IFMBE Proceedings

Volume 88

The IFMBE Proceedings Book Series is an official publication of *the International Federation for Medical and Biological Engineering* (IFMBE). The series gathers the proceedings of various international conferences, which are either organized or endorsed by the Federation. Books published in this series report on cutting-edge findings and provide an informative survey on the most challenging topics and advances in the fields of medicine, biology, clinical engineering, and biophysics.

The series aims at disseminating high quality scientific information, encouraging both basic and applied research, and promoting world-wide collaboration between researchers and practitioners in the field of Medical and Biological Engineering.

Topics include, but are not limited to:

- Diagnostic Imaging, Image Processing, Biomedical Signal Processing
- Modeling and Simulation, Biomechanics
- Biomaterials, Cellular and Tissue Engineering
- Information and Communication in Medicine, Telemedicine and e-Health
- Instrumentation and Clinical Engineering
- Surgery, Minimal Invasive Interventions, Endoscopy and Image Guided Therapy
- Audiology, Ophthalmology, Emergency and Dental Medicine Applications
- Radiology, Radiation Oncology and Biological Effects of Radiation

IFMBE proceedings are indexed by SCOPUS, EI Compendex, Japanese Science and Technology Agency (JST), SCImago.

Proposals can be submitted by contacting the Springer responsible editor shown on the series webpage (see "Contacts"), or by getting in touch with the series editor Ratko Magjarevic.

More information about this series at https://link.springer.com/bookseries/7403

Simona Vlad · Nicolae Marius Roman
Editors

7th International Conference on Advancements of Medicine and Health Care through Technology

Proceedings of MEDITECH-2020,
13–15 October 2020

Springer

Editors
Simona Vlad
Faculty of Electrical Engineering
Technical University of Cluj-Napoca
Cluj-Napoca, Romania

Nicolae Marius Roman
Faculty of Electrical Engineering
Technical University of Cluj-Napoca
Cluj-Napoca, Romania

ISSN 1680-0737 ISSN 1433-9277 (electronic)
IFMBE Proceedings
ISBN 978-3-030-93563-4 ISBN 978-3-030-93564-1 (eBook)
https://doi.org/10.1007/978-3-030-93564-1

This Springer imprint is published by the registered company Springer Nature Switzerland AG
The registered company address is: Gewerbestrasse 11, 6330 Cham, Switzerland

Preface

The 7th "Conference on Advancements of Medicine and Health Care through Technology" – MediTech 2020 took place in Cluj-Napoca in October 13th–15th, 2020 was and held online.

The *MediTech conferences* are part of a series of conferences organized by the "***Romanian National Society for Medical Engineering and Biological Technology***" under the auspices of the "***International Federation for Medical and Biological Engineering***," which take place every two years, on even years, in Cluj-Napoca, Romania, being one of the most important scientific events in the region, in this field.

The MediTech 2020 conference was co-organized by representatives of the academic and medical environment, namely the ***Technical University of Cluj-Napoca***, the "***Iuliu Haţieganu***" ***University of Medicine and Pharmacy of Cluj-Napoca*** and "***Dr. Constantin Papilian***" ***Military Emergency Hospital Cluj-Napoca, Romania***. It was attended by over 75 guests from Romania and abroad, who presented their latest research results in medical and technical universities, in clinics and hospitals, or from the field of clinical engineering and medical devices.

We were honored to have Kang-Ping Lin, Distinguished Professor, at the Chung-Yuan Christian University, Taiwan, and IFMBE Secretary-General attend our conference for the third time in a row and delivered a speech during the opening ceremony. We were also honored to welcome Dr. Christoph Baumann, Editorial Director, Springer Nature, among our guests.

We proved once again that this conference is a good opportunity for all the participants from the academic, technical and medical fields to exchange ideas, knowledge and achievements in the extremely important field of medical engineering.

I would especially like to thank Distinguished Professor Ratko Magjarevic, Vice President of IFMBE, for his support regarding the publication of the conference's papers and for his kind words regarding our conference, organized by the Romanian National Society for Medical Engineering and Biological Technology, the technical and medical universities and a renowned hospital unit.

In the end, we would like to kindly thank all the participants, speakers, academics, researchers, doctors, sponsors and members of the scientific and organizing committees for their hard work and dedication.

Please visit our Web site for more information https://snimtb.ro/.

Nicolae Marius Roman
Conference Chair, President of R.N.S.M.E.B.T.

Organization

Organizers

Romanian National Society for Medical Engineering and Biological Technology
Technical University of Cluj-Napoca, Romania
"Iuliu Haţieganu" University of Medicine and Pharmacy Cluj-Napoca, Romania

Partner

Dr. Constantin Papilian — Military Emergency Hospital, Cluj-Napoca, Romania

Conference Chair

Nicolae Marius Roman — Romanian National Society for Medical Engineering and Biological Technology

Honorary Chair

Radu Vasile Ciupa — Romanian National Society for Medical Engineering and Biological Technology

Scientific Advisory Committee

Ancuta-Coca Abrudan (RO)
Laura Bacali (RO)
Doina Baltaru (RO)
Maria Beudean (RO)
Lelia Ciontea (RO)
Radu Ciorap (RO)
Radu V. Ciupa (RO)
Mariana Constantiniuc (RO)
Cecilia Cristea (RO)
Paul Farago (RO)
Anca Galaction (RO)
Monica Gavris (RO)
Stefan Gergely (RO)
Zoltan German-Sallo (RO)

Laura Grindei (RO)
Sorin Hintea (RO)
Rodica Holonec (RO)
Ioan Jivet (RO)
Mihaela-Ruxandra Lascu (RO)
Mircea Leabu (RO)
Kang-Ping Lin (TW)
Angela Lungu (RO)
Eugen Lupu (RO)
Dan Mandru (RO)
Alma Maniu (RO)
Winfried Mayr (A)
Amalia Mesaros (RO)
Bogdan Micu (RO)
Dan D. Micu (RO)
Dan L. Milici (RO)
Alexandru Morega (RO)
Mihaela Morega (RO)
Marius Muji (RO)
Calin Munteanu (RO)
Mihai S. Munteanu (RO)
Radu A. Munteanu (RO)
Anca I. Nicu (RO)
Maria Olt (RO)
Tudor Palade (RO)
Doina Pasla (RO)
Sever Pasca (RO)
Alessandro Pepino (IT)
Petre G. Pop (RO)
Dan V. Rafiroiu (RO)
Nicolae-Marius Roman (RO)
Corneliu Rusu (RO)
Adrian Samuila (RO)
Dan I. Stoia (RO)
Mihai Tarata (RO)
Marina Topa (RO)
Vasile Topa (RO)
Doru Ursutiu (RO)
Liliana Verestiuc (RO)
Radu C. Vlad (RO)
Simona Vlad (RO)
Dan Zaharia (RO)

Local Organizing Committee

Radu V. Ciupa
Levente Czumbil
Rodica Holonec
Angela Lungu
Alma Maniu
Bogdan Micu
Calin Munteanu
Mihai S. Munteanu
Anca I. Nicu
Maria Olt
Sever Pasca
Dan V. Rafiroiu
Nicolae Marius Roman
Adrian Samuila
Marina Topa
Simona Vlad

Speakers at the Opening Ceremony

Radu Vasile Ciupa, Honorary chair - Romanian National Society for Medical Engineering and Biological Technology

Mihaela Băciuţ, vice-rector "Iuliu Haţieganu" University of Medicine and Pharmacy, Cluj-Napoca, Romania

Doina Pîslă, vice-rector Technical University of Cluj-Napoca, Romania

Doina Baltaru, commander of "Dr. Constantin Papilian" Emergency Military Hospital Cluj-Napoca, Romania

Kang-Ping Lin, IFMBE Secretary-General, Chung-Yuan Christian University, Taiwan

Christoph Baumann, Editorial Director, Springer Nature

Nicolae Marius Roman, Conference Chair, Romanian National Society for Medical Engineering and Biological Technology

Invited Speakers

Kang-Ping Lin, IFMBE Secretary-General, Chung-Yuan Christian University, Taiwan,
Potentials and Challenges of Healthcare AI Technology for Medical Devices

Winfried Mayr, Medical University of Vienna, Austria,
The roll of electrical parameters and electrodes in FES supported motor control

Ramon Pallàs Areny, Universitat Politècnica de Catalunya (BarcelonaTech), Spain,
Capacitive electrodes for bioelectric signals: solution or chimera?

Lucio Tommaso De Paolis, University of Salento, Italy,
How Augmented Reality is Transforming Medical Imaging

Doru Ursuţiu, "*Transilvania*" *University of Braşov, Romania,*
Virtual Instrumentation, Reconfigurable Systems, from IoT to Medical IoT

Valentin Oprea, "Dr. Constantin Papilian" Military Emergency Hospital Cluj-Napoca, Romania,
A rationale for tension measurement in abdominal wall hernia repair

Dan L. Dumitraşcu, "Iuliu Haţieganu" University of Medicine and Pharmacy, Cluj-Napoca, Romania,
Breath test applications in gastroenterology

Alessandro Pepino, University of Naples "Federico II," Italy,
How Simulation can support covid-19 challenges in medical education

Thierry Marchal, Ansys, Inc. & Avicenna Alliance,
The role of in silico technologies for medical innovation, testing and regulatory approval: illustrations with implantable devices, wearables and cancer treatments.

Rahul Bharadwaj, VP of Engineering and Business Development at ESSS,
Multiscale and Multiphysics Simulation Approach for Increased Productivity in Tablet Manufacturing

Tony Ward, University of York, England,
Can Higher Education ever meet the requirements industry has in graduates?

Mircea Gelu Buta, "Babeş-Bolyai" University of Cluj-Napoca, Romania,
The Disenchantment of Medicine

Paula Pavel, "Dr. Constantin Papilian" Military Emergency Hospital Cluj-Napoca, Romania,
Available Diagnostic Methods of Deformities in the Diabetic Foot

Sponsors

INAS SA
Farmec SA
Tehno Industrial SA
Laitek Medical Software, Romania HQ
Medical Technologies Infinity SRL
Cefmur SA

Chairmen

Radu Ciorap
Radu Vasile Ciupa
Laura Grindei
Rodica Holonec
Eugen Lupu
Dan Mandru
Mihai Stelian Munteanu
Radu Adrian Munteanu
Calin Munteanu
Anca Iulia Nicu
Tudor Palade
Andra Păstrăv
Nicolae Marius Roman
Adrian Samuila
Marina Topa
Simona Vlad

Media Partners

TVR Cluj
Radio Romania Cluj

Contents

Telemedicine and Health Care Information Systems

Medical Devices, Measurements and Instrumentation; Biomedical Signal Processing

A Matlab Simulation Platform for a Biometric Identification System Based on ECG Fiducial Features

Claudia Cordoş, Paul Faragó(✉), Sorin Hintea, and Călin Fărcaş

Technical University of Cluj-Napoca, Cluj-Napoca, Romania
paul.farago@bel.utcluj.ro

Abstract. Traditional data security methods, such as passwords or numeric codes, have been replaced by biometric methods of identification. In this paper, an ECG biometric recognition system was developed and simulated under the Matlab environment. The application allows the selection of two ECG recordings from a database, which are compared based on the extracted morphological features. Based on the comparison results, it will be established whether the two recordings belong to the same subject. The operation of comparing the two ECG recordings can be done only after the fiducial points of the PQRST signature have been identified for an RR interval, normalized in the time and magnitude domains. The implemented algorithm was validated on a data set of 14 recordings taken from the PhysioNet platform.

Keywords: Electrocardiogram · Fiducial features · Feature extraction · Biometric identification

1 Introduction

It is a well-known fact that security plays an important role in our lives, being found in a wide variety of fields. Starting with securing an ordinary account on social media or a mobile phone and continuing with online payments or even physical access into a building. This is an important finding in the understanding of the improvement of security methods, that became increasingly advanced with the passing of the years.

In spite of the fact that alphanumeric passwords are the most popular security methods, a major problem is that the users choose weak passwords and malicious persons can hack them. If a few years ago it was enough to secure our phone with an alphanumeric key, nowadays it is not sufficient anymore, more sophisticated methods being needed such as fingerprint recognition or facial recognition. The prestigious institutions do not limit to previously mentioned authentication methods. These take in account authentication methods based of biometrics recognition like iris recognition, palm, shape of the hand and voice recognition. New security methods are based on biometrics recognition because biometric features are unique and difficult to hack.

Biometric recognition can be explained as a pattern recognition problem: if a user wants to be authenticated, he must provide a set of physiological and/or behavioral characteristics which must match with a previously registered signature [1].

S. Vlad and N. M. Roman (Eds.): MEDITECH 2020, IFMBE Proceedings 88, pp. 3–8, 2022.
https://doi.org/10.1007/978-3-030-93564-1_1

In the last few years, the use of electrocardiogram (ECG) in biometric authentication systems has grown. ECG signals have many advantages, such as: uniqueness, are measurable, can be acquired continuously, are available for living human beings, contains information about the person's clinical condition, their collection is non-invasive, are difficult to falsify, their acquisition for long period of time does not require specific actions from the user.

Between 2001–2010 the processing of biosignals progressed in the field of diagnosis, so that their use in biometric systems became easy. Therefore, the development of a non-clinical application received increased attention from the scientific community engaged in the development of intelligent systems [2].

Biel et al. demonstrated the uniqueness of the ECG signal in 2001. Furthermore, they presented the first proof of identifying a person using features extracted from a single lead of the ECG signal [3]. Wübbeler et al. (2007) used 234 ECG recordings which were preprocessed. They applied a moving median width, a low-pass filter and a threshold procedure to calculate the positions of the R-peaks [4]. Shen et al. (2010) determined R-R intervals using Pan and Tompkins method and detected PQRST fiducial points using digital filtering [5]. Salloum and Kuo (2017) used recurrent neural networks (RNNs) with LSTM (Long Short-Term Memory) architecture and a segmented heartbeat sequence. They obtained 100% accuracy for the ECG-ID database, which contains 90 ECG recordings [6].

2 Biometric Identification Based on Fiducial Points

The main ECG components of a heartbeat are presented in Fig. 1: the P wave represents the depolarization of the atrium; the QRS complex consisting of the Q, R and S waves respectively, illustrates ventricular depolarization; the T wave marks ventricular repolarization [2]. Intervals are then measured between the ECG waves.

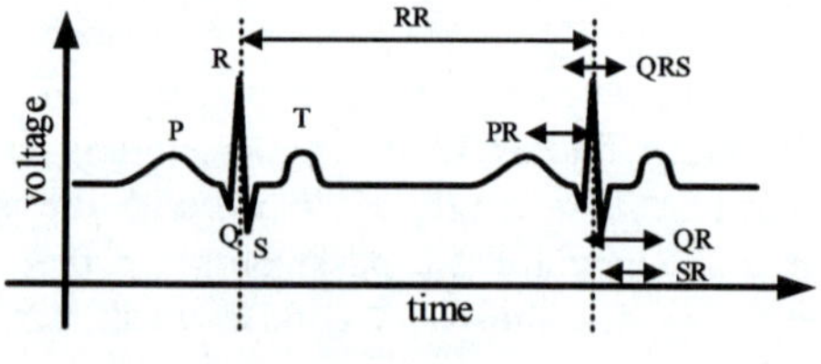

Fig. 1. The standard ECG waveform [7]

The implemented biometric recognition system based on the ECG signal compares two ECG recordings and determines whether or not they belong to the same subject. To be noticed is that the ECG is sensitive to interferences, the real amplitude and duration of the signal can be changed by additive noise or by different types of artifacts. As such, a denoising and artifact removal stage in advisable. In this work, we have employed previously preprocessed recordings. These are amplified and filtered using a moving median width to obtain a smooth, noise-free signal.

In this paper the ECG signal is analyzed from the point of view of the fiducial points contained in the PQRST signature. The detection of fiducial points of PQRST signature is done for an RR interval normalized at unit amplitude and 1 s duration. An RR interval is defined as the time difference between two consecutive R peaks, is equivalent to a heartbeat.

The amplitude and time intervals extracted from the ECG are unique morphological features of each person, so they can be used for identification. Thus, we measured the time intervals between the determined fiducial points, as listed in Table 1, and the amplitudes of the P, Q, S, T waves in relation to the unit R wave, as listed in Table 2. The difference of the measurements obtained for the recordings of the same subject are compared with an empirically determined threshold to decide upon identification.

3 Graphical User Interface

For an interactive application, a user-friendly graphical user interface (GUI) has been created using Matlab development environment. The user will not need training to be able to interact with the interface, because the controls are intuitive. The graphical user interface allows the user to choose two ECG signals from the database to establish the person's identity. Once the ECG recordings to be compared have been chosen, the P, Q, R, S, T waves on the RR interval will be displayed on the main page of the interface. The RR interval is extracted from both the reference signal and the signal with which it will be compared. At the first analysis on the overlapping of RR intervals, the user can make assumptions about the membership of the records, whether or not they belong to the same individual (see Fig. 2). When the Check button is pressed, a message box will appear approving whether the ECG recordings belong to the same subject.

Table 1. The results obtained by measuring time intervals

Time intervals										
Database	*1st Recording*					*2nd Recording*				
	PQ(s)	QRS(s)	QT(s)	ST(s)	RT(s)	PQ(s)	QRS(s)	QT(s)	ST(s)	RT(s)
Subject 1	0.18	0.07	0.33	0.26	0.29	0.14	0.07	0.33	0.25	0.28
Subject 2	0.12	0.05	0.30	0.24	0.27	0.09	0.05	0.25	0.20	0.23
Subject 3	0.17	0.07	0.38	0.30	0.34	0.14	0.07	0.37	0.30	0.34
Subject 4	0.12	0.08	0.34	0.25	0.30	0.11	0.08	0.34	0.26	0.31
Subject 5	0.10	0.05	0.28	0.22	0.25	0.10	0.05	0.28	0.22	0.25
Subject 6	0.07	0.06	0.26	0.19	0.22	0.11	0.06	0.26	0.19	0.23
Subject 7	0.15	0.06	0.39	0.32	0.36	0.10	0.06	0.36	0.30	0.32

Table 2. The results obtained by measuring amplitudes

Amplitudes								
Database	*1st Recording*				*2nd Recording*			
	RP	RQ	RS	RT	RP	RQ	RS	RT
Subject 1	0.98	1.17	1.26	0.62	1.03	1.18	1.21	0.64
Subject 2	0.93	1.21	1.76	0.45	0.96	1.23	1.74	0.29
Subject 3	0.90	1.08	1.03	0.52	0.96	1.20	1.15	0.76
Subject 4	0.92	1.25	1.22	0.69	0.95	1.22	1.22	0.85
Subject 5	0.96	1.30	1.22	0.42	1.00	1.31	1.42	0.44
Subject 6	1.02	1.11	1.09	0.82	1.01	1.15	1.07	0.79
Subject 7	0.92	1.18	1.37	0.67	0.95	1.21	1.40	0.74

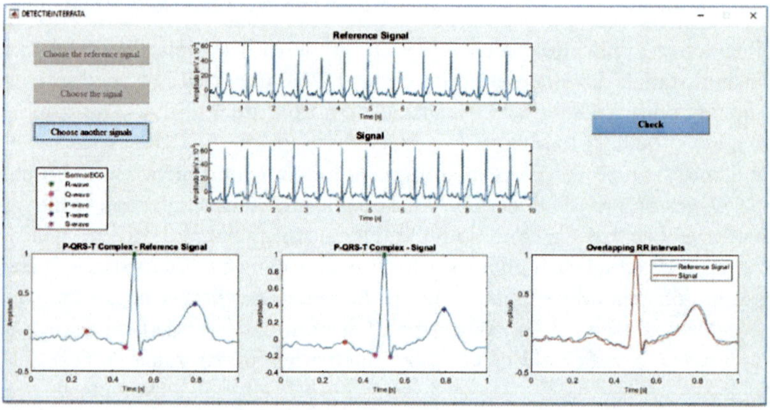

Fig. 2. The main page of the user interface. The P-QRS-T Complex graphs contain the P, Q, R, S, T waves. The overlap of the RR intervals is represented in the figure on the bottom right side

4 Simulation Results

The algorithm was tested on 14 recordings of 10 s each, two recordings for the same subject. The data set was taken from the PhysioNet platform and ECG recordings are chosen randomly from the ECG-ID database. The identification of all subjects based on the implemented algorithm was done successfully. Considering the rather small dataset, i.e. 7 individuals, we were able to obtain a 0% False Rejection Rate and 0% False Acceptance Rate. Although the dataset is relatively small, the simulation results illustrate the strength of the implemented approach. Indeed, individuality of the ECG signature allowed for the correct identification after the PQRST morphology was extracted and identified.

In the following images, there are the results obtained for some of the database's ECG recordings. For the result obtained in Fig. 3, the rec7_01 reference ECG recording was compared with the rec7_02 ECG recording. The message box indicates that both ECG recordings are acquired from the same person, which is true, therefore, the identification is done successfully.

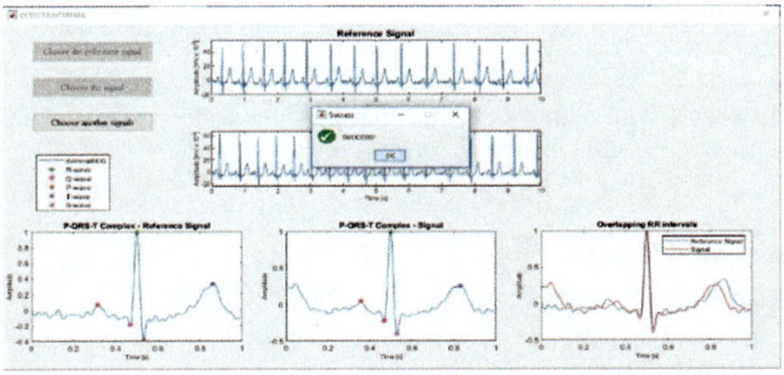

Fig. 3. The result obtained for the rec7_01 reference signal and the rec7_02 signal

The message box, shown in the Fig. 4, indicates that the two recordings do not belong to the same subject, so that the identification of the two people is done correctly. The rec3_02 reference signal was compared with the rec6_02 signal.

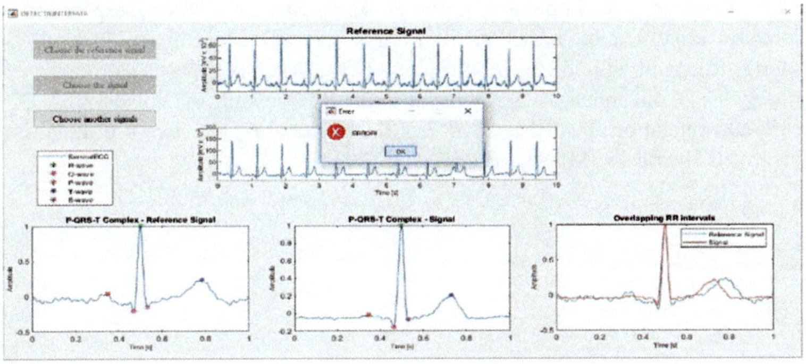

Fig. 4. The result obtained for the rec3_02 reference signal and the rec6_02 signal

5 Conclusions

This paper presented a solution for ECG biometric identification, using the comparison of two ECG recordings. The application was implemented in the Matlab development environment.

The ECG signals were preprocessed, then an RR interval was extracted from each signal. The detection of P, Q, R, S, T fiducial points was performed for a normalized RR interval for each of the two compared ECG recordings. Time intervals and amplitudes were measured taking into account the PQRST signature. These unique morphological features were correlated and compared with an error.

The identification was successfully performed for a set taken from 7 individuals, two filtered ECG recordings for each individual. The 14 tested ECG recordings were taken from the PhysioNet platform.

As future work, we aim to extend the data set and to analyze the fiducial points for a larger number of RR intervals and to calculate RR interval's average prior to feature extraction.

Conflict of Interest. The authors declare that they have no conflict of interest.

References

1. Ivanciu, L., Farago, P., Hintea, S.: A review of ECG based biometric systems. Acta Technica Napocensis **59**, 1–4 (2018)
2. Merone, M., Soda, P., Sansone, M., Sansone, C.: ECG databases for biometric systems: a systematic review. Expert Syst. Appl. **67**, 189–202 (2017)
3. Biel, L., Pettersson, O., Philipson, L., Wide, P.: ECG analysis: a new approach in human identification. IEEE Trans. Instrum. Meas. **50**(3), 808–812 (2001)
4. Wübbeler, G., Stavridis, M., Kreiseler, D., Bousseljot, R.D., Elster, C.: Verification of humans using the electrocardiogram. Pattern Recognit. Lett. **28**(10), 1172–1175 (2007)
5. Shen, T.-W.D., Tompkins, W.J., Hu, Y.H.: Implementation of a one-lead ECG human identification system on a normal population. J. Eng. Comput. Innov. **2**(1), 12–21 (2010)
6. Pinto, J.R., Cardoso, J.S., Lourenço, A., Arya, K.V., Bhadoria, R.S.: Deep neural networks for biometric identification based on non-intrusive ECG acquisitions. In: The Biometric Computing: Recognition and Registration, pp. 217–234 (2019)
7. Choudhary, T., Manikandan, M.S.: A novel unified framework for noise-robust ECG-based biometric authentication. In: IEEE 2015 2nd International Conference on Signal Processing and Integrated Networks (SPIN), pp. 186–191 (2015)

A Study on ECG Denoising Using Wiener Filter

Laura Mihăilă, Paul Faragó(✉), Călin Fărcaş, and Sorin Hintea

Technical University of Cluj-Napoca, Cluj-Napoca, Romania
paul.farago@bel.utcluj.ro

Abstract. The electrocardiogram (ECG) is an essential signal in the medical field, with the help of which doctors manage to identify the electrical functions of the heart. From the interpretation of the ECG signal it is possible to diagnose and measure the presence of abnormal heart rhythms, as well as the presence of indicators for future abnormalities. The captured ECG signal is usually contaminated with noise. Therefore, it is important to eliminate the noise to prevent any error in the subsequent analysis of the signal. This paper presents a study of ECG noise reduction via adaptive filtering using the Wiener filter. Simulation results obtained in Matlab provide a numeric assessment of the approach.

Keywords: Electrocardiogram · Filtering · Noise · Wiener · Signal to noise ratio · Processing time

1 Introduction

The electrocardiogram (ECG) is an electrical recording of the heart activity and is one of the most widely used diagnostic tools in medicine. The ECG frequency band is 0.05–100 Hz and its amplitude is in the range of 4 μV–4 mV. The ECG signal is used to identify the electrical and muscular functions of the heart and reveals if there are various heart conditions, effects of various medications or implants [1]. From the interpretation of the signal, heartbeat deficiencies can be identified, thus any deviation of the ECG signal parameters (P wave, ORS complex and a T wave) indicates abnormalities present in the activity of the heart. During the acquisition of the ECG signal, various noises may occur over the useful signal leading to its disturbance and misinterpretation [2]. In principle, there are two types of ECG noises: high frequency noise (e.g. White Additive Gaussian Noise, power line interference, electromyogram noise, etc.) and low frequency noise (Baseline Wander). All these noises can be produced by different sources of artificial or biological nature and all aim to error the ECG signal.

This paper studies Wiener filtering as a method for ECG noise suppression. This is an adaptive filter, being considered in the literature a powerful tool that has changed the traditional approaches to signal processing [3]. The Wiener filter aims to minimize the mean square error and subtract the noise from the signal based on the frequency spectrum. This method is optimal and efficient in removing unwanted interference from the ECG signal [4].

S. Vlad and N. M. Roman (Eds.): MEDITECH 2020, IFMBE Proceedings 88, pp. 9–17, 2022.
https://doi.org/10.1007/978-3-030-93564-1_2

2 The Wiener Filter

The Wiener filter belongs to the class of adaptive filters and operates on the basis of static properties of both signal and noise, in order to be able to reconstruct the useful signal and possible from a large noise flow [4]. The filter coefficients are re-computed periodically for each signal segment, thus adapting the filter to the characteristics of each segment and becomes adaptable to the segment (block). Therefore, Wiener filters are important in a variety of applications, ranging from echo cancellation to linear prediction, signal restoration, system identification and channel equalization. A Wiener filter can be with infinite impulse response (IIR) - the result is a set of nonlinear equations, or with finite impulse response (FIR) - the result is a set of linear equations.

This paper proposes a method that uses a Wiener filter with finite impulse response (FIR), because FIR filters are simple to calculate, stable and practical. The disadvantage of FIR filters is that they require a larger number of coefficients to be able to approximate the desired response [5].

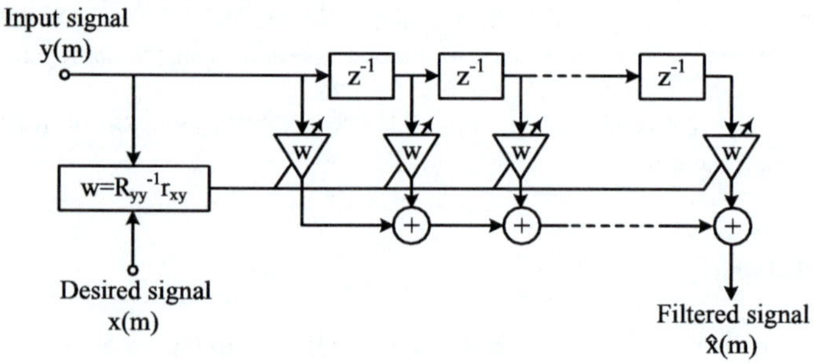

Fig. 1. Wiener filter block diagram

Figure 1 illustrates the block diagram of the Wiener filter as a function of the coefficient vector *w*. The filter has an input signal *y(m)* which is the distorted variant of the signal *x(m),* and the output signal is $\hat{x}(m)$ and is calculated so as to obtain a minimum square mean between this signal and the desired one. Time-domain and frequency-domain operation of the Wiener filter is then formulated as follows [6–8].

2.1 Time-Domain Equations

The equation which describes Wiener filter operation is

$$\hat{x}(m) = \sum_{k=0}^{P-1} w_k y(m-k) = w^T y \tag{1}$$

where *m* represents the index of the discrete-time, y^T = [y(m), y(m − 1), … , y(m − P − 1)] is the input signal and vector $w = [w^T]$ represents the Wiener coefficients. The error signal of the Wiener filter is defined as the difference between the desired signal *x(m)* and the filter output signal $\hat{x}(m)$:

$$\begin{aligned} e(m) &= x(m) - \hat{x}(m) \\ &= x(m) - w^T y \end{aligned} \tag{2}$$

Equation (2) can be rewritten to matrix form in order to highlight the relation between the Wiener coefficient vector *w* and the error signal *e(m)*

$$\begin{pmatrix} e(0) \\ e(1) \\ e(2) \\ \cdots \\ e(N-1) \end{pmatrix} = \begin{pmatrix} x(0) \\ x(1) \\ x(2) \\ \cdots \\ x(N-1) \end{pmatrix} - \begin{pmatrix} y(0) & y(-1) & y(-2) & \cdots & y(1-P) \\ y(1) & y(0) & y(-1) & \cdots & y(2-P) \\ y(2) & y(1) & y(0) & \cdots & y(3-P) \\ \cdots & \cdots & \cdots & \cdots & \cdots \\ y(N-1) & y(N-2) & y(N-3) & \cdots & y(N-P) \end{pmatrix} \begin{pmatrix} w_0 \\ w_1 \\ w_2 \\ \cdots \\ w_{P-1} \end{pmatrix} \tag{3}$$

Using vector notation, the relation (3) may be written as:

$$e = x - Yw \tag{4}$$

where *e* is the error vector, *x* is the desired signal vector and $Yw = \hat{x}$ is the filtered signal.

We are interested in the dependence of the solution of Eq. (3) on the number of input samples. Thus, we will have 3 distinct cases:

- **N = P**, in this case we have N equations with N unknowns, which leads to a determined system with unique solution w, and zero estimation error;
- **N < P**, in this case we have a subdetermined system, an infinity of solutions w;
- **N > P**, in this case we have an overdetermined system, a unique solution, but non-zero estimation error.

Thus, the coefficients of the Wiener filter will be calculated to minimize the mean error, respectively the mean square error $E[|\ e^2(m)]$, where by *E[.]* the expectation operator was noted, in this case the mediation.

$$\begin{aligned} E[e^2(m) &= E\left[\left(x(m) - w^T y\right)^2\right] \\ &= E\left[x^2(m)\right] - 2w^T E[yx(m)] + w^T E\left[yy^T\right] w \\ &= r_{xx}(0) - 2w^T r_{yx} + w^T R_{yy} w \end{aligned} \tag{5}$$

where, $Ryy = E\ [y(m)\ \cdot y^T(m)]$ is the autocorrelation of the input, and $r_{yx} = E[y(m) \cdot x(m)]$ is the intercorrelation vector of the input signal and the desired signal. If we develop Eq. (5) can be obtained as:

$$E\left[e^2(n)\right] = r_{xx}(0) - 2\sum_{k=0}^{P-1} w_k r_{yx}(k) + \sum_{k=0}^{P-1} w_k \sum_{j=0}^{P-1} w_j r_{yy}(k-j) \tag{6}$$

where $r_{yy}(k)$ and $r_{yx}(k)$ are the autocorrelation elements of matrix Ryy and the cross-correlation vector r_{xy} respectively. From Eq. (5), the mean square error function gradient with respect to the filter coefficient vector is determined by:

$$\frac{\partial}{\partial w}E[e^2(m) = -2E[y(m) \cdot x(m)] + 2w^T E[y(m) \cdot y^T(m)] \tag{7}$$

where, the gradient vector is defined as:

$$\frac{\partial}{\partial w} = \left[\frac{\partial}{\partial w_0}, \frac{\partial}{\partial w_1}, \frac{\partial}{\partial w_2}, \ldots, \frac{\partial}{\partial w_{P-1}}\right]^T \tag{8}$$

The minimum mean square error Wiener filter is obtained by setting Eq. (7) to zero:

$$R_{yy}w = r_{yx} \tag{9}$$

or, equivalently,

$$w = R_{yy}^{-1} r_{yx} \tag{10}$$

2.2 Frequency-Domain Equations

The output signal of the Wiener filter in the frequency domain is the product of the input signal Y(f) and the filter frequency response W(f):

$$\hat{X}(f) = Y(f) \cdot W(f) \tag{11}$$

The estimation error signal E(f) is defined as:

$$\begin{aligned} E(f) &= X(f) - \hat{X}(f) \\ &= X(f) - Y(f) \cdot W(f) \end{aligned} \tag{12}$$

The mean square error at a frequency *f* is given by:

$$E\left[|E(f)|^2\right] = E[(X(f) - Y(f) \cdot W(f)) * (X(f) - Y(f) \cdot W(f))] \tag{13}$$

To obtain the least mean square error filter we set the complex derivative of Eq. (13) with respect to filter W(f) to zero:

$$\frac{\partial E\left[|E(f)|^2\right]}{\partial W(f)} = 2 \cdot W(f) \cdot P_{yy}(f) - 2 \cdot P_{xy}(f) = 0 \tag{14}$$

where, $P_{yy}(f) = E[Y(f)Y^*(f)]$ is the power spectrum of Y(f) and $P_{xy}(f) = E[X(f)Y^*(f)]$ is the cross-power spectrum of Y(f) and X(f) respectively.

Thus, the least mean square error Wiener filter in the frequency domain is given as:

$$W(f) = \frac{P_{xy}(f)}{P_{yy}(f)} \tag{15}$$

3 Implementation

The Wiener filter, as given in Eq. (15), and the filter test scenarios were implemented in Matlab. The block diagram for the implementation is illustrated in Fig. 2 and is explained as follows.

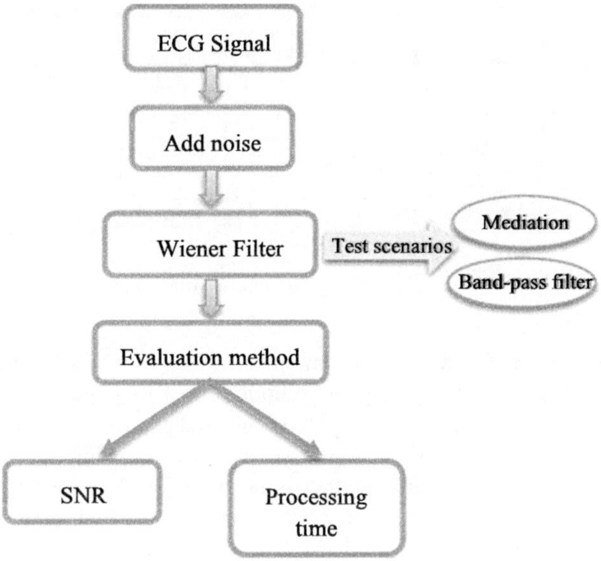

Fig. 2. Flow diagram of the implemented solution

The signals used were imported from the PhysioNet platform. ECG recordings were obtained from volunteers aged between 13 and 75 years. Each signal represents a 20 s recording, sampled at 500 Hz with a resolution of 12-bits over a nominal voltage range of ±10 mV.

Over the imported ECG signal, colored noise (pink, purple, blue, Gaussian white) is added and the noisy signal is filtered using a Wiener filtering algorithm. This method follows a series of test scenarios and comparison to a 5-sample local signal mediation and 4th order bandpass filtering with Butterworth approximation, and finally, criteria for objective evaluation of the filtered signal are applied.

4 Simulation Results

The implemented algorithm was simulated on several signals taken from different patients. Also, the implementation was simulated both on low noise signals and on high noise signals. From the results obtained, the Wiener filtering method produces satisfactory results on the ECG signal, eliminating noise considerably (Fig. 3).

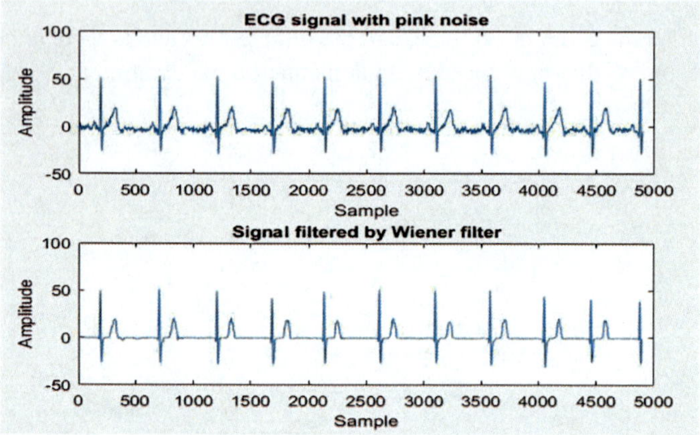

Fig. 3. Noisy signal and filtered signal using the Wiener filter, amplitude is measured in mV

During several tests, it was noticed that at a very high level of noise on the ECG signal, the filter no longer recognizes all components of the signal and produces significant signal distortion after filtering. Thus, the Wiener filter is limited to a large amount of noise present in the ECG signal as can be seen in Fig. 4.

By applying classical filtering methods, such as local mediation or 4th order bandpass filtering with Butterworth approximation, the noise in the ECG signal is eliminated, but not completely, as plotted in Fig. 5.

From Table 1 it is observed that the Wiener filter has a long processing time compared to the other methods. This is explained by the implementation method because this adaptive filter requires a much more complex implementation, which leads to a longer processing time. From diagram it can also be seen graphically the increase of the SNR value after Wiener filtering at all types of noise. The filtering approaches were tested and compared on added noise, namely pink noise, blue noise and white noise respectively. The bar-graph plotted in Fig. 6 illustrates the SNR gain via Wiener filtering for each noise color respectively (Tables 2 and 3).

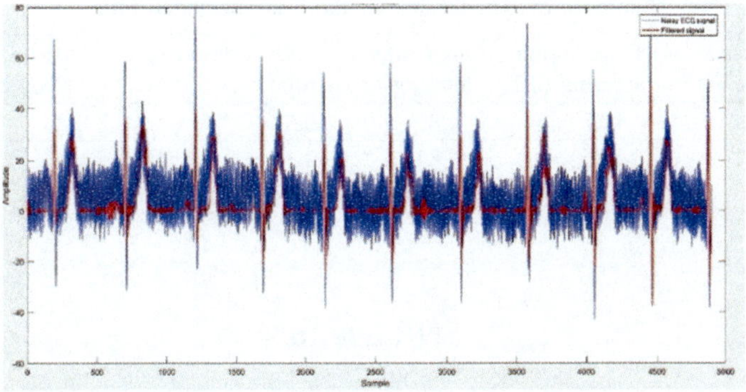

Fig. 4. Signal with high level of noise (blue) and filtered signal using the Wiener filter, amplitude is measured in mV

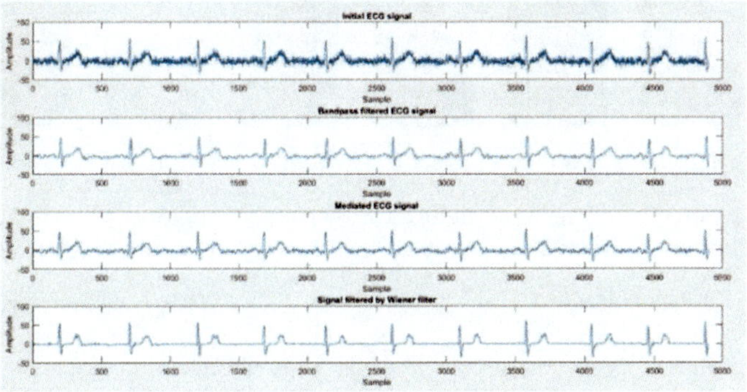

Fig. 5. ECG waveforms to illustrate denoising via bandpass filtering, averaging and Wiener filter

Table 1. SNR measures for the Wiener filter

Signal	Processing time (s)	SNR before filtering (dB)	SNR after filtering (dB)	SNR improvement (dB)	SNR improvement (SNR/sec)
Pers. 1	2,24	5,34	22,12	16,76	7,4
Pers. 2	2,26	5,79	27,13	21,34	9,3
Pers. 3	2,41	13,42	27,35	3,93	5,7

Table 2. SNR measures for the bandpass filter

Signal	Processing time (s)	SNR before filtering (dB)	SNR after filtering (dB)	SNR improvement (dB)	SNR improvement (SNR/sec)
Pers. 1	0,7	4,95	6,51	1,56	2,22
Pers. 2	0,93	5,28	7,01	1,73	1,86
Pers. 3	0,66	14,08	15,28	1,2	1,81

Table 3. SNR measures for the local mediation filter

Signal	Processing time (s)	SNR before filtering (dB)	SNR after filtering (dB)	SNR improvement (dB)	SNR improvement (SNR/sec)
Pers. 1	0.19	4,27	4,85	0,58	3,05
Pers. 2	0.21	4,46	4,84	0,38	1,8
Pers. 3	0.20	11,73	12,22	0,49	2,45

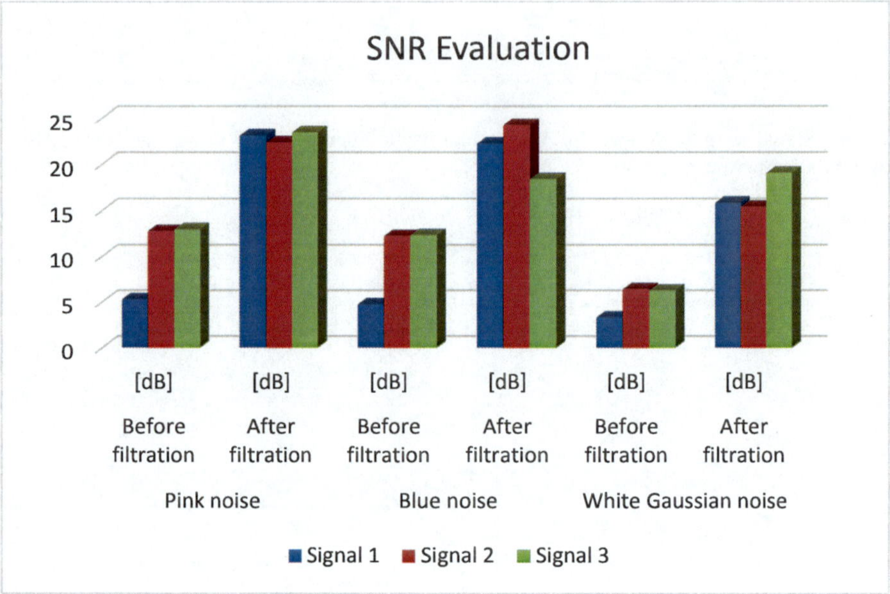

Fig. 6. SNR values of the ECG signal to illustrate denoising performances via bandpass filtering, averaging and Wiener filter

5 Conclusions

The ECG is an essential signal in the medical field for the analysis and interpretation of heart activity. As such, the integrity of the ECG signal is crucial for diagnosis and follow-up. Given the ECG amplitude and baseband frequency range, it is affected by a variety of noises and artifacts, the suppression of which requires ECG-specific algorithms and approaches.

This paper studied the applicability of optimal numeric filtering, in this work Wiener filter, for ECG noise reduction. ECG denoising results obtained with the Wiener filter were compared to those obtained with traditional approaches: bandpass filter and local averaging.

As expected, in comparison to the traditional approaches, Wiener filter operated towards a considerably larger improvement of the SNR, but with the expense of a longer execution time. Also, the Wiener filter was tested for ECG denoising with various noise colors.

It was demonstrated that the Wiener filter is suitable for ECG denoising, provided runtime is not an issue. As such, alternative means are advisable for real-time applications.

Conflict of Interest. The authors declare that they have no conflict of interest.

References

1. Becker, D.E.: Fundamentals of electrocardiography interpretation. Anesth. Prog. **52**(2), 53–64 (2006)
2. Velayudhan, A., Peter, S.: Noise analysis and different denoising techniques of ECG signal - a survey. IOSR J. Electron. Commun. Eng. (IOSR-JECE), 40–44. e-ISSN 2278-2834, p-ISSN 2278-8735
3. Chandrakar, C., Kowar, M.K.: Denoising ECG signals using adaptive filter algorithm. Int. J. Soft Comput. Eng. (IJSCE) **2**(1) (2012). ISSN 2231-2307
4. Smith, S.W.: Digital Signal Processing. California Technical Publishing, San Diego (1997–1999)
5. Avalos, G., Sanchez, J.C., Velazquez, J.: Applications of Adaptive Filtering. Accessed 5 July 2011
6. Vaseghi, S.V.: Multimedia Signal Processing Theory and Applications in Speech, Music and Communications. Wiley (2007). ISBN 978-0-470-06201-2
7. Smith, S.W.: The Scientist and Engineer's Guide to Digital Signal Processing. http://www.dspguide.com/
8. Lascu, M., Lascu, D.: Electrocardiogram compression and optimal ECG filtering algorithms. WSEAS Trans. Comput. **7**(4), 155–164 (2008). ISSN 1109-2750

Correlation of Electric EEG and VEP Signals in Normal Neuro-Physiological Brain Activity

Nicolae Stelian Stanciu(✉)

Faculty of Electric Engineering, University Politehnica Bucharest,
Spaiul Independentei nr 313, 060042 Bucharest, Romania
sdie@upb.ro
http://www.electro.pub.ro

Abstract. Correlation analysis of EEG signals and VEP signal components, brings new information of normal brain activity. An EEG analyzer Zwőnitz 2000 was used. Relative power of brain spectral activity is obtained. Analysis used the Pearson correlation. The correlation procedure is applied both on the EEG power values in steady state and white flash stimulation and on VEP components amplitudes, in normal subjects. The VEP amplitudes are correlated and then compared with each other. Analysis indicates that new features of electrical processing of visual signal appear involving both higher levels of linearity on minimized space areas of brain activity and patterns of electric conduction and transport. The electric brain activity tends to construct pathways on which the signal is not distorted on its shape. The electric energy transport of signal is minimal and optimized in normal brain activity. By VEP correlation are obtained the links between VEP structures generators.

Keyword: EEG signal · Linearity · Correlation · VEP signal

1 Introduction

Worldwide the humans try to understand deeply their nature, especially the way the human brain works and activates. Synthetically, this paper analyzes the correlation between EEG activity power in different brain areas and the correlation of visual evoked potentials components acquired in steady state and white flash stimulation. The aim of the work is the identification of neuronal features and electric features of the signal conduction by revealing the correlated processes implied in different conditions of stimulation for generation the visual perception. The procedure of recording of signals and then correlate these values by each other is done by standard procedure in normal functional cases, and no pathology situation is included [3, 5, 6]. The Results Sect. 3 (notice Sect. 3, A, B, C) highlight both new and specific electric features of the signal conduction and new neuronal features which appear in different conditions of flash stimulation in normal status. The Discussion Sect. 4 clearly explains the correlation results presented in Sect. 3. As such, the goals and the claims included in the paper's abstract presented above are met.

S. Vlad and N. M. Roman (Eds.): MEDITECH 2020, IFMBE Proceedings 88, pp. 18–26, 2022.
https://doi.org/10.1007/978-3-030-93564-1_3

2 Data Base and Methods

The database used in this study was obtained in Neurophysiology Department of a Neurology Hospital, [3] by selecting 30 normal cases in correlation analysis of the values of weighted electric EEG power, and 10 normal ones in correlation analysis study of VEP amplitude components. The study was done using an EEG analyzer Bioscript - Zwonitz 2000 with EEG QA (Quellen Ableintung) method of recording.

The records have of a length of 3 min each, both in stand-by and flash stimulation conditions, with time base of 250 ms, with 25 sweeps number for VEP recording, frequency stimulating of 1 Hz, signal filters between 0.3–70 Hz, with Map EEG 2000 analysis signal software. The electric EEG signal is digitized with resolution at 8 bits over 50 mV. The white flashes for VEP signal were emitted from 30 cm in front of the eyes, in 30 normal subjects. Each record for VEPs contains of two simultaneously recorded electric signals taken from Occipital left-right, O1-O2 in flash condition. The EEG records consists of 19 recording areas spread over the scalp in standard schema of 10–20 electrodes and the null electrode connected to the ear [1, 3, 5]. In EEG analysis the electric signals after the recording on each brain area, are then correlated on each other and results obtained are mapped for each of the 19 recording areas, both in steady state and in flash stimulation condition. The VEP amplitude signals are correlated by each other and then mapped both for Left and Right Hemisphere [2]. Correlation analysis used Pearson correlation coefficient [4]:

$$\rho_{x,y} = \rho = \frac{Cov(X, Y)}{\sigma_x \sigma_y} \tag{1}$$

with its propriety

$$-1 \leq \rho_{x,y} \leq 1 \tag{2}$$

The covariance coefficient is:

$$Cov(X, Y) = \frac{1}{n} \sum_{j=1}^{n} (x_j - \mu_x)(y_j - \mu_y) \tag{3}$$

where μ_x, μ_y are the averages of variables x and y; n = number of samples.

The standard deviation σ used for x and y is calculated by the following formula:

$$\sigma_x = \sqrt{\frac{n\sum x^2 - (\sum x)^2}{n(n-1)}}; \sigma_y = \sqrt{\frac{n\sum y^2 - (\sum y)^2}{n(n-1)}} \tag{4}$$

The relation between coefficients a, b in the linearity expression $y = ax + b$ where with a is the slope and b the intersection of regression line with OY axe, and $\rho_{x,y}$, the Pearson correlation coefficient between X and Y variables is the following [4]:

$$a = \rho \cdot \frac{\sigma_y}{\sigma_x} \text{ and } b = \mu_y - \rho \cdot \frac{\sigma_x}{\sigma_y} \cdot \mu_x = \mu_y - a \cdot \mu_x \tag{5}$$

3 Results

A) The correlation of EEG electric weighted power in normal individuals in steady-by state for Delta, Theta. Alpha and Beta is indicated in maps below:

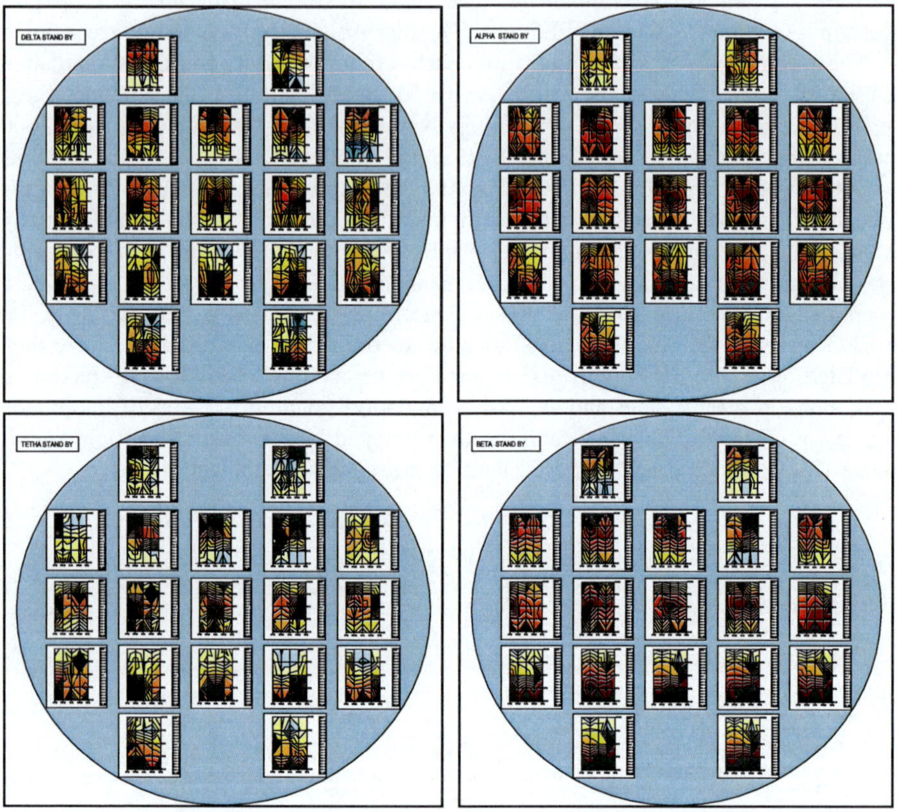

Fig. 1. The distribution of the correlation values of weighted electric EEG power in stand by state for Delta, Theta, Alpha, Beta is indicated in maps above [2]. Generally, the values of correlation decrease slowly from activation area for all frequency domains. Some patterns tend to be formed in Delta F_3-F_4,C_3-C_4,P_3-P_4,O_1-O_2; on Theta in PF_1-PF_2,Pz-P_4-T_6, O_1-O_2; on Alpha the tendency is noticed in F_3-F_4,T_3-T_5, T_5-P_3-P_z-P_4-T_6-O_1-O_2; on Beta PF_1-PF_2, F_7-F_3,C_3-C_4, T_5-P_3-P_z-P_4-T_6-O_1-O_2; in these patterns values gradually decrease, (look inside article), [2].

B) The correlation of EEG electric weighted power in normal individuals by flash stimulation for Delta, Theta, Alpha and Beta is indicated in maps below:

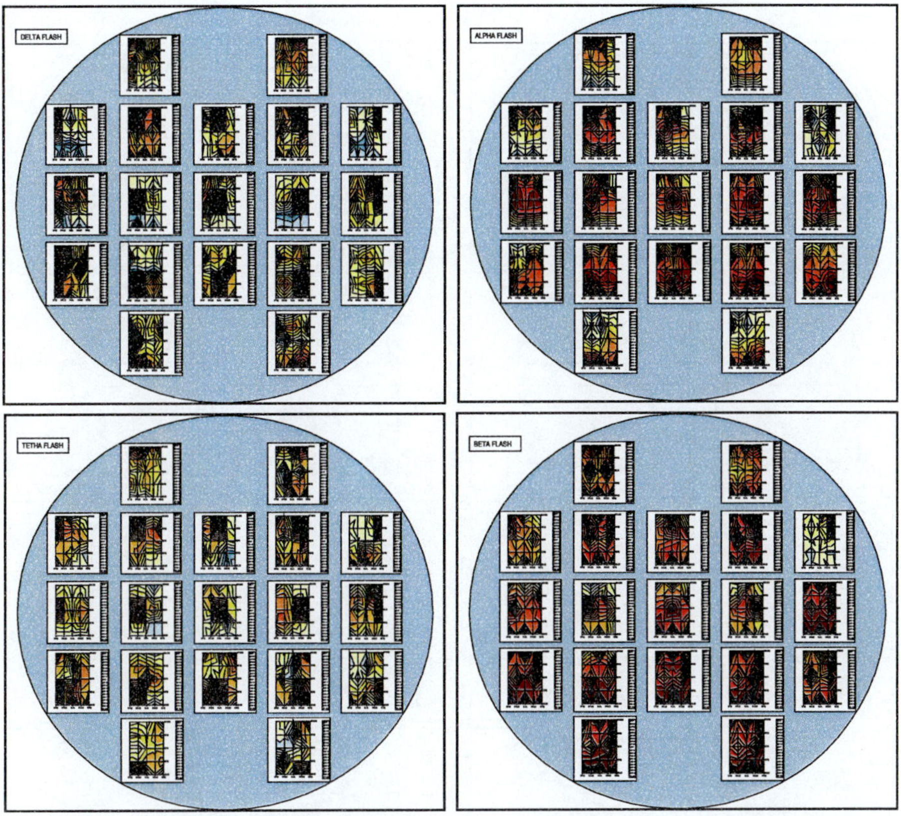

Fig. 2. The distribution of the correlation values of weighted electric EEG power by flash stimulation for Delta, Theta, Alpha and Beta is indicated in maps above [2]. On Alpha, Flash in F_3-F_4, T_3-T_4, P_3-P_4, O_1-O_2 it's noticed the constructed patterns. The Alpha domain is connected with basic cortical activation. On Beta domain at Flash, it's noticed the presence of pattern formation in PF_1-PF_2, T_3-T_4, T_5-P_3-P_z-P_4-T_6-O_1-O_2, C_3-C_4. To be noticed that the concentration of correlation values around activation areas at flash are higher than in stand-by state, mostly in Delta and Alpha, (look inside article), [2].

C) The VEP standard components correlated to each other, and the average correlation of the VEP components related to the flash stimulation energy, in normal individuals at white flash stimulation [2].

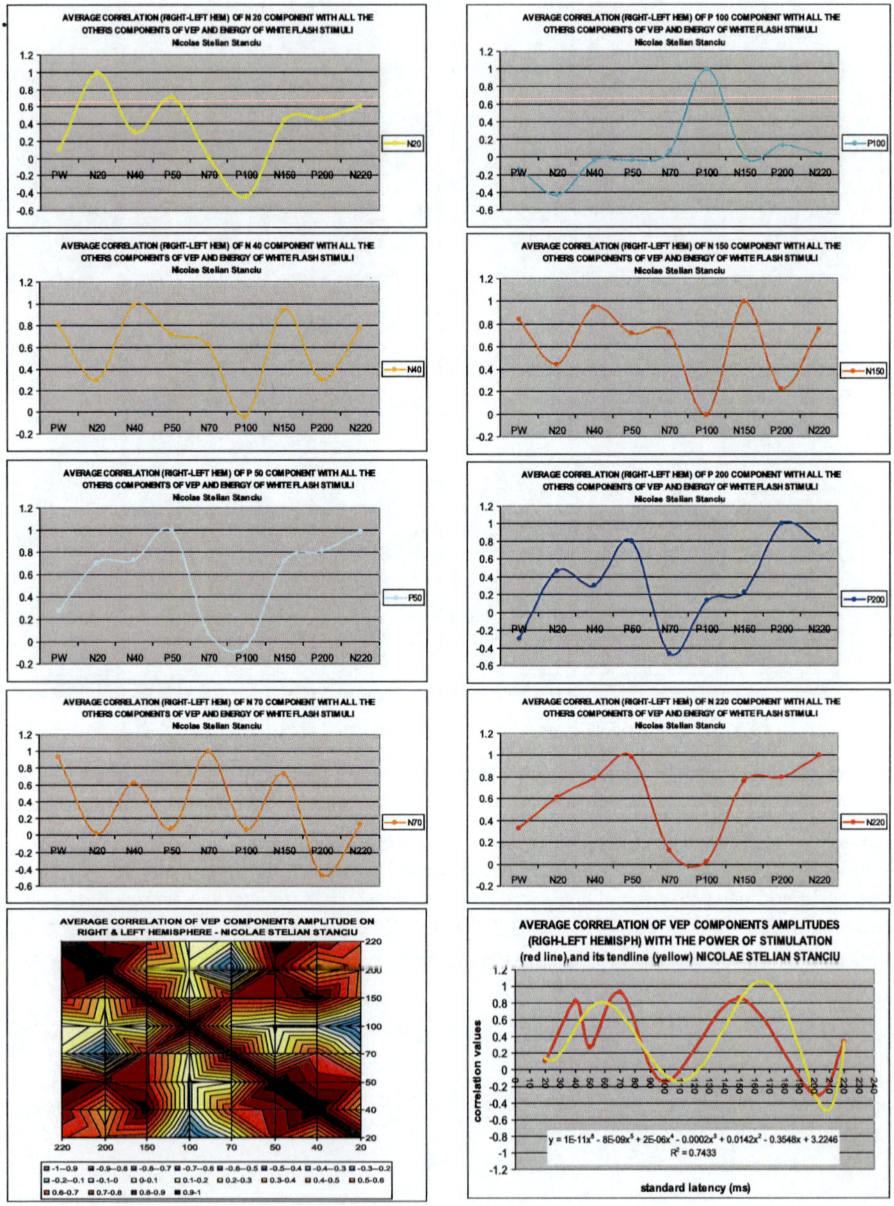

Fig. 3. Correlation of VEP components with each other and with the energy of flash [2]. To be noticed the correlation and the relation between inhibitory and excitatory VEP components

4 Discussion

EEG Correlation Analysis

In steady state Delta correlation values at stand by state, appears as being spread bilaterally, toward symmetrical hemispheric areas. The spreading of correlation values is gradually done with an average spatial gradient indifferent of recording position. The prefrontal and frontal areas are clearly delimited of occipital areas. and they have opposite correlation values in steady state. Theta correlation at steady state tends to spread bilaterally symmetric to activation area and appear to be restrained spatially on prefrontal areas. On parietal-occipital areas tends to form correlation patterns and have opposite signs as in Frontal areas F3-Fz-F4. On Alpha, correlation values get higher values and spread bilaterally. The temporal-parietal-occipital areas tends to form correlation patterns. On Beta domain the correlation values are continued with higher values in parietal and occipital areas. The Beta values of correlation are signed opposite in prefrontal and occipital areas (see Fig. 1). At stand-by state, the values of correlation decrease slowly from activation area for all frequency domains. As a synthesis, some patterns tend to be formed in Delta domain: F_3-F_4, C_3-C_4, P_3-P_4, O_1-O_2; on Theta domain in PF_1-PF_2, Pz-P_4-T_6, O_1-O_2; on Alpha domain, the tendency is noticed in F_3-F_4, T_3-T_5, T_5-P_3-P_z-P_4-T_6-O_1-O_2; and on Beta domain pattern tendency is seen in PF_1-PF_2, F_7-F_3, C_3-C_4, T_5-P_3-P_z-P_4-T_6-O_1-O_2. Inside these patterns, the values of correlation gradually decrease [2].

By Flash stimulation, at Delta it's noticed the reducing values of spatial domains far away from activation zone, on which the correlation values get higher than in steady state. At Theta Flash the correlation with higher values is restricted mostly to the activation areas. It tends to form correlation patterns in temporal T3–T4. In Alpha Flash it exists patterns of correlation widely spread all over the cortical areas, mostly symmetric on frontal-parietal-temporal areas. In Beta domain at Flash exists lots of patterns of correlation spread all over the cortical areas, with higher values in temporal-parietal-occipital areas. These patterns tend to avoid central areas C3–C4, (see Fig. 2). As a synthesis at flash stimulation, on Alpha, pattern construction is observed in F_3–F_4, T_3–T_4, P_3–P_4, O_1-O_2. The Alpha domain is connected with basic cortical activation (thalamic activation). On Beta domain at Flash, it's noticed the presence of pattern formation in areas: PF_1-PF_2, T_3-T_4, T_5-P_3-P_z-P_4-T_6-O_1-O_2, C_3-C_4. To be noticed that the concentration of correlation higher values around activation areas at flash is greater than in stand-by state, mostly in Delta and Alpha, and the tendency for pattern formation increase as the frequency increase, by flash stimulation [2].

VEP Correlation Analysis:

It shows different structural and electric functional links in between neuronal structures involved into the electric signal processing along the visual pathways [1, 2, 6]. The excitatory VEP components have a positive correlation values with the energy (power) of stimulation, and the inhibitory ones have negative correlation values with the energy of flash stimuli. There are specific links between different VEP components which show the tendency in producing a migration of correlation values between neuronal VEP electric generators, which depends on stimulation energy (see Fig. 3). As a synthesis, the EEG and VEP signal amplitudes are signals of electric activity, then some average correlation on left-right hemisphere in normal subjects, and diagrams of VEP components amplitude can be obtained. The Fig. 3, indicates that the inhibitory VEP's component P_{100} is negatively correlated with all the others excitatory components of the visual response N_{20}, N_{150}, and forms a coupled system with N_{70}. Inhibitory component P_{50} is positively correlated with the excitatory Hippocampus component N_{220}. To be noticed the negative correlation value of P_{100} with excitatory component N_{70} and its correlation values with the level of stimulation flash power. The excitatory component N_{70} correlates positively with all the others excitatory components both on the primary and secondary pathway N_{150}, N_{220}. To be noticed, the N_{70} positive correlation value with the level of power of flash stimulation and its almost zero correlation value with retinal response N_{20}. The dependency of power of flash stimulation shows that both the primary pathways VEP components N_{20}-P_{30}-N_{40}-P_{50}-N_{70} and the second VEP excitatory components N_{150}, N_{220} have positive correlation values with stimulation flash power. To be noticed that the inhibitory components P_{100} and P_{200} have negative correlation value with the flash stimulation power (see Fig. 3, the yellow curve of trend, down right). So correlation analysis highlights a control system of excitatory and inhibitory VEP components, which is connected to the neuronal structures which generate the visual signal.

5 Conclusion

The correlation of EEG and VEP is a measure of functional linear linkage in between the neuronal structures implied in electric visual conduction. It is noticed by correlation analysis that exists in EEG activity the tendency for electric pattern activity construction. Some of the features of pattern formations (see Sect. 3 and 4) are further on exposed:

1 - These patterns (to be noticed Fig. 1 and Fig. 2, Sect. 3 and 4) have different shapes related to the activation areas, frequencies and condition of stimulation. These patterns in EEG correlation activity are clearly viewed in Alpha and Beta by Flash, placed symmetrically, mostly in Parietal – Occipital – Temporal areas.

2 - These patterns highlight the necessity of construct a structure for electric conduction in excitation phenomena by creating pathways which go through different neuronal areas, including both structures with grey matter and with white matter, with different levels of morphology and functionality. The way these correlation values appear reveals how the electric field interacts with the neuronal white or grey substance [7] and finally, how the patterns are constructed [2].

3 - Inside these patterns, the correlation values of EEG weighted power indicate that exists a strong linear relation between the power of electric signal distributed on different cortical areas, fact which give birth to the pathways formation based on the law of linearity.

4 - These patterns appear to be necessary, both for construction of electric pathways and sustain an optimal electric energy consuming [1] on neuronal coherent behavior and for generating linear responses in neuronal structures [2].

5 - All these happens in order to preserve the shape of VEP [2]. The VEP correlation analysis shows electric functional links between neuronal structures situated in visual system and show a tendency of migration of correlation values in between VEP generators. [2] This feature is sustained also by the model of conduction pathways in between neuronal structures implied in the Visual Perception indicated in [5, 6].

6 - The correlation analysis applied on EEG and VEP signals highlights a control system of excitatory and inhibitory VEP components, which is connected to the neuronal structures which generate the visual signal and use linearity, in order to preserve the signals shapes and energy.

7 - Moreover, as a linkage and possible application to the present paper's subject are some of the EEG correlation studies which can reveal connections between some features of EEG oscillatory activity and inter or intra hemispheric correlation [8] or to point the sex differences in inter hemispheric correlations and spectral power of EEG activity [9]. Knowing better and better this domain of the EEG electric activity correlation generates studies connected to the present subject which imply optimizing the cognitive performance with inter-hemispheric EEG correlation [10], an now a day priority subject in education domain.

Acknowledgment. Many thanks to all in the research team of Neurophysiology Department of Medical Academy for the EEG and VEP data recording, and for professional advices and scientific discussions that implied the neurophysiologic themes attained, and for the work editing and presentation advices. Many thanks to those who has helped me to obtain an optimal status for editing this article.

Conflict of Interest. The author declare that he has no conflict of interest.

References

1. Stanciu, N.S.: Biophysics of Neuronal Excitation. EstFalia Press, Bay Village (2017)
2. Stanciu, N.S.: Analiza de corelatie de semnal in cazuri normale si patologice. EstFalia Press, Bay Village (2020)
3. Nicolae, S., Mihaela, M., Dan, P.M.: Coherence between neocortical areas. Electrical activity recorded in monopolar and source derivation. Rom. J. Neurol. **44**(1–2), 109 (2006)
4. Constantin, A., Tudorel, A., Stelian, B.L.: Bazele teoretice ale Statisticii, Universitatea Spiru Haret, Editura Fundatiei "Romaina de Maine" (2000)
5. Mihai, P.D., Mihaela, M.: Neural generators of visual evoked potentials components. Rom. J. Neurol., Nr 35 (1997)
6. Stanciu, N.S.: Aspecte Biofizice ale Propagarii Influxului Nervos in Sistemul Visual. EstFalia Press, Bay Village (2013)
7. Bhavsar, R., Sun, Y., Helian, N., Davey, N., Mayor, D., Steffert, T.: The correlation between EEG signals as measured in different position on the scalp varying with distance; The University of Hertfordshire, Hatfield; Open University London, UK, 8th Annual International Conference on Biologically Inspired Cognitive Architecture, BICA 2017. Elsevier

8. Hermann, C.S., Struber, D., Helfrich, R.F., Engel, A.K.: EEG oscillations: from correlation to causality. Int. J. Psychophysiol. **103**, 12–21 (2016)
9. Juarez, J., Cabrera, M.C.: Sex differences in inter hemispheric correlation and spectral power of EEG activity. Brain Res. Bull. **38**(2), 149–151 (1995)
10. Cabrera, M.C., Gutierez, S., Ramos, J., Arce, C.: Interhemisferic correlation of EEG activity during successfull and unsuccessfull cognitive performance. Int. J. Neurosc. **39**(3–4), 2253–259 (1988). Published on line 07 July 2009

Information Theory-Based Methods in HRV Analysis

Zoltán Germán-Salló(✉)

"George Emil Palade" University of Medicine, Pharmacy, Sciences and Technology of Tirgu-Mures, Târgu Mureş, Romania
zoltan.german-sallo@umfst.ro

Abstract. HRV signals can be treated as discrete time series. Healthy hearts usually have a more complex beat rate, therefore a measured or estimated complexity indicator of the HRV signal can be used as diagnosis tool. Complexity measures are related to signal's information content and estimates have been made in many ways in last years. This paper uses information theory-based estimates of complexity and tries to find a relationship between the introduced method offered results and clinical observations using annotated signals from specific databases. The proposed procedure uses the permutation entropy as complexity measure, in order to estimate the complexity of HRV signals.

Keywords: Heart rate variability (HRV) · Entropy · Complexity

1 Introduction

Heart rate variability is actually the variance in time between the beats of a human heart and is strongly related to the activity of autonomic nervous system (ANS). Heart rate, recorded as time duration between two heart-beats or as a distance R-R (R being the peak of QRS complex on a surface electrocardiogram (ECG)), is a measure of human body's response to parasympathetic and sympathetic inputs. This is a sign that the nervous system is balanced, and the organism is adequate of adapting to its environment. Heart rate variability is an extremely sensitive metric which varies greatly throughout the day, from one day to the next, and from one person to another. Lately, scientific research has shown a relationship between low HRV and different cardiovascular diseases. Quantifying the complexity of a Heart Rate Variability signal is an important task in order to have a deeper insight into the ANS mechanisms governing the human body. This paper introduces the permutation entropy as an information theory-based parameter, in order to evaluate the complexity of HRV signals. Both of healthy and unhealthy signals are analyzed, the evaluations are performed on annotated signals from specific databases. The used test signals are obtained from PhysioNet, a research resource for studies of complex physiologic and biomedical signals, namely from Normal Sinus Rhythm RR Interval Database and Congestive Heart Failure RR Interval Database, obtained from long-term ECG recordings. This method offered results are compared with other measures given by sample entropy based approach. This paper is organized as follows: the second chapter offers a review of the most used entropy metrics and presents the permutation entropy as a signal processing procedure. In chapter three is

S. Vlad and N. M. Roman (Eds.): MEDITECH 2020, IFMBE Proceedings 88, pp. 27–34, 2022.
https://doi.org/10.1007/978-3-030-93564-1_4

presented the proposed procedure, chapter four brings the experimental results. Finally, concluding remarks and possible future work ideas are presented.

2 Information Theory-Based Indices in HRV Analysis

2.1 Entropy Indices Overview

Since Shannon's generic definition, a lot of other entropy interpretations and definitions appeared in nowadays scientific researches. Basically, the entropy is strongly related to the information content of a discrete event (or signal) due to his probability density function [1]. The concept of Shannon entropy is a measure of the average information content of a discrete information source, defined by the following expression

$$S(x) = -\sum\nolimits_{i=1}^{n} p(x_i) log_2[p(x_i)] \tag{1}$$

where *X(xi)* is a discrete random variable, *p(xi)* is the probability.

Some of lately introduced entropies are the Approximation Entropy (AE) and the Sample Entropy (SE) relatively simple indices which estimate the conditional probability that two sequences of given number successive samples in a signal that are similar to each other (within a given tolerance) will remain similar when one more successive sample is included. The AE was first proposed by Pincus [2], and it shows a good performance in the characterization of randomness even when the data sequences are not very long [3]. These entropies asses the irregularity or randomness of a time series.

The sample entropy (SE) as a modification of AE was introduced by Richman and Moorman [4] and is the negative natural logarithm of the conditional probability that two sequences similar for a given number of points remain similar when one more point is added, where self-matches are eliminated in calculating the probability.

The probability that two sequences match for m points

$$P_m(r) = \frac{1}{N-m+1} \sum\nolimits_{i=1}^{N-m+1} C_i^m(r) \tag{2}$$

the number of matching templates for each template within a time series [6].

$$C_i^m(r) = \frac{1}{N-m} \sum\nolimits_{j=1, x \neq 1}^{N-m+1} H(r - d_{ij}) \tag{3}$$

H is the Heaviside function, defined as

$$H(z) = \begin{cases} 1\, if\; z > 0 \\ 0\, if\; z \leq 0 \end{cases} \tag{4}$$

For $m+1$, $C_i^m(r)$ and $\Phi^m(r)$ are calculated. Finally, sample entropy is:

$$SE(n,r) = -lg\frac{P_{m+1}(r)}{P_m(r)} \tag{5}$$

The Fourier entropy is another complexity index defined in frequency domain. To obtain this entropy, a power-spectral-density (PSD) of a time series is computed. The PSD is normalized over the definition domain *S*, to produce a probability-like distribution, the corresponding Shannon entropy is called the Fourier entropy [5].

$$FE = -\sum\nolimits_S PSD \cdot logPSD \tag{6}$$

Fourier entropy seems to be a reference in comparison with other methods, but often is the easiest approximation in case of large data sets.

The wavelet entropy (WE) uses orthogonal wavelet transforms (continuous or discrete) to characterize different discrete signals. According to the characteristic of orthogonal wavelet transform, at a certain time window the total signal energy E is the sum of energy E_j of each component obtained from a signal x(k) at m scales. The wavelet energy entropy is defined as

$$WE = -\sum\nolimits_j p_j \cdot logp_j, j = 1, \cdots m \tag{7}$$

It carries information about the degree of order/disorder associated with a multi-frequency signal response Usually, in this case the complexity measure is related to the number of wavelet components needed to approximate the analyzed signal [6].

The generalized entropies of Rényi allow to explore the information content enclosed by time-frequency representations (TFR) of biomedical signals. The Rényi entropy which is a generalized form of Shannon entropy can be evaluated in many ways, obtaining different outcomes. The definition of Rényi entropy

$$R(x) = \frac{1}{1-\alpha}log_2\left(\sum\nolimits_{i=1}^{n} p(x_i)\right) \tag{8}$$

where p(xi) is the probability that a random variable takes a given value out of n values, and α is the order of the entropy measure. The best choice depends upon the uniqueness of the considered application [7].

2.2 The Permutation Entropy

The permutation entropy describes complexity of a time series or discrete signals through phase space reconstruction, and takes into account non-linear behavior of it. It was introduced by C. Bandt and B. Pompe in 2002 [8]. As signal processing procedure, the algorithm will partition the one-dimensional time series into a matrix of overlapping column vectors using two parameters, the embedding time delay τ which controls the number of time periods between elements of each of the new column vectors and the embedding dimension D which controls the length of each of the new column vector. The embedding time delay τ any positive integer, the embedding dimension D will be between 3 and 7 [8]. In the new matrixeach column vector has a number of elements equal to the embedding dimension, D. In addition, the time period between each element in the vectors is equal to the embedded time delay.

The permutation entropy is a natural encoding which reflects the rank order of successive x_i in sequences of length n and thus defining permutation entropy by

$$PEn = \sum_{i=1}^{n} \pi_i \cdot log_2 \pi_i \tag{9}$$

π_i represent the relative frequencies of the possible patterns of symbol sequences, named permutations. These relative frequencies are obtained as appearance probabilities in the created patterns. One of the most important parameters in computing permutation entropy is the pattern length n which provide n! possible permutation patterns build from the numbers 1, 2, … , n. Their dynamical representation is shown in Fig. 1.

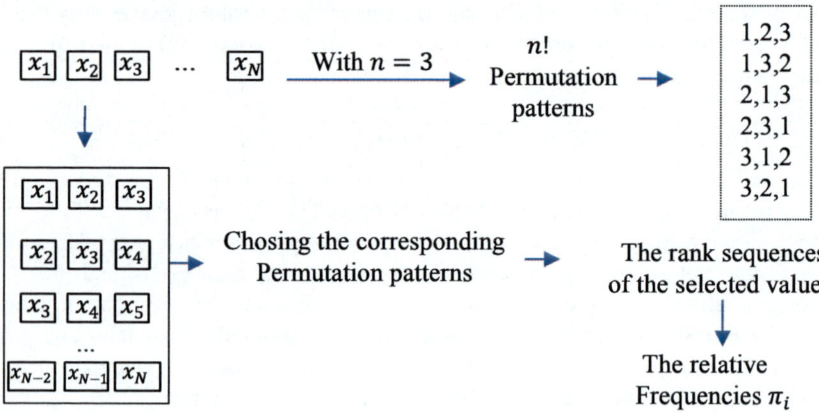

Fig. 1. Exemplification of permutation pattern probability computation for pattern length 3

The ranks are the indices of the values in ascending sorted order. As it can be seen, the permutation entropy is based on the notions of entropy and symbolic dynamics, assures time causality and is non-parametric meaning that is free of restrictive parametric model expectations. The permutation entropy has shown its usefulness in many signal analysis field as HRV analysis [11].

3 The Proposed Analysing Procedure

In this paper is performed a permutation entropy based estimation of HRV signal complexity. A higher complexity of HRV signal will be always a sign of a better response of the human body exposed to external excitations. A lower complexity will emphasize a worse response and eventually will be assosciated with increased risk of diseases. The used signals are taken from from PhysioNet, a research resource for studies of complex physiologic and bio-medical signals. Three signals are from Normal Sinus Rhythm RR Interval Database, they are the reference signals and other three are taken from Congestive Heart Failure RR Interval Database, all of them are obtained from long-term ECG recordings [9]. The six RR interval time series used in this analysis were preprocessed [10] by the removal of non-valid values caused probably by data aquisition errors. Figure 2 presents the applied procedure.

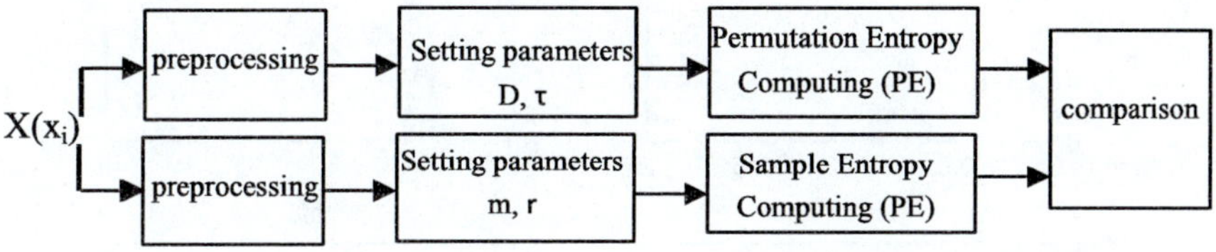

Fig. 2. The proposed procedure to compute the permutation entropy.

In the preprocessing phase, gaussian white noise can be added in order to evaluate the robustness against additive noise. Different levels of added noise can figure out this characteristic.

4 Experimental Results

The procedure is carried out under Matlab environment, the signals are normed in order to have an uniform appreciation over them. The normal sinus rythm signals are noted nsr1 (length 117690), nsr2 (length 102253), nsr3 (length 102001) and the others with chf1 (length 112181), chf2 (length 109296), chf3 (length 99382). They have various lengths, all of them are preprocessed and processed with their original sizes.

After preprocessing and normalization the histograms of these signals are computed and presented on Fig. 3. A first visual investigation of these histogram reveals the dynamics of the signals, the distribution of consecutive heartbeat intervals presents serious differences between normal and non-normal HRVs.

The values of obtained permutation entropies are presented Fig. 5, for both normal (Pen) and non-normal (Peb) signals. There is easy to notice the existing permutation entropy differences between them. These values are obtained for unity time delay and variable embedding dimensions which varies between 3 and 6.

m	3	4	5	6
PEn1	1.7433	3.0537	4.5358	6.1274
PEn2	1.74	3.0474	4.5367	6.1712
PEn3	1.7117	2.9415	4.3158	6.1712
PEb1	1.6797	2.8997	4.2817	5.7913
PEb2	1.7397	3.0567	4.5597	6.2134
PEb3	1.6474	2.8683	4.2167	5.6272

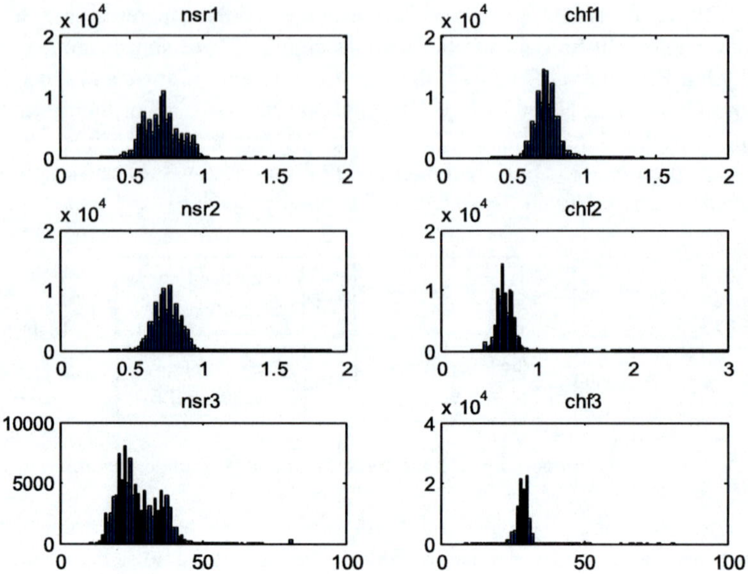

Fig. 3. The histograms of the used signals.

The plot clearly indicates differences between them, where, compared to a normal patient, the CHF patient has a lower variability. Figure 4 presents the obtained permutation values, a more complex signal as a healthy HRV has greater permutation entropy value. Also the differences between the permutation entropies of two signals increase with the embedding length, meaning that this parameter is important in complexity evaluation.

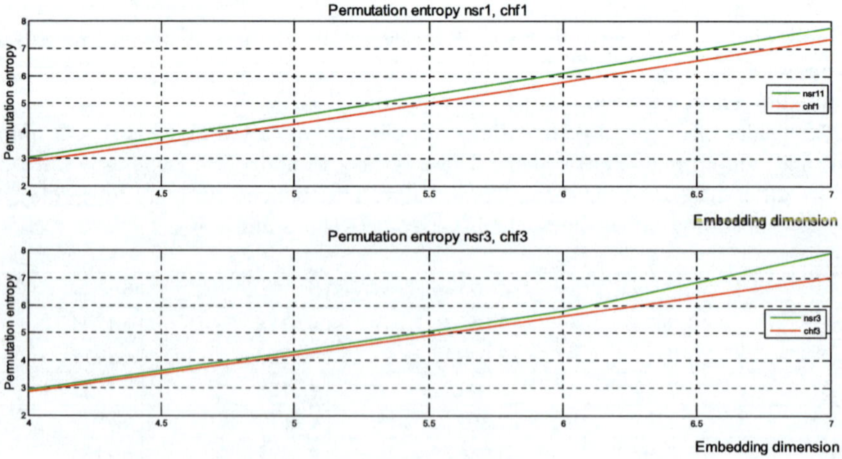

Fig. 4. Permutation entropy values obtained for different signal with different embedded dimensions.

In order to compare the results offered by permutation entropy computing, also sample entropies for embedding dimensions and a given tolerance (usually 20% of standard deviation) for the same two signals were calculated. The obtained values are synthesized on Fig. 5, the results show a similarity with the previous evaluation.

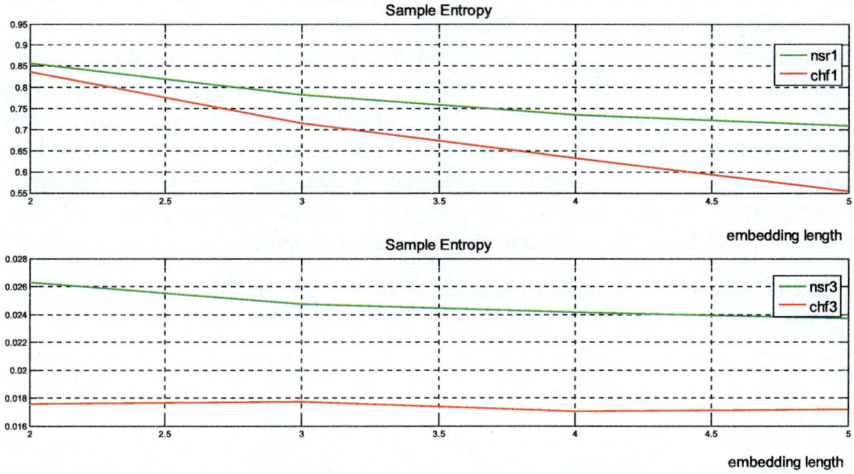

Fig. 5. The sample entropy for the studied signals with the standard tolerance

The effect of an additive noise can be observed on Fig. 6, where one of the normal sinus rythm signal was studied through the permutation entropy with different levels of noise. The noise level is relatively low, at higher level the algorythm of permutation entropy does'nt work properly. The graph shows that after a certain level of an gaussian random noise, the permutation entropy does not change because the noise level becomes comparably with the difference between two consecutive samples.

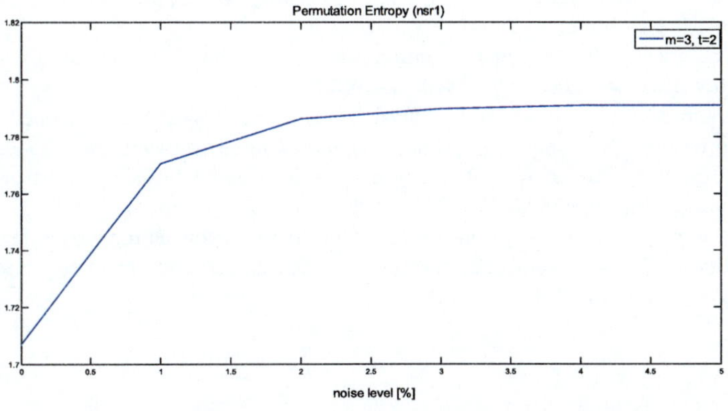

Fig. 6. The dependence of the permutation entropy (PE) on additive noise level

5 Conclusions

Investigation and analysis of HRV is a growing field of research. In this study permutation entropy of HRV sequences were calculated. This parameter obtained by a nonlinear method is a trustable complexity index, the experiments show that a HRV corresponding to the normal sinus rythm has a greater permutation entropy than other corresponding to some diseases. The one of disadvantages is that requires a lot of computational resources, especially with relativeley great embedding dimensions and embedding delays. The experiments emphasized also the fact that after a certain level of added noise this parameter could be almost invariant, therefore is recommendable to use previously denoised signals. Also, the amplitude information is discarded; the., equal values in the analysed signal are not properly dealt with. The sample entropy and the permutation entropy are different parameters which evaluate the complexity of signals, they can be used correlated with the signal properties.

Conflict of Interest. The author declare that he has no conflict of interest.

References

1. Shannon, C.E.: A mathematical theory of communication. Bell Syst. Tech. J. **27**, 379–423, 623–656 (1948)
2. Pincus, S.M.: Approximate entropy as a measure of system complexity. Proc. Natl. Acad. Sci. USA Math. **88**, 2297–2301 (1991)
3. Jiang, Y., Mao, D., Xu, Y.: A fast algorithm for computing sample entropy. Adv. Adapt. Data Anal. **03**(01), 167–186 (2011)
4. Richman, J.S., Moorman, J.R.: Physiological time-series analysis using approximate entropy and sample entropy. Am. J. Physiol.-Heart Circ. Physiol. **278**(6), H2039–H2049 (2000)
5. Fefferman, R.A.: A theory of entropy in Fourier analysis. Adv. Math. **30**(3), 171–201 (1978)
6. Rossoa, O.A., et al.: Wavelet entropy: a new tool for analysis of short duration brain electrical signals. J. Neurosci. Methods **105**(1), 65–75 (2001)
7. Renyi, A.: On measures of entropy and information. In: Proceedings of the Fourth Berkeley Symposium on Mathematical Statistics and Probability, vol. 1, pp. 547–561. University of California Press, Berkeley (1960)
8. Bandt, C., Pompe, B.: Permutation entropy: a natural complexity measure for time series. Phys. Rev. Lett. **88**, 174102:1–174102:4 (2002)
9. Goldberger, A., et al.: PhysioBank, PhysioToolkit, and PhysioNet: components of a new research resource for complex physiologic signals. Circulation **101**(23), e215–e220 (2000)
10. Unakafova, V.A., Keller, K.: Efficiently measuring complexity on the basis of real-world data. Entropy **15**(10), 4392–4415 (2013). Physical review letters
11. Ravelo-García, A.G., et al.: Application of the permutation entropy over the heart rate variability for the improvement of electrocardiogram-based sleep breathing pause detection. Entropy **17**, 914–927 (2015)

Margin Pruning Adaboost Classification with Nature Inspired Optimization for Alcoholic EEG Classification

Harikumar Rajaguru[1](✉) and Sunil Kumar Prabhakar[2]

[1] Department of ECE, Bannari Amman Institute of Technology, Sathyamangalam, India
[2] Department of Brain and Cognitive Engineering, Korea University, Seoul, Korea

Abstract. The diagnosis of alcoholism is of utmost importance to all the human beings in the world. Severe pathological effects of alcoholism can be witnessed in heart, brain, lungs, immune system, and liver. The alcoholic patients can be analyzed with the help of Electroencephalography (EEG) signals and with the advent of both optimization schemes and pattern recognition techniques, they aid the medical community for the automated diagnosis of this disorder by providing a good classification accuracy. In this work, initially features are extracted with the help of Support Vector based Non Linear Regression (SVNLR) technique and then it is optimized with two optimization algorithms such as Brain Storm optimization and Sine Cosine optimization before proceeding to classification with the help of a Margin Pruning Adaboost classifier. This kind of a different and interesting approach provided results with a high classification accuracy of 97.33% when SVNLR features and Brainstorm optimization is utilized to classify with Margin Pruning Adaboost Classifier.

Keywords: EEG · Alcoholism · Optimization · Regression · Classification

1 Introduction

The number of individuals consuming alcohol throughout the world is increasing day by day [1]. Alcoholism if consumed in moderation does not cause too much damage to the human body instead can enhance the sociability and mood of a person. But too much of alcohol consumption can lead to a condition called alcoholism. Alcoholism is a psychiatric condition where the consumption of alcoholic beverages is very high and can lead to adverse consequences to the health such as high blood pressure, stroke, cancer, and liver cirrhosis [2]. The functions of the central and peripheral nervous system too can get affected easily by this disorder. Alcoholism affects the social lives of the person and their respective family members to a great extent. Some of the famous techniques which aid the doctors to study about the chronic alcoholism are EEG, Computed Tomography (CT) and Position Emission Tomography (PET). The effect of alcohol on the cortical regions of the brain are easily analyzed by the EEG signals.

S. Vlad and N. M. Roman (Eds.): MEDITECH 2020, IFMBE Proceedings 88, pp. 35–44, 2022.
https://doi.org/10.1007/978-3-030-93564-1_5

The usage of EEG signals can work efficiently when compared to other techniques due to its ease in the acquirement and cost effectiveness. With the development of a lot of pattern recognition techniques, it is easy to classify EEG signals in medical practice as it provides a good assistance to the doctors in automated diagnosis of this disorder. The strategies utilized to assess and classify the EEG signals have become more versatile and is nearing to perfection. As the diagnosis becomes more reliable with less cost, EEG signals feasibility for the alcoholism diagnosis has become very high.

2 Materials and Methods

The database utilized in this paper was obtained from UCI-KDD dataset [3]. The dataset has EEG recordings of 122 normal and alcoholic EEG patients. The electrodes position was placed in a standard manner. For various stimuli, all the subjects have undergone about 120 trials. 256 Hz was the sampling rate of the system and the resolution range was 12 bits. To eliminate the artifacts, the EEG signal was band pass filtered between 0.5 Hz and 70 Hz. Thirty such data recordings were considered for the alcoholic EEG data set and it was subdivided into 120 files each comprising a total length of about 2560 samples. The block diagram of the work is shown in Fig. 1.

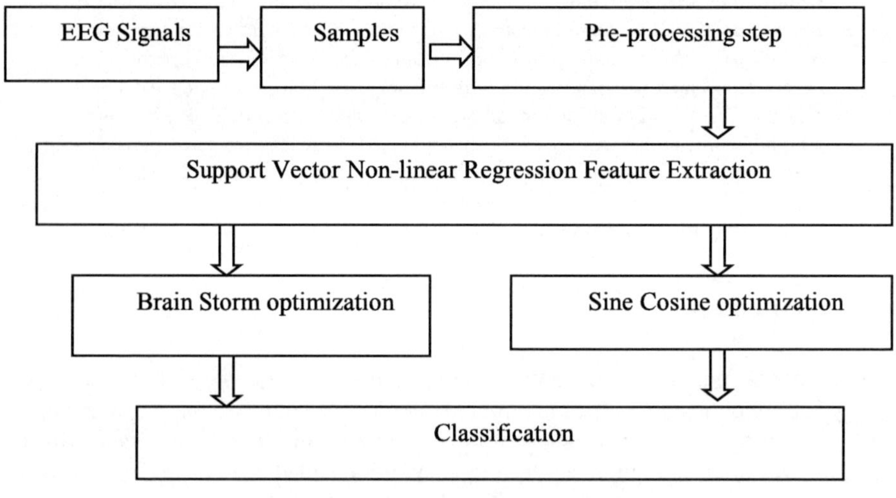

Fig. 1. Block diagram of the work

3 SVNLR Based Feature Extraction

Instead of extracting the standard statistical type of features like mean, variance, standard deviation, kurtosis, entropy etc… here the type of features extracted is SVNLR features [4]. For Support Vector Regression (SVR), between the matrix of the wavelength signals W and the corresponding concentration Z, a linear regression can be performed as

$$Z : Z = qW + A \tag{1}$$

where the matrix of the weight coefficient is expressed as q and the bias vector is represented as A. Based on the Lagrange multiplier and Karush-Kuhn-Tucker (KKT) condition, SVR is expressed as

$$q = \sum_{i=1}^{l} (\alpha_i - \alpha_i^*) w_i \tag{2}$$

where a variable of matrix W is expressed as w_i. The corresponding Lagrange coefficients are expressed as α_i and α_i^*, the total number of samples are expressed as l. Therefore, the linear regression equation is expressed as

$$Z = \sum_{i=1}^{l} (\alpha_i - \alpha_i^*) w_i^T w_j + A \tag{3}$$

The calculation of the regression function is done by the concentration of the component $Z = v\emptyset(w) + A.\emptyset(w)$ and is a high dimensional mapping that is initiated to finish the non-linear regression. w is an input variable of the wavelength signals, v and A acts as the same role in SVR. The kernel function $K = \emptyset\emptyset^T$ is defined [5]. Using the following equation, the regression model's component concentration is expressed as

$$\begin{gathered} Z = \sum_{i=1}^{l} (\alpha_i - \alpha_i^*) \o(w_j)^T \emptyset(w_j) + A \\ Z = \sum_{i=1}^{l} (\alpha_i - \alpha_i^*) K_y + A \end{gathered} \tag{4}$$

where for the validation set Z denotes an output variable. The matrix of validation variable feature space mapping is expressed by $\emptyset_y$. K_y expresses the matrix comprised of $K_{ij} = K(w_i, w_j)$, where the input variable to the validation set is expressed as w_i and the input variable to the calibration set is expressed as w_j. The Kernel based SVR leads to the attainment of the SVNLR features. The original alcohol EEG data has samples of about [2560 × 64], that is for 64 channels with each of them having a duration of 10 s. The sampling rate is 256 samples per second. Then it is reduced to SVNLR features as [256 × 64], that is ten times reduction per channel. Then these features are further optimized with the two optimization algorithms to have an eight-time reduction of only [2560 × 1] feature length per patient.

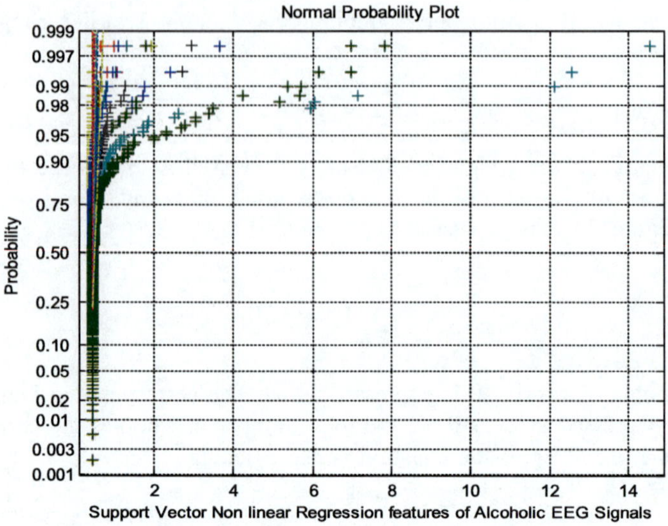

Fig. 2. Normal plot for Support Vector Non linear Regression features

Figure 2 shows the normal plot of SVNLR features extracted from alcoholic EEG signals. The presence of skewness and non-Gaussian representation in the graph indicates the need for additional optimization techniques so that it could ease the burden of the classifier. Therefore, Brain Storm optimization and Sine Cosine optimization are utilized in this work and is explained in Sect. 4.

4 Optimization Techniques

The extracted features are further optimized with the two optimization techniques used here such as Brain Storm optimization algorithm and Sine Cosine optimization algorithm.

4.1 Brain Storm Optimization Algorithm

Some problems which cannot be solved by person can be solved with the intervention of many humans and based on this inspiration, Brain Storm optimization algorithm was developed [6]. It is actually a type of clustering algorithm where the solutions are clustered into several types. To produce the new offspring, one or three clusters are selected. The new off springs are then compared with the best solution of the entire population present in the same cluster. This is done so that the population is updated easily. The strengthening of the local search ability is done if parents are chosen from one or two neighboring clusters. The global search ability is heightened if the new offspring generated by parents are selected from three random clusters. The procedure is explained in Pseudocode 1.

Pseudocode 1:

Initialize: Generate 'M' solutions in a random manner.

Problem-specific method is used to generate N solutions

Evaluate these solutions

While the stopping criteria is not satisfied **do**

Clustering process: M solutions are clustered into Q clusters by a clustering algorithm (In our work K- Nearest Neighbour (KNN) is utilized for clustering).

Generation of new solutions: To produce new solutions, select one or two clusters randomly

Updating: The new solution is compared with the existing solution present in the same cluster.

The best is kept and the other is discarded.

4.2 Sine Cosine Algorithm

With a random set of search agents, this algorithm initiates the optimization process [7]. Using a fitness function, the quality of each agents is evaluated. After the fitness is evaluated, the algorithm traces and finds the best search agents obtained and then assigns it to the far away destination points. Based on this, the updation of the other solution is done. By undergoing a lot of iterations, this algorithm can reach the expected solution. Once the termination criteria are satisfied, the algorithm terminates. The procedure is explained in Pseudocode 2.

Pseudocode 2:

Step 1: A random set of search agents is generated

Step 2: The fitness of every search agents is evaluated

Step 3: The best solution achieved so far is updated

Step 4: $q_1, q_2, q_{3,} q_4$ are updated

Step 5: The search agents' position is updated

Step 6: While (t <maximum number of generations)

Step 7: The best solution obtained so far is returned as the global optimum.

As the number of iteration increases, the updation of the ranges of the sine and cosine function A_j^t shall be done. For both the phases, this update in position is expressed as:

$$A_j^{t+1} = A_j^t + q_1 \times \sin(q_2) \times \left| q_3 M_j^t - A_j^t \right| \tag{5}$$

$$A_j^{t+1} = A_j^t + q_1 \times cos(q_2) \times \left| q_3 M_j^t - A_j^t \right| \tag{6}$$

The above equations are combined and used in the following equation as:

$$A_j^{t+1} = \begin{cases} A_j^t + q_1 * \sin(q_2) * \left| q_3 M_j^t - A_j^t \right|, q_4 < 0.75 \\ A_j^t + q_1 * cos(q_2) * \left| q_3 M_j^t - A_j^t \right|, q_4 \geq 0.75 \end{cases} \quad (7)$$

According to the above equation, there are four main parameters such as q_1, q_2, q_3 and q_4. These are helpful to determine the position of the new solution. The next position's region is dictated by the parameter q_1 which could be either in the space between the destination and solution. The progressive movement towards or away from the destination is determined by q_2. A random weight for the destination is brought out by q_3 in order to define the effect of destination from the distance. The switching between sine and cosine components is done with the help of the parameter q_4. To have a good balance between the exploration and exploitation phases, the sine and cosine ranges is adaptively changed using the following equation as

$$q_1 = b - t\frac{b}{T} \quad (8)$$

where the current iterations is expressed as t and the maximum number of iterations is expressed as T, b is a constant and in our experiment the value of it is expressed as 4 to get the best value after going through several rounds of trial and error basis with other values. After the fitness is evaluated, a better solution is produced by the initial population which helps to update the position. To achieve a better exploitation capability the periodic behavior of the sine and cosine function necessitates the solution to be generated. Between the obtained current solution and the best solution, if the new solution is generated, then the global search capability is easily generated during this optimization method.

In time domain, one of the ways to indicate the statistical property of a signal is by the usage of Hjorth parameter and the most important parameters here are the mobility and complexity [8]. The Hjorth parameters are calculated for both the optimizations as expressed in Table 1.

Table 1. Hjorth parameters for both the optimization techniques.

S. no	Optimization techniques	Complexity	Mobility
1	Brain Storm optimization	1.6428	0.0439
2	Sine cosine optimization	1.6877	0.1623

5 Margin Pruning Classification

Instead of using the conventional classifiers like neural networks and before proceeding into advanced deep learning methods an attempt to use a variant of the standard Adaboost classifier called Margin Pruning Boost Classifier [9] is done. To mitigate the influence of noise like instances, Margin Pruning boost was designed for our work.

(1) The initial weights are set as

$$w_{l,1} = 1/2, \text{ where } l = 1, 2, \ldots., N \tag{9}$$

(2) The following tasks for $t = 1, 2, \ldots., T$ is performed.

a) Based on the weighted instances, a weak classifier in trained and then W_{t+}^k and W_{t-}^k is calculated for every partition P_t^k.

b) The weak hypothesis $f_t^k(x)$ for every partition P_t^k is computed by

$$f_t^k(x) = \frac{(W_{t+}^k - W_{t-}^k)}{(W_{t+}^k + W_{t-}^k)} \tag{10}$$

c) The weights of instances are updated as

$$w_{l,t+1} = \exp\left[-y_l \sum_t f_t(x_l)\right] \tag{11}$$

d) A threshold Z_{t+1} is set by the following equation

$$Z_{t+1} = \max_l\{w_{l,t+1}\} - \frac{\max_l\{w_{l,t+1}\} - \min_l\{w_{l,t+1}\}}{50} \tag{12}$$

For any instance x_l, if $w_{l,t+1} > Z_{t+1}$, then reset $w_{l,t+1} = 1$ and $\sum_t f_t(x_l) = 0$. Compute $A_t = \sum_i w_{l,t+1}$, then the normalization is done by letting $w_{l,t+1} = \frac{w_{l,t+1}}{A_t}$.

(3) Assume $F_T(x_l) = \sum_{t=1}^{T} f_t(x_l)$, the output of the strong classifier $G(x_l)$ is expressed as

$$G(x_l) = sign[F_T(x_l)] \tag{13}$$

6 Results and Discussion

In this work, a K-fold cross-validation methodology was used. At the start, in this methodology the splitting of the dataset into 'K' equal size points are done. To train the classifiers, $K - 1$ groups are used, and the remaining part is utilized for testing. The validation is repeated for K number of times. The computation of the classifier performance is obtained based on the K results. Various values of K are selected and in our work, $K = 10$. Therefore, 90% of the data was used for training and 10% for testing in the 10-fold cross-validation method. The procedure was repeated 10 times for each fold of process. The performance metrics parameters utilized here are TP, TN, FP and FN and they denote True Positive, True Negative, False Positive and False Negative, respectively. The Tables 2 and 3 shows the consolidated result analysis of SVNLR features with both the Brain Storm optimization and sine cosine optimization along with Margin Pruning Adaboost classifier. The performance metrics such as classification accuracy, sensitivity, specificity, and Performance Index (PI) are explained as follows.

$$Accuracy = \frac{TP + TN}{TP + TN + FP + FN} \times 100 \tag{14}$$

$$Sensitivity = \frac{TP}{TP + FN} \times 100 \tag{15}$$

$$Specificity = \frac{TN}{TN + FP} \times 100 \tag{16}$$

$$PI = \left(\frac{(TP + TN) - FN - FP}{(TP + TN)} \right) \times 100 \tag{17}$$

Table 2. Consolidated result analysis of SVR non-linear features and Brain Storm optimization with Margin Pruning Adaboost Classifier.

Classifiers	PI (%)	Sensitivity (%)	Specificity (%)	Accuracy (%)
Margin Pruning Adaboost Classifier	94.49	94.79	100	97.33

Table 3. Consolidated result analysis of SVR non-linear features and Sine Cosine optimization with Margin Pruning Adaboost Classifier

Classifiers	PI (%)	Sensitivity (%)	Specificity (%)	Accuracy (%)
Margin Pruning Adaboost Classifier	88.38	89.59	100	94.795

Table 2 shows that a higher accuracy of 97.33% is accomplished in the case of Brain Storm optimization method with Margin Pruning Adaboost Classifier and it is due to higher PI and perfect specificity with a less sensitivity. As depicted in the Fig. 3. the scatter plot among Brain Storm optimization and Sine Cosine optimization, the Brain Storm optimization (blue colour) covers the entire region of the classification rather than a small, clustered area as shown for Sine Cosine optimization (red colour). This leads to lower value of 94.79% accuracy in the Margin Pruning Adaboost classifier for Sine Cosine optimization with a less sensitivity value as shown in the Table 3.

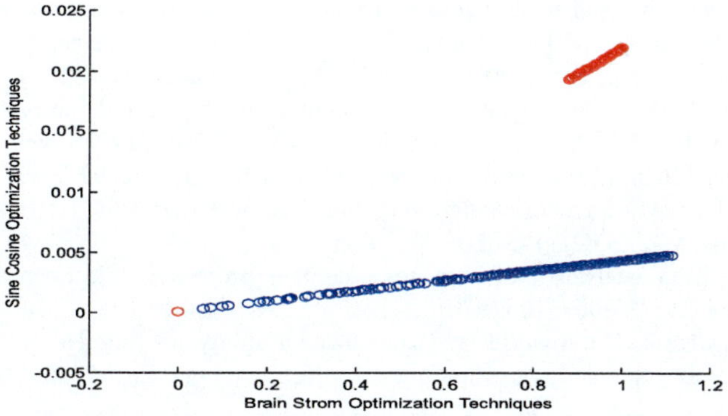

Fig. 3. Scatter plot among Brainstorm optimization and Sine cosine optimization

7 Conclusion

Around the world, one of the serious neurological disorder affecting many people is alcoholism and because of it, many people are suffering from deadly diseases like heart attacks, liver cirrhosis, cancer etc. Therefore, a computer aided automated analysis of alcohol classification from EEG signals is of crucial importance. Here in our work, the approach developed seems to be quite effective, easy, and useful for the alcoholic EEG signal classification. The results prove that for SVNLR features when optimized with Brain Storm optimization and classified with Margin Pruning Adaboost classifier, a higher classification accuracy of 97.33% is obtained along with a PI of 94.49% when compared with that of utilizing Sine Cosine optimization and classified with Margin Pruning Adaboost classifier which reported an accuracy of 94.79% and a PI of 88.38%. Future works aim to work with other optimization techniques and other feature extraction and advance deep learning techniques for classification of alcohol from EEG signals.

Conflict of Interest: The authors have no conflict of interest.

References

1. Zhang, L., Begleiter, H., Porjesz, B., Litke, A.: Electrophysiological evidence of memory impairment in alcoholic patients. Biol. Psychiat. **42**, 1157–1171 (1997)
2. Kousarrizi, M.R.N., Ghanbari, A.A., Gharaviri, A., Teshnehlab, M., Aliyari, M.: Classification of alcoholics and non-alcoholics via EEG using SVM and neural networks. In: 2009 3rd International Conference on Bioinformatics and Biomedical Engineering, pp. 1–4 (2009)
3. UCI KDD Database. https://archive.ics.uci.edu/ml/datasets/eeg+database
4. Smola, J., Scholkopf, B.: A tutorial on support vector regression. J. Stat. Comput. **1**, 199–222 (2002)

5. Platt, J.C.: Fast training of support vector machines using sequential minimal optimization. In: Schölkopf, B., Burges, C.J.C., Smola, A.J. (eds.) Advances in Kernel Methods: Support Vector Learning, chap. 12, pp. 185–208. MIT Press, Cambridge (1999)
6. Xie, L., Wu, Y.: A Modified multi-objective optimization based on brain storm optimization algorithm. In: Tan, Y., Shi, Y., Coello, C.A.C. (eds.) ICSI 2014. LNCS, vol. 8795, pp. 328–339. Springer, Cham (2014). https://doi.org/10.1007/978-3-319-11897-0_39
7. Yong, L., Liang, M.: Sine cosine algorithm with nonlinear decreasing conversion parameter. Comput. Eng. Appl. **53**(2), 1–5 (2017)
8. Hjorth, B.: EEG analysis based on time domain properties. Electroencephalogr. Clin. Neurophysiol. **29**(3), 306–310 (1970)
9. Wu, S., Nagahashi, H.: Analysis of Generalization ability for different AdaBoost variants based on classification and regression trees. J. Electr. Comput. Eng. **2015** (2015). Article ID 835357

Hand Movements Monitoring Device for Post Paresis Recovery Process

Titus E. Crisan(✉), Madalin I. Ardelean, Bogdan Tebrean, and Radu A. Munteanu

Electrical Engineering and Measurements Department, Technical University of Cluj-Napoca, Cluj-Napoca, Romania
titus.crisan@ethm.utcluj.ro

Abstract. The objective of this paper is to present a device designed for the monitoring of hand movements and grip force in specific situations namely in case of an impaired hand, due to either paresis or trauma situations. Such a device will allow the real time monitoring of the hand during the recovery process and its data logging capabilities will permit a better following and response during recovery. The device is based on several force sensors included in a form which allow the hand to rest on it in a relaxed position thus permitting to relatively small/weak movements to be measured. It is possible to measure either each finger individually or the entire hand. The same device allows further on the treatment to measure the grip force of the recovered/recovering hand or fingers. The device is microcontroller based and the software used for the application is *LabVIEW*. The force sensors – FSR type – were calibrated and the actual mass-voltage diagram was experimentally determined.

Keywords: FSR · Paresis · Force measurement

1 Introduction

The current paper is expressing the results of an experimental study for the fine moves of a human hand. First of all, we will monitor the normal movement of the human hand and the gripping force. After defining the default normal values, we will start experimenting the minimal moves detected by the system. Only after determining the facts mentioned above we'll start monitoring cycles of recovery from paresis or other mobility impairing hand injures.

The main idea is to follow a series of exercises meant to determine the basic hand and fingers movement that improve in time. In the first few weeks of recovery, most of the patients don't move their hands at all or the values are insignificant.

For this experiment we will use a specifically constructed device designed for a simple use and tailored for a specific or a specific group of patients. It consists of a base and a handhold molded using the shape of a human palm. On the mold will be placed six force resistive sensors, one for each finger and one for the palm pressing point. Two more bending sensors will be placed on the wrist, one placed above and one below.

The force sensors will measure the force applied by the human hand that is monitored, while the bending sensors will notice the very first weak moves of the hand (Fig. 1).

S. Vlad and N. M. Roman (Eds.): MEDITECH 2020, IFMBE Proceedings 88, pp. 45–54, 2022.
https://doi.org/10.1007/978-3-030-93564-1_6

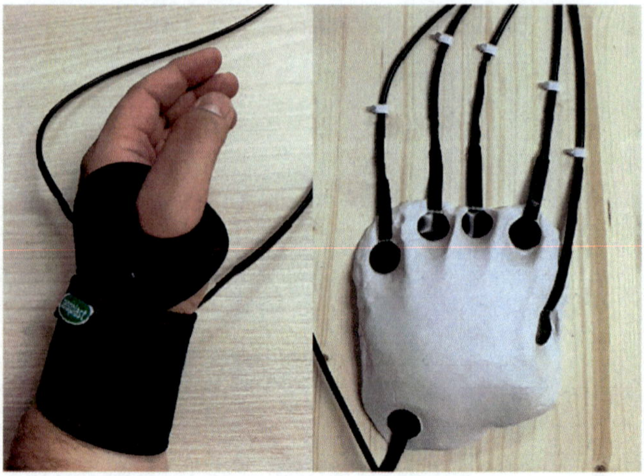

Fig. 1. Hand relaxed position/Molded part of the device with sensors

2 Theoretical Considerations

2.1 Hand's Movement

Natural hand movement is defined by the ability of a human hand to execute a wide range of movements consisting of rotations, various types of bending and flexing within a certain portion of space. Among many of this movements we consider the gripping movement of being important both in everyday use and in recovery monitoring.

Human hand has multiple flexing points, it is important to mention the wrist horizontal and vertical moving angles, and the finger joints conferring mobility to the hand. As natural examples for this particular study we are mentioning grabbing, squishing, even keyboard typing. All this simple reflex requires neurological capabilities, otherwise will be noticed symmetry defects.

2.2 Paresis

Even if this device can be used in a large number of situations involving impaired hand movements for the purpose of this paper, we chose to exemplify using the paresis condition.

Considering that paresis is partial paralysis of some nerve damaging or disease, we may say that hand paresis is affecting natural grip force reflexes and movement range. In most cases patients are highly affected and they cannot practice even the most basic exercises.

Paresis as an affection evolves from a sudden effect, manifesting slow improvements step by step in the recovery process. Recovery treatment and monitoring is the same for most of the cases, while time and results may vary depending on the patient.

3 Practical Implementation

The actual construction of the experimental stand is presented below.

For this stand we used: one Arduino Uno R3 micro-controller, one breadboard, connection wires, one 16 Channel Analog/Digital Multiplexer, six force resistive sensors, two bending sensors, eight 10 kΩ resistors, one mold and the base.

A special consideration was given to the insulation aspect thus all wire ends were insulated from the start so the values could not be influenced by the direct interaction with the hand.

Considering the expected results, the most important part of the experimental stand is the mold, being made from a special material giving the ability to replicate any hand shape. It is necessary to discuss hand dimensions and finger shapes when realizing the mold in order to obtain accurate results, this topic will be approached further in the paper.

3.1 Construction Design

The experimental stand was designed to be practical, targeting the ease of use.

Positioning the hand on the mold being one of the top priorities, therefore after testing the sensors individually and confirming as supposed that the best way to measure is to apply the force as close as possible to the center. Thus, the finger placement was designed to be as close as possible to the center of each sensor (Fig. 2).

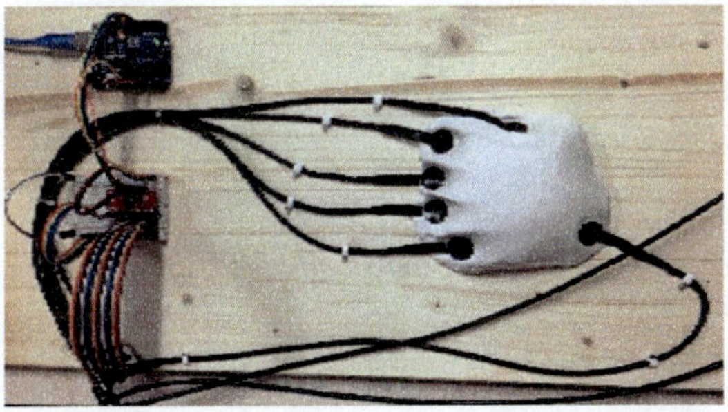

Fig. 2. Experimental device, molded part and electronic circuit

It is essential to model the form following the basic steps, starting by defining the highest point on the palm while the hand is standing in a natural relaxed position.

For this specific experiment we used as the main testing hand, the one presented below (see Fig. 3).

The first step as mentioned is defining the highest point, noted in this case with B. Next step is to measure the distance between point B and the fingertip of the middle finger, noted in this case with A, and distance between point B and the lowest point of the palm, noted in this case with C.

Therefore $D1 = B \rightarrow A$ and $D2 = B \rightarrow C$.

In this particular case D1 is equal to 12 cm and D2 is equal to 6 cm. We may approximate that the test hand has a length of 18 cm.

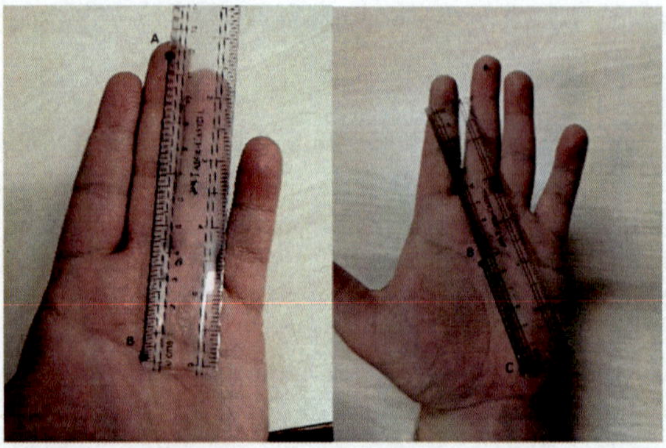

Fig. 3. Specific hand dimensions for mold design

3.2 Sensor Calibration

The date on the available FSR used is a little scarce so it was deemed important to perform a calibration process for the sensors.

The calibration setting was simple, and it was performed off-mold for a better accuracy of data. In the molded form of the design the sensors are placed so that the finger force vector is orthogonal to the sensor thus the same model was used for calibration (Fig. 4).

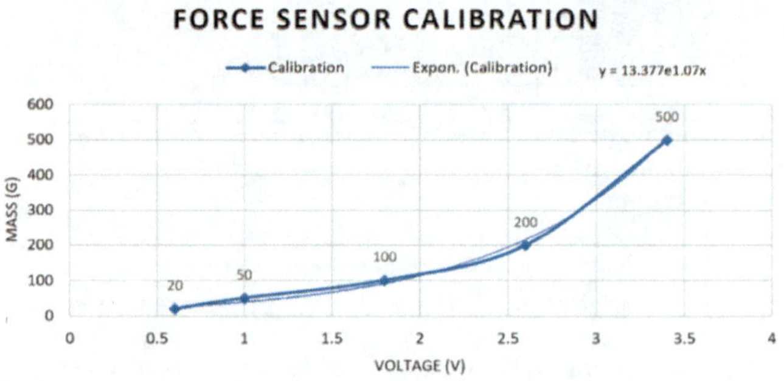

Fig. 4. Calibration diagram of the FSR

Calibration was done with a series of metrological weights of 500 g, 200 g, 100 g, 50 g and 20 g. For each of these weights, the displayed voltage acquired thus permitting to compute the closes function to the measured values. The resulting function is an exponential one as presented below:

$$\text{Calibrated Value} = 13.377 * e^{1.07 * \text{Un}} \quad (1)$$

The calibration for the bending sensors was made using the free 90-degree reference on a protractor. The voltage values obtained from the 40, 50, 60, 70, 80, 90, 100,

110, 120, 130, and 140° of gravity were measured by bending the sensor from the center point.

Similar to the FSR the exponential function was found to be:

$$\text{Calibrated Value} = 21.233 * e^{1.2618*\text{Un}} \quad (2)$$

From the actual experimental data, it occur that due to the small forces acquired in early stages of recovery it might be possible to us a linear approximation – within the range of 0 … 100 g (0 … 1 N) – but the actual exponential function was preferred.

3.3 Data Acquisition

The LabView software and an Arduino microcontroller were used to make the data acquisition setup. The Arduino microcontroller was used to retrieve the sensor data, and the LabView software was used to process data acquisition.

To process all data as required, the virtual instrument was structured in 4 different steps: Increment sequence for channel readout, Digital writing and analog reading, converting into vector type data, Graphic representation.

The electrical diagram of the system is presented in Fig. 5.

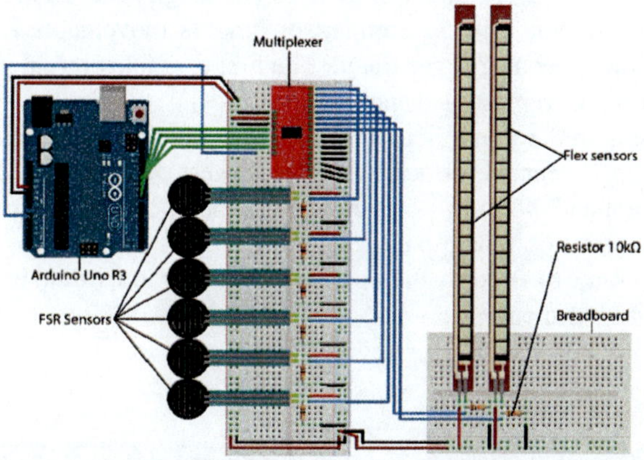

Fig. 5. Electrical diagram of the system

4 Experimental Results

Several sets of measurements were taken from which the most relevant were presented. Interpretation of the results was based on normal and erroneous use of the experimental stand.

The first significant aspect is the difference in force between the fingers of the hand. The fingers that exert the greatest force in the tightening process are the Thumb, the Index Finger, and the Middle Finger.

The movements of the joints are visible from the first pulses, the signal then oscillating in intensity and speed (Fig. 6).

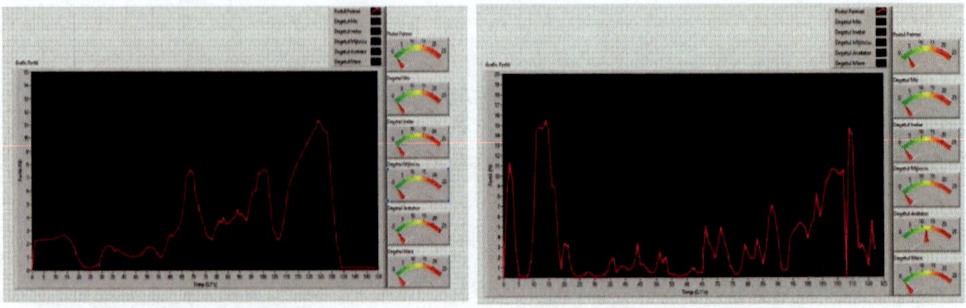

Fig. 6. Finger movement measured as force variation on sensors

After studying the peripheral clamping points, a test cast on which the sensors were mounted was modeled. After the tightening process was measured the difference in signal received from different hand sizes, noticing that sensors must be placed different on the mold for every patient.

More cycles of tests were taken with different length and intensity to study the behavior of the hand. For a better comparison fingers movements and actions were monitored both simultaneous and separated. The first example presented is the one with the force of each finger represented individually.

The thick line is in the graph is representing the value taken from the force sensor placed in the palm. It can be noticed that the lowest value is given by the thumb because in a free hand holding position the thumb is pressing from side without influencing the contraction of the palm.

Differences in force exerted by the opposite sex in the individual tightening process is about 20% higher at men.

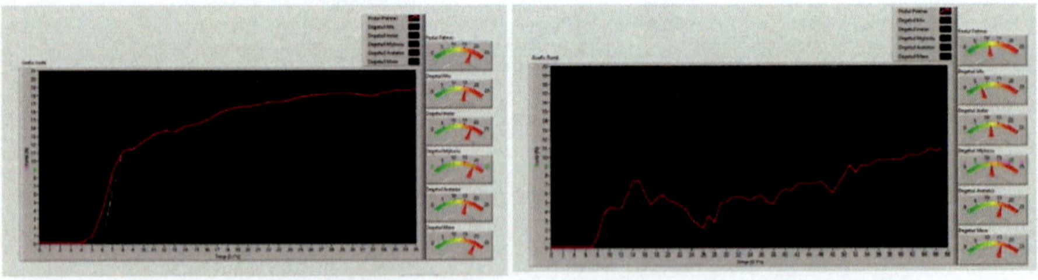

Fig. 7. Maximal gripping force for men and women

The graph above is a forceful tightening exercise of the hand of an apt person. When exercising the maximum hand force, the physical difference between the fingers of the hand can be observed. Using the values of analog indicators, it can be calculated an average hand force of approximately 19 N.

According to experimental measurements from a sample of 20 people, 10 male and 10 female resulted in a male tightening is within the range of 18 N and 22 N, and women between 10 N and 16 N. (see Fig. 7).

Testing of the measuring system was performed both under normal operating conditions and under increased sensitivity conditions. The shock method refers to touching the sensors with each independent finger as fast and powerful as possible. This method reflects sensor response time and finger dexterity in the recovery process (Fig. 8).

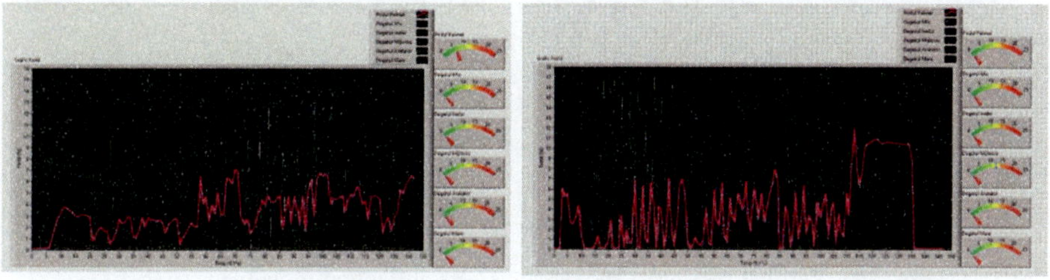

Fig. 8. Dexterity (left)/sensitivity (right)

The response time of the sensors is favorable to the monitoring of the recovery process as finger movements can be observed from tightening exercises to fine fingers movements.

The nonlinear form of the arm prevents the sensors from being fixed to the 90° position. Several types of exercises were performed to determine the results of the wrist movement.

The first type of exercise is the vertical bending of the hand at a normal pace (Fig. 9).

Fig. 9. Normal wrist movement (left) versus rest state (right)

The resting state of the wrist in the right position shows a difference between the two signals. The difference decreases gradually depending on the bending due to the sensor's return to the natural shape.

The most relevant bending exercise is the broad hand movement at an alert pace. Thus, one can observe the natural and fluid movement of the hand of a person without disabilities, which is taken as a standard for normal values.

Unlike force sensors, the bending movement was similar in both male and female.

In the case of horizontal bending, the measured values are less clear due to sensor location points (Fig. 10).

Fig. 10. Vertical fast wrist movement (left) horizontal movement (right)

The physical limits of the experimental device are determined by the greatest extent of the difficulty of the exercises. As a result of repeated tests carried out on different persons, the fluidity of the signal acquired in the normal use of the measuring system.

The experimental stand in question was assembled as all purpose stand but futher measurements proved that for different hands – dimensionally different – significant errors may occur thus user needs a mold, modeled after his own hand.

Using a general molding can result in erroneous results due to different measuring points (Fig. 11).

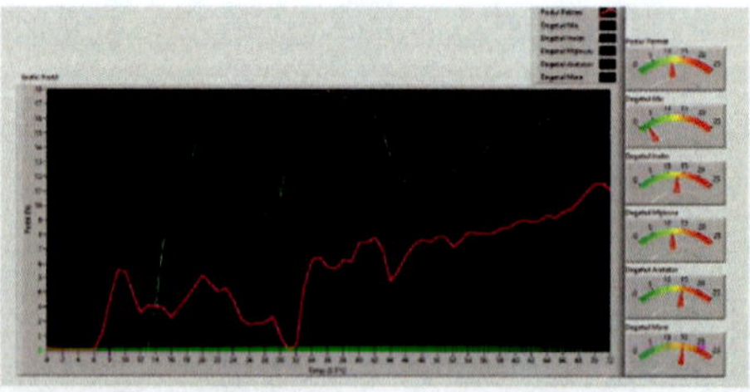

Fig. 11. Errors due to incorrect mold use or big differences in hand dimensions

Referring to the initial design idea it is recommended to use a fit mold for every patient. From our test trials it is shown that a hand size that differs with more than 5% from the main testing hand will return false or unwanted results.

For this measuring trial, the testing hand is smaller than the main one, D1 being 10.5 cm and D2 being 5 cm.

Size differing with more than 5% doesn't allow the fingers to be placed on the center of the sensor.

Diameter of the FSR sensor is 1.27 cm and for the most reliable result, the finger must press on the whole surface. Having a case with a smaller hand, the fingers won't be able to press on the required surface, therefore the values won't be reliable.

Real-time short timed measurements while useful may be relevant only for a short-term diagnosis at high intensity.

In the post-paresis recovery process, it is recommended to log the data either in numeric or graphic form both for tightening and for mobility so that the basis and evolution of the movement/recovery can be suitably monitored.

5 Conclusions

The paper presents the development of a system for monitoring fine hand movements in the post-paresis recovery process. The completion of the study required the theoretical considerations on hand anatomy, FSR sensors and the 16 channels multiplexer.

Following the assimilation of the theoretical considerations, the practical implementation part was taken, consisting of the physical construction of the experimental stand, the connection of the electronic components and the data acquisition.

Data acquisition was performed using LabView and Lynx Maker hub software, taking the signals from the Arduino Uno microcontroller.

After completion of the practical implementation, different types of experiments were simulated to obtain the experimental data. The study of the experimental data revealed the operation of the experimental stand according to the initial hypothesis.

The response time of the sensors is small enough to sense any kind of motion or hand pressure.

Data analysis is based on values wrote into a vector and graphic representation in LabView. For individual measurement point, an analog indicator panel is provided displaying the last values read from the vector.

According to simulated exercises, we proved the possibility of monitoring the entire post-paresis recovery process (dexterity, tightening, mobility, fatigue).

Initial trials have been successfully completed, and perspective proposals have been considered for future implementation such as (Storing results in a database, Adjusting the structure to create a portable device, Setting comparison values for each user, Extending the bending part measurements to the fingertips).

Conflict of Interest. The authors declare that they have no conflict of interest.

References

1. Damian, R.O.: Mişcări coreice
2. Voll, M., Wesker, K.: Atlas de anatomie. Thieme, USA (2010)
3. Dragomir, N.D.: col. - Mãsurarea electricã a mãrimilor neelectrice, vol. I. Mediamira Publishers, Romania (1998)

4. Bogdan, L.: Traductoare De Poziţie Şi Deplasare
5. Testemitnu, N.: Anatomia Topografică a Mâinii
6. Morris, R., Whishaw, I.: Arm and hand movement: current knowledge and future perspective. Front. Neurol. **6**, 19 (2015). https://doi.org/10.3389/fneur.2015.00019
7. Ju, Z., Liu, H.: Human hand motion analysis with multisensory information. IEEE/ASME Trans. Mechatron. **19** (2013). https://doi.org/10.1109/TMECH.2013.2240312
8. *** - https://www.robofun.ro/analog-digital-multiplexor

Monitoring of Respiratory Parameters During Wearing Medical Masks

Doru Andritoi, Radu Ciorap(✉), and Catalina Luca

"Grigore T. Popa" University of Medicine and Pharmacy of Iasi, Iasi, Romania
radu.ciorap@umfiasi.ro

Abstract. To prevent the spread of infections, hospitals required their employees and visitors to wear a mask at all times when in the facility. Wearing masks for a prolonged amount of time causes a host of physiologic and psychological burdens and can decrease work efficiency. In most cases the persons subjected to such measurements perform various activities and most of the time without being constantly monitored by a member of the medical staff. What happens if such a subject quickly develops respiratory problems and is unable to oxygenate effectively? How long does it take for another person to achieve this and especially how long does it take until the specialized intervention in these cases? Starting from these questions, we set out to create a device designed to have a dual role: monitoring respiratory activity and the second role is to alert the person in question in the event that he develops breathing problems and thus oxygenation.

Keywords: Respiratory rate · Mask · Respiratory monitoring

1 Introduction

Wearing a mask is one of the measures to prevent and limit the spread of certain respiratory diseases, including that caused by the new coronavirus. However, the use of a mask alone is not sufficient to ensure an adequate level of protection and other equally relevant measures should be taken. The use of the mask must be combined with proper hand hygiene and other measures to prevent and control disease, especially from person to person.

The mask is also an essential element used by medical staff for protection, regardless of specialization or location. Medical personnel are advised to use a respiratory mask whenever performing procedures that may generate aerosols (e.g., tracheal intubation, noninvasive ventilation, tracheotomy, cardiopulmonary resuscitation, manual ventilation before intubation, and bronchoscopy) [1].

Wearing a mask, when not indicated, can cause unnecessary costs, burden on purchases and create a false sense of security that can lead to neglect of other essential measures, such as hand hygiene practices. Moreover, the incorrect use of a mask can affect its effectiveness in reducing the risk of transmission. Some of the studies reported headaches [2], others discussed skin problems due to prolonged wearing of masks [3].

S. Vlad and N. M. Roman (Eds.): MEDITECH 2020, IFMBE Proceedings 88, pp. 55–60, 2022.
https://doi.org/10.1007/978-3-030-93564-1_7

2 System Architecture

In this experiment we try to evaluate changes in respiratory reactivity and the response of people forced to wear a mask in pandemic conditions with the new coronavirus and to make a correlation of different parameters obtained by non-invasive methods to determine what can lead to early diagnosis followed by actions appropriate therapeutic.

The respiratory system includes the airways (leading to the lungs), the lungs, and the thoracic structures responsible for the circulation of air through the respiratory system. The airways begin at the nostrils, which is the interface with the air in the atmosphere. During a normal resting breathing cycle, the only type of movement is the diaphragm up and down movement. The inspiratory stage of the respiratory cycle causes the diaphragm to move down along with the lungs. During the expiration stage, the diaphragm relaxes and the elastic retraction of the lungs determines the return to the initial position. In case of forced exhalation, the abdominal muscles contract, causing the abdominal organs to reach the diaphragm. These movements of the diaphragm muscle and the rib cage can be recorded and the sequence of inspiration/expiration determined during one minute represents the respiratory rate.

Using this monitoring method for a long time can lead to discomfort or in some cases prevent the subject from performing activities [4].

Respiratory rate control ensures efficient exchange of respiratory gases, helping to maintain the body's pH and minimize energy consumption.

Considering the above mentioned, we opted for respiratory rate monitoring through a thermistor small positioned effectively for this purpose.

Respiratory rate measurement is based on temperature variations of inhaled/exhaled air. An SMD thermistor (10 kohm at 25°) is used as the temperature/voltage transducer. The signal provided by it is amplified and subsequently filtered in order to retain efficient data.

Thermography is a noninvasive imaging method of investigation which implies no side effects and no contraindications for the patient. Unlike other imaging methods of investigation, medical thermography is completely without risk for the patient and physician.

The thermistor should be placed near the nostrils of the subject using the facemask. It measures the temperature difference between exhaled and inhaled air which translates into a variation in electrical resistance. This variation taken with a resistive divider and amplified with an operational amplifier provides an analog voltage proportional to the temperature at the output connector (Fig. 1).

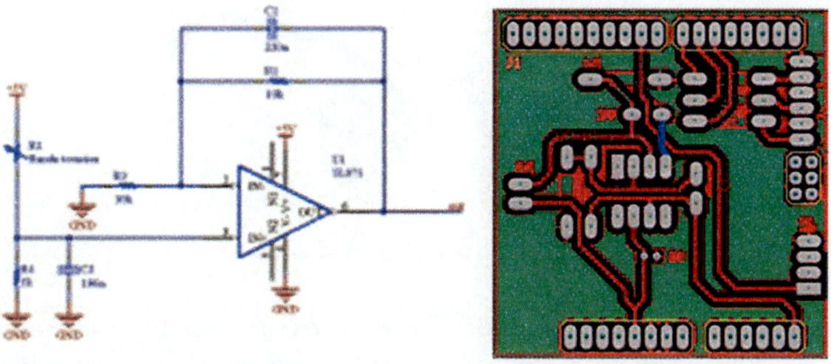

Fig. 1. Wiring diagram and wiring for the amplification and filtering part of the respiratory signal.

The prototype system does provide options for power-up using batteries and has access to a USB 2.0 port for transferring conversion data, and programming the microcontroller.

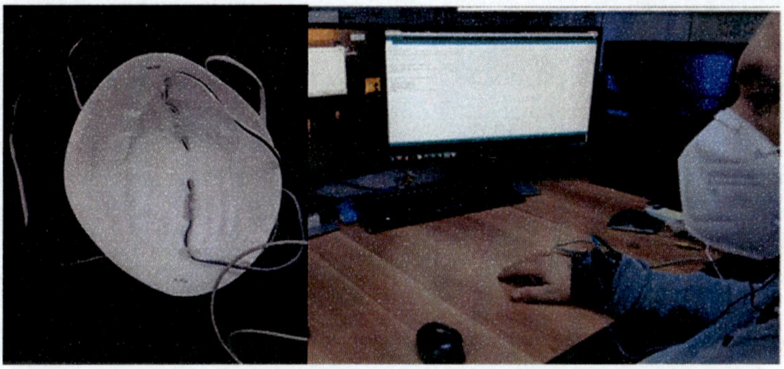

Fig. 2. First prototype of the equipment (left) and final setup (right).

In order to facilitate its placement at the level of the mask, the thermistor is fastened with the help of a metal clip fixed at the level of the nasal support of the mask as can be seen in Fig. 2. This system allows easy removal of the sensor when you want to replace the mask. Also, the capsule thermistor allows disinfection using standard solutions for medical use, in order to comply with safety rules.

The pulse oximetry module allows by a non-invasive method the monitoring of the oxygen level from the hemoglobin level in the patient's blood A small device was designed, that integrates the pulse sensor using Autodesk FUSION 360 [5] (Fig. 3).

Fig. 3. Design of PPG sensor

Although pulse oximetry is used to monitor oxygen levels, it can also provide information about oxygen metabolism or the amount of oxygen the subject/patient needs [6, 7]. In this case it is necessary to correlate the two parameters. It can also be helpful in detecting ventilation abnormalities.

3 Functionality of the System

The microprocessor microsystem used is the Arduino Uno development system, based on the Atmel AtMega324 microcontroller, with 20 I/O ports, 32 kB flash memory, 2.5 kB SRAM, 1 kB EEPROM, 16 MHz CPU frequency, 10-resolution ADC bits and integrated USB support. The characteristics of the microcontroller but also the format of the development system facilitated the development of the prototype, allowing the implementation of a system with small dimensions.

The information coming from the 2 sensors is transmitted to the microcontroller for processing. The system being developed in order to monitor the parameters related to respiration analyzes this information and compares the values obtained with some standards, considered optimal. If deviations from these normal intervals occur, the system carrier receives a warning by coloring an LED from green (normal) to red (danger). In this way the subject is warned of certain changes in the ventilatory parameters.

There is also the possibility to connect to a PC in order to visualize and graphically represent these parameters, in the future we want the possibility to attach a storage system in order to obtain a holter type system for the previously mentioned parameters (Fig. 4).

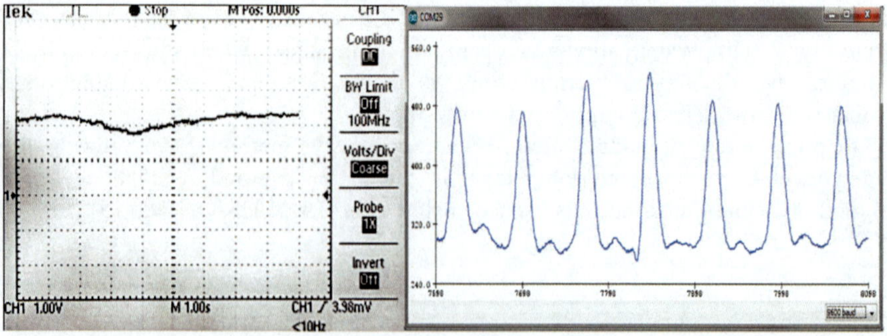

Fig. 4. Signal representation on oscilloscope (left) and using a graphic interface (right).

4 Conclusions

An experimental model of the device was implemented and tested. The signal acquisition and user alerting part was implemented as a shield for the Arduino Uno development system with the possibility of further development by including new types of acquisition signals but also communication modes.

The proposed device offers the subject in question an opportunity to learn about possible episodes of respiratory problems that may occur while wearing a mask and can give him or a loved one a chance to mitigate their effects. Another possible use could be its integration into a system of diagnosis or treatment of diseases related to respiratory problems, to trigger or record these events or to optimization the therapy in case of chronic diseases [8, 9].

In the future, we will focus on a better detection of the parameters but also on an integration of a protocol for wireless transmission of the acquired data that would allow the recording of the time moments in which the targeted events occurred.

Conflict of Interest. The authors declare that they have no conflict of interest.

References

1. Johnson, A.T.: Respirator masks protect health but impact performance: a review. J. Biol. Eng. **10**, 4 (2016)
2. Lim, E.C.H., Seet, R.C.S., Lee, K.-H., Wilder-Smith, E.P.V., Chuah, B.Y.S., et al.: Headaches and the N95 face-mask amongst healthcare providers. Acta Neurol. Scand. **113**, 199–202 (2006)
3. Foo, C., Goon, A.T.J., Leow, Y.-H., Goh, C.-L.: Adverse skin reactions to personal protective equipment against severe acute respiratory syndrome- a descriptive study in Singapore. Contact Derm. **55**, 291–294 (2006)
4. Rosner, E.: MSN, RN-C (2020), adverse effects of prolonged mask use among healthcare professionals during COVID-19, Rosner. J. Infect. Dis. Epidemiol. **6**(3),130, 1–5 (2020)
5. Ciorap, R., Andritoi, D., Ciorap, M., Munteanu, M.: Monitoring system for home-based hand rehabilitation. In: International Conference and Exposition on Electrical and Power Engineering (EPE 2018), pp. 836–839 (2018). https://doi.org/10.1109/ICEPE.2018.8559682
6. Andriţoi, D., David, V., Ciorap, R.: An portable device for ECG and photoplethysmographic signal acquisition. In: 2014 International Conference and Exposition on Electrical and Power Engineering (EPE 2014), pp. 16–18, October 2014, Iasi, Romania (2014). https://doi.org/10.1109/ICEPE.2014.6969967
7. Hritcu-Luca, C., Corciova, C., Ciorap, R.: Wireless device for SpO_2 monitoring. In: Proceedings of the International Conference on Advancements of Medicine and Health Care Through Technology, Cluj-Napoca, Romania 23–26 September 2009, IFMBE 26, pp. 133–136 (2009). https://doi.org/10.1007/978-3-642-04292-8_30

8. Andriţoi, D., David, V., Ciorap, R.: Dynamics analysis of heart rate during magneto-therapy session. In: 2014 International Conference and Exposition on Electrical and Power Engineering (EPE 2014), 16–18 October 2014, Iasi, Romania (2014). https://doi.org/10.1109/ICEPE.2014.6969961
9. Ciorap, R., Andritoi, D., Luca, C., Corciova, C.: Monitoring of cardiovascular parameters during rehabilitation after stroke event. In: Vlad, S., Roman, N.M. (eds.) 6th International Conference on Advancements of Medicine and Health Care through Technology; 17–20 October 2018, Cluj-Napoca, Romania. IP, vol. 71, pp. 103–107. Springer, Singapore (2019). https://doi.org/10.1007/978-981-13-6207-1_17

The Incubation Time Effects on the Precision and Accuracy of the Glycated Hemoglobin

Manole-Stefan Niculescu[1,2](✉), Adriana Florescu[1], and Pasca Sever[1]

[1] University POLITEHNICA of Bucharest, 060042 Bucharest, Romania
[2] Ana Concept SRL, Dorneasca, 11, 051713 Bucharest, Romania

Abstract. Health can be easily monitored by performing routine blood assays. The blood science offers complete information about the clinical state of the body from a sample to which the concentration of the parameter of interest is measured using specialized medical equipment. In order to maintain or to improve the medical judgement, the measured results for the assay parameters must be precise and accurate. If this condition is not fulfilled, the quality of the medical diagnosis can be affected, which is translated into improper therapy that has negative impact over the patient's health. In this paper is presented the incubation time effect over the precision and the accuracy of the concentration of one of the most important parameter used in the self-monitoring procedure of diabetic patients, which is the glycated hemoglobin. Regarding the experiment, HbA1c lateral flow immunofluorescence test strips were used in order to measure the amount of glycated hemoglobin for different incubation periods from an artificial quality control material with known manufacturer assigned target values.

Keywords: Immunofluorescence · Incubation time · Lateral flow analyzer

1 Introduction

The blood science represents a step in the development of a particular path of pathology [1]. Health is the most important state of a human being and in order to properly monitor and diagnose it, periodical medical evaluation must be performed. The most simple and accessible medical monitoring procedure is represented by the blood assays.

The complete blood assay profile consists in a set of parameters that will provide important information about the vital functions and the chemical and metabolic processes that take place in the human body.

Blood is a complex suspension of cells in plasma proteins that flows through arteries, veins and capillaries and transports the nutrients along with the oxygen to the cells [2]. The analytical chemistry measures quantitatively the concentration of a particular parameter from the blood, supplying important information that will improve the clinical judgement when the diagnostic procedure is performed. The quantitative measurements must be precise and accurate in order not to affect the quality of the medical act. It is desired to obtain results as close as possible to the real clinical state of the tested patient. The collection and the assay procedure must be performed following strictly the medical protocol in order to minimize the operation errors that can be

S. Vlad and N. M. Roman (Eds.): MEDITECH 2020, IFMBE Proceedings 88, pp. 61–68, 2022.
https://doi.org/10.1007/978-3-030-93564-1_8

induced in the measurement process. Any operation issue involved in this process will conduct to false positive or false negative results that can generate an improper medical diagnosis translated into severe repercussions over the clinical state of the tested patient.

In this paper is presented the effect of the incubation time over the accuracy and precision of the results obtained on lateral flow immunoassay glycated hemoglobin (HbA1c) tests. The paper is organized as it follows: in Sect. 2 are described the precision and the accuracy of a blood assay analyzer, followed by Sect. 3, where the HbA1c lateral flow immunoassay measurement technique is presented, ending with the experimental results in Sect. 4.

2 The Precision and Accuracy of the Medical Assays

The medical record should include blood assay results in order to provide a correct medical diagnostic. It is very important that the blood results to be precise and accurate and in accordance with the clinical state of the patient because depending of the values obtained during the assay procedure, the patient's diagnostic will be framed in normality or in a specific pathology. Every assay has its specific normal range and any other deviation from this limit represents a low or a high abnormality.

In the case of blood analyzers, the assay limits are specific to each type and model and are provided by the manufacturer, with the emendation that the limits can be modified according to every laboratory's own targets. In Fig. 1 is illustrated the measurement domain of a blood analyzer.

The normal range corresponds to the limit in which a result is reported as normal, followed by the low and high range, where the result is outside the normal limit, being situated in a specific pathology [3]. Any results that are outside the pathological range will not be reported due to the measurement limitation of the instrument.

Fig. 1. The measurement domain of a blood analyzer.

There are cases when the instrument is not properly maintained or operation issues are induces generating measurement errors. A tested patient with a normal health state can be easily misplaced in a pathological case with improper medical treatment that can degenerate its health due to a false positive result. In the same manner, a patient with health issues can be declared normal if the measured result is false negative, meaning that there is no need for medical treatment, increasing the risk for the disease to progress.

In order to establish the accuracy and the precision of a medical analyzer, the standard deviation (SD) must be calculated using (1), where n represents the number of measured results, x_n are the measured results and $\overline{x}$ is the mean of the measured values:

$$\overline{x} = \sum_{i=1}^{n} \frac{x_n}{n} \tag{1}$$

$$SD = \sqrt{\frac{\sum_{i=1}^{n} (x_i - \overline{x})^2}{n - 1}} \tag{2}$$

The coefficient of variation (CV) can be calculated from (1) and (2):

$$CV = SD/\overline{x} \tag{3}$$

One of the most important assay that needs to be constantly monitored is the glucose. The capillary blood glucose is an important indicator in the diabetes care management system [4].

This substance is stored as glycogen in the skeletal muscle and in the liver cells [5] and it is regulated by insulin, which is a hormone produced by the pancreas. The blood sugar levels are mostly affected by external factors, such as bad gastronomic habits, stress, sports, but also by internal factors, like illness or slow metabolism.

Along with the blood glucose, a more relevant parameter involved in the diabetes care management system is the glycated hemoglobin (HbA1c), which is a standard for the clinical evaluation of the glycemic control, providing complete informational resources in the diabetes treatment.

In 2019, all over the world were reported approximately 463 million diabetic adults with ages from 20 to 79 years old. The number is increasing each year and this is a very alarming fact because diabetes caused 4.2 million deaths so far [6].

In order to have a better clinical understanding about the patient's glycemic management and to monitor if his metabolic control is in the normal range, HbA1c should be tested every 3 months.

This indicator is primarily used to monitor the effectiveness of a treatment or to modify the glycemic therapy [7]. This means that HbA1c is a blood assay that needs to be precise and accurate and should be taken into account to be measured when the routine blood assays are performed.

3 The Lateral Flow Immunoassay Principle

The monitoring procedure of the HbA1c can be performed in specialized laboratories using automated or semi-automated equipment or with the help of point-of-care (POC) analyzers. In the case of POC instruments, the assay can be performed also by non-medical personnel for self-monitoring at home, since the POC instruments are mainly Clinical Laboratory Improvement Amendments (CLIA) – waived equipment [8].

For the self-monitoring procedure are used rapid tests, which are easy to use, quick and cheap, useful for home or for on-site applications. The main disadvantage in the case of the rapid tests is that they are qualitative and do not provide a measured and concise value that will help in the therapy management system.

The POC instruments use tests that are quantitative, providing a measurable value for the concentration of the analyzed parameter. In this paper, for the experiment are used HbA1c immunofluorescence quantitative tests that are preprocessed in a time variable incubator at 37 °C and after the incubation is finished, they are measured on a lateral flow immunofluorescence instrument.

The immunofluorescence lateral flow principle is based on the reaction that takes place in the test strip between the antigens and the antibodies. The blood sample is placed on the application pad from the test strip, which has the role of transporting the sample to the conjugate pad. It is pretreated to separate the sample components, to adjust the pH of the sampled substance and to remove any other interferences involved in the measurement process.

The conjugate pad is the segment on the test strip, where the fluorescent labeled bio-recognition molecules trigger the labeled conjugate when interacts with the sample transported from the application pad.

The next layer consists in a nitrocellulose membrane, where the analyte-labeled antibody complex is transported by capillary action to the primary antibodies test line, forming a labeled antibody-analyte-primary complex. The excess labeled antibodies travel to the control line and bind with the secondary antibodies. When the complex reacts with the secondary antibodies, the control line is activated, which means that the test was performed in accordance with the testing procedure and it is valid. The color intensity of the test line is direct proportional with the concentration of the measured analyte. In order to obtain a quantitative value, the test strip is measured with an optical strip reader [9]. In Fig. 2 is presented the composition of a lateral flow immunofluorescence test strip.

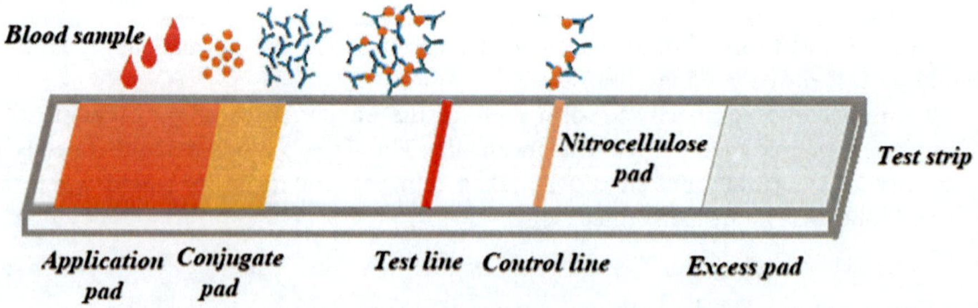

Fig. 2. The composition of a lateral flow immunofluorescence test strip.

4 Experimental Results

The experiment was performed on the FA-160 immunofluorescence lateral flow analyzer using the bi-level CFPO-96 quality control (QC) for the HbA1c from Boditech.

The manufacturer assigned reference interval values from the QC's insert sheet for the HbA1c are 5.40% and 6.21% corresponding to the 1st level and 8.84%–11.96% corresponding to the 2nd level.

The HbA1c testing procedure consists in analyzing 10 µl of control blood diluted in 1 ml of sample diluent, letting the mix to harmonize for 1 min and transferring the mix to the application pad from the test strip, incubating it for 3, 5, 10, 15, 20, 25, 30, 35 and 40 min in order to observe if the results suffer any modifications for different incubation periods.

In order to maintain a safe and correct assay procedure, the test strips were unsealed only when the sample was about to be applied and for every incubation period a new test was used.

In Fig. 3 is presented the FA-160 analyzer along with the volumetric pipette used for transferring the sample into the diluent, the HbA1c tests and the HbA1c Boditech control material.

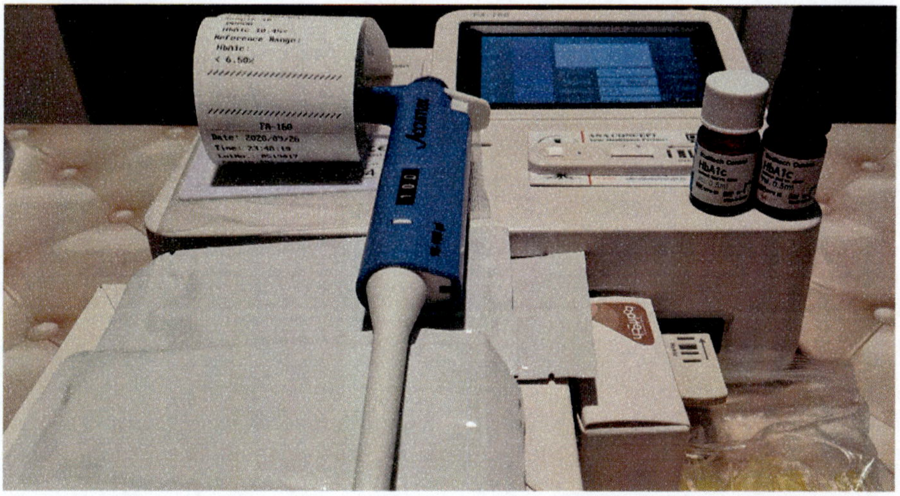

Fig. 3. The experimental setup containing the FA-160 immunofluorescence analyzer and the testing materials.

In Table 1 are presented the measured results when the 1st level of QC was analyzed at the predefined incubation intervals and in Table 2 are presented the measured results corresponding to the 2nd level of QC for the same incubation intervals as when QC 1 was analyzed. The target value represents the mean between the low and the high value of the quality control material assigned in its insert sheet.

Table 1. The experimental results for QC level 1.

Incubation time [min]	Measured HbA1c [%]	Target value [%]	SD	CV [%]
3 min	6.01	5.80	0.15	2.51
5 min	6.05	5.80	0.18	2.98
10 min	6.08	5.80	0.20	3.33
15 min	6.14	5.80	0.24	4.03
20 min	6.11	5.80	0.22	3.68
25 min	6.51	5.80	0.50	8.16
30 min	6.78	5.80	0.69	11.02
35 min	6.89	5.80	0.77	12.15
40 min	7.11	5.80	0.93	14.35

Table 2. The experimental results for QC level 2.

Incubation time [min]	Measured HbA1c [%]	Target value [%]	SD	CV [%]
3 min	10.37	10.40	0.02	0.20
5 min	10.37	10.40	0.02	0.20
10 min	10.38	10.40	0.01	0.14
15 min	10.45	10.40	0.04	0.34
20 min	10.40	10.40	0.00	0.00
25 min	12.82	10.40	1.71	14.74
30 min	14.05	10.40	2.58	21.11
35 min	14.12	10.40	2.63	21.46
40 min	14.38	10.40	2.81	22.71

In Fig. 4 is presented the plot corresponding to the coefficient of variation in relation with the incubation time. There are two series for each QC level from which it can be stated that after approximately 20 min of incubation, the measured results are slightly increased, being falsely elevated.

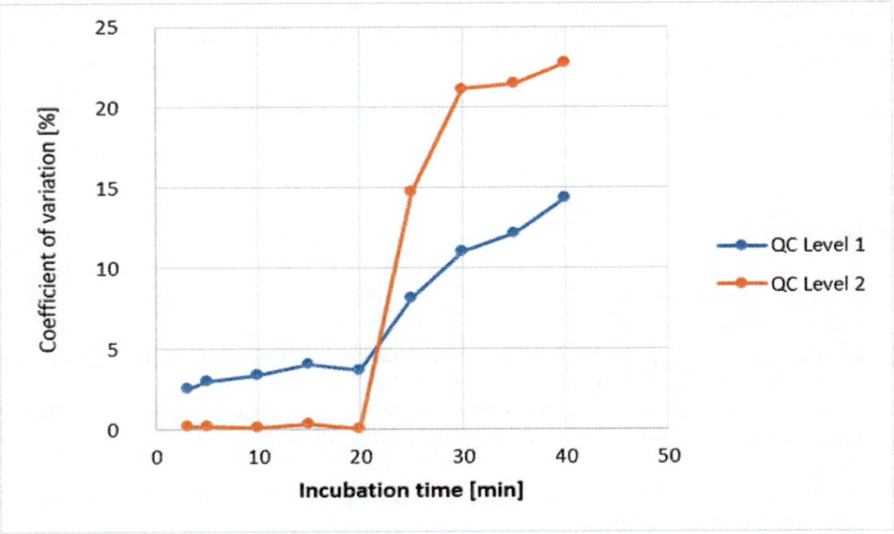

Fig. 4. The coefficient of variation for each level of QC depending of the incubation time.

5 Conclusion

The blood assay science offers complete information about the clinical state of a patient. It is the simplest method to diagnose potential diseases from a small blood sample by analyzing it on a specialized medical equipment.

The improper collection, handling, assaying procedures or the abnormal functionality of the medical instrument can affect the precision and the accuracy of the measured results. This fact will decrease the quality of the medical diagnostic procedure and can generate severe health issues due to improper medication therapy for the patient from who the blood sample was collected and analyzed. There are situations when the health state must be continuous monitored, as in the case of diabetic patients, where the glycemic control must be maintained in normal limits with insulin compensation therapy.

An efficient long term diabetes indicator is the glycated hemoglobin (HbA1c), which must be periodically monitored. The patient can use a immunofluorescence lateral flow POC HbA1c analyzer that is CLIA-waived because it is cheap, simple and reliable. In order for the reaction between the antigen and the labeled antibody to take place when a HbA1c sample is analyzed, the incubation of the test must be performed. As can be observed from the presented experiment, a longer incubation time (after approximately 20 min) in the case of a HbA1c test strip, affects the precision and the accuracy of the measured results by increasing with almost 1% for the low levels and 3% for the higher pathological levels. This fact is possible because longer exposure in the incubator will dry the nitrocellulose membrane containing the labeled antibody-analyte-primary complex and will modify the color intensity of the test line.

Acknowledgement. The work has been funded by the Operational Programme Human Capital of the Ministry of European Funds through the Financial Agreement 51675/09.07.2019, SMIS code125125.

Conflict of Interest. The authors declare no conflict of interest.

References

1. Blann, A., Nessar, A.: Blood Science – Principles and Pathology, pp. 10–12. Wiley Blackwell, Hoboken (2014)
2. Handin, R.I., Lux, S.E., Stossel, T.P.: Blood – Principles and Practice of Hematology, pp. 15–17. Lippincott Williams & Wilkins, Philadelphia (2003)
3. Niculescu, M.S.: Optical method for improving the accuracy of biochemical assays. In: 6th IEEE International Conference on E-Health and Bioengineering, Sinaia, Romania, pp. 1–2 (2017)
4. Jia, W.: Continuous Glucose Monitoring, pp. 97–99. Springer, Singapore (2018). https://doi.org/10.1007/978-981-10-7074-7
5. Martini, F.H., Ober, W.C., Welch, M.D., Hutchings, R.T., Ireland, K.: Anatomy and Physiology, pp. 463–467. Prentice Hall, Singapore (2005)
6. International Diabetes Federation. https://www.idf.org/aboutdiabetes/what-is-diabetes/facts-figures.html. Accessed 01 Sept 2020
7. Jameson, J.L., De Groot, L.J., Kretser, D.M., Giudice, L.C., et al.: Endocrinology – Adult and Pediatric, 7th edn, p. 877. Elsevier Saunders (2016)
8. Warekois, R.S., Robinson, R., Primrose, P.B.: Phlebotomy – Worktext and Procedures Manual, 5th edn, p. 258. Elsevier (2020)
9. Wong, R., Tse, H.: Lateral Flow Immunoassay, p. 59, 74, 95–97. Humana Press (2009)

Consideration Regarding the Implementation of a System for Monitoring a Person in the Office

Mihaela Pavăl[1,2](✉), Mihai Cenușă[1], and Dan Laurențiu Milici[1]

[1] Faculty of Electrical Engineering and Computer Science, University Ștefan cel Mare of Suceava, Suceava, Romania
{mpoienar,dam}@usm.ro

[2] Integrated Center for Research, Development and Innovation in Advanced Materials, Nanotechnologies, and Distributed Systems for Fabrication and Control, University Ștefan cel Mare of Suceava, Suceava, Romania

Abstract. The paper presents a study on the performance of system of monitoring the activity of a person in the office. The equipment proposed by the authors is based on a continuous evaluation, using data collection and processing systems with wireless transmission, so that the monitored individual is not uncomfortable in the activities they perform and it is performed based on the evaluation of the position of the monitored person during the period in which he carries out his activity sitting on the chair, but also during the rest period.

Keywords: Microcontroller · Monitoring · Wireless · Device · Sensor · Acquisition system

1 Introduction

In contemporary societies, more and more companies are interested in studying the degree of office comfort of employees that has effects on their productivity.

Spending a number of hours at the office, in front of a computer without the existence of breaks leads to a decrease in their performance and the degree of involvement.

Concerns about evaluating the activity of a person who works in the office has attracted over time countless researchers in the field of psychology and beyond, countless solutions have been developed based on sets of tests grouped by type of activity, symptoms or emotions [1–3].

The European Union made a priority o support the prevention of work-related diseases. EU directives and campaigns at Member State level (e.g. the Healthy Workplaces - Stress Management Campaign, Healthy Workplaces Lighten the Load, Healthy Workplaces for All Ages) on occupational health make more and more companies look for solutions to assess working conditions and take actions which can improve health [1, 4–7].

According to a study conducted by the market research company 4Service Group Romania, in November 2019, 53% of employees in our country believe that the companies they work for do not give any or very little importance to their physical and

S. Vlad and N. M. Roman (Eds.): MEDITECH 2020, IFMBE Proceedings 88, pp. 69–76, 2022.
https://doi.org/10.1007/978-3-030-93564-1_9

emotional health, and 7 of 10 employees complain that their managers do not provide them with constructive support and feedback [8–10].

The equipment proposed by the authors is based on a continuous evaluation, using data collection and processing systems with wireless transmission, so that the monitored individual is not uncomfortable in the activities they perform and it is performed based on the evaluation of the position of the monitored person during the period in which he carries out his activity sitting on the chair, but also during the rest period [11–13].

The system of monitoring the activity of a person in the office (Fig. 1), is constituted mainly of a chair 1, which has the seating area made of two overlapping plates 2, 3 and, separated by four resorts 4, 4′, 4″, 4‴ between which are placed, on the corners of the plates, four force sensors 5, 5′, 5″, 5‴, connected to an acquisition microsystem with microcontroller 6 that transmits the acquired data to a computer 8 to be saved and processed [13] (Fig. 2).

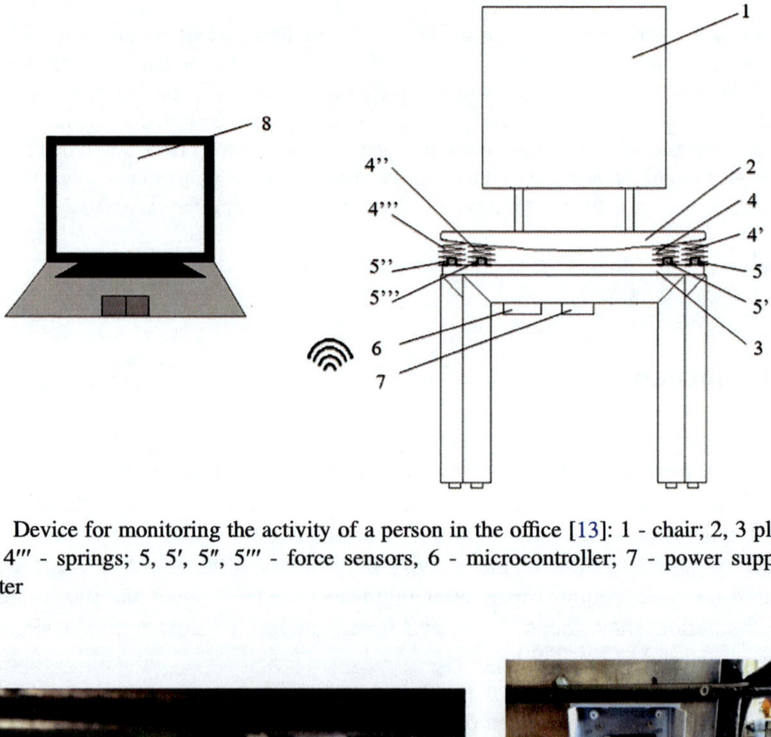

Fig. 1. Device for monitoring the activity of a person in the office [13]: 1 - chair; 2, 3 plates; 4, 4′, 4″, 4‴ - springs; 5, 5′, 5″, 5‴ - force sensors, 6 - microcontroller; 7 - power supply; 8 - computer

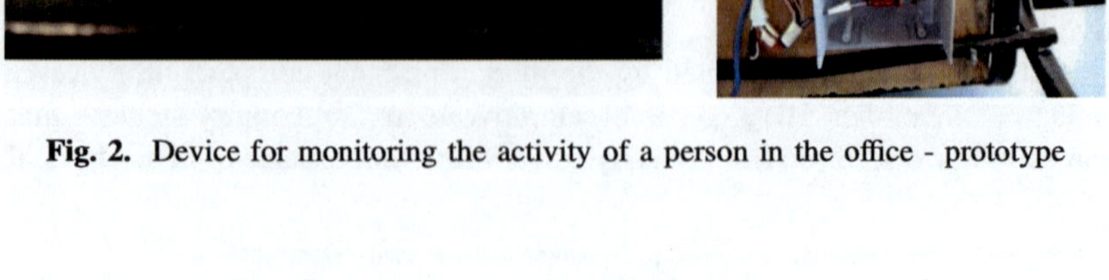

Fig. 2. Device for monitoring the activity of a person in the office - prototype

The system takes information about the distribution of body weight in the four points and through the microcontroller can be determined by analyzing the variation of signals from each sensor, with its own algorithm adapted to the activity of the monitored person, weight, period of sitting, position and its activity (leaning forward towards the sheets or a screen, left on its back in a relaxed position, leaning sideways towards another person, with feet resting on the floor or not, sudden movements, comfort level, continuous movements of the legs, etc.), information that is transmitted to the computer where they are saved and can generate warning messages regarding weight change over long periods of time, the need for relaxation breaks in case of long periods of work, activities not suitable for work tasks, overload states, stress or anxiety of the monitored person [13] (Fig. 3).

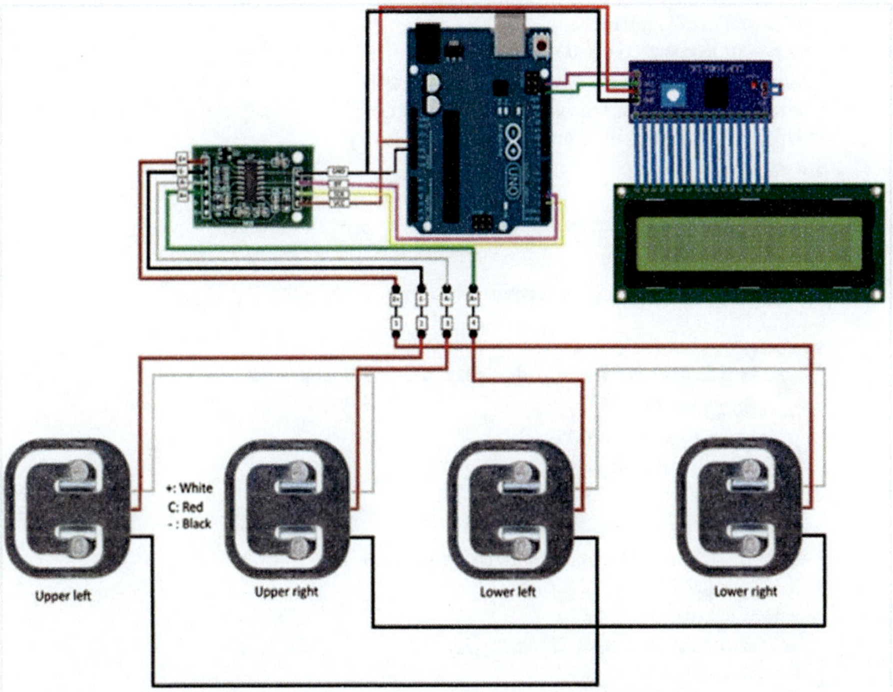

Fig. 3. Circuit diagram for experimental assembly

Data retrieved from the system is processed in a software application. Initially, the system is calibrated, and the unit of measurement (kg or lbs) can be selected, then the user's weight distribution is analyzed.

The program for analyzing the position of the activity of the person at the office is presented below.

```
'include "HX711.h"
"include <Hire.h>
'include <LiguidCrystalI2C.h>
HX711 scale(2, 3);
LiguidCrystal_I2C lcd(0x27, 2, 1, 0, 4, 5, 6, 7, 3, POSITIVE);
float greutate; // definire paramentru greutate
float factor_calibrare=7240.00; // The sensor calibration factor for the kg unit of measurement // The calibration factor for the Lbs unit is 15950.00

void setup()

  Serial.begin (9600 );
  Serial.println ("HX711 calibrare");
  Serial.println ("indepartati orice greutate de pe scaun");
  Serial.println ("Dupa ce incepe o citire, puteti pune o greutate");
  Serial.println ("Apasa tasta a sau c pentru a create factorul de calibrare");
  Serial.println ("Apasa tasta x sau v pentru a scidea factorul de calibrare");
  scale.set_scale ();
  scale.tare ();
  long zero_factor=scale.read_average ();
  Serial.println(zero_factor );
  lcd.begin (16,2); // Initialize the LCD display connection
  lcd.backlight ();
  void loop()
    scale.set_scale(factor_calibrare ); // Adjustment to the calibration factor
    Serial.print("Citire: ");
    greutate= scale.get_units(5)*0.453592;
    Serial.print( greutate,1 ); // display weight
    Serial.print(" Kg");
    Serial.print(" Factor_calibrare: ");
    Serial.print(factor_calibrare); // Display calibration factor
    Serial.println();
    lcd.setCursor(0, 0);
    lcd.print("Greutate[kg] :"); // LCD data display
    lcd.setCursor(0, 1); /
    lcd.print(greutate,1 );
    if(Serial.available ()) // change calibration factor
      char temp = Serial.read();
      if(temp == '+' II temp = 'a')
       factor_calibrare += 10;
   else if(temp ==II temp == 'z')
   factor_calibrare -= 10;
                 if(temp == 'c')
                 factor_calibrare += 50;
   else if(temp == 'v' )
   factorul de calibrare factor_calibrare -= 50;
    }
```

2 Experimental Testing

In order to verify the functionality of the device, a series of tests were performed on the weight variation over time for a person with a weight of 84 kg seating at the desk: performing a simple sitting and stationary operation, without any movement from the sitting position and with the feet touching the ground (Fig. 4), the variation of weight according to time, for performing a stressed activity and with office support (Fig. 5).

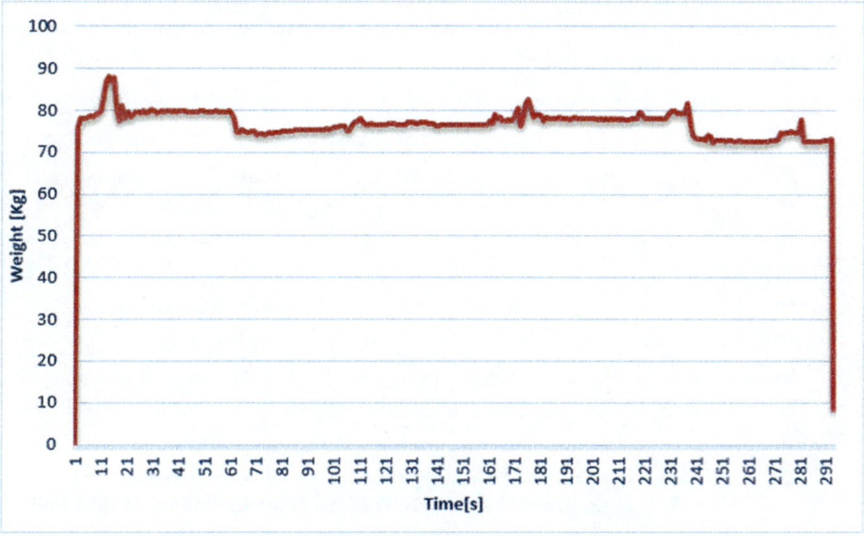

Fig. 4. Weight variation when sitting operation

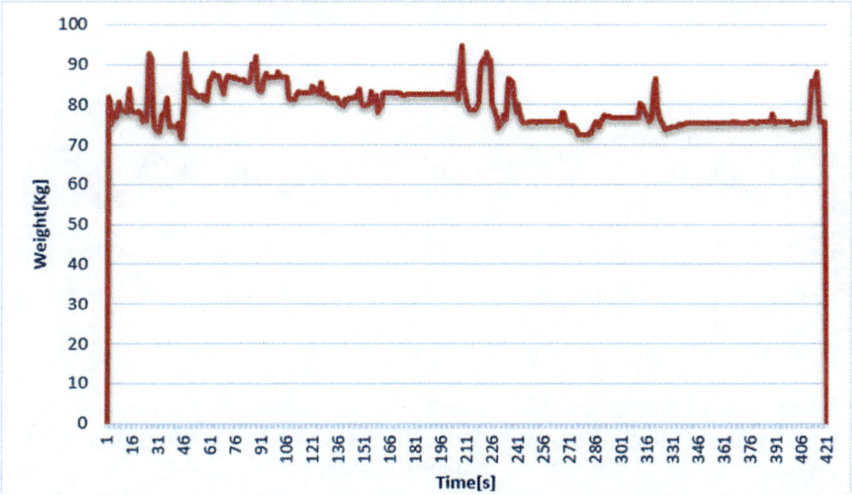

Fig. 5. Weight variation when sitting at the desk and performing a stressing activity

According to the data obtained, a minor increase in the first part of the position over the weight of the experienced person can be observed, indicating the moment of stabilization of the position on the chair, and throughout the station there are certain fluctuations representing small position changes but without significant fluctuation.

Compared to the graph obtained in the situation where the person sits relaxed at the office, in the second graph it can be seen that in the first 4 min there is a variation in weight between −15 kg and +10 kg compared to normal weight, indicating an increased degree of stress during this period that generates repeated movements of the person in the chair. In the next 3 min, a stabilization in time of the sitting position and the appearance of smaller fluctuations in amplitude can be identified, indicating a certain relaxation of the monitored person.

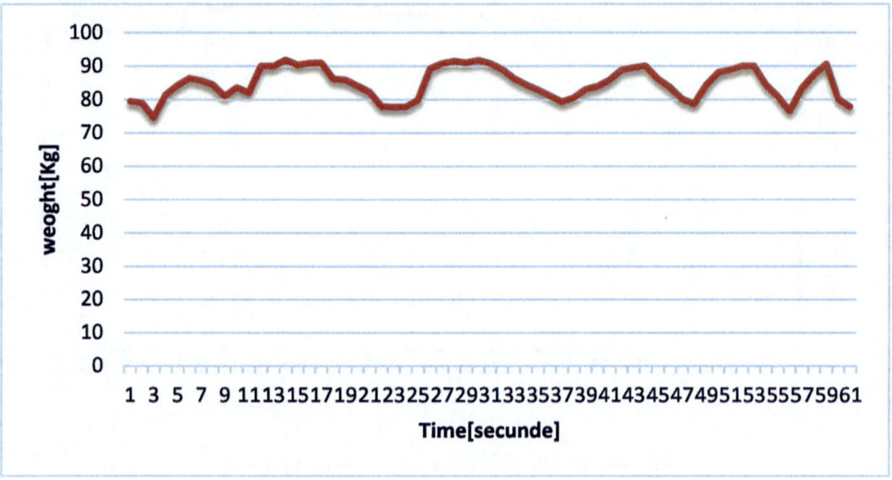

Fig. 6. Variation of weights on left-right movement

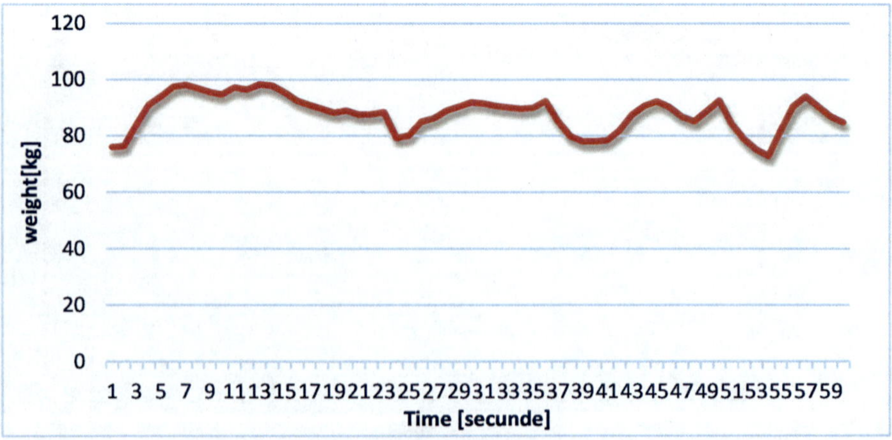

Fig. 7. Variation of weights on front-back movement

The influence of an activity can be observed by the continuous variation of the weight and movement that the analyzed person performs during the time of sitting on the chair. In order to be able to observe the variation of the movement on the experimental assembly, certain movements were performed from left to right and in front of the back which are shown in Figs. 6 and 7.

The oscillations that can be observed on the graph from Fig. 6 represent movements either to the left or to the right. The minimum values represented on the graph representing resting positions and the maximums represent the movements to the left or to the right. The maximum weight variation between rest and movement in the case of the 84 kg subject is 10–12 kg for each movement.

In the case of back-and-forth movements, Fig. 7, a double weight increase can be observed compared to the left-right movement, this is due to changes in the center of gravity of the body by pressing, only on a certain portion of the seat emphasizing weight in the direction of movement.

To determine a differentiation between the positions and movements performed by the person using the chair, the relative errors lower and higher than the value of the subject's weight are calculated.

3 Conclusions

The system presents the possibility to determine the activity performed by the monitored person, as well as his degree of involvement or stress [13].

The importance of developing systems for monitoring a person's office activity is highlighted by the fact that at European level there are various programs to improve the quality of life at work, as well as the increased interest of companies for integrated systems to provide an evaluation of employee activity of working conditions in order to be able to adopt measures that can improve the state of health.

The system of monitoring the activity of a person in the office has the following advantages:

- the possibility to identify the type of activity carried out by the person sitting on the chair;
- the possibility to evaluate the level of involvement in activities of the person sitting on the chair;
- the possibility of evaluating the time intervals in which the person sat on the chair;
- accuracy of processed data.

Acknowledgement. This paper has been financially supported within the project entitled "DECIDE - Development through entrepreneurial education and innovative doctoral and post-doctoral research, project code POCU/380/6/13/125031, project co-financed from the European Social Fund through the 2014–2020 Operational Program Human Capital".

Conflict of Interest. The authors declare that they have no conflict of interest.

References

1. Milici, D., Milici, M., Cernomazu, D., Popa, C.: Modeling of physical and psyhological human performance evolution. In: 2015 9th International Symposium on Advanced Topics in Electrical Engineering (ATEE), pp. 965–968 (2015)
2. Milici, M.R., Mihai, I., Milici, L.D.: Aspects of engineering education in signal technology using virtual instrumentation. Elektronika ir Elektrotechnika **94**(6), 113–116 (2009)
3. Milici, D.: Computerized system for testing and formation the speed of backward pusf of sportmen. In: 13th International Symposium on Measurements for Research and Industry Applications organisating by International Measurement Confederation IMEKO TC4, pp. 673–677 (2004)
4. Pavel, P., Cenuşă, M., Cernuşcă, D.: Human performance assessment system in sport trening and physiotherapeutic rehabilitation. In: Buletinul ştiinţific al academiei Forţelor Terestre Nicolae Bălcescu Sibiu, Sibiu (2016)
5. Milici, L.D., Graur, A., Poienar, M., Cenuşă, M., Barascu, A.A.: Study of a tiredness identification system. In: 11th International Conference on Electromechanical and Power Systems (SIELMEN 2017), Chisinau, Moldova, 12–13 October 2017, pp. 493–496 (2017). ISSB/ISBN: 978-1-5386-1845-5
6. Milici, L.D., Milici, M.R.: Consideration about the evolution of performance in nature and technology. Eur. J. Sci. Theol. **9**(1), 157–166 (2013)
7. Milici, L.D., Milici, M.R., Pentiuc, S.G.: E-learning application for the modeling and studying of data acquisition system's working. In: 4th IEEE Workshop on Intelligent Data Acquisition and Advanced Computing Systems: Technology and Applications, pp. 545–549 (2007)
8. Chiuchişan, I., Costin, H.N., Geman, O.: Adopting the Internet of Things technologies in health care systems. In: International Conference and Exposition on Electrical and Power Engineering (EPE 2014), Iasi, Romania, 16–18 October 2014 (2014)
9. Paradiso, R., Loriga, G., Taccini, N.: A wearable health care system based on knitted integrated sensors. IEEE Trans. Inf. Technol. Biomed. **9**(3), 337–344 (2005)
10. Barros, A.K., Vigario, R., Jousmachi, V., Ohnishi, N.: Extraction of event-related signals from multichannel bioelectrical measurements. IEEE Trans. Biomed. Eng. **47**(5), 583–588 (2000)
11. Bhattacharya, A., Basu, D., Goswami, B., Ghosh, R.: Measuring human biopotentials to ascertain parameters for health. In: Communication and Industrial Application (ICCIA) (2011)
12. Milici, L.D., Plăcintă, V.M., Bujor, L., Milici, M.R.: System for highlighting the emotional states, used in assessing the teaching methods. In: 9th International Symposium on Advanced Topics in Electrical Engineering (ATEE), pp. 965–968 (2015)
13. Milici, D., et al.: Sistem de monitorizare a activităţii unei persoane la birou. EPO pattent application no. 20464011.4

The Influence of Electromagnetic Waves Emitted by PIFA Antennas on the Human Head

Claudia Constantinescu(✉), Claudia Pacurar, Adina Giurgiuman, Calin Munteanu, Sergiu Andreica, and Razvan Gliga

Technical University of Cluj-Napoca, Cluj-Napoca, Romania
Claudia.Constantinescu@ethm.utcluj.ro

Abstract. The electromagnetic waves emitted by antennas are found everywhere around us. The antennas are very frequently used nowadays, having many fields of applicability. Thus, in order to offer very good quality services, an increasing number of antennas are used. Many of the devices currently used by humans contain antennas that emit electromagnetic waves, which, if they exceed certain permissible limits, become harmful to health. This paper treats the influence of electromagnetic waves emitted by PIFA (planar inverted-F antenna) antennas on the human head. The PIFA antenna will be tridimensional numerical modelled to determine its parameters and its fields emissions. The influence of the PIFA antenna on the human head is modelled. The antenna is placed parallel with the ear, simulating the position of the mobile phone. The electromagnetic wave emissions are analysed, and the results are checked with the admitted exposure limits from the standards in force. Also, the presence of a foreign body, such as the earring, is considered. This case is also tridimensional numerically modelled for the analysis of the influence of electromagnetic waves emitted on the human head. The results of the three modelled cases are finally compared and are treated in report with the human exposure to the electromagnetic waves emissions in 0–10 GHz frequency range. We want to check if the admissible limits provided in the standards in force are exceeded or not in these three cases.

Keywords: PIFA antenna · Electromagnetic waves emissions · Human head

1 Introduction

The PIFA antenna is the focus of this paper. Antenna is an element connected to the outside or inside of devices that communicate via radio waves. The role of the antenna for a device that communicates through radio waves is understood only if we have a clear notion of electromagnetic field. The electromagnetic field that propagates in space is called the electromagnetic wave. Radio signals are electromagnetic waves through which information is transmitted remotely, wirelessly, but the antenna effectively provides the connection. Radio waves as well as coloured light differ in wavelength (λ) and frequency (f), being linked by the formula: $\lambda = c/f$. Electromagnetic waves have two components: electric field (E) and magnetic field (H) in mutually perpendicular, inseparable planes. The mechanism of propagation of electromagnetic waves is based on the phenomenon of mutual generation of electric and magnetic fields. In the RF antenna (Radio Frequency) an electric field is generated from the voltage and magnetic field variations to the current

S. Vlad and N. M. Roman (Eds.): MEDITECH 2020, IFMBE Proceedings 88, pp. 77–91, 2022.
https://doi.org/10.1007/978-3-030-93564-1_10

variations and thus an electromagnetic field will appear propagating in space in the form of sinusoidal waves. Similarly, when an RF antenna is in an electromagnetic field, changes in the magnetic field occur with changes in current, and changes of the electric fields occur with variations of the voltage in the antenna circuit. To obtain broad bandwidth characteristics, the PIFA antenna was introduced [1, 6]. The PIFA is widely used in nowadays mobile handheld devices. It is a self-resonating antenna with purely resistive impedance at the frequency of operation. This makes it a practical candidate for mobile handheld design since it does not require a conjugate circuit between the antenna and the load, reducing both cost and losses [2]. The PIFA antennas have many advantages: low volume and weight, low manufacturing costs, easy integration into circuits, robustness from a mechanical point of view when are placed on a rigid surface and low profile planar configuration that can conform to the surface on which are applied.

2 PIFA Antenna High Frequency Modelling

The PIFA antenna is increasingly used in the mobile phone market. This type of antenna resonates to a quarter-wavelength (thus reducing the space required on the phone) and usually has good SAR properties. This antenna looks like an inverted F, which explains the name PIFA. The PIFA antenna is popular because it has a low profile and an omnidirectional pattern [8]. In Fig. 1 is presented the geometry of the PIFA antenna that is analysed by three-dimensional modelling.

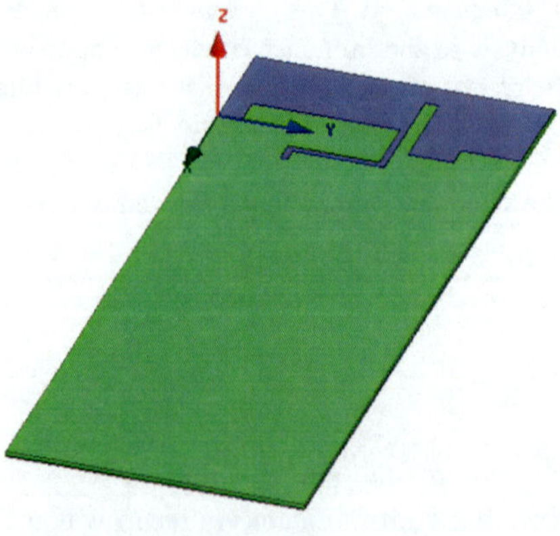

Fig. 1. The PIFA antenna geometry

These antennas have a low profile, are compliant with planar or non-planar surfaces, their manufacturing process is simple and inexpensive using modern printed circuit technology, they are mechanically robust when mounted on rigid surfaces, and when a particular shape of the patch is chosen and a certain mode of power supply, they are very versatile in terms of resonance frequency, polarization and impedance. In addition, by adding loads between the patch and the ground plane, such as pins or diodes, adaptive elements with resonant frequency, impedance, variable polarization can be designed [5].

PIFA is one of the most promising types of antennas, because it is small and has a low profile, which makes it suitable for mounting on portable equipment. Wireless communication was intended for voice and short messaging services, but from time to time some new services are integrated into the existing infrastructure to make communication more and more lively. With the current advancement there is a trend of thin mobile phones, with a functionality almost equivalent to a computer. Now, if mobile phones need to be thin, then there is a high requirement for a single antenna to be multifunctional because using multiple antennas for multiple function leads to a bulky device. To reduce the size of the conventional microstrip antenna for this purpose, a flat inverted F antenna (PIFA) can be used. A PIFA antenna can reduce the size of the PATCH by half.

The modelled PIFA antenna is composed of the FR-4 Epoxy substrate; the perfect E ground plane; the perfect E patch with dimensions from (Fig. 2a) and the coaxial supply wire of cooper (Fig. 2b) [7, 9].

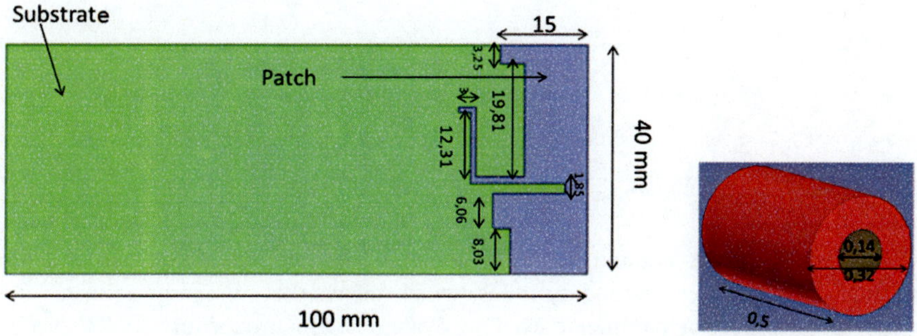

a) PIFA antenna b) coaxial supply wire

Fig. 2. The PIFA antenna components and dimensions

The PIFA antenna was modelled at 0,835 GHz, to determine its parameters (S-Parameters, directivity) and field quantities (SAR, electric and magnetic field). A variable frequency is considered, between 0 to 10 GHz.

In Fig. 3 is represented the S-Parameters variation versus frequency.

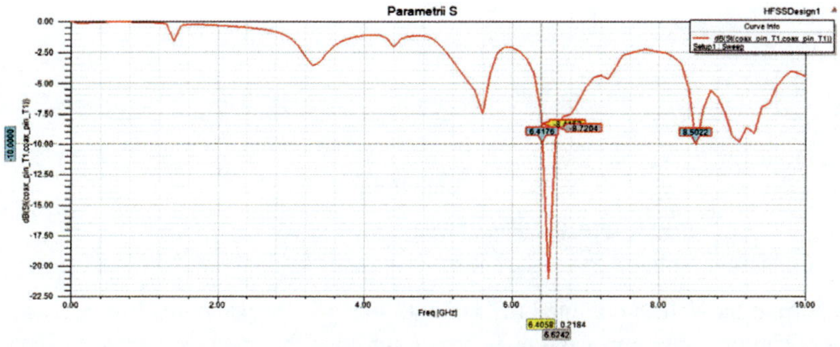

Fig. 3. The S-Parameters of the PIFA antenna

The graph shows that the antenna begins to radiate at 6.41 GHz, resonates at 6.51 GHz and begins to reflect part of the wave at 6.62 GHz. The average calculated frequency is 6.49 GHz. The calculated band width is 2.61% [3].

In Fig. 4 is presented the PIFA antenna directivity.

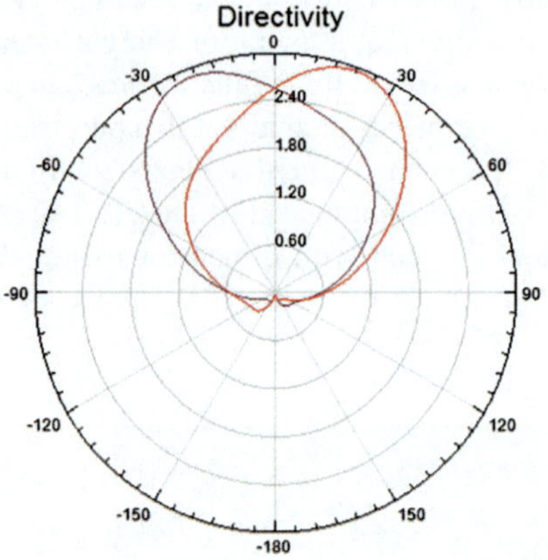

Fig. 4. The directivity of the PIFA antenna

The modelled PIFA antenna emits in one main direction. We have a main radiation lobe and we do not have secondary lobes, which means that no energy is lost.

The most important parameter used to evaluate human exposure to RF electromagnetic fields emitted by mobile phones is the specific absorption rate (SAR).

The Local SAR and Average SAR for the PIFA antenna modelled are presented in Fig. 5.

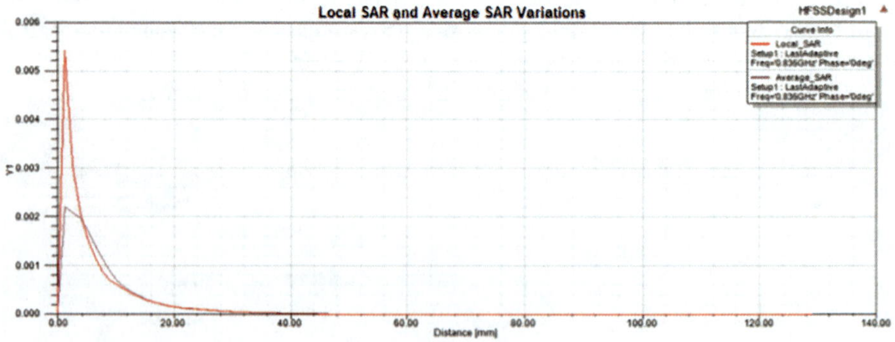

Fig. 5. PIFA antenna Local_SAR and Average_SAR

Reporting the values obtained (maximal 0.0054) to the standards in force, either to the American or to the European ones, it is found that the antenna does not exceed the legal limits.

Figure 6a highlights the electric field produced by the antenna in the YZ plane, in the direction where the head will be placed. The electric field will be represented in the domain 0–100 V/m, to be compared with the case of antenna with head present near it.

The magnetic field is presented in Fig. 6b in a cross-section (YZ plane), as considered for the electric field. The values are represented in the limits 0–0.5 A/m in order to compare them with the antenna magnetic field in the presence of the head [4].

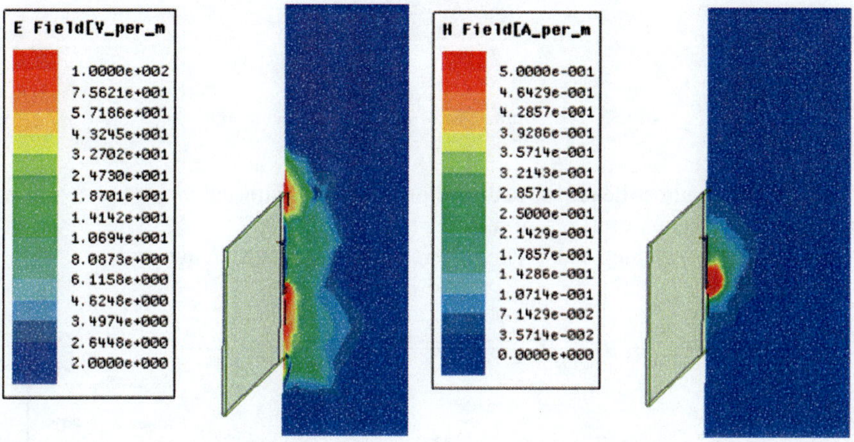

a) E field b) H field

Fig. 6. PIFA antenna with the electric and magnetic field produced by it in a YZ cross section

It can be observed that the values are closer to the limits stated in the standards in force only in the vicinity of the antenna, where the patch is present.

3 The Influence of the Electromagnetic Waves Emitted by the PIFA Antenna on the Human Head

The PIFA antenna is modelled near the human head to study the influence of electromagnetic emissions on its tissues. Two cases are detailed in this paper, the first one with the human head and PIFA antenna, and the second one with human head, gold earring and antenna.

3.1 The PIFA Antenna Influence on the Human Head

The head implanted in the ANSYS-HFSS program is presented in Fig. 7.

Fig. 7. The human head and the PIFA antenna

The head is a ghost head, formed of the skin layer, with radius of 111.5 mm and bone layer, with radius of 106.5 mm.

In Fig. 8 is represented the S-Parameters variation versus frequency.

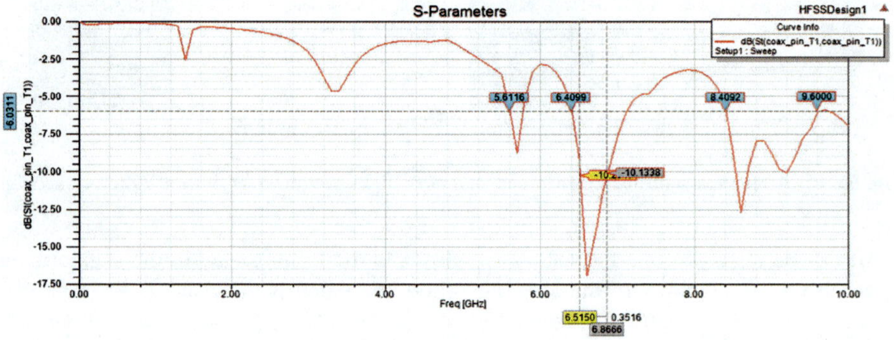

Fig. 8. S-Parameters for the PIFA antenna with the human head

The antenna is functioning on three different frequency ranges, between 5.6–5.8 GHz, 6.5–6.86 GHz and 8.4–9.6 GHz [3]. Compare with S Parameters of the PIFA antenna (Fig. 3) can be seen that the bandwidth increased with the appearance of the head in the action area of the antenna.

The PIFA antenna directivity is also influence by the human head presence in its emission direction. In Fig. 9 is presented the polar and 3D diagram of the directivity.

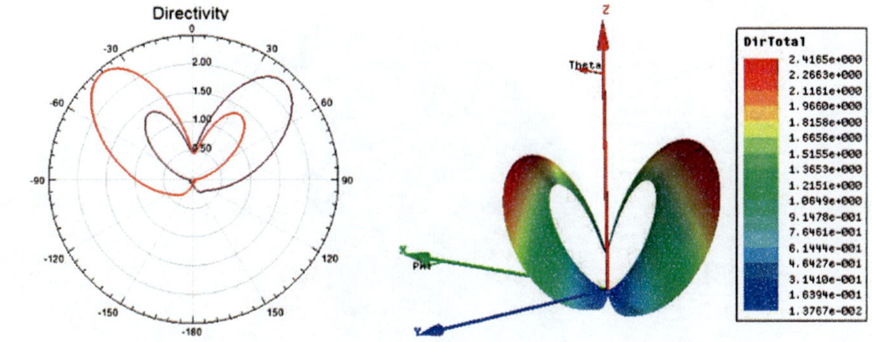

a) polar diagram b) 3D diagram

Fig. 9. The directivity for the PIFA antenna with the human head

We have two main radiation lobes but also secondary lobes, which means that a small part of the energy is lost to do the head presence.

The Local SAR and Average SAR for the PIFA antenna modelled with human head are presented in Fig. 10. The SAR reaches a maximum value of 0.0086, so on the surface of the antenna the values are below the imposed limits.

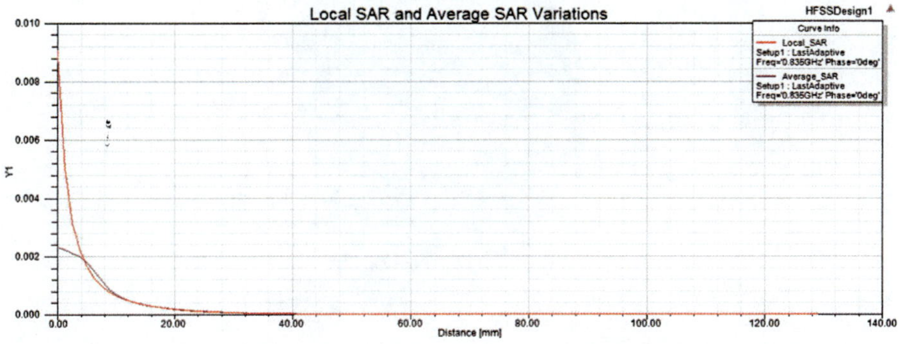

Fig. 10. The PIFA antenna with human head Local_SAR and Average_SAR

Figure 11 highlights the impact of the electric field on the human head. The limit values are higher only in the skin layer and on a small surface, where the patch is placed near the head. In the bone layer the electric field is present, but with small values.

The magnetic field is presented in Figs. 12. It can be observed how it penetrates inside the head, the values being represented in the limits 0–0.5 A/m, in order to compare them with the antenna magnetic field. The magnetic field values do not exceed the limits and is present only in the skin layer.

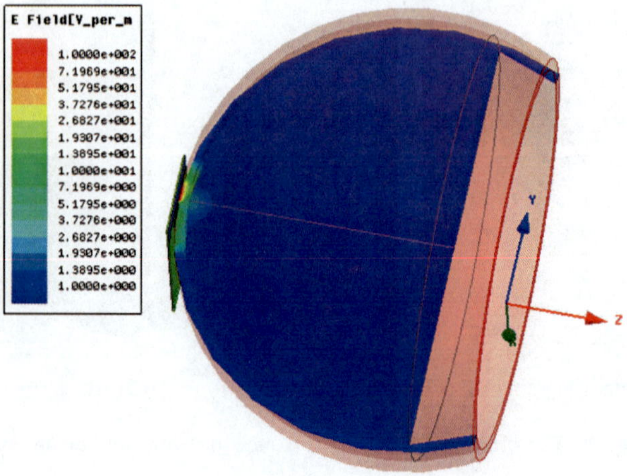

Fig. 11. The electric field on skin and bone

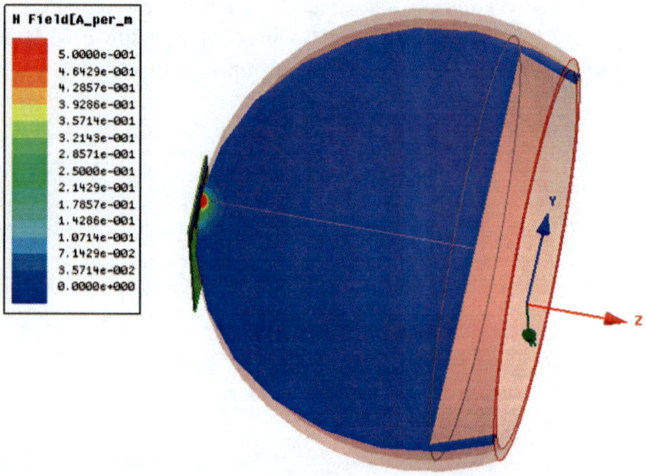

Fig. 12. The magnetic field on skin and bone

3.2 The Gold Earring Presence Influence

The PIFA antenna is modelled near the human head to study the influence of electromagnetic emissions in case of gold earring presence. An earring of gold of 9.5 mm exterior diameter and width of 2 mm is drawn (Fig. 13).

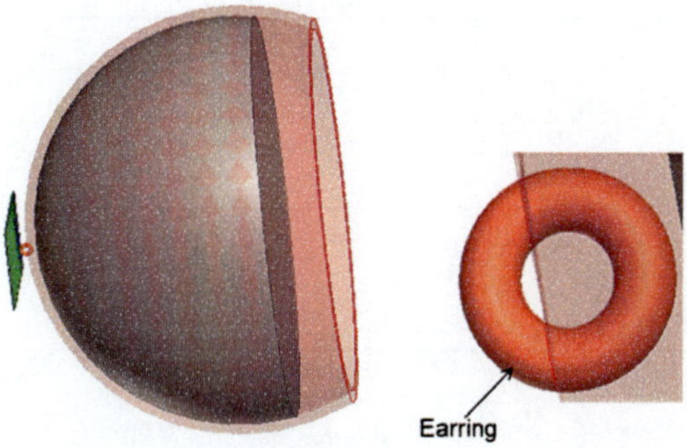

Fig. 13. The electric field on skin layer

The PIFA antenna is positioned at three distances relative with the human head. The S parameters for distance of 0 mm are presented in Fig. 14a), for 10 mm in Fig. 14b) and for 20 mm in Fig. 14c).

The presence of an earring, no matter the material, influences the S parameters, as it can be observed comparing Fig. 14a and Fig. 8.

It can be observed that the distance of the antenna from the head influences the antenna functionality. The antenna does not have the same operating frequency ranges and it better functions when closer to the head.

The PIFA antenna and human head with gold earring directivity is presented also for the three considered position in Fig. 15, as polar diagram, and 3D diagram.

As it can be seen above, the antenna's directivity becomes bidirectional in the presence of the human head. The main lobes are influenced by the distance between the head and the antenna, the smaller lobe increasing with the increase of the distance.

The PIFA antenna and human head with gold earring Local SAR and Average SAR variations are presented also for the three considered position in Fig. 16.

Considering the three graphs, it is determined that the values of the local and average SAR is higher when the antenna is closer to the head, decreasing when the distance between the antenna and the head is decreased.

The PIFA antenna and human head with gold earring model's SAR Field is presented for the three considered position also locally colour coded in Fig. 17.

The conclusions remain the same as for the graphs, but this representation gives us a better view at the fact that the maximum values are present where the earring and the patch are present.

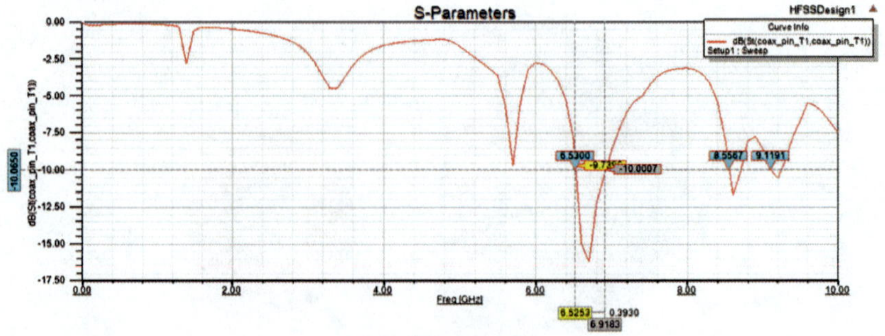

a) 0 mm

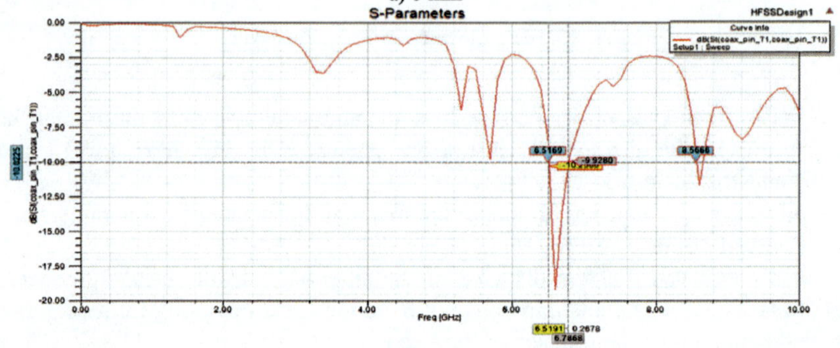

b) 10 mm

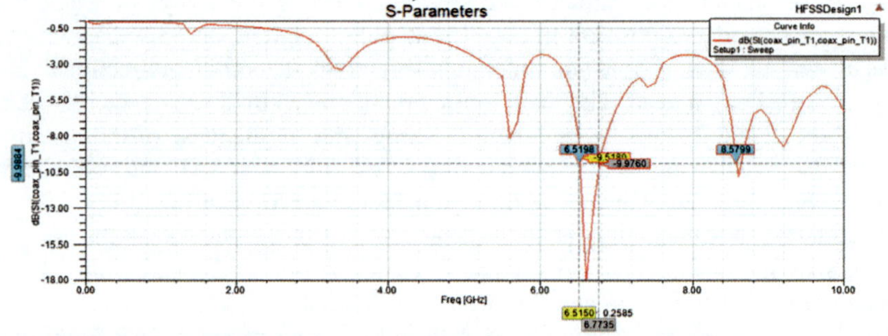

c) 20 mm

Fig. 14. S-Parameters for the PIFA antenna and human head with gold earring

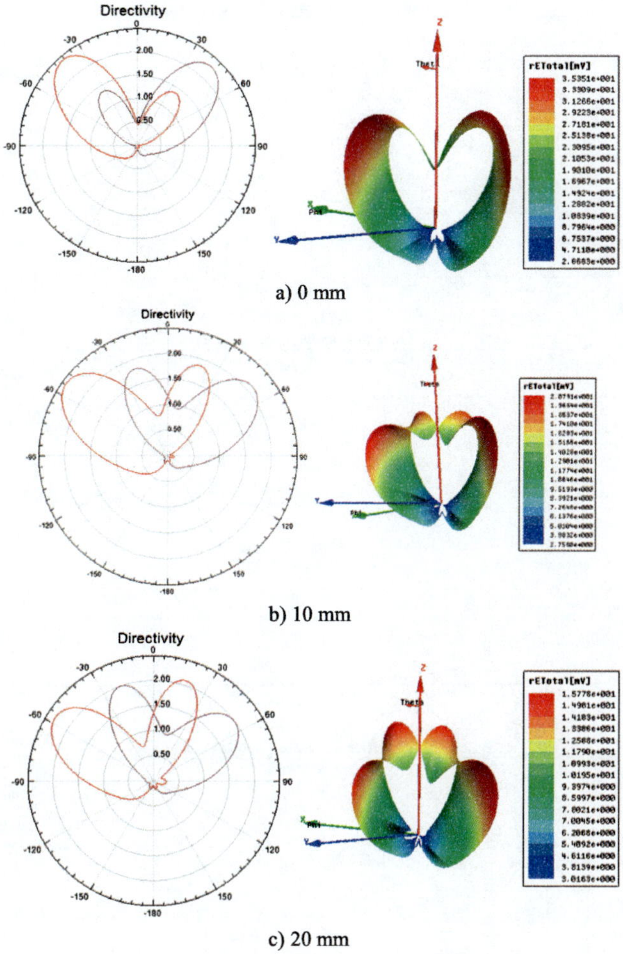

Fig. 15. The directivity for the PIFA antenna and human head with gold earring

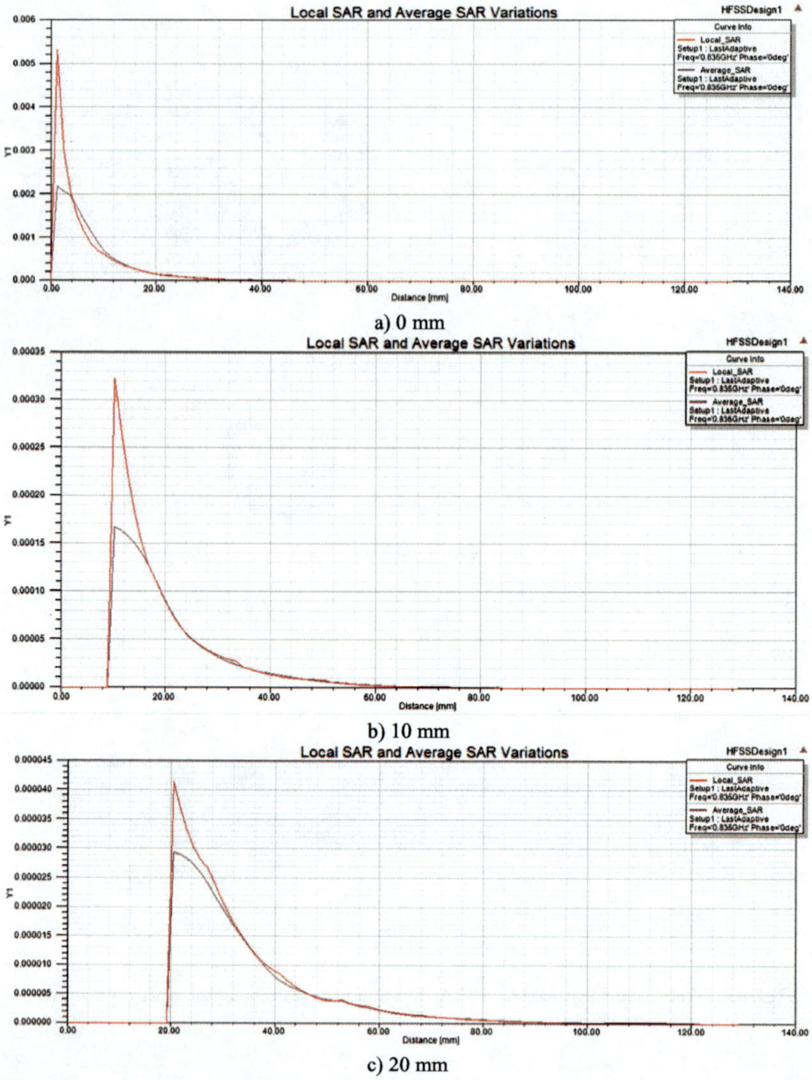

Fig. 16. The local and average SAR for the PIFA antenna and human head with gold earring

The electric field emitted by the PIFA antenna for the head with earring is represented for the limits 0 to 100 V/m in Fig. 18.

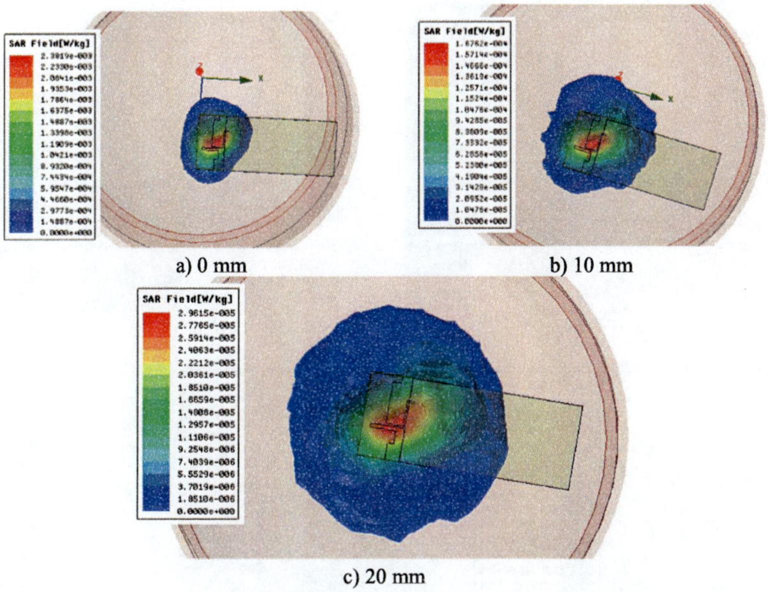

Fig. 17. The SAR-field for the PIFA antenna and human head with gold earring

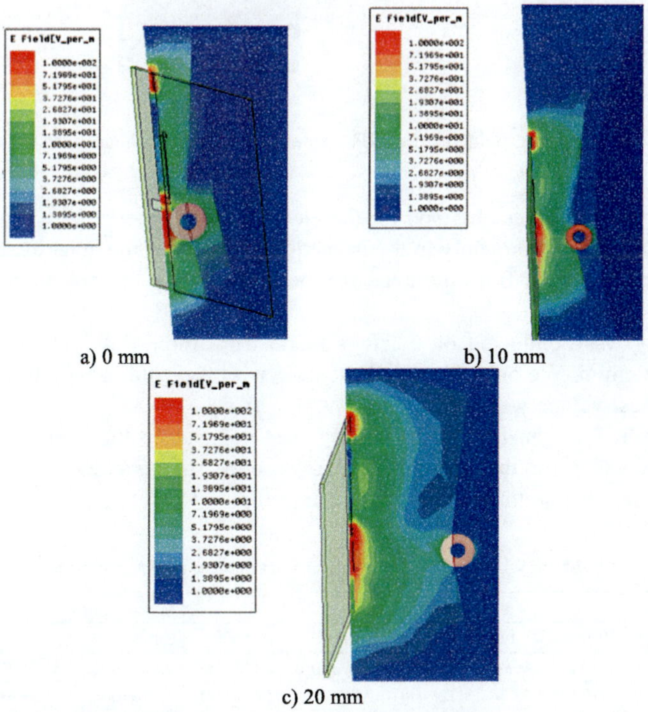

Fig. 18. The electric field for the PIFA antenna and human head with gold earring

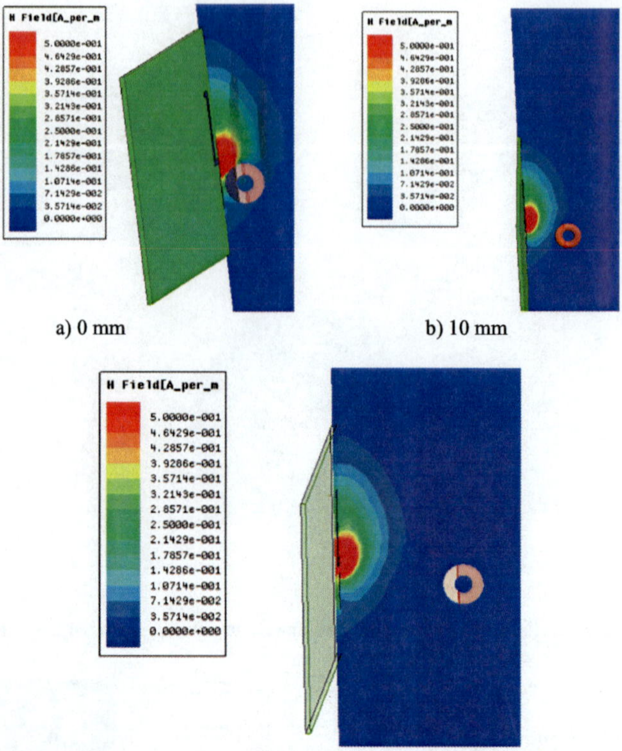

a) 0 mm b) 10 mm

c) 20 mm

Fig. 19. The magnetic field for the PIFA antenna and human head with gold earring

Considering the values obtained for the electric and magnetic field, it can be stated that they closer to the limits only in the immediate vicinity of the antenna. It is better to keep the antenna further from the head for the higher values to not be present on the skin surface (Fig. 19).

The study was conducted on earrings made from three different materials, gold, silver, and titanium. We only presented the case with the gold earring because for this case the highest values were obtained.

From Table 1, it can be said that SAR does not exceed the imposed limits. The bandwidth is wider for the antenna in the presence of the golden earring, but the differences between the three cases are not that big.

Table 1. Values of the specific parameters of the antenna.

Materials	Bandwidth			Directivity [mV]			Specific absorption rate [W/kg]		
	0 mm	10 mm	20 mm	0 mm	10 mm	20 mm	0 mm	10 mm	20 mm
Gold	5,84%	4,06%	3,89%	3,53e+001	2,07e+001	1,57e+001	2,38e−003	1,67e−004	2,96e−005
Silver	5,56%	4,04%	3,78%	3,52e+001	2,06e+001	1,56e+001	2,37e−003	1,66e−004	2,96e−005
Titanium	5,54%	4,02%	3,74%	3,52e+001	2,06e+001	1,56e+001	2,37e−003	1,66e−004	2,96e−005

4 Conclusions

This paper focuses on the PIFA antenna and its influence on the human head, considering its position and the presence of foreign bodies like different material earrings. It was observed that the values obtained for the electric and magnetic field limits are closer to the higher limits of the standards in force in the near vicinity of the antenna, in the analysed models.

When the antenna is present next to the human head the limit values are higher only in the skin layer and on a small surface, where the patch is placed near the head. In the bone layer the electric field is present, but with small values. The magnetic field values do not exceed the limits and is present only in the skin layer.

The presence of an earring influences the S parameters. Regarding the directivity, the main lobes are influenced by the distance between the head and the antenna. SAR is higher when the antenna is closer to the head. All the values do not exceed the standard limits and are higher where the patch and the earring are present.

Considering the values obtained for the electric and magnetic field, it can be stated that they have higher values only in the immediate vicinity of the antenna and the surface on which they have greater values is negligible. It is better to keep the antenna further from the head for the higher values to not be present on the skin surface.

Conflict of Interest. The authors declare that they have no conflict of interest.

References

1. Taga, T., Tsunekawa, K.: Performance analysis of a built-in inverted-F antenna for 800 MHz band portable radio units. IEEE J. Sel. Areas Commun. **5**(5), 921–929 (1987)
2. Geyi, W., Rao, Q., Ali, S., Wang, D.: Handset antenna design: practice and theory. Prog. Electromagn. Res. J. (PIER) **80**, 123–160 (2008)
3. Constantinescu, C., Munteanu, C., Pacurar, C., Racasan, A., Gliga, M., Andreica, S.: High frequency analysis of bowtie antennas. In: 2019 11th International Symposium on Advanced Topics in Electrical Engineering, ATEE 2019, Bucharest, Romania, 28–30 March 2019 (2019). https://doi.org/10.1109/ATEE.2019.8724972. ISBN: 978-147997514-3
4. ORDER no. 1193 of 29 September 2006 for the approval of the Norms regarding the limitation of the exposure of the general population to electromagnetic fields from 0 Hz to 300 GHz
5. Council Recommendation of 12 July 1999 on the limitation of exposure of the general public to electromagnetic fields (0 Hz to 300 GHz) (1999/519/EC)
6. Pacurar, C., et al.: Analysis of the patch antennas at high frequency. Acta Electrotehnica J. **55** (3–4), 169–173 (2014)
7. User's Guide – High Frequency Structure Simulator (2005)
8. Decision No. 520/2016 of 20 July 2016 on the minimum safety and health requirements regarding the exposure of workers to risks generated by electromagnetic fields
9. Constantinescu, C., Munteanu, C., Pacurar, C., Racasan, A.: Influence of the patch antenna feeding on their parameters. In: 10th International Conference and Exposition on Electrical and Power Engineering, Iasi, Romania, pp. 235–240 (2018)

Biomedical Imaging and Image Processing

Fractal Dimension Box-Counting Algorithm Optimization Through Integral Images

Mircea-Sebastian Şerbănescu(✉)

University of Medicine and Pharmacy of Craiova,
Petru Rarş St., No. 2-4, 200349 Craiova, Romania
mircea_serbanescu@yahoo.com

Abstract. The fractal dimension is an index used in computer vision for characterizing fractal patterns, by quantifying their complexity as a ratio of the change in detail to the change in scale. A common approach for fractal dimension estimation in binary images is the box-counting algorithm that counts the number of boxes needed to cover the object at different scales. A new approach for the counting task of the algorithm is proposed using integral images (summed-area tables). We experimented with our proposed algorithm on two distinct datasets (a texture dataset and a medical image dataset). The experiments have proven that our algorithm statistically outperforms the standard approach.

Keywords: Fractal dimension · FD · Box-counting · Integral image · Summed-area table · Optimization

1 Introduction

The fractal dimension (FD) is an index for characterizing fractal patterns or sets by quantifying their complexity as a ratio of the change in detail to the change in scale [1]. The first concept around the fractal behavior cannot be traced in time, but the name itself and current use was introduced in 1967 [2].

The FD is used in various computer vision applications, ranging from texture description to medical image analysis [3]. In the medical field it is used as a feature useful to characterize images that are too complex to be described by other features.

Let D_0 be the box-counting dimension, a numerical approximation of the FD:

$$D_0 = \lim_{e \to 0} \frac{\log N(e)}{\log \frac{1}{e}}$$

where e is the scale and $N(e)$ represents the detail at e-sized scale.

The practical numerical approach approximates scaling and detail using the limits gathered from regression lines over log vs log plots of detail vs scale.

For a binary image, a good approximation of the FD can be obtained by the box-counting algorithm that considers as detail the number of boxes of e-size needed to cover the object represented in the image.

S. Vlad and N. M. Roman (Eds.): MEDITECH 2020, IFMBE Proceedings 88, pp. 95–101, 2022.
https://doi.org/10.1007/978-3-030-93564-1_11

The applications of FD computation in biology and medical fields are numerous [4], practically any signal or image that can be acquired, with proper preprocessing, can be described by its FD. The FD has been successfully used in bio-signals [5], echography [6], computerized tomography [7], nuclear resonance [8], microscopic images [9], ophthalmoscopy images [10], and the list could continue.

1.1 Box-Counting Algorithm

A box counting algorithm consists of three steps: (1) image preprocessing, (2) box-counting at different scales (3) estimating the best fit of a line between the box-count pairs and scale pairs computed at the previous step.

Image preprocessing implies that the image is square shaped and its size is set as the smallest power of 2 that the original image can fit in. After estimating the smallest power of 2 that is larger than each of the image dimensions, the original image is padded with zeros to obtain the required square shape.

Box-counting at different scales implies generating the scale (box size) and then counting the boxes that contain at least a pixel of the object. For the first scale, each image dimension is split in two halves, thus dividing the original image in four equally-shaped sub-images. The box-count and the scale are saved. Next each sub-image is split again in 4 sub-images, and the process continues until the image can no longer be split. Two vectors of images are obtained: one is a power of two series of numbers and the other is the box-count of the squares of that scale that contained at least one pixel of the object.

The FD is the slope of the best fit line between equivalent points of the vectors computed at the previous step. It can be determined using polynomial fit.

1.2 Integral Images

Summed-area tables were introduced in 1984 [11], but they became known to computer vision applications in 1995 [12] as "integral images". For example, they were used within Viola–Jones object detection framework [13] in 2001 as a common approach to real-time face recognition.

In order to obtain an integral image, the value of each pixel is replaced with the sum of all the pixels that are located above and to the right of it, including itself.

Let

$$I(x,y) = \sum_{\substack{x' < x \\ y' < y}} i(x',y')$$

where x, y are the coordinates of a pixel in the integral image (I) computed from the original image (i).

The advantage of the technique is that after transforming an (n x n) image into an (n x n) integral image using an $O(n^2)$ algorithm, computation of the sum of pixels in any sub-image of the original image can be done in $O(1)$ using 4 simple algebraic operations, regardless of the size of the sub-image, where here and in what follows, n denotes the height/weight of the square shaped image.

If x_1, y_1 and x_2, y_2 are the top-left and bottom-right corners of a sub-image then the sum of the pixels in sub-image (s) is:

$$s = I(x_2, y_2) - I(x_1, y_2) - I(x_2, y_1) + I(x_1, y_1)$$

The complexity of any sub-image evaluation, including the whole image, is $O(1)$. The area of the marked sub-image can be computed from the integral image using 4 algebraic operations (for example in Fig. 1 this sum is computed as: $46 - 22 - 20 + 10 = 14$).

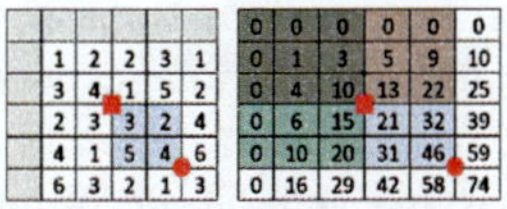

Fig. 1. Computing the sum of a sub-image in $O(1)$ complexity using integral images approach. Original image on the left, computed integral image on the right, red square is x_1, y_1, red circle is x_2, y_2.

Current work proposes a hybrid algorithm for speed optimization of FD estimation in binary image using integral images for the counting task.

2 Material and Method

2.1 Integral Image (Summed-Area Table) Box-Counting Algorithm

The standard algorithm for the counting task is:

1. Set current scale to half of image size.
2. For each sub-image with current scale size, search for a value larger than 0 (if found break loop).
3. Save the scale and the resulted non-zero sub-image count.
4. While current scale is larger than 1, set current scale to:
   $scale = \frac{current\ scale}{2}$
   and repeat step 2.

This algorithm computes two vectors representing the scales and their corresponding box counts.

The hybrid integral image approach for the counting task, that replaces the iterative counting task with the sum obtained from the integral sub-image is:

1. Compute the integral image
2. Set current scale to image size/2.
3. Count the sub-images with current scale size that have a summed value larger than 0 (meaning that they have at least one object pixel) with complexity $O(1)$.
4. Save the scale and the resulted non-zero sub-image count.

5. While current scale is larger than 1, set current scale to:
   $scale = \frac{current\,scale}{2}$
   and repeat step 4.

This algorithm determines two vectors, similarly to those obtained by the standard algorithm.

Note that image preprocessing and FD estimation of the FD remain unchanged in both approaches.

2.2 Dataset 1

Dataset 1 is a set of real medical images consisting in 460 histological images stained with Masson's trichrome procedure. Binary images were obtained using the algorithm presented in [14].

A sample image and its binarized version are presented in Fig. 2.

2.3 Dataset 2

Dataset 2 is a public set consisting in 111 grayscale texture images offered by Trygve Randen [15] and used in other FD applications [16]. The binary images were obtained using a 0.5 (128/255) threshold. From the 643 × 643 images a crop of 512 × 512 was made starting from the pixel at position (5, 5), thus removing some artifact of the dataset. Image 14 (D14.gif) from the dataset is missing.

An image sample and its binary computed mask are presented in Fig. 3.

2.4 Statistical Validation

Mean computation time between the standard and hybrid algorithm for both datasets were assessed with the Wilcoxon test, since the data distribution in the samples did not have a normal distribution (Lilliefors test, p value > 0.05).

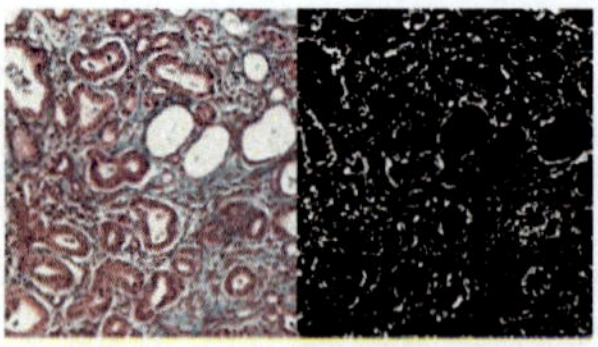

Fig. 2. Sample image from Dataset 1 (left), and resulted binary image (right); computed FD = 1.334.

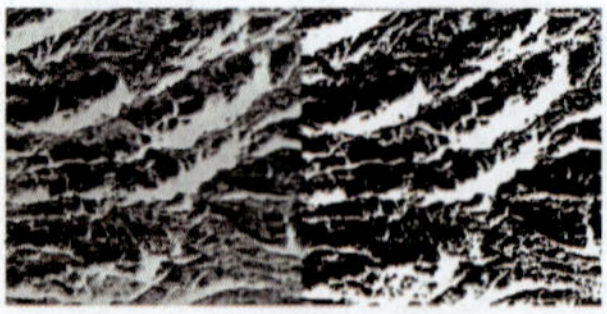

Fig. 3. Crop of image 13 (D13.gif) from Dataset 2 (left), and resulted binary image (right); computed FD = 1.888.

3 Results and Discussion

Computational Time for the Standard Algorithm (CTSA) and Computational Time for the Hybrid Algorithm (CTHA) were measured. Statistics of Dataset 1 computation time is presented in Table 1. Individual values and statistics of Dataset 2 computation time are presented in Table 2.

For both datasets the FD estimation for each individual image was identical for the two algorithms.

Table 1. Statistics and computed values for FD, CTSA, and CTHA on Dataset 1.

	FD	CTSA	CTHA
Average	1.381	0.015	0.006
Standard deviation	0.210	0.002	0.001

Table 2. Statistics and computed values for FD, CTSA, and CTHA on Dataset 2.

#	FD	CTSA	CTHA	#	FD	CTSA	CTHA	#	FD	CTSA	CTHA	#	FD	CTSA	CTHA
1	1.986	0.010	0.007	29	1.781	0.013	0.007	57	1.816	0.014	0.007	85	1.929	0.014	0.014
2	1.918	0.012	0.007	30	1.862	0.022	0.008	58	1.924	0.012	0.008	86	1.784	0.013	0.009
3	1.963	0.012	0.008	31	1.841	0.014	0.007	59	1.682	0.017	0.007	87	1.889	0.014	0.012
4	1.900	0.014	0.008	32	1.673	0.014	0.008	60	1.956	0.019	0.008	88	1.905	0.013	0.007
5	1.916	0.012	0.007	33	1.561	0.015	0.007	61	1.792	0.017	0.006	89	1.941	0.012	0.012
6	1.643	0.013	0.006	34	1.549	0.014	0.006	62	1.837	0.015	0.008	90	1.848	0.014	0.011
7	1.949	0.013	0.007	35	1.923	0.013	0.008	63	1.946	0.019	0.008	91	1.826	0.016	0.007
8	1.891	0.014	0.009	36	1.912	0.013	0.007	64	1.713	0.014	0.007	92	1.909	0.015	0.011
9	1.917	0.013	0.010	37	1.890	0.013	0.007	65	1.696	0.017	0.006	93	1.932	0.015	0.009
10	1.981	0.011	0.008	38	1.858	0.013	0.010	66	1.632	0.016	0.011	94	1.784	0.015	0.009
11	1.908	0.015	0.009	39	1.725	0.014	0.007	67	1.839	0.014	0.007	95	1.960	0.013	0.008
12	1.893	0.013	0.008	40	1.732	0.014	0.007	68	1.757	0.014	0.007	96	1.983	0.015	0.010
13	1.888	0.015	0.008	41	1.669	0.015	0.006	69	1.765	0.017	0.008	97	1.946	0.014	0.008
14	-	-	-	42	1.793	0.013	0.007	70	1.546	0.015	0.007	98	1.754	0.016	0.011
15	1.943	0.014	0.010	43	1.982	0.011	0.007	71	1.922	0.029	0.008	99	1.792	0.016	0.008
16	1.872	0.016	0.009	44	1.985	0.011	0.006	72	1.819	0.026	0.008	100	1.831	0.020	0.008
17	1.847	0.014	0.009	45	1.978	0.010	0.008	73	1.891	0.014	0.008	101	1.941	0.013	0.007
18	1.861	0.014	0.010	46	1.915	0.013	0.007	74	1.907	0.015	0.007	102	1.807	0.013	0.007
19	1.915	0.015	0.009	47	1.647	0.021	0.006	75	1.944	0.014	0.009	103	1.886	0.013	0.007
20	1.851	0.013	0.007	48	1.869	0.013	0.006	76	1.812	0.016	0.009	104	1.921	0.012	0.007
21	1.836	0.013	0.013	49	1.941	0.012	0.009	77	1.862	0.014	0.009	105	1.868	0.015	0.009
22	1.940	0.013	0.007	50	1.780	0.016	0.008	78	1.819	0.014	0.008	106	1.935	0.015	0.010
23	1.865	0.013	0.007	51	1.860	0.013	0.011	79	1.872	0.014	0.009	107	1.813	0.014	0.007
24	1.928	0.015	0.009	52	1.805	0.013	0.007	80	1.881	0.016	0.009	108	1.968	0.012	0.011
25	1.479	0.019	0.007	53	1.877	0.019	0.007	81	1.879	0.015	0.015	109	1.705	0.022	0.008
26	1.673	0.017	0.008	54	1.893	0.020	0.009	82	1.925	0.013	0.009	110	1.967	0.011	0.009
27	1.864	0.014	0.007	55	1.900	0.013	0.007	83	1.918	0.015	0.015	111	1.847	0.014	0.007
28	1.896	0.016	0.008	56	1.881	0.012	0.008	84	1.880	0.014	0.008	112	1.939	0.014	0.007
Average													1.851	0.015	0.008
Standard deviation													0.104	0.002	0.001

Average computation time, in 100 runs, for Dataset 1 was 0.015 s for the standard algorithm and 0.006 s for the hybrid algorithm. Average computation time for Dataset 2 was 0.015 s for the standard algorithm and 0.008 s for the hybrid algorithm. Statistical assessment of the mean differences shows significant differences between the two groups (Wilcoxon test p value < 0.05) for both datasets. Tests were carried out on a computer equipped with an I7-7700K processor with 16 GB of RAM using MATLAB 2018a.

Integral image computation is done in $O(n^2)$ complexity. Since the resulted integral image can be used to compute other features its determination can be moved from the counting task to the preprocessing task thus resulting in a faster algorithm.

For large-scale images the summed-area pixel values at the end of a grayscale image can overcome the possibilities of the standard integer formats, but, since the algorithm is proposed for binary images this event in highly unlikely for standard applications.

For grayscale images other algorithms for feature extraction have been proposed [17], with possible optimization with integral images.

Integral images have already been used in various applications like: very fast template matching [18], detecting small objects in high resolution images [19], local adaptive thresholding techniques for optical character recognition [20] and real-time visual attention system using smart feature computation techniques based on integral images [21].

4 Conclusion

A fast hybrid algorithm for FD estimation through box-counting using integral images (summed-area tables) for the counting task was developed.

The resulted algorithm outperforms the standard one, even when the integral image computation is included in the benchmark, with statistical significance. Taking into consideration that for some applications the integral image can be reused for other tasks, thus saving the time required for its re-computation (complexity $O(n^2)$), this might result in the further improvement of applications based on box-counting.

5 Grant Support

This work was supported by the University of Medicine and Pharmacy of Craiova, research grant number 519/20.09.2019.

Conflict of Interest. The author declares that he has no conflict of interest.

References

1. Fractal dimension – Wikipedia. https://en.wikipedia.org/wiki/Fractal_dimension. Accessed 01 Sept 2020
2. Mandelbrot, B.: How long is the coast of Britain? Statistical self-similarity and fractional dimension. Science **156**(3775), 636–638 (1967)

3. Serbanescu, M.S., Plesea, R.M., Covica, V., Plesea, I.E.: Fractal dimension of stromal fibrillar network: a new approach to prostate carcinoma architectural assessment. In: Proceedings of the 27th European Congress of Pathology, p. S235. Virchows Archiv, Belgrad (2015)
4. Przemysław Leszczyński, P., Sokalski, J.: The use of fractal analysis in medicine: a literature review. Dent. Med. Probl. **54**(1), 79–83 (2017)
5. Tapanainen, J.M., et al.: Fractal analysis of heart rate variability and mortality after an acute myocardial infarction. Am. J. Cardiol. **90**, 347–352 (2002)
6. Fiz, J.A., et al.: Fractal dimension analysis of malignant and benign endobronchial ultrasound nodes. BMC Med. Imaging **14** (2014). Article number: 22. https://doi.org/10.1186/1471-2342-14-22
7. Torres, S.R., Chen, C.S., Leroux, B.G., Lee, P.P., Hollender, L.G., Schubert, M.M.: Fractal dimension evaluation of cone beam computed tomography in patients with bisphosphonate-associated osteonecrosis. Dentomaxillofacial Radiol. **40**, 501–505 (2011)
8. Uemura, K., Toyama, H., Baba, S., Kimura, Y., Senda, M., Uchiyama, A.: Generation of fractal dimension images and its application to automatic edge detection in brain MRI. Comput. Med. Imaging Graph. **24**(2), 73–85 (2000)
9. Plesea, R.M., et al.: The study of tumor architecture components in prostate adenocarcinoma using fractal dimension analysis. Rom. J. Morphol. Embryol. **60**(2), 501–519 (2019)
10. Frydkjaer-Olsen, U., Soegaard Hansen, R., Pedersen, K., Peto, T., Grauslund, J.: Retinal vascular fractals correlate with early neurodegeneration in patients with type 2 diabetes mellitus. Invest. Ophthalmol. Vis. Sci. **56**, 7438–7443 (2015)
11. Crow, F.: Summed-area tables for texture mapping. In: Proceedings of the 11th Annual Conference on Computer Graphics and Interactive Techniques, DBLP Computer Science Bibliography, Minneapolis, pp. 207–212 (1984)
12. Lewis, J.P.: Fast template matching. In: Vision Interface 1995, 15–19 May 1995, pp. 120–123 (1995)
13. Viola, P., Jones, M.: Robust real-time object detection. Int. J. Comput. Vis. **57**(2), 1–25 (2001)
14. Şerbănescu, M.S., et al.: Automated collagen segmentation from masson's trichrome stained images—preliminary results. In: Vlad, S., Roman, N.M. (eds.) 6th International Conference on Advancements of Medicine and Health Care through Technology, 17–20 October 2018, Cluj-Napoca, Romania. IP, vol. 71, pp. 163–167. Springer, Singapore (2019). https://doi.org/10.1007/978-981-13-6207-1_26
15. Brodatz Textures. http://www.ux.uis.no/~tranden/brodatz.html. Accessed 01 Sept 2020
16. Li, J., Du, Q., Sun, C.: An improved box-counting method for image fractal dimension estimation. Pattern Recogn. **42**(11), 2460–2469 (2009)
17. Serbanescu, M.S., Plesea, I.E.: R-VA a new fractal parameter for grayscale image characterization. Ann. Comput. Sci. Ser. **13**(1), 9–14 (2015)
18. Schweitzer, H., Bell, J.W., Wu, F.: Very fast template matching. In: Heyden, A., Sparr, G., Nielsen, M., Johansen, P. (eds.) ECCV 2002. LNCS, vol. 2353, pp. 358–372. Springer, Heidelberg (2002). https://doi.org/10.1007/3-540-47979-1_24
19. Leyva, R., Sanchez, S.V., Li, C.T.: Detecting small objects in high resolution images with integral fisher score. In: 25th IEEE International Conference on Image Processing (ICIP), pp. 316–320. University of Warwick Publications Service & WRAP, Athens (2018)
20. Shafait, F., Keysers, D., Breuel, T.M.: Efficient implementation of local adaptive thresholding techniques using integral images. In: Yanikoglu, B.A., Berkner, K. (eds.) Document Recognition and Retrieval XV. SPIE, vol. 6815, pp. 681510-1–681510-6. SPIE Digital Library (2008)
21. Frintrop, S., Klodt, M., Rome, E.: A real-time visual attention system using integral images. In: 5th International Conference on Computer Vision Systems (ICVS), pp. 1–10. Applied Computer Science Group, Bielefeld (2017)

Exploring Two Deep Learning Based Solutions for Improving Endoscopy Artifact Detection

Radu Razvan Slavescu[1(✉)], Ioan Catalin Sporis[1], Kriszta Gombos[1], and Kinga Cristina Slavescu[2]

[1] Technical University of Cluj-Napoca, Cluj-Napoca, Romania
Radu.Razvan.Slavescu@cs.utcluj.ro, {Ioan.Sporis, Kriszta.Gombos}@student.utcluj.ro

[2] Iuliu Hatieganu University of Medicine and Pharmacy, Cluj-Napoca, Romania
Cristina.Slavescu@umfcluj.ro

Abstract. This paper presents two approaches for the task of multi-class object detection at the 2019 Endoscopy Artifact Detection competition. In the first approach, we successively combined a Faster R-CNN neural network (NN) with two different object detectors (VGG16 and ResNet50 respectively). An ensemble of NNs of this type was also investigated. The second approach consists of a deep learning architecture based on the DarkNet framework. The Faster R-CNN based solution obtained a better Intersection over Union score for most of the artifact classes, while the DarkNet one proved better in terms of mean Average Precision. The performances are competitive with the other solutions submitted to the competition.

Keywords: Faster R-CNN · VGG16 · ResNet50 · DarkNet · Endoscopy artifact detection · Ensemble learning

1 Introduction

Endoscopy is a medical procedure used both for disease investigation and treatment, in which an instrument (endoscope) introduced into the patient's body helps gain a view of the internal structures. Even if high quality frames are captured from the instrument, these might be corrupted with different artifacts, such as motion blur, debris or various reflections. For this reason, visual diagnosis becomes difficult, so frame restoration algorithms are needed.

To address this, the endoscopy artifacts detection (EAD) competition [1] has been set. It asks the participants to provide solutions for detection and classification of multiple objects in the same frame. Three tasks are proposed. First, the bounding box localization of multi-class artifacts requires that any artifact belonging to one of the seven classes specified to be precisely identified together with its spatial coordinates in a set of images. Then, semantic segmentation of artifacts, which means correctly identifying the class the object belongs to and highlighting the region of interest for that artifact. Last, the generalization task is concerned with applying Machine Learning (ML) techniques across different endoscopic datasets.

S. Vlad and N. M. Roman (Eds.): MEDITECH 2020, IFMBE Proceedings 88, pp. 102–112, 2022.
https://doi.org/10.1007/978-3-030-93564-1_12

In this paper, we present our approaches to the first of these tasks. The first approach relies on the object detection method from the family of R-CNN [2], Fast R-CNN [3] and Faster R-CNN [4], which gets combined with two different object detectors (VGG16 [5] and ResNet50 [6] respectively). Another solution consists of the NN architectures based on the Darknet framework (pjrreddie.com/darknet), which performs well in terms of speed. We also report on using an ensemble of NNs whose performances are competitive with the other solutions submitted to the competition.

The paper is organized as follows. Section 2 positions our work among similar approaches. Section 3 details our solutions. Section 4 and 5 describe the experiment setup and the results obtained. Section 6 concludes and sketches future work lines.

2 Related Work

EAD contest challenged competitors to come up with systems able to correctly identify the 7 classes and their bounding box coordinates. One such systems [7] is based on RetinaNet [4], which basically is a single NN, consisting of a backbone and two other task-specific sub-networks. For the backbone NN, the authors of paper [8] adapted RetinaNet to work with ResNet101 [6], obtaining a maximum final score equal to 49.26 when running the model on their test dataset.

Another system is described in [9]. Based on the unbalanced sample distribution of the classes in the dataset, a model centered on Cascade R-CNN architecture [10] is proposed. In this approach, the original Cascade R-CNN NN is modified by adding a Feature Pyramid Network (FPN) [11] module during feature extraction [9]. The same backbone NN, ResNet101, as in [7] is used. To avoid sharp Mean Average Precision (mAP) dropping when the Intersection over Union (IoU) score is high, three cascade stages of object detection NNs are applied. The best result mentioned is 36.10.

In [12] the authors propose an ensemble containing 7 RetinaNet models, for which they varied the hyperparameters, data augmentation techniques, backbone NNs and the subset of images they used. Finally, all 7 architectures were combined based on an effective voting scheme [12], obtaining a final score of 34.51.

3 Proposed Models

Our first approach uses a combination of two models, each relying on the Faster R-CNN architecture. In brief, Faster R-CNN is a two-stage object detector having three main components: the convolutional layers (backbone of the NN), the region proposal network (RPN) and the module for predicting classes and the bounding boxes' coordinates. Its architecture is presented in Fig. 1.

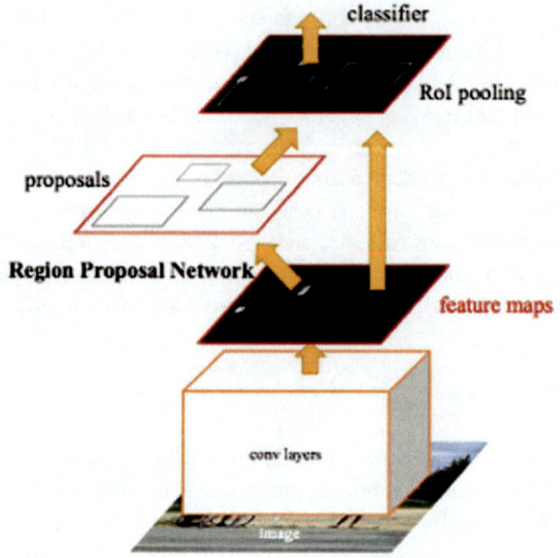

Fig. 1. Faster R-CNN architecture [4]

The backbone is basically another NN that handles the features extraction and feeds them to the other layers of the model. After analyzing the dataset, we noticed that the object instances are usually small, many of which could easily either be mistaken with the background (e.g., contrast) or be interpreted as belonging to a different class (e.g., bubbles, specularity). For the backbone, we chose VGG16 and ResNet50.

The RPN is a small NN whose purpose is to slide over the last feature map of the convolutional layers and predict whether that region may contain an object or not. It takes care of generating anchor boxes, classifying anchors into background or foreground components and learning the offsets of boxes to fit the objects.

The last component of Faster R-CNN is yet another NN which processes the outputs of RPN and predicts the label (classification) and bounding box's coordinates (regression) of an object from the image.

Figure 2 presents our architecture which relies on two similar models trained with different backbones, VGG16 and ResNet50. The first model was trained by fine-tuning the pre-trained Faster R-CNN model from Tensorflow Object Detection API. We used a batch size of 1 together with the Momentum optimizer, training the model for 70000 steps (the policy for Tensorflow models). We started with a learning rate of 0.02, and reduced it down to 0.0002 during training. The ResNet50 backbone was initialized with the weights trained on COCO dataset (cocodataset.org). The second model was trained using a Keras implementation of Faster R-CNN and a VGG16 backbone. This was done for 27 epochs, each of size 1000. We used Momentum optimizer for the first model and Adam Optimizer with a learning rate of 0.00001 for the second one. Due to the inconsistent size of the image, they were rescaled during the training pipeline so that the dimensions remain constant throughout the whole dataset.

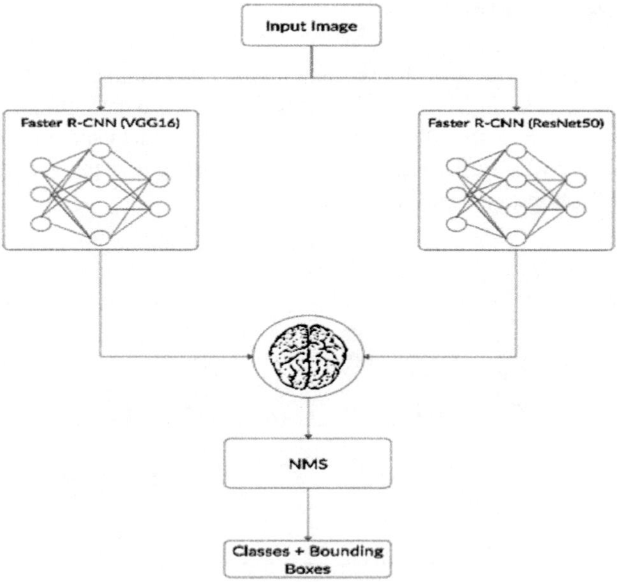

Fig. 2. Ensemble architecture

Once these NNs trained, they were combined in an ensemble where the bounding boxes were obtained with an affirmative strategy, meaning that all predictions coming from the models are kept. To keep the number of false positive results low, we set the threshold of these predictions to 0.85.

The last step of our design was to apply the Non-Maximum Suppression (NMS) algorithm. This way, we ensure that only the most relevant bounding boxes are kept and we do not have duplicates for the same object in the image. After testing several NMS thresholds, we found that the value 0.36 would return the best results.

Since the actual training dataset was small, we included an image augmentation phase into the training pipeline. We applied vertical and horizontal flips, small angle rotations and small random adjustments to the HSV values.

The second architecture we tried consists of 77 convolutional and 3 YOLO layers, implemented using Redmon's Darknet framework. The number of iterations in the training process (*max_batches*) was 14000 (in line with the usual practice to set it at 2000 * number of classes). The learning rate was initially set to 0.001 and diminished after 80% and 90% of *max_batches* iterations respectively. Each time, its value was divided by 10. Finally, the *batch* and *subdivisions* hyperparameters, which determine how many images are processed în parallel by the GPUs, were set to 64 and 8 respectively.

4 Experimental Setup

The dataset was the one provided for training by the EAD competition. Each image can contain more artifacts in the 7 specified classes. We split the dataset into two parts: one used for testing, containing 209 samples and another one used for training (1871 images). The latter was further split, so that 1684 images were used for the actual training and 188 samples for validating the model.

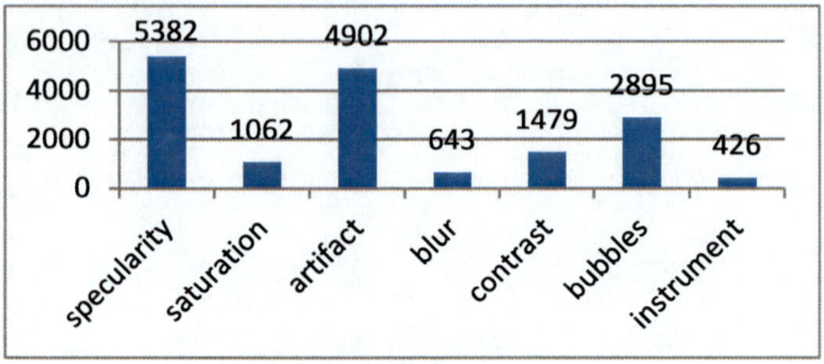

Fig. 3. Class instances in the entire dataset

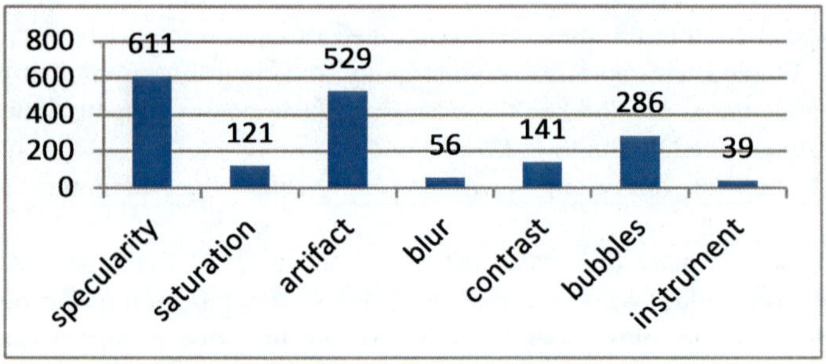

Fig. 4. Class instances in the test dataset

One can see in Fig. 3 that the available dataset was imbalanced, with 5382 specularity instances and only 426 instrument objects. An overview of the test data is presented in Fig. 4.

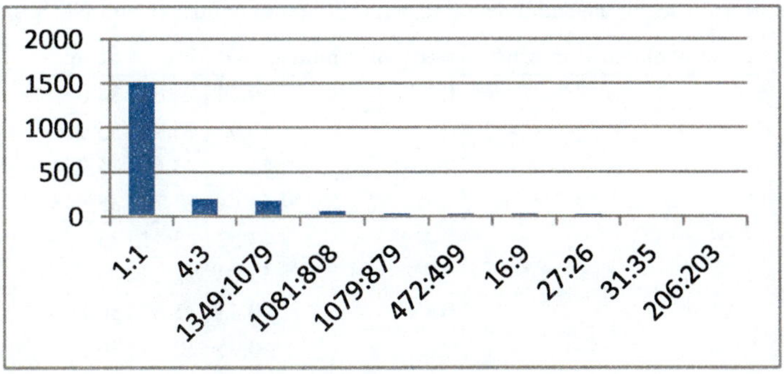

Fig. 5. Aspect ratio examples

During training each image was resized to a constant size. Even though most of the images have a ratio of 1:1, we can see in Fig. 5 that exceptions might occur.

For model evaluation, we used the metric in [1].

$$\text{Score} = 0.6 * \text{mAP} + 0.4 * \text{IoU} \quad (1)$$

The mAP is computed as a mean of each class's average precision, while IoU represents the ratio of overlap between the predicted and ground truth bounding boxes.

5 Results

Table 1 contains the results of running each model on the test dataset. The most significant score difference was obtained in the case of ResNet50 backbone, for which the score value jumped from 30.57 to 34.17 due to augmentation. For the VGG16 backbone, the difference was much lower (33.6 and 34.66). DarkNet performed better from this perspective (44.62).

Another interesting outcome obtained when experimenting with the two NN configurations was the high value of IoU in the case of Faster R-CNN combined with the ResNet50 backbone. This actually means a better match between the prediction and the ground truth.

Table 1. Independent models' results

	mAP	IoU	Score
Faster R-CNN + VGG16 (no augmentation)	27.26	42.27	33.26
Faster R-CNN + ResNet50 (no augmentation)	18.73	48.35	30.57
Faster R-CNN + VGG16 (with augmentation)	29.43	42.52	34.66
Faster R-CNN + ResNet50 (with augmentation)	23.45	50.25	34.17
DarkNet (with augmentation)	44.26	45.16	44.62

Table 2. Results using ensemble learning with different confidence and NMS thresholds

Min confidence threshold	NMS threshold	mAP	IoU	Score
0.6	0.17	39.64	25.67	34.052
	0.3	43.96	23.95	35.956
	0.33	44.11	23.67	35.934
	0.36	43.6	23.72	35.648
	0.4	43.52	23.41	35.476
	0.5	43.03	22.48	34.81
0.7	0.17	40.43	31.52	36.866
	0.3	40.95	30.07	36.598
	0.33	41.13	29.68	36.55
	0.36	41.25	29.71	36.634
	0.4	40.61	29.41	36.13
	0.5	40.53	28.09	35.554
0.8	0.17	36.71	38.91	37.59
	0.3	37.23	38.14	37.594
	0.33	37.41	37.48	37.438
	0.36	37.37	37.49	37.418
	0.4	37.11	37.01	37.07
	0.5	36.76	36.56	36.68
0.85	0.17	33.66	43.23	37.488
	0.3	33.89	41.92	37.102
	0.33	34.62	42.76	37.876
	0.36	34.8	42.77	**37.988**
	0.37	34.61	42.57	37.794
	0.4	34.34	42.22	37.492
	0.5	33.76	41.45	36.836
0.9	0.36	29.15	46.1	35.93

We tried different parameters combinations for our ensemble system. We focused on varying two such parameters. The first one is the minimum confidence value for which a bounding box would be taken into consideration. To eliminate similar detections and keep only the best match for each object in the image, we altered the value of the NMS threshold, which is in fact our second parameter. Table 2 and Fig. 6 presents a comparison of the results we obtained by using different combinations. We started with a confidence threshold of only 0.6 and obtained just a small improvement in the final score, besides our best single models' results. After several experiments, we concluded that for this problem our system would reach the highest score value of almost 38 (37.988) when the minimum confidence level is 0.85 and the NMS threshold 0.36.

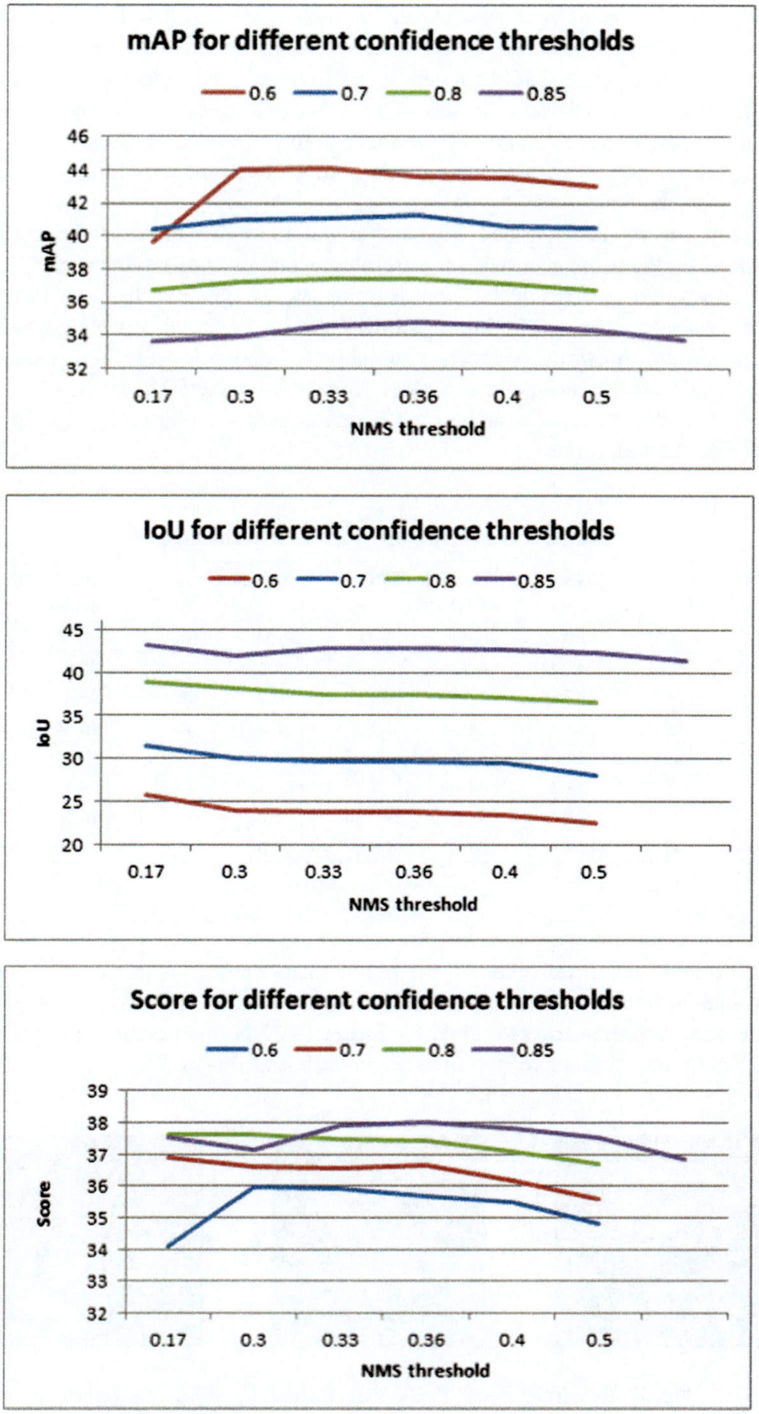

Fig. 6. Charts of the experimental results from Table 2

In [12], the best result of a system with 7 models run on the test dataset provided by EAD was 34.51. Due to the lack of the same test data, we only ran the models on our 209 selected images. From Table 1 it can be inferred that our model using VGG16 with augmented images has already surpassed the other model, being closely followed by the one using ResNet50. Not only did we manage to outperform the best results in [12] by testing on our dataset with a single model, but we even increased the final score up to 37.98 using the ensemble.

One can see in Table 2 and Fig. 6 that the more we increase the confidence threshold value, the more accurate our ensemble becomes, thus having the highest IoU when the threshold is 0.9. On the contrary, the system obtained the best mAP value when the threshold is set to 0.6. Similar values for the two parameters were obtained by using a minimum threshold of 0.8, for which mAP and IoU were both close to 37. Considering all three models, one can observe that the ResNet50 based model obtained the best IoU, with the cost of a smaller mAP value, having a total score slightly smaller than the VGG16 based one.

Table 3. mAP and IoU values for each class.

	Faster R-CNN + VGG 16		Faster R-CNN + ResNet50		Ensemble		DarkNet based model	
Class	mAP	IoU	mAP	IoU	mAP	IoU	mAP	IoU
Specularity	5.14	41.41	6.52	40.58	8.35	42.98	35.5	37.18
Saturation	35	45.59	5.94	39.99	39.24	46.23	40.59	43.08
Artifact	23.71	41.43	13.86	53.05	28.93	44	40.54	49.01
Blur	4.46	20.26	5.35	25.95	8.37	28.9	59.6	60.34
Contrast	50.56	58.2	45	58.22	61.58	49.32	25.32	27
Bubbles	5.69	25.71	0.58	40.54	6.44	26.65	18.04	32.01
Instrument	81.42	64.97	53.84	80.09	90.63	61.26	90.19	67.48

Table 3 presents the overall results for each of the seven classes. As one can see, the ResNet50 based model has obtained the highest IoU values for instruments, bubbles, contrast and artifacts, among all models. Also, considering the mAP values, our ensemble has overcome the two singular Faster R-CNN architectures for each class, however, obtaining weaker results then DarkNet based model (Fig. 7).

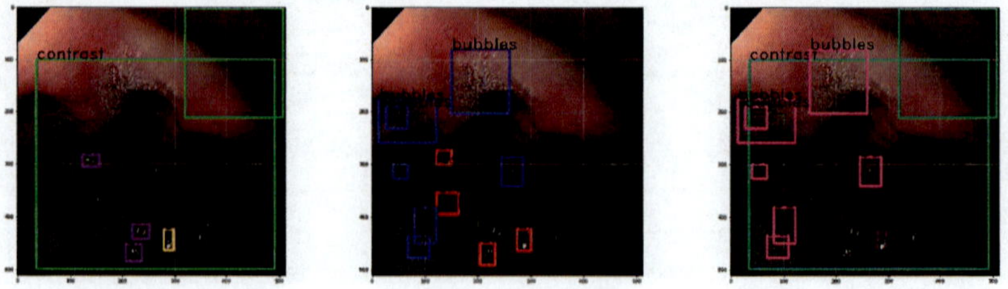

Fig. 7. ResNet50 model (left), VGG16 (center), ensemble (right)

Judging by the images in Fig. 6, we observed that for this problem, Faster R-CNN with VGG16 has been capable of better identifying the small objects then the second model we used. The ensemble proved useful for bigger objects detection, like contrasts or instruments.

From the classification speed perspective, DarkNet performed better. For a Google Colab implementation (12 GB NVIDIA Tesla K80, Intel® Xeon® CPU @ 2.2 GHz, 56 MB cache), classification was done in 20 to 25 ms.

6 Conclusions and Future Work

We proposed two approaches for the task of multi-class object detection at the 2019 Endoscopy Artifact Detection competition. The first one was based on combining a Faster R-CNN neural network with two different object detectors (VGG16 and ResNet50 respectively). An ensemble of NNs of this type was also investigated. The second approach consists of a deep learning architecture based on the DarkNet framework. Both our models obtained promising scores, comparable with the other architectures proposed at the competition. Our ResNet50 based model performed better in terms of accuracy, always having high IoU values. More, we improved further the performance of the system by introducing an ensemble followed by a NMS layer. Experiments with different thresholds proved that the final score could reach the value of 38.

DarkNet based architecture was the best in terms of mAP and overall performance according to the metric used in competition. Finally, integrating the models into a single system proved that they could compensate each other in terms of the objects they detect, resulting in a robust and reliable endoscopic system.

In the future, experimentation with more advanced backbone NNs could be investigated. Another challenge would be to propose a real time detector for which DarkNet based architectures could be considered.

Acknowledgement. This work was supported by the Computer Science Department of the Technical University of Cluj-Napoca, Romania.

Conflict of Interest. The authors declare they have no conflict of interest.

References

1. Ali, S., et al.: endoscopy artifact detection (EAD 2019) challenge dataset (2019)
2. Girshick, R., Donahue, J., Darrell, T., Malik, J.: Rich feature hierarchies for accurate object detection and semantic segmentation. In: IEEE Conference on Computer Vision and Pattern Recognition 2014, ICCV, pp. 580–587. Columbus, OH (2014)
3. Girshick, R.: Fast R-CNN. In: IEEE International Conference on Computer Vision 2015, pp.1440–1448. ICCV, Santiago (2015)

4. Ren, S., He, K., Girshick, R., Sun, J.: Faster R-CNN: towards real-time object detection with region proposal networks. In: IEEE Transactions on Pattern Analysis and Machine Intelligence 2017, vol. 39, pp. 1137–1149 (2017)
5. Simonyan, K., Zisserman, A.: Very Deep Convolutional Networks for Large-Scale Image Recognition (2014)
6. He, K., Zhang, X., Ren, S., Sun, J.: Deep Residual learning for image recognition. In: IEEE Conference on Computer Vision and Pattern Recognition 2016, pp. 770–778, CVPR, Las Vegas (2016)
7. Khan, M., Choo, J.: Multi-class artefact detection in video endoscopy via convolutional neural networks. In: Challenge on Endoscopy Artefacts Detection 2019, EAD, Venice (2019)
8. Lin, T., et al.: Focal loss for dense object detection. In: IEEE International Conference on Computer Vision 2017, pp. 2999–3007, ICCV, Venice (2017)
9. Yang, S., Cheng, G.: Endoscopic artefact detection and segmentation with deepconvolutional neural network. In: Challenge on Endoscopy Artefacts Detection 2019, EAD, Venice (2019)
10. Cai, Z., Vasconcelos, N.: Cascade R-CNN: delving into high quality object detection. In: IEEE/CVF Conference on Computer Vision and Pattern Recognition 2018, pp. 6154–6162, Salt Lake City (2018)
11. Lin, T., Dollár, P., Girshick, R., He, K., Hariharan, B., Belongie, S.: Feature pyramid networks for object detection. In: IEEE Conference on Computer Vision and Pattern Recognition 2017, pp. 936–944, CVPR, Honolulu (2017)
12. Kayser, M., Soberanis-Mukul, R., Albarqouni, S., Navab, N.: Focal loss for artefact detection in medical endoscopy. In: Challenge on Endoscopy Artefacts Detection 2019, EAD, Venice (2019)

HCC Recognition Within B-Mode and CEUS Images Using Traditional and Deep Learning Techniques

Delia Mitrea[1(✉)], Cosmina Mendoiu[1], Paulina Mitrea[1], Sergiu Nedevschi[1], Monica Lupşor-Platon[2,3], Magda Rotaru[3], and Radu Badea[2]

[1] Technical University of Cluj-Napoca, 26-28 Baritiu Street, 400027 Cluj-Napoca, Romania
delia.mitrea@cs.utcluj.ro
[2] Regional Institute of Gastroenterology and Hepatology, Iuliu Hatieganu University of Medicine and Pharmacy, Cluj-Napoca, 19-21 Croitorilor Street, 400162 Cluj-Napoca, Romania
[3] I. Haţieganu University of Medicine and Pharmacy of Cluj-Napoca, 8 V. Babeş Street, 400012 Cluj-Napoca, Romania

Abstract. Hepatocellular Carcinoma (HCC) is the most often met malignant liver tumor and one of the most frequent causes of death worldwide. The most reliable method for HCC diagnosis is the needle biopsy, but this is invasive, dangerous, leading to the spread of the malignity through the body, as well as to infections. We develop noninvasive, computerized techniques for performing HCC diagnosis based on ultrasound images. In this research work, we combined the information provided by both B-mode ultrasound images and Contrast Enhanced Ultrasound (CEUS) images, in order to improve the HCC recognition performance. We computed classical and advanced textural features from both types of images, we applied dimensionality reduction methods, and then we provided the resulted features at the entrances of both traditional and deep learning classifiers. Similar experiments were conducted using the pixel values of the regions of interest. We obtained a recognition accuracy of 90%.

Keywords: Hepatocellular Carcinoma (HCC) · Contrast Enhanced Ultrasound (CEUS) images · Dimensionality reduction · Stacked Denoising Autoencoders (SAE)

1 Introduction

HCC is one of the most often met malignant liver tumors, appearing in 70% of the liver cancer cases. It usually evolves on the top of the cirrhotic parenchyma. The golden standard for HCC diagnosis is the needle biopsy, but this is invasive and dangerous [1]. Ultrasonography is an investigation technique which is non-invasive, cheap, repeatable. The CEUS imaging assumes the injection into the blood of a specific contrast agent, consisting of gas filled microbubbles, which spreads through the human body and highlights the vessel structure. Within B-mode images, focal HCC appears as a

S. Vlad and N. M. Roman (Eds.): MEDITECH 2020, IFMBE Proceedings 88, pp. 113–120, 2022.
https://doi.org/10.1007/978-3-030-93564-1_13

well-defined region, of 3–5 cm, being hyperechogenic and heterogeneous [1]. In CEUS images, HCC appears more highlighted, due to the specific dense and complex vessel structure. However, in many cases, within both B-mode and CEUS images, HCC is hardly distinguishable from the cirrhotic parenchyma [2]. In our research, we develop non-invasive, computerized methods, for the automatic diagnosis of HCC based on ultrasound images. Regarding *the state of the art*, many approaches perform tumor recognition within medical images [3, 4]. Also, multiple approaches perform tumor automatic diagnosis based on contrast enhanced medical images or on the combination between multiple image types [5–7]. In [5] the authors aimed to recognize the liver tumors using the textural parameters resulted from classical, as well as from contrast enhanced CT images. The textural features computed from the classical CT image were concatenated with those derived from the contrast enhanced image. After feature selection, a C4.5 classifier was applied, leading to an accuracy over 90%. Regarding the application of the deep learning techniques for the considered objective, in [7] the segmentation of the sarcoma tumors was performed, by fusing Magnetic Resonance Images (MRI), CT and Positron Emission Tomography (PET) images. Three approaches were taken into account: (a) the fusion of the features individually determined on each image type was achieved, then the result of this fusion was provided at the input of a CNN (feature-level fusion); (b) only the parts performing convolution were separated within the CNN, so that the images of each category were provided in parallel, to identical convolution units, the results being processed by a single fully connected network, which determined the class (classifier-level fusion); (c) the images of various types were provided as inputs to different CNNs, the class resulting through majority voting (decision-level fusion). The feature level fusion yielded the best results, the accuracy being around 95%. However, no relevant approach exists, which performs HCC recognition within both B-mode and CEUS images. In the current work, we exploit the information provided by both B-mode and CEUS images, in order to get an increased recognition accuracy. The textural features derived from the B-mode and CEUS images, as well as the fused features, are provided at the entrances of conventional classifiers. These features are also provided at the input of the SAE deep learning classifier [9]. Also, the raw image data was provided at the entrances of these classifiers.

2 Proposed Methods

2.1 Texture Analysis Methods

Classical, as well as advanced texture analysis methods were employed in our work. As classical techniques, second order gray level statistics, such as the Haralick features derived from the Gray Level Cooccurrence Matrix (GLCM), as well as the autocorrelation index were computed. The Hurst fractal index was also calculated, besides the edge based statistics, respectively the frequency and density of the textural microstructures obtained after applying the Laws' filters [9]. Multiresolution features such as the Shannon entropy, computed after applying the Wavelet transform, recursively, twice [9], were employed as well. Concerning the advanced textural features, the Haralick parameters derived from the third order GLCM [3] were added to the classical feature set. The second and third order GLCM matrices and the Haralick features were computed as described in [3]. All the textural features were determined on rectangular regions of interest, after the median filter application, independently on orientation, illumination and region of interest size.

2.2 Dimensionality Reduction Techniques

As *feature selection techniques*, we considered the Correlation based Feature Selection (CFS) and the Information Gain Attribute Evaluation (IGA) methods, which provided the best results in our experiments [10]. *CFS* is a feature selection method that computes, for each attribute subset, a merit, in order to retain those features correlated with the class parameter and uncorrelated between themselves [10]. The *IGA technique* calculates the entropy of the class ***C***, before and after observing the attribute ***A*** [5]. The gain due to the attribute ***A*** is provided by the measure in which the attribute ***A*** leads to the decrease of the entropy of the class ***C***. As for CFS, the subset of features corresponding to the highest merit was taken into account, while for IGA, the first ranked features, having a relevance score above 0.1, were considered. Finally, the union of these subsets provided by each individual method was assessed. *Principal Component Analysis (PCA)* projects the initial data on a lower dimensional space where the main variation modes are highlighted [11]. *Kernel Principal Component Analysis (KPCA)*, the generalized version of PCA, represents the transposition of PCA in a superior space, built using a kernel function having different forms, such as Gaussian, polynomial and linear, in the last case being equivalent with the classical PCA [11]. In our work, all the three versions of KPCA were experimented. The dimensionality reduction methods were applied as follows: (a) on the features resulted from the B-mode images; (b) on the features resulted from the CEUS images; (c) on the concatenated features, resulted from the B-mode and CEUS images.

2.3 Classification Methods

For assessing the classification performance due to each image type, respectively to the combination between the B-mode and CEUS images, the textural features, as well as the pixel values, were provided at the entrances of the following *traditional classifiers*: Support Vector Machines (SVM), Multilayer Perceptron (MLP) and AdaBoost combined with the C4.5 technique [8]. These classification methods provided the best results in our experiments [8]. Regarding *the deep learning techniques*, Stacked Denoising Autoencoder (SAE) presents an architecture based on stacking multiple Denoising Autoencoder units. A Denoising Autoencoder consists of a neural network that receives $x \in [0; 1]^d$ as input and maps it into a hidden representation $y \in [0; 1]^d$, through the mapping provided by Eq. (1):

$$y = s(Wx + b) \tag{1}$$

In Eq. (1), s is a non-linear transfer function, as the sigmoid, W is the weight matrix, b being a constant, assigned by designer. The auto-encoder performs the encoding of the input, preserving also the information and removing the noise. The code y contains a distributed representation, capturing the main data variation modes, which makes this method almost similar to PCA. However, due to the hidden layers, the autoencoder is able to capture more complex information than classical PCA. Multiple denoising autoencoders can be stacked, yielding a deep autoencoder, each autoencoder on the layer above receiving the output of the unit one layer below. Firstly, an unsupervised, pre-training process takes place, by minimizing the reconstruction error. A second, supervised training stage can be performed, by adding a *logistic regression* or *softmax* layer on the network top [8]. In our work, both the relevant textural features, as well as the pixel values were provided as inputs to SAE, for the B-mode and CEUS images separately, as well as for the combined information, resulted for the two types of images. The fusion of the information from the two image types was performed as follows: (a) feature level fusion, by concatenating the feature vectors resulted from the two image types; (b) classifier level fusion, by providing the data corresponding to each image type to an individual SAE architecture, the resulting features being provided to a unified *softmax* layer; (c) decision level fusion, by employing separated SAE classifiers for each image type, performing the arithmetic mean of the class probabilities.

3 Experiments and Results

3.1 The Dataset and the Experimental Settings

During the experiments, the B-mode and CEUS images corresponding to 35 patients affected by HCC were considered. All these patients underwent biopsy or CT for diagnostic confirmation. The images were acquired using a GE Logiq E9 XDclear 2.0 (General Electric, USA) ultrasound machine, under the same settings: Frequency of 6 MHz, Gain of 58, Depth of 16 cm, and Dynamic Range (DR) of 111. For the CEUS images, the Sonovue (Bracco®) agent was employed, the images corresponding to the arterial phase being considered. Pairs of B-mode and CEUS images simultaneously acquired for each patient were taken into account. From these images, regions of interest of 51×51 pixels, for HCC and cirrhotic parenchyma where HCC evolved, were automatically extracted, using a sliding window algorithm, being augmented by rotations with 90° and 180°. These two classes were compared, as the tissue aspect within ultrasound images is almost similar in many situations. The data was uniformly split, 1000 patches/class being considered, 70% of the data representing the training set, while 30% of the data constituting the test set. Both the pixels values and the textural features (41 attributes) computed by our Visual C+ + software modules were involved in our experiments. The feature selection techniques and the traditional classifiers were implemented using the Weka 3.8 library [12]. For feature selection, the CfsSubsetEval (CFS) method was implemented with BestFirst search, and also InfoGainAttributeEval (IGA) was employed with the Ranker search method. Regarding the conventional classifiers, the John Platt's Sequential Minimal Optimization (SMO) algorithm [12], standing for SVM, was experimented, the best results being obtained for the polynomial kernel, of third or fifth degree. Also, the MultilayerPerceptron (MLP) technique was employed, with a learning rate of 0.2, a momentum of

0.8 and a training time of 500 epochs. Correspondingly, various architectures, consisting of one, two or three hidden layers, each of them having the number of nodes equal with the arithmetic mean between the number of features and the number of classes, were assessed. The AdaBoost metaclassifier with 100 iterations, with the J48 technique, the Weka equivalent of C4.5, was also implemented. The KPCA and the SAE deep learning techniques were employed in Matlab 2020. KPCA was implemented using the Matlab-Kernel-PCA toolbox [12]. The Gaussian, the third degree polynomial and the linear versions were considered. The SAE technique was employed using the Matlab Deep Learning toolbox [14]. The number of hidden layers, respectively the number of nodes within each hidden layer were tuned in order to achieve the best performance, a *softmax* layer being added on the top. The input data was normalized. The other parameters were also tuned, the values leading to the best performance being: 800 for the number of epochs, 0.0004 for weight regularization, 4 for sparsity regularization, 0.14 for sparsity proportion, the transfer function being the logistic sigmoid.

3.2 Results and Discussions

Classification Performance Assessment Using Textural Features

Concerning the results obtained by employing traditional classifiers, after performing feature selection on B-mode images, the best performance resulted for AdaBoost: accuracy (recognition rate) of 72.33%, true positive rate (sensitivity) of 86.1%, true negative rate (specificity) of 58.6%, area under ROC (AUC) of 75.9%. After applying traditional classifiers on the relevant features obtained from CEUS images, the best accuracy, of 80.94%, the highest sensitivity, of 90.6%, the highest specificity, of 71.3% and the best AUC, of 86%, resulted for AdaBoost, as well. The classification performance obtained after applying traditional classifiers on the concatenated textural features, resulted from both B-mode and CEUS images, after feature selection, is illustrated in Table 1. As we can notice, the results overpassed those obtained when employing only CEUS images, as well as those obtained when employing only B-mode images. The best accuracy (86.01%), the best sensitivity (97.5%) and the best AUC (92.4%), resulted for AdaBoost, while the best specificity (75%) was achieved for MLP. We observe that the classification performance is higher than that obtained previously. Concerning the relevant textural features, we observed in this case an increased density of the attributes resulted from the CEUS images, such as those derived from the GLCM matrices, the autocorrelation index, the Hurst index and the entropy computed after applying the Wavelet transform, indicating the complex, chaotic structure of HCC and granularity differences between the two considered tissue classes.

Table 1. The performance of the traditional classifiers on CEUS + B mode textural features

Classifier	Rec. rate (%)	TP rate (%)	TN rate (%)	AUC
SMO	82.123	93.0	71.3	82.1
MLP	80.48	86	**75**	89.1
AdaBoost + J48	**86.01**	**97.5**	73.5	**92.4**

The results obtained by employing SAE after feature selection are provided in Table 2. The best accuracy (90.1%) the best specificity (94.3%) and the best AUC (90.17%), resulted for feature level fusion, while the best sensitivity (92%) was achieved for both classifier and decision level fusion.

Table 2. Results obtained with the SAE technique on the relevant textural features

Fusion type	Rec. rate (%)	TP rate (%)	TN rate (%)	AUC
B-mode	76.6	86.7	66.6	77.7
CEUS	79.9	90.3	69.4	81.21
CEUS + B-mode (feature level)	**90.1**	85.4	**94.3**	**90.17**
CEUS + B-mode (classifier level)	83	**92**	74	84.11
CEUS + B-mode (decision level)	84	**92**	76	84.89

KPCA was also applied on the original textural feature sets, the best performance resulting for the Gaussian kernel. For the B-mode and CEUS images, the first 20 components were considered. For the B-mode images, the highest accuracy, of 87.29%, the highest sensitivity, of 75.4% and the highest AuC, of 89.4%, resulted for AdaBoost, while the highest specificity, of 99.4%, was obtained for SMO. For the CEUS images, the best accuracy, of 87.32%, the best sensitivity, of 75.4%, the best specificity, of 99.2% and the best AUC, of 89.9%, were achieved for AdaBoost. In the case of the concatenated features, the first 40 components were considered. The best accuracy, of 86.68%, the best sensitivity, of 74.4% and the best AUC, of 99.2% were obtained for AdaBoost, the best specificity, of 89.2%, resulting for AdaBoost and SMO.

Classification Performance Assessment When Employing Directly the Pixel Values
The classification performance obtained when providing the pixel values to SAE are depicted in Table 3. The best accuracy (68.4%), the best specificity (59.9%) and the best AUC (74.63%), resulted for the combination between CEUS and B-mode images, for classifier level fusion, the best sensitivity (89.1%), corresponded to feature level fusion.

Table 3. Results obtained with the SAE technique on the grayscale and CEUS images

Fusion type	Rec. rate (%)	TP rate (%)	TN rate (%)	AUC
B-mode	64.9	83.8	55.9	71.53
CEUS	65.3	83.8	56.8	71.90
CEUS + B-mode (feature level)	65.1	**89.1**	50.1	73.12
CEUS + B-mode (classifier level)	**68.4**	86	**59.9**	**74.63**
CEUS + B-mode (decision level)	65.6	82.9	57.2	71.47

The KPCA method was also applied on the pixel values, the Gaussian kernel leading to the best results. For the B-mode, as well as for the CEUS images, the first 250 components were considered. For the B-mode images, the best accuracy (83.55%)

the best sensitivity (75.6%), the best specificity (99.6%) and the best AUC (93.1%) resulted for AdaBoost. For CEUS images, the highest values for the accuracy (79.77%), sensitivity (61.6%), specificity (97.8%) and AuC (93%) were obtained for MLP. In the case of the concatenated images, the first 400 components were considered and the best accuracy (85.4%) the best sensitivity (70.2%) the best specificity (99.6%), respectively the best AuC (93.1%) were achieved for AdaBoost.

Discussions

A comparison of the accuracy values obtained by applying the corresponding methodologies, upon B-mode images, upon CEUS images and upon the combined information is illustrated within Fig. 1. We notice that the information provided by the CEUS images and especially that provided by the combination between B-mode and CEUS images always led to an accuracy improvement. The classification performance was generally better when employing the textural features, compared with the case when using the pixel values. The time was also considerably improved when employing the textural features (under 5 min), compared with the other case (30–60 min), on a computer with an i5 processor, 2.70 GHz frequency and 4 GB of RAM. The best overall accuracy was achieved for SAE, when receiving as input the concatenated relevant textural features resulted from CEUS and B-mode images. The SAE classifier led to classification accuracy improvement compared with the approach based on feature selection followed by traditional classifiers, also compared with the linear PCA.

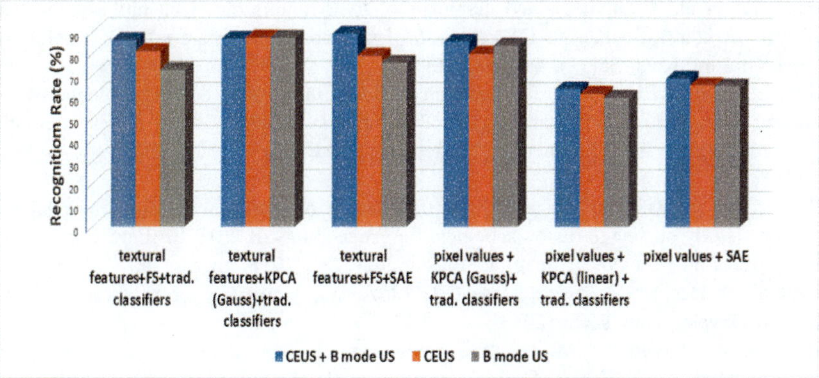

Fig. 1. The classification accuracy improvement due to the CEUS images

4 Conclusions

As our experiments illustrate, the information obtained from the CEUS images, as well as from the combination between B-mode and CEUS images, led to an improvement of the classification performance, in comparison with the case when only the B-mode images were employed, the best obtained classification accuracy being above 90%. The results obtained when providing the textural features at the classifier inputs overpassed those achieved when the pixel values where directly used. The currently achieved results are also comparable with the state of the art results [5–7]. In our future research, we intend to improve the experimental dataset, by acquiring images corresponding to more HCC cases, as well as to benign liver tumors. We also aim to employ other deep learning methods, such as Deep Belief Networks (DBN), respectively CNN.

Acknowledgement. This work was founded by the Romanian National Authority for Scientific Research and Innovation under Grant number PN-III-P1-1.2-PCCDI2017-0221 Nr.59/1st March 2018.

Conflicts of Interest. The authors declare they have no conflict of interest.

References

1. American liver foundation, Online: https://liverfoundation.org/for-patients/about-the-liver/diseases-ofthe-liver/liver-cancer/. Accessed 10 Sep 2020
2. Sherman, M.: Approaches to the diagnosis of hepatocellular carcinoma. Curr. Gastroent. Reports **7**(1), 11–18 (2005)
3. Brehar, R., Mitrea, D., et al.: Comparison of deep-learning and conventional machine-learning methods for the automatic recognition of the hepatocellular carcinoma areas from ultrasound images. Sensors **20**(11), 22 (2020)
4. Ker, J., Wang, L., et al.: Deep learning applications in medical image analysis. IEEE Access **6**, 9375–9389 (2017)
5. Duda, D., et al.: Computer aided diagnosis of liver tumors based on multi-image texture analysis of contrast-enhanced CT selection of the most appropriate texture features. Stud. Logic Gramm. Rhet. **35**(12), 49–70 (2013)
6. Guo, L., et al.: CEUS-based classification of liver tumors with deep canonical correlation analysis and multi-kernel learning. In: Proceedings of 2017 39th Annual International Conference of the IEEE Engineering in Medicine and Biology Society, 2017, pp. 1748–1751 (2017)
7. Guo, Z., et al.: Medical image segmentation based on multi-modal convolutional neural network: study on image fusion schemes. In: Proceedings of 2018 IEEE 15th International Symposium on Biomedical Imaging, Washington, DC, 2018, pp. 903–907 (2018)
8. Tutorial of Deep Learning, Release 0.1, LISA lab, University of Montreal, Copyright Theano Development Team (2015)
9. Materka, A., Strzelezki, M.: Texture analysis methods – a review, Technical Report, University of Lodz (1998) Online: http://www.eletel.p.lodz.pl/programy/cost/pdf_1.pdf. Accessed 30 Aug 2020
10. Hall, M.: Benchmarking attribute selection techniques for discrete class data mining, IEEE Trans. Knowl. Data Eng. **15**, 1–16 (2003)
11. Van der Maaten, L.J.P., et al: Dimensionality reduction: a comparative review (2009). https://lvdmaaten.github.io/publications/papers/TR_Dimensionality_Reduction_Review_2009.pdf. Accessed 30 Aug 2020
12. Weka 3. (2020). http://www.cs.waikato.ac.nz/ml/weka/. Accessed 10 Aug 2020
13. Kitayama, M.: Matlab-Kernel-PCA toolbox (2017). https://it.mathworks.com/matlabcentral/fileexchange/71647-matlab-kernel-pca. Accessed 15 Sep 2017
14. Deep Learning Toolbox for Matlab (2020). https://it.mathworks.com/help/deeplearning/index.html?s_tid=CRUX_lftnav. Accessed 17 Sep 2017

Modified Adam to Update Control Parameters of Crow Search Algorithm in Transformation Technique Based Dementia MRI Image Classification

N. Bharanidharan[1(✉)] and Harikumar Rajaguru[2]

[1] Dayananda Sagar University, Bangalore, India
[2] Bannari Amman Institute of Technology, Sathyamangalam, India

Abstract. This research paper focuses on classifying brain MRI images into two classes: Non-Dementia & Dementia. Crow Search Algorithm is used as transformation method for converting non-linearly separable data points into linearly separable data points. The accuracy enhancement in Support Vector Machine and K-Nearest Neighbor classifiers through the usage of this transformation technique is analyzed in this paper. To increase the efficiency of Crow Search Algorithm as transformation technique, modified Adam method is newly proposed. Support Vector Machine and K-Nearest Neighbor classifiers provide Balanced Classification Accuracy of 72% and 68% respectively while usage of Crow Search Algorithm with modified Adam as transformation technique in Support Vector Machine and K-Nearest Neighbor classifiers provide Balanced Classification Accuracy of 90% and 86%.

Keywords: MRI · Crow search algorithm · Modified Adam

1 Introduction

Dementia is a collective term used to represent the impairments originated in the human brain that is responsible for the gradual decline in memory, reasoning and other cognitive functions of human [1]. Currently around 50 million humans are affected with dementia worldwide and 10 million humans are getting affected by dementia every year [2]. Magnetic Resonance Imaging (MRI) is used extensively for diagnosing dementia. Computer aided automated classification will be very supportive to the clinician during diagnosis since manually inspecting large number of MRI images will be tedious task and sometimes it may leads to error in diagnosis [3].

Meta-heuristic algorithms are preferred for solving real-world problems due to its adaptability, derivation free structures and capability to avoid local optima [4]. Some of the popular meta-heuristic algorithms are Genetic Algorithm (GA), Particle Swarm Optimization (PSO), Crow Search Algorithm (CSA), Harmony Search Algorithm (HSA), Bat Algorithm (BA), etc. In general these meta-heuristic algorithms are used mainly for cracking numerical optimization problem, training neural networks and

S. Vlad and N. M. Roman (Eds.): MEDITECH 2020, IFMBE Proceedings 88, pp. 121–131, 2022.
https://doi.org/10.1007/978-3-030-93564-1_14

feature selection in classification problem. In this research work, CSA will be used as transformation technique that helps to classify the input MRI images into two groups: Dementia (DEM) & Non-Dementia (ND).

Experimentally it is found that control parameters present in the equations of meta-heuristic algorithms plays a significant role and it affects the accuracy to a greater extent when used in classification problem. Instead of fixing the control parameters through Trial and Error Method (TEM), the control parameters are updated using the modified Adam method. Adam is one of the popular techniques for updating the weights in neural networks. It will be novel to use this technique for updating the control-parameters of CSA. In addition to this, modified Adam is proposed to improve its efficiency in transforming non-linearly separable data points into linearly separable data points.

Many research articles can be found regarding the usage of Support Vector Machine (SVM) and K-Nearest Neighbor (KNN) classifiers for classification of diseases using medical images [5]. These two popular classifiers are used to prove the accuracy enhancement attained through the usage of CSA based transformation technique with modified Adam for control parameters updation.

The remaining paper is organized as follows: The second section deals with the overall methodology; Fundamentals of CSA is presented in third section whereas the procedure for implementing CSA using Trail and Error Method is detailed in fourth section; Fifth section deals with the implementation of CSA transform using Modified Adam; Cumulative results are discussed in the sixth section and concluded in the last section.

2 Overall Methodology

Brain MRI images related to dementia is openly available in Open Access Series of Imaging Studies. From this database, MRI images of 117 patients with different age groups are considered for this research work. Among these 117 patients, 65 patients are ND and 52 patients are DEM. MRI images of 40 ND patients and 40 DEM patients are considered for training while the MRI images of remaining 25 ND and 12 DEM patients are used for testing.

The overall methodology is depicted in Fig. 1. Each input MRI image has 65536 pixels (256 * 256) and these 65536 intensity values are directly given as input to the CSA. Modified Adam is used to update the control parameters of CSA in each iteration. The output of CSA will be given to the SVM or KNN classifier and then input image will be categorized as either ND or DEM. Finally the confusion matrix shown in Table 1 is constructed and widely used performance metrics given in Eqs. (1–6) are used to evaluate the performance of classification.

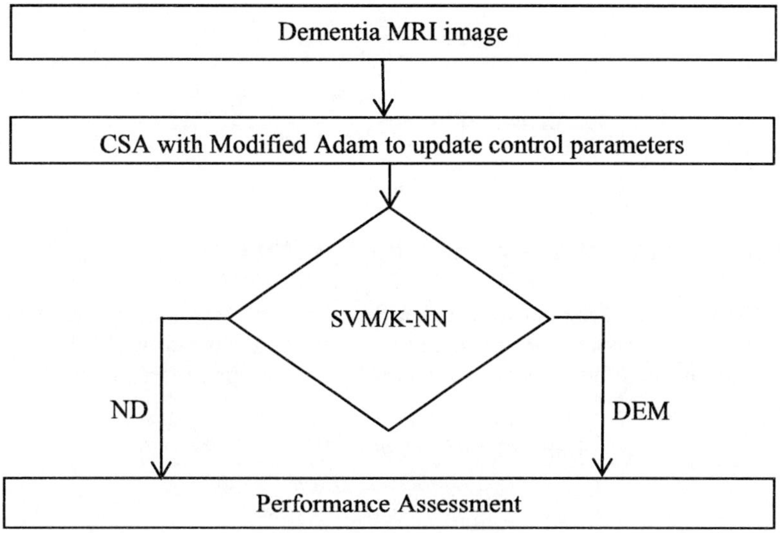

Fig. 1. Overall methodology

Table 1. Confusion matrix of dementia MRI image classification

Confusion matrix	Class	Predicted	
		ND	DEM
Actual	ND	TN	FP
	DEM	FN	TP

In Table 1, True Positive (TP) symbolizes the total number of DEM images properly classified as DEM while True Negative (TN) denotes the total number of ND images correctly classified as ND; False Positive (FP) denotes the sum of ND images which are wrongly classified as DEM and False Negative (FN) represents sum of DEM images which are wrongly classified as ND.

$$\text{Balanced Classification Accuracy (BCA)} = \frac{1}{2}\left(\frac{\text{TP}}{\text{TP}+\text{FN}} + \frac{\text{TN}}{\text{TN}+\text{FP}}\right) \times 100\% \quad (1)$$

$$\text{F1 score} = \frac{2\text{TP}}{2\text{TP}+\text{FP}+\text{FN}} \times 100\% \quad (2)$$

$$\text{Mathews Correlation Coefficient (MCC)} = \frac{\text{TP} \times \text{TN} - \text{FP} \times \text{FN}}{\sqrt{(TP+FP)(TP+FN)(TN+FP)(TN+FN)}} \times 100\% \quad (3)$$

$$\text{Jaccard Index (JI)} = \frac{\text{TP}}{\text{TP}+\text{FN}+\text{FP}} \times 100\% \quad (4)$$

$$\text{Sensitivity (SENS)} = \frac{\text{TP}}{\text{TP} + \text{FN}} \times 100\% \quad (5)$$

$$\text{Specificity (SPEC)} = \frac{\text{TN}}{\text{TN} + \text{FP}} \times 100\% \quad (6)$$

3 Fundamentals of Crow Search Algorithm

Inspired by the smart behavior of crows, CSA is formed. Generally a crow will make efforts to steal the resources present in hiding places of another crow [6]. The positions of crows are updated iteratively using this behavior as represented in Eq. (7).

$$x_i^{t+1} = \begin{cases} x_i^t + r1 * FL_i^t * \left(m_j^t - x_i^t\right) & if \quad r2 > AP_j^t \\ a \quad random \quad position & otherise \end{cases} \quad (7)$$

Here x_i^{t+1} denote the position of ith crow in the iteration (t+1) i.e., x_i^{t+1} represents the new location and x_i^t represents the previous position of ith crow. FL_i^t denotes to the flight length of ith crow and indicates thAwaressbability (AP) of jth crow. m_j^t denotes the best place of the jth crow obtained till iteri t. The best place of jth crow is identified based on the fitness function. $r1$ and $r2$ denotes the random numbers in the range 0 to 1. For any Swarm Intelligence algorithm, exploration and exploitation process are vital to get the optimum solutions. Exploration is responsible for searching in a large solution space and exploitation is responsible for fine tuning of the solution. In simple words, the amount of exploration will decide the global searching ability. If Flight Length (FL) value is small, then there will more exploitation ability and if FL value is large, then it has more exploration ability [7].

4 Transformation Technique Using CSA with Trial and Error Method for Control Parameters Updation

When CSA is used to solve feature selection and numerical optimization problems, the position of crows will be initialized randomly. But for using CSA as a transformation technique for solving classification problem, the position of crows are initialized using 65536 intensity values directly obtained from the brain MRI image. These intensity values will be varying in the range 0–255. Fitness function is used to identify the hiding place (best position) of each crow. Fitness function of each crow will be equal to the Euclidean distance between the target value and current position of that crow. Target value is fixed as 300. In each iteration after finding the fitness function, the position of crows will be updated using Eq. (7).

AP and FL are the two control parameters of CSA. Usually the value of AP will vary from 0 to 1 and the value of FL will vary from 0 to 2 [8]. More than these two control parameters, Maximum Number of Iterations (MAXIT) will play a vital role since CSA is an iterative algorithm. Hence these two control-parameters are fixed at their mid-points: AP = 0.5, and FL = 1 while searching the optimum value for MAXIT using Trial and Error Method (TEM) as shown in Fig. 2.

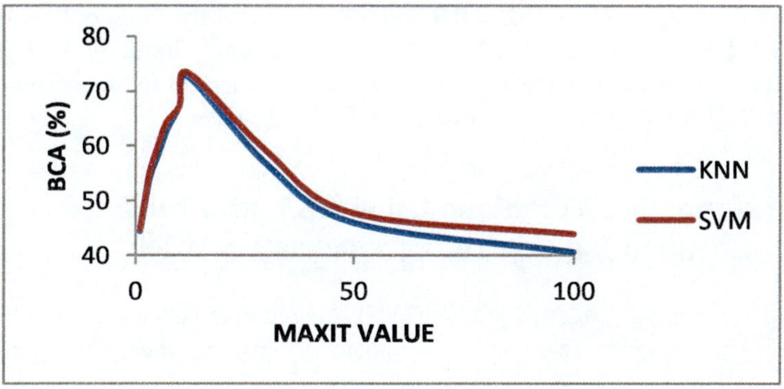

Fig. 2. Finding the optimum value for MAXIT

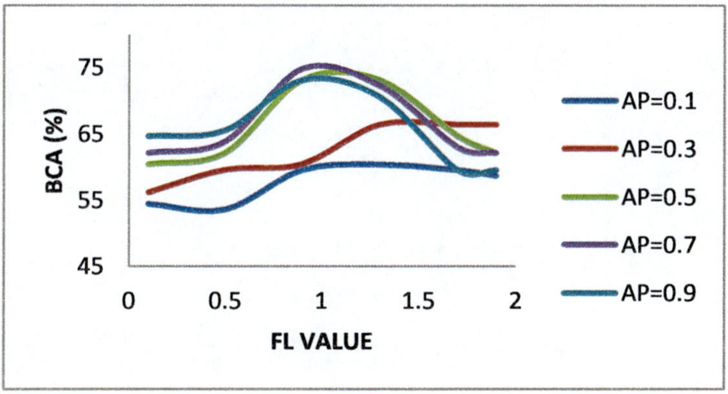

Fig. 3. Finding the optimum values of AP & FL in CSA-KNN classifier

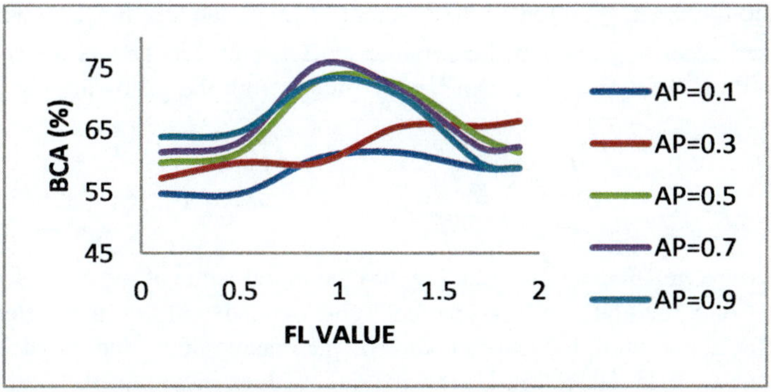

Fig. 4. Finding the optimum values of AP & FL in CSA-SVM classifier

When MAXIT = 12, the maximum BCA of 86% and 87% are attained for CSA-KNN and CSA-SVM classifiers respectively. Then the optimum values for FL and AP are identified by keeping the value of MAXIT at optimum value as depicted in Fig. 3 and Fig. 4 for CSA-KNN and CSA-SVM classifiers respectively. Through Fig. 3, the optimum values are found as FL = 0.9 and AP = 0.7 and for these optimum values, maximum BCA of 88% is achieved for CSA-KNN classifier. Similarly through Fig. 4, the optimum values are found as FL = 0.9 and AP = 0.7 and for these optimum values, maximum BCA of 90% is achieved for CSA-KNN classifier.

5 Transformation Technique Using CSA with Modified Adam Method for Control Parameters Updation

Exponential moving averages of gradients and squared gradients are used in the Adam method [9–12]. The control parameters are updated according to the following equations:

$$w_{t+1} = w_t - \frac{Lrate}{\in + \sqrt{\widehat{S}_t}} * \widehat{V}_t \tag{8}$$

$$\widehat{V}_t = \frac{v_t}{1 - Bet_1^t}. \tag{9}$$

$$\widehat{S}_t = \frac{s_t}{1 - Bet_2^t} \tag{10}$$

$$v_t = Bet_1 * v_{t-1} + (1 - Bet_1) * \frac{\partial L}{\partial w_t} \tag{11}$$

$$s_t = Bet_2 * s_{t-1} + (1 - Bet_2) * \left(\frac{\partial L}{\partial w_t}\right)^2 \tag{12}$$

In the above equations, w_t& w_{t+1} denotes to past and new control parameters; Lrate denotes the learning rate of Adam method; $\in$ denotes a small constant used to avoid division by zero; $\frac{\partial L}{\partial w_t}$ refers to the gradient of L i.e., the loss function which has to be minimized according to control parameter w; Bet_1 and Bet_2 denote the constants of Adam algorithm. The gradient L can be computed using the following equation:

$$\frac{\partial L}{\partial w_t} = \begin{cases} \frac{ER_t}{w_{init}} & if\, t = 1 \\ \frac{ER_t - ER_{t-1}}{w_t - w_{t-1}} & if\, t > 1 \end{cases} \tag{13}$$

Here Error Rate(ER) = 1-BCA and w_{init} is the initial value of the control parameter. In original Adam method, the loss gradient value depends only on the current ER and previous ER along with the current control parameter value and previous control parameter value. It is neglecting all the previous values including the best value. To increase the efficiency of Adam method, the modified version of Adam is proposed in which additional term is introduced as represented in Eq. (14).

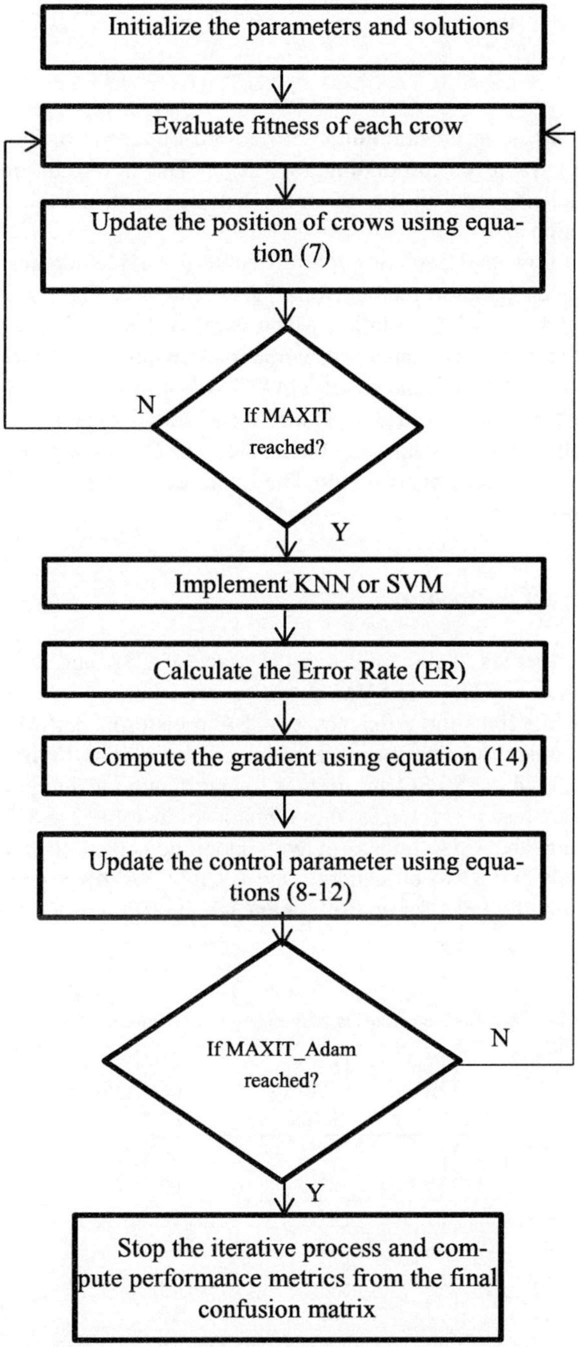

Fig. 5. Flow chart depicting the implementation of CSA-KNN or CSA-SVM classifier with modified Adam to update the control parameters of CSA

$$\frac{\partial L}{\partial w_t} = \begin{cases} \frac{ER_t}{w_{init}} \; if \; t = 1 \\ \frac{ER_t - ER_{t-1}}{w_t - w_{t-1}} + \frac{ER_t - ER_{min}}{w_t - w_{best}} \quad if \; t > 1 \end{cases} \tag{14}$$

In Eq. (14), ER_{min} is the minimum Error Rate obtained so far and w_{best} is the control parameter value corresponding to, $ER_{min.}$ The newly introduced term will compute the gradient based on the difference between current ER and minimum ER attained till iteration *t*.

The common flowchart depicting the procedure for implementing modified Adam as control parameter updation method for CSA is shown in Fig. 5. In Fig. 5, Control parameter *w* is updated using modified Adam iteratively and *t* is the current iteration number; *MAXIT* denotes the maximum number of iterations for the CSA and from Fig. 2, its optimum value is found as 12; *MAXIT_Adam* denotes the maximum number of iterations for the modified Adam algorithm; Through experiments, the optimum value for the following constants are found as: *MAXIT_Adam* = 35, *Lrate* = 0.001, $\in$, = 10^{-8}, $Bet_1 = 0.9$ and $Bet_2 = 0.999$. The hyper-parameter is initialized randomly (w_{init}) in the range [0, 1].

6 Results and Discussion

The performance metrics attained through the usage of CSA and its variants as transformation technique in KNN and SVM classifiers are tabulated in Table 2 and Table 3 respectively. To highlight the efficiency of CSA transform, performance metrics of individual KNN and SVM classifiers are presented. In addition to that Principal Component Analysis (PCA) and PSO are used as transformation technique instead of CSA and those results are also presented for the comparison. In Table 2 and Table 3, CSA (M-ADAM) represents the CSA transform with modified Adam to update the control parameters in KNN and SVM classifiers respectively; CSA-KNN represents the usage of CSA transform with TEM for control parameters updation in KNN classifier.

Table 2. Performance metrics of various transforms in KNN classifier

	KNN	PSO-KNN	PCA-KNN	CSA (TEM)-KNN	CSA (ADAM)-KNN	CSA (M-ADAM)-KNN
BCA	68	72	70	75	80	86
F1	52	57	55	62	69	77
MCC	31	40	37	47	58	69
JI	35	40	38	44	53	63
SENS	67	67	67	67	75	83
SPEC	69	77	74	83	86	89

Table.3. Performance metrics of various transforms in SVM classifier

	SVM	PSO-SVM	PCA-SVM	CSA (TEM)-SVM	CSA (ADAM)-SVM	CSA (M-ADAM)- SVM
BCA	72	75	73	76	82	90
F1	57	62	59	64	72	81
MCC	40	47	44	51	62	75
JI	40	44	42	47	56	69
SENS	67	67	67	67	75	92
SPEC	77	83	80	86	89	89

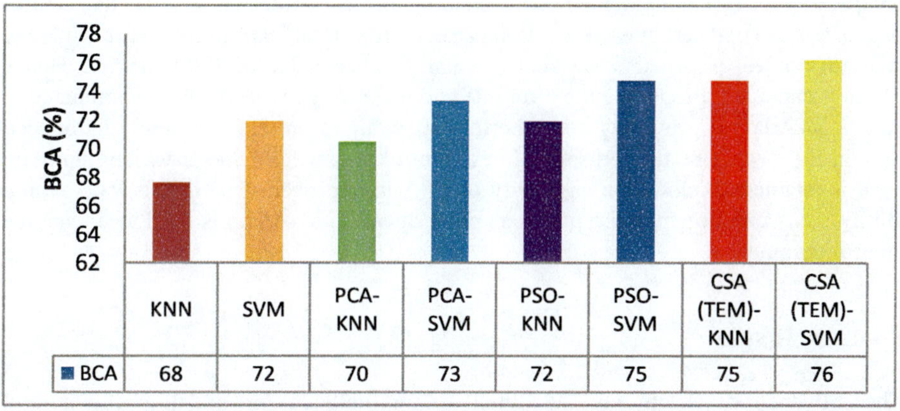

Fig. 6. BCA comparison of KNN and SVM classifiers with various transformation techniques

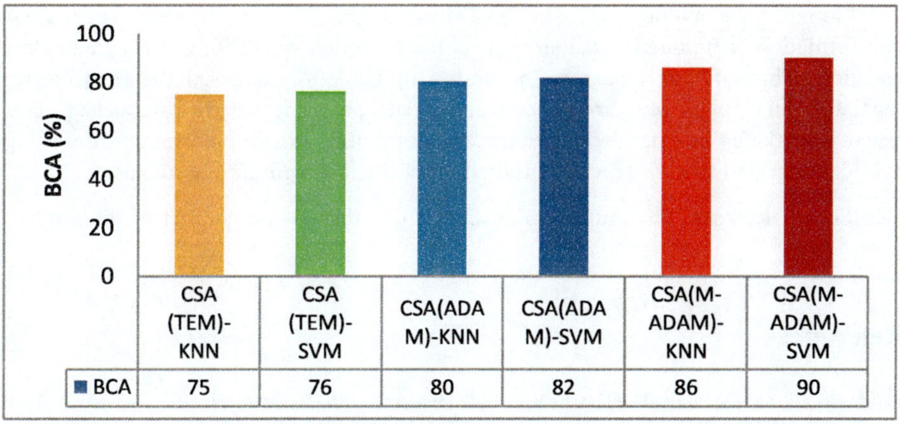

Fig. 7. BCA comparison of KNN and SVM classifiers with CSA and its variants

The efficiency of CSA as transformation technique is proved through the comparison with various transforms as depicted in Fig. 6. BCA is mainly used in this research work since the test dataset is highly imbalanced. When compared with KNN and SVM classifiers without any transform, 11% and 6% increase in BCA is witnessed if CSA (TEM) is used as transform in KNN and SVM classifiers respectively. Also higher performance can be witnessed for CSA (TEM) transform when compared to PCA and PSO based transformation techniques in both the KNN and SVM classifiers. The primary reason behind the superior performance of CSA (TEM) is the good balance between the exploration and exploitation ability since the value of FL is fixed approximately at the mid-point (FL = 0.9).

The significant performance of Adam and Modified Adam (M-Adam) as control parameter updation methods for CSA can be justified through the comparison with CSA that uses TEM for control parameters updation as depicted in Fig. 7. When compared to TEM, Adam method is yielding better performance since the control parameters are updated in each iteration based on the classification Error Rate while the value of control parameters are kept constant at ideal value in TEM method. Hence Adam is capable for reducing the error rate and increasing the BCA. When compared to Adam, M-Adam is providing better performance due to presence of newly introduced term in Eq. (14). Due to the usage of best control parameter value at which minimum error is attained, exploitation capability of CSA is improved. Due to this local search ability, CSA can fine tune the positions properly when M-Adam is used to update the control parameters.

7 Conclusion

The influence of Adam and Modified Adam methods for updating the control parameters of CSA based transformation technique in KNN and SVM classifiers are analyzed. The results clearly depict the superior performance of the above mentioned techniques through comparison with TEM method and other transformation techniques such as PCA and PSO. When compared to the KNN and SVM classifiers without any transformation technique, 27% and 25% increase in BCA is observed when CSA transformation technique is used along with the Modified Adam for control parameters updation. Through this research, the ability of CSA to transform the non-linearly separable data points into linearly separable data points is clearly established. This research work can be extended in the direction of building more precise classifier using hybridization and modification of original algorithms in appropriate manner.

Conflict of Interest. The Author(s) declare(s) that there is no conflict of interest.

References

1. International Statistical Classification of Diseases and Related Health Problems, 10th Revision. World Health Organization, Geneva (1992)

2. Paraskevaidi, M., Martin-Hirsch, P.L., Martin, F.L.: Progress and challenges in the diagnosis of dementia: a critical review. ACS Chem. Neurosci. **9**(3), 446–461 (2018). https://doi.org/10.1021/acschemneuro.8b00007
3. Murray, A.D.: Imaging approaches for dementia. Am. J. Neuroradiol. (2011). https://doi.org/10.3174/ajnr.A2782
4. Mirjalili, S., Mirjalili, S.M., Lewis, A.: Grey wolf optimizer. Adv. Eng. Softw. **69**, 46–61 (2014)
5. Xie, T., Yao, J., Zhou, Z.: DA-Based parameter optimization of combined kernel support vector machine for cancer diagnosis. Appl. Data Model. Anal. Tech. Cancer Therapy. **7**, 263 (2019). https://doi.org/10.3390/pr7050263
6. Oliva, D., Hinojosa, S., Cuevas, E., Pajares, G., Avalos, O., Gálvez, J.: Cross entropy based thresholding for magnetic resonance brain images using crow search algorithm. Exp. Syst. Appl. **79**, 164–180 (2017)
7. Liu, D., et al.: ELM evaluation model of regional groundwater quality based on the crow search algorithm. Ecol. Indic. **81**, 302–314 (2017)
8. Askarzadeh, A.: A novel metaheuristic method for solving constrained engineering optimization problems: crow search algorithm. Comput. Struct. **169**, 1–12 (2016)
9. Kingma, D., Ba, J.: Adam: a method for stochastic optimization. arXiv preprint arXiv:1412.6980 (2014)
10. HaideWang, J.Z., et al.: Optimization algorithms of neural networks for traditional time-domain equalizer in optical communications. Appl. Sci. **9**, 3907 (2019). https://doi.org/10.3390/app9183907
11. Prabowo, A.S., Sihabuddin, A., Azhari, S.N.: Adaptive moment estimation on deep belief network for rupiah currency forecasting, Indonesian J. Comput. Cybernet. Syst. **13**(1), 31–42 (2019). https://doi.org/10.22146/ijccs.39071
12. Voronov, S., Voronov, I., Kovalenko, R.: Comparative analysis of stochastic optimization algorithms for image registration. In: IV International Conference on Information Technology and Nanotechnology (ITNT 2018) (2018)

Automatic Detection of Diabetic Retinopathy Using Image Processing Techniques

Mihai Ciobancă(✉), Simina Emerich, and Mircea Giurgiu

Technical University of Cluj-Napoca, 26 Barițiu Street 400027, Cluj-Napoca, Romania

Abstract. Diabetic retinopathy manifests in the case of diabetic people and in many cases the detection of this disease is too late. As the number of people with diabetes doubles every ten years, the ophthalmologists will be inevitably overloaded, which will lead to an increased risk of human errors in the diagnosis. Therefore, we propose in this paper a solution for automatic detection of this disease even in its early stage by image analysis of the retina, particularly the optic nerve and the veins structure. The originality consists in a new processing pipeline of the images and a careful feature selection to reduce the high dimensional image representation. Then a number of automatic classification models, typically used for image recognition (e.g. K-nearest neighbor [KNN], Support Vector Machines [SVM] and Multilayer Perceptron [MLP]), have been tested. Finally, based on these simulation results, we designed and delivered a software application to support the doctors for a fast analysis process and decision making. The application has been tested with doctors in real life scenarios and it proved its efficiency.

Keywords: Diabetic retinopathy · Automatic classification · Image processing

1 Introduction

All blood vessels of diabetic people are affected, but especially small capillaries. Capillary changes are the precursor to retinopathy in such patients. In some cases the blood vessels are obliterated, in others to compensate for the lack of blood supply the vessels dilate, appearing certain deviations, bulges, of the capillary wall. These swellings represent the so-called micro aneurysms, which upset the permeability of the vascular wall, thus infiltrating the plasma in the retinal tissue, causing the retina edema. Prolonged existence of retinal edema causes compression and loss of nerve cells. The exudates represent albino complexes, lipid formations on the retina.

Pathological changes are not only in the capillaries, large blood vessels, veins in the retina are also affected. Thus, visual acuity begins to decrease with bleeding and the number and size of exudative foci. Non-proliferative diabetic retinopathy is the term used by ophthalmologists to classify the period until the appearance of newly formed pathological vessels and scar changes. When the blood glucose level is high, there is a thickening of the blood vessels. At the level of the retina when the blood vessels are thickened, at first, the vision becomes blurred and without proper treatment can lead to complete blindness, this process of retinal damage is called ophthalmic diabetic retinopathy (Fig. 1).

S. Vlad and N. M. Roman (Eds.): MEDITECH 2020, IFMBE Proceedings 88, pp. 132–142, 2022.
https://doi.org/10.1007/978-3-030-93564-1_15

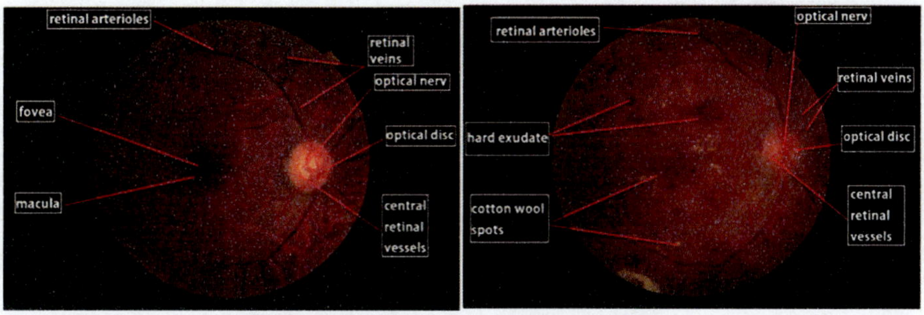

Fig. 1. Difference between normal retina and retina with diabetic retinopathy

Diabetic retinopathy is defined by damage to the structure of blood vessels and degradation of normal functions. At the level of the blood veins on the retina, proper irrigation is no longer performed, the veins becoming weaker, producing ruptures and blood leaks that flood the retina, causing exudates and retinal hemorrhages, thus reducing vision [1]. According to a study launched by the World Health Organization, it is shown that the number of diabetic patients will increase from 130 million to 350 million in the next twenty years [2].

The ophthalmological examination is a complex one, doctors do an analysis, a research of the type of disease, age and other risk factors, after this the visual acuity is determined with correction, an examination takes place, using a device called a biomicroscope, of the anterior pole of the eye, is made a measurement of the tension of the eyeball [3]. Early diagnosis of diabetic retinopathy can play a key role in maintaining vision for as long as possible. Currently, ophthalmologists analyze the images of the retina directly, following certain aspects of the tissue behind the eye, the method being a difficult one that can suffer due to the human factor and the diagnosis is subjective. Given the complexity of this medical analysis, we propose in this paper an automatic detection of the retinopathy using image processing techniques in order to fast the analysis process and to support the doctors in their decision.

2 Proposed Method

2.1 Image Database

In order to test the proposed method one of the largest image databases with the bottom of the eyeball from the Internet was used in our research, the MESSIDOR database [4]. The database is part of a French research program conducted by the Ministry Research and Defense from the Techno-Vision program, aimed to compare and evaluate different techniques for segmentation and indexing of retina images. The 1200 images from MESSIDOR were taken by three ophthalmology clinic departments: Service d'Ophtalmologie - Hôpital Lariboisière Paris, Service d'Ophtalmologie - E3063 Faculté de Médecine St Étienne şi LăŢIM - CHRU de Brest. In the clinical imaging it was used a 3CCD video camera mounted on an analysis device Topcan TRC NW6 non-mydriatic retinograph with an inclination of 45 of degrees (Fig. 2) [4].

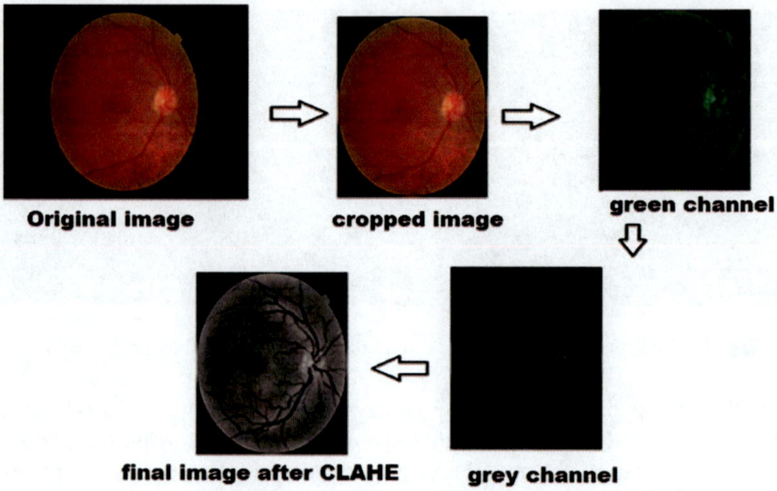

Fig. 2. The preprocessing stages

2.2 Features Extraction

In the pre-processing part, the image is cropped, using some predefined cropping windows, to a window containing only the retina and as little additional space as possible. After cropping, using the predefined functions, the image is divided into the main RGB color channels (Red, Green, Blue) and the green channel is used. Then, using the rgb2gray function a conversion is performed to transform the green channel image to gray [5]. It removes the hue and saturation information while retaining luminance, thus transforming the RGB image into a grayscale, in which case the function only transforms the Green channel you are working with. The image must be turned gray because the features in the image are extracted using the Gray Level Co-occurrence Matrix (GLCM) [6] in which the inserted image must be gray. Finally, we apply a method of equalization of contrast-limited adaptive histograms (CLAHE) [7] to improve the local contrast of an image. At the end, we apply a feature selection process, Gray Level Co-occurrence Matrix, in which the most important features are retained. There are 13 features to be used, the most relevant features in the process of image classification: contrast, correlation, energy, homogeneity, mean, standard deviation, and entropy, Root Mean Square-RMS, variance, smoothness, kurtosis, skewness and Inverse Difference Movement-IDM. Smoothness and IDM have been removed because they did not showed contribution for the classification process, having similar values, then only 11 features are used.

2.3 The Training Process

The features extracted from images are labeled by the user to a class. This class is used to classify the image, showing the state of the retina. Class 0 is for a retina without diabetic retinopathy. Class 1 is for the identification of Non-Proliferative Diabetic Retinopathy - NPDR in its infancy. Class 2 is for the identification of retinopathy non-proliferative diabetics who are in a middle condition and Class 3 to identify retinopathy non-proliferative diabetics who are in a severe condition. All annotated feature vectors are saved in a training database. Further on, a part of this database is used to train 3 classification models (KNN, SVM, MLP) and the other part is used for testing. The training process is implemented with the tools in WEKA, because it was not our aim to develop new classification algorithms, rather than a robust feature extraction from the image and finally an integrated software tool for the doctors.

2.4 The Testing Process

In the testing process of analyzing an image, an image not used before in the training process is loaded in to the program, in order to be tested and subsequently classified. The selected image goes through the same type of pre-processing, the image is automated cropped, using some predefined cropping windows, so that the extracted information is taken from a window that contains retina and as little additional space as possible. After cropping, the image is divided into the main RGB color channels (Red, Green, Blue) and the green channel is used, as previously explained, this channel provides the best contrast. After the entire pre-processing stage takes place, 13 features are extracted using the Grey Level Cooccurrence Matrix (GLCM).

These extracted features are compared with the database obtained in the training part, following the result obtained from the classification algorithm, between the data obtained from the image and the data from the database a class is assigned to the entered image and thus a diagnosis. Class 0 is for a retina without diabetic retinopathy, a healthy retina. Class 1 for the identification of non-proliferative Diabetic Retinopathy (NPDR) in the early stages. Class 2 for the identification of non-proliferative diabetic retinopathy which is in a moderate condition and Class 3 for the identification of non-proliferative diabetic retinopathy which is in a severe state.

3 Experimental Results

We have selected for the experiment a total of 413 images (125 from Class 0, 85 from Class 1, 74 from Class 2, and 129 from Class 3), respecting so the proportions with the entire MESSIDOR dataset. In the first scenario 34% of the images are retained for testing and in the second scenario 20% of the images are used for testing.

3.1 Results for the 1st Scenario

For the KNN classifier we have selected the value of k to 1, batchSize is 100 and numDecimalPlaces is 2. Following this classification, a correct classification of 75.7%

was obtained. For the SVM classifier it was chosen the batchSize of 100, cacheSize 40, with a cost of 200, 3°, loss of 0.1 and gamma 0.01. Following this classification, a correct classification of 72.8% was obtained. For the MLP classifier default values from WEKA have been maintained. Following this classification, a correct classification of 56.4% was obtained. This poor rate is explained by the fact that not enough data was provided for the training. However, our aim was to train the classifiers with as little data as possible and still to obtain good performance. Figure 3 presents the confusion matrix for these 3 classifiers.

3.2 Results for the 2nd Scenario

In this scenario the parameters of the 3 classifiers have been maintained the same as for the 1st scenario, only the size of the training data has been increased (80% of the total database), so using 20% of images for testing. The classification performance is increased to 78,3% for the KNN, to 76% for the SVM, and to 61,4% for the MLP. Figure 4 presents the confusion matrix for these 3 classifiers. Overall, we may conclude that the KNN classifier, although very simple, is the most efficient for our task.

The performance of the proposed classification method is compatible with current state of the art results on similar task, even with those employing Deep Neural Networks (DNN). More, our method is more efficient both in what concerns a small size of the feature vector, as well as the simplicity of the classification method.

```
=== Confusion Matrix ===

 a  b  c  d   <-- classified as
26  0  5  7 |  a = 0
 2 21  5  1 |  b = 1
 4  4 20  1 |  c = 2
 4  0  1 39 |  d = 3
```

```
=== Confusion Matrix ===

 a  b  c  d   <-- classified as
27  0  7  4 |  a = 0
 3 20  2  4 |  b = 1
 7  6 15  1 |  c = 2
 3  0  1 40 |  d = 3
```

```
=== Confusion Matrix ===

 a  b  c  d   <-- classified as
26  1  7  4 |  a = 0
 8 16  4  1 |  b = 1
11  3 12  3 |  c = 2
11  0  8 25 |  d = 3
```

Fig. 3. Confusion matrix (from left to right) to: KNN, SVM, MLP (1st scenario)

```
=== Confusion Matrix ===

 a  b  c  d   <-- classified as
20  0  1  3 |  a = 0
 1 11  2  1 |  b = 1
 4  4 11  0 |  c = 2
 0  1  1 23 |  d = 3
```

```
=== Confusion Matrix ===

 a  b  c  d   <-- classified as
16  0  5  3 |  a = 0
 2  9  2  2 |  b = 1
 3  4  6  6 |  c = 2
 2  0  3 20 |  d = 3
```

```
=== Confusion Matrix ===

 a  b  c  d   <-- classified as
20  0  2  2 |  a = 0
 1 11  1  2 |  b = 1
 3  6  9  1 |  c = 2
 0  1  1 23 |  d = 3
```

Fig. 4. Confusion matrix (from left to right) to: KNN, SVM, MLP (2nd scenario)

4 Practical Implementation of an Assistive Software Tool

Given the promising results obtained in the experimental phase, we have been advised by the collaboration doctors to develop for them a dedicated software tool to help them in image analysis, as well as an automated support in the decision making. Therefore, a desktop application was developed in MATLAB with intuitive interfaces for image processing and classification.

The user interface for the annotation of images (Fig. 4) is divided into three parts: left side, for image selection, central part for the annotation of the input image with the label of the class type, and the right side for the description of the classes with the retina followed by two buttons to clear the image and to close the application. The footnote on the left has the utility of specifying the meaning of various actions, so better informing the user about his activity.

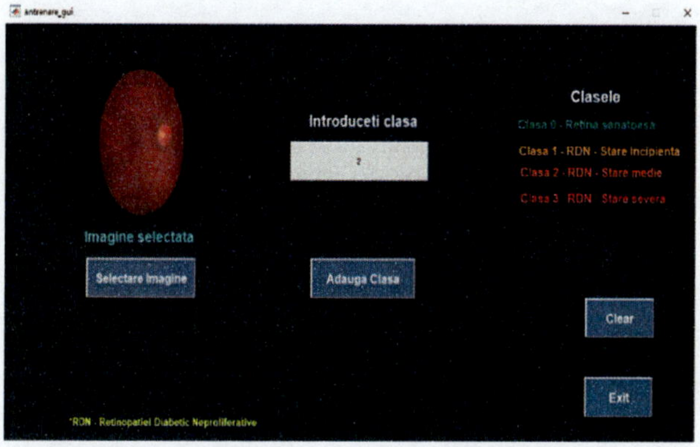

Fig. 5. The user interface for the process of annotation of images

The user interface for image analysis and classification (Fig. 5) implements the proposed image processing algorithms for feature extraction and for automatic classification. The interface respects the classification steps described above in the detection process. The interface is divided into three components: buttons, images and features. The left side is the component part with the buttons that aim to achieve the whole algorithm operation process. In order to use the whole process as intuitively as possible is represented in descending order, from top to bottom, each button having the functionality described on it. The central part is the component part used for the purpose of a better view of the results and in the presentation of the diagnosis. The results are presented in chronological order, being represented from left to right from top to bottom the whole process through which it is passed image from loading to diagnosis. In the last tab, on the right side are presented the features extracted from the image processed. These features are the features used by the classification algorithm to provide an estimation for the diagnosis after being compared with the previously trained database.

To demonstrate this practical implementation we present here two testing scenarios. The first one (left side in Fig. 6) is with an image of a healthy eye, and the second scenario is with an image corresponding to diabetic retinopathy (right side in Fig. 6).

For the first scenario, the image "20051201-38041_0400_PP.tif" of a healthy eye from the MESSIDOR database was used. After the pre-processing process, the image looks like in Fig. 6(b), left side, and the values of the extracted features are presented in Table 1.

Table 1. Features for the first scenario

Features	Values
Contrast	0.165374
Correlation	0.911675
Energy	0.324627
Homogeneity	0.923914
Mean	73.3008
Standard deviation	26.9088
Entropy	6.17078
RMS	15,9683
Variance	469,421
Smoothness	1
Kurtosis	2.92782
Skewness	−0.302063
IDM	255

The edge detection process (Fig. 6c-left) reveals only the optic nerve and veins, not aneurysms or areas to present diseases. The segmentation (Fig. 6-left) accentuates the optic nerve and the veins around it, not aneurysms or yellow areas that have ailments. The classification result indicates a "healthy" diagnosis with a condition "normal". Checking the excel file with the original annotations provided by the ophthalmology clinic (Fig. 7) it is noticed a correct classification, the image coming from a healthy patient (Class 0).

For the second scenario, the image "20051214_51953_ 0100_PP.tif" corresponding to a diabetic retinopathy was selected from the MESSIDOR database. After pre-processing, the image looks like in Fig. 6(b), right side, and the values of the extracted features are presented in Table 2.

Table 2. Features for the second scenario

Features	Values
Contrast	0.205721
Correlation	0.917664
Energy	0.251231
Homogeneity	0.909828
Mean	74.9558
Standard deviation	32,857
Entropy	6,07844
RMS	15.9687
Variance	800.23
Smoothness	1
Kurtosis	3.06385
Skewness	−0.110852
IDM	255

The edge detection (Fig. 6c-right indicates the optic nerve and the veins, and additionally the aneurysms or areas presenting diseases and secretions of plasma and fat. The segmentation (Fig. 6d-right) accentuates not only the optic nerve, but also the yellow areas, cotton wool spots, that fat and plasma deposition who have diseases. The classification result indicates "Retinopathy Non-Proliferative Diabetics" with a "Severe" condition. This is fully verified with the original annotations of doctors from the excel file (Fig. 8 and Fig. 9), "Class 2".

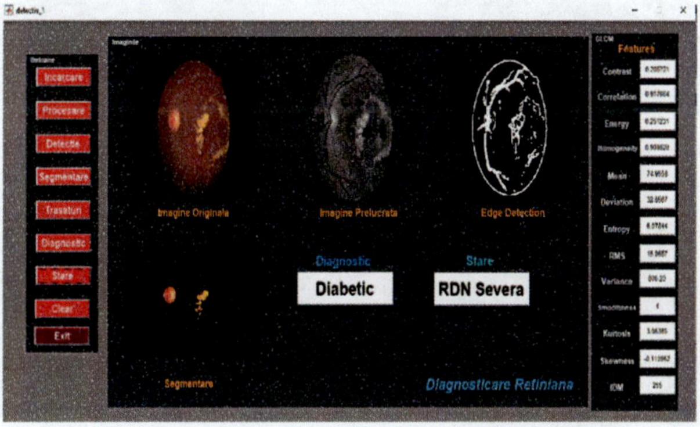

Fig. 6. The user interface for image analysis

Left side: test scenario for a healthy eye **Right side**: test for diabetic case

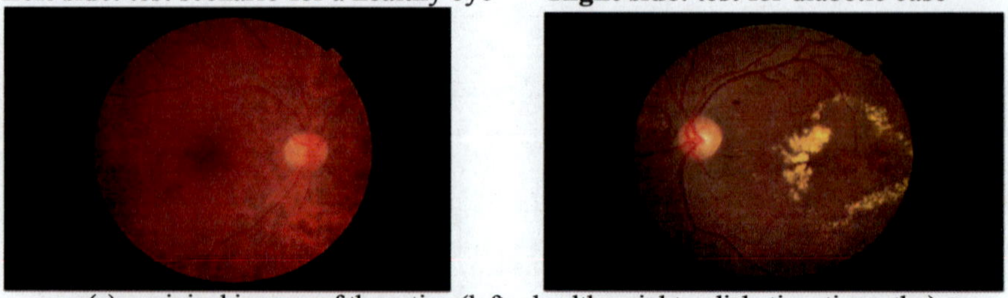

(a) original images of the retina (left – healthy, right – diabetic retinopathy)

(b) the images after the pre-processing stage

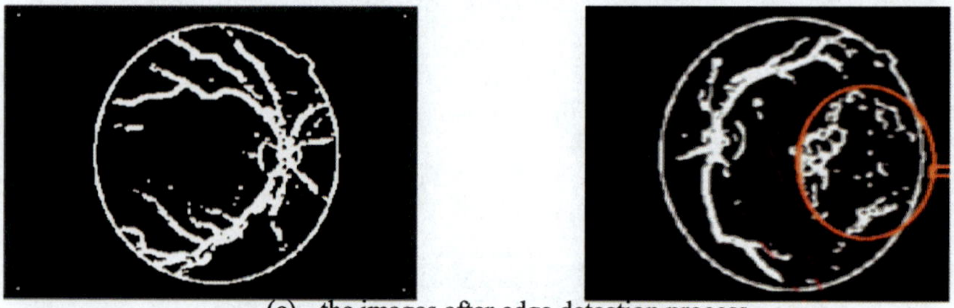

(c) the images after edge detection process

(d) the images after the segmentation process

Fig. 7. Demonstration of the processing flow in the software application

84	20051201_37557_0400_PP.tif	LaTIM - CHU de BREST	0	0
85	20051201_38041_0400_PP.tif	LaTIM - CHU de BREST	0	0
86	20051201_38081_0400_PP.tif	LaTIM - CHU de BREST	0	0

Fig. 8. The classification result indicates a correct classification (Class 0) for Scenario 1

75	20051214_51921_0100_PP.tif	Service Ophtalmologie Lariboisière	3	2
76	20051214_51953_0100_PP.tif	Service Ophtalmologie Lariboisière	3	2
77	20051214_52115_0100_PP.tif	Service Ophtalmologie Lariboisière	0	0

Fig. 9. The classification result indicates a correct classification (Class 2) for Scenario 2

5 Conclusions

The analysis of the retina images in the case of diabetic retinopathy reveals that the entire network of blood vessels is affected, including small blood vessels, small capillaries which are most prone to changes in pathological order. Capillary changes are the precursor to diabetic retinopathy in humans who have insulin deficiency. In some cases the blood vessels are obliterated, in others (to compensate for the lack of blood supply) the vessels dilate, appearing certain bumps, bulges, capillary wall.

In this paper we have proposed an efficient and robust retina images processing pipeline in order to extract a minimal set of 13 features which are able to describe such complex variability of eye's blood vessels. These features have been further used to train a number of well known classification models in order to automatically detect the diabetic retinopathy. Still, there is room for improvement of the classification performance and we look for the following solutions: to process a more large dataset possible recorded in severe conditions (e.g. such as noisy images, different illumination), to improve the processing pipeline adding new features that may be more relevant for this task, improving the software application and integrating it with the medical devices from the ophthalmology clinics.

Conflict of Interest. The authors declare that they have no conflict of interest.

References

1. Wu, G.: Diabetic Retinopathy: The Essentials. Lippincott Willia & Wilkins, Philadelphia (2010)
2. Cunha-Vaz, J.: Diabetic Retinopathy. University of Coimbra, Coimbra (2018)
3. WestEye Hospital website. https://www.westeyehospital.ro/ro/afectiuni-si-tratamente/retinopatia-diabetica. Accessed 12 Apr 2019
4. MESSIDOR DataBase Site. http://www.adcis.net/en/third-party/messidor/. Accessed 28 Jan 2020
5. Mathworks web site. https://www.mathworks.com/help/matlab/ref/rgb2gray.html. Accessed 20 Apr 2020

6. Brynolfsson, P., et al.: Haralick texture features from apparent diffusion coefficient (ADC) MRI images depend on imaging and pre-processing parameters. Sci. Rep. **7**(1), 4041 (2017). https://doi.org/10.1038/s41598-017-04151-4. PMID: 28642480; PMCID: PMC5481454
7. ImageJ web site. https://imagej.net/Enhance_Local_Contrast_(CLAHE). Accessed 5 Apr 2020

Improvement of the *in vitro* Pig Mandible Periodontal Tissue Images Quality Recorded by an OCT Handheld in 1200–1400 nm Infrared Domain: A Preliminary Study

Florin Toadere[1], Nicoleta Tosa[1(✉)], Mindra Eugenia Badea[2(✉)], Andreea Miron[2], Victor Cebotari[1], Valer Tosa[1], Ioan Turcu[1], and Radu Chifor[2]

[1] Molecular and Biomolecular Department, National Institute for Research and Development for Isotopic and Molecular Technologies, Donat 67-103, 400293 Cluj-Napoca, Romania
nicoleta.tosa@itim-cj.ro

[2] Department of Preventive Dentistry, "Iuliu Hațieganu" University of Medicine and Pharmacy Cluj-Napoca, Avram Iancu 8, 400349 Cluj-Napoca, Romania
mebadea@umfcluj.ro

Abstract. In this study we proposed the improvement of the quality of the periodontal images recorded *in vitro* on a defrozen pig mandible using a spectral domain optical coherent tomography (SD-OCT) handheld operating in 1200–1400 nm infrared domain. Anatomical elements of periodontal tissue necessary to establish the diagnosis of periodontal disease by OCT imaging as a non-contact and non-invasive alternative to periodontal probing could be identified. The anatomical elements could be clearer visualized by using a dedicated imaging processing algorithm, which applied a package of proper filters to remove the noise and the blur, but also to enhance the original images contrast.

Keywords: Spectral domain - optical coherent tomography · Periodontal tissue imaging · Imaging processing algorithm

1 Introduction

Optical coherent light tomography (OCT) is a non-invasive and non-destructive imaging technique that allows the visualization of biomedical samples in the form of two-dimensional (2D) and three-dimensional (3D) images. The operating principles of the first OCT systems were presented in 1991 [1]. The first OCT image was recorded with an axial resolution of 15 μm and a scan depth of 2 mm. These parameters are lower than those obtained using ultrasound [2], where they depend on the ultrasonic wave's frequency, and higher than those obtained by confocal microscopy [3], where they depend on the microscope objective numerical aperture.

In OCT, the microscopic objective influences the transverse resolution value, and the axial resolution depends on the scanning source. Therefore, an axial resolution of 1–2 μm was obtained with the help of high-resolution OCT systems [4]. In OCT

S. Vlad and N. M. Roman (Eds.): MEDITECH 2020, IFMBE Proceedings 88, pp. 143–150, 2022.
https://doi.org/10.1007/978-3-030-93564-1_16

scanning, the lateral resolution is 1–2 μm and the depth of penetration into biological tissues is about 1–2 mm. In clinical applications, OCT is used to obtain *in vivo* images of the human eye, but there is an increasing interest in its application in dermatology, dentistry and biological tissues [5].

In OCT the detected interferometric signal is correlated in phase and amplitude with the light reflected from the given depth of the sample. Thus, a reflective axial profile is obtained at a point in the tissue (type A or A-scan). Different OCT types are already well described [6]. Nowadays, major research areas in OCT regards to swept source and spectral (Fourier) domain OCT, respectively [7, 8]. Spectral-domain OCT (SD-OCT) methodology deals with the spectra analysis at the output of the interferometer. SD-OCT is attractive because it eliminates the need for depth scanning, but the obtained depth resolution depends on the illumination source bandwidth [6]. The quality of digital images of OCT type can be also improved by using additional filters [9–12]. Therefore, their effect among a processed digital image reveals little details which are not distinctly visualized in the original image.

In our study is proposed the improvement of the quality of the periodontal images recorded *in vitro* on a defrozen pig mandible using an SD-OCT handheld in 1200–1400 nm range. The anatomical elements of periodontal tissue necessary to establish the diagnosis of periodontal disease by OCT imaging as a non-invasive alternative to periodontal probing could be identified. These anatomical elements could be clearer visualized by using a dedicated imaging processing algorithm, which applied a package of proper filters to remove the noise and the blur, but also to enhance the original images contrast.

2 Experimental

2.1 Materials and Methodology

A defrozen pig mandible from a butcher's shop was used as sample within the *in vitro* OCT scanning examination of the periodontal tissue for our preliminary study.

The SD-OCT measurements were recorded using a fiber-based portable handheld scanner OCTH-1300 from Thorlabs (see Fig. 1b).

The light source used was a superluminescent diode operating in the 1200–1400 nm range with 5.6 mW power output. The OCTH-1300 is based on an interferometer with optically matched sample and reference arms, to minimize the image distortion caused by dispersion. The reference arm and the sample arm are housed in a common path design, ensuring the sample arm phase stability. The handheld scanner has an internal MEMS scanner with a 2D scanning mirror with 3 mm clear aperture diameter (see Fig. 1a) and an integrated color CMOS camera and LED illumination for sample real-time visualization during the OCT scanning. The OCT scanner operation is completely automated and controlled by a computer using the ThorImageOCT software.

The OCT investigations were done in the 2D and 3D scanning modes with a speed of 76 kHz at a refractive index set to 1.5, which ensures a depth of penetration of 2.4 mm and an axial resolution of 5.5 μm in air and 4.4 μm in water, respectively. The light was focused into the sample using a compatible lens kit OCTH-LK30 with focal length of 30 mm, working distance of 22 mm, field of view of 10 mm (diameter) and lateral resolution of 24 μm. A compatible working distance holder OCTH-AIR30 with focal length of 30 mm and working distance of 22 mm is attached to the lens kit, preventing both the direct contact of the lens with the sample and ensuring the keeping of the sample in focus (see Fig. 1b).

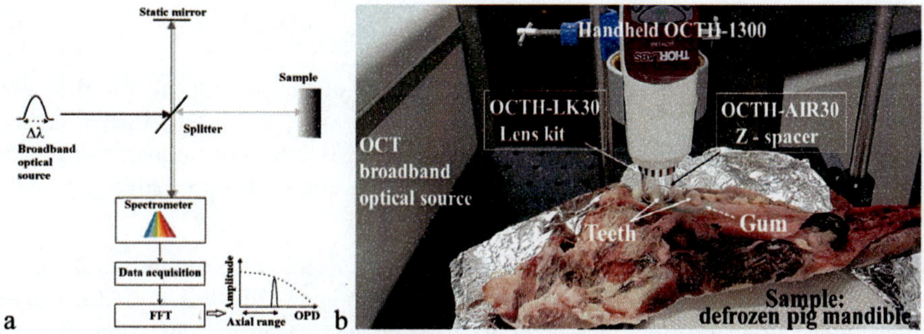

Fig. 1. SD-OCT setup: a. operating principle [6]; b. experimental, on periodontal tissue of a defrozen pig mandible using the OCTH-1300 handheld operating in 1200–1400 nm range.

2.2 Image Processing Algorithm

Acquired periodontal tissue OCT images by *in vitro* scanning of a defrozen pig mandible are blurred and faded in terms of contrast and luminosity. Therefore, a four steps image processing algorithm [9–12] that filters the OCT images to improve the quality of the original images is developed as follows:

1. The use of a sharpening filter, to diminish the image blur without affecting fine details and edges,
2. The use of an anisotropic diffusion filter, to reduce the image noise without removing significant edges, lines or other fine details,
3. The use of a homomorphic filter, to correct the non-uniform illumination of the images,
4. The enhancement of the image quality in terms of contrast and brightness.

Sharpening Filter. The OCT images are faded and blurred. Blur removal requires the employment of a sharpening filter. Such filter reduces the blur and enhances the filtered image content by revealing its details that were not visible previously filtering.

Performing sharpening of the OCT images requires:

1. Frequency low pass filtering by a first order Butterworth low-pass filter (Eq. 1) [9, 10]:

$$H(u,v) = \frac{1}{1 + [D(u,v)/D_0]^{2n}} \tag{1}$$

where: D_0 is the cutoff frequency and is a positive constant having the values in the range (0, 1), $n = 1$ is the first order of the Butterworth, u and v are the image frequency coordinates and *D*(*u*,*v*) is the Euclidean distance from any pixel having the coordinates (*u*, *v*) to the center of the frequency plane, i.e., $D(u, v) = \sqrt{u^2 + v^2}$.

2. Image sharpening is achieved by subtracting the low pass filtered image from the initial image (Eq. 2):

$$S(x, y) = \frac{c}{2c - 1} O(x, y) - \frac{1 - c}{2c - 1} L(x, y) \quad (2)$$

where: *S*(*x*,*y*), *O*(*x*,*y*) and *L*(*x*,*y*) represent the values of the pixel located at the coordinate (*x*,*y*) in the filtered image, original image and low pass filtered image, respectively. The values of the constant *c* range between 0.556 and 1. Sharpening filter enhances the visibility of the filtered image content by emphasizing its edges. Once the blur decreases, the noise reveals itself.

Bilateral Filter. Noise presented in the OCT images needs to be filtered. In the noise filtering process, an anisotropic diffusion filter is employed. This is a gradient filter, which filters along image edges without crossing them. Isotropic diffusion equation (Eq. 3) characterizes the operation of the anisotropic diffusion filter as follows [11, 12]:

$$\frac{\partial}{\partial t} I(x, y, t) = div[c(x, y, t) gradI(x, t)] \quad (3)$$

where *I*(*x*,*y*,*t*) is the image's intensity, *x* and *y* are image's coordinates, *t* is the time parameter, *c(x,y,t)* is the diffusion constant.

Filtering OCT image's noise using Eq. 3 is similar to filtering the image's noise when apply a Gaussian filter. Parameter *c(x,y,t)* controls the diffusion rate of the bilateral filter, so that the edges are preserved. Its mathematical expression is given in Eq. 4:

$$c(x, y, t) = \exp\left(-\left(\frac{\nabla I(x, y, t)}{k}\right)^2\right) \quad (4)$$

Empirically, we found that the value of the diffusion rate is k = 0.2 and the number of iterations is 100.

Homomorphic Image Filter. The homomorphic filter corrects the non-uniform illumination of the images [9, 10]. In the image formation model, the intensity of a pixel is the percent of the light reflected by a point located in the scene [12]. Thus, the reflected intensity is the product between the illumination of the scene *L(x,y)* and the reflectance of the objects *R(x,y)* located inside the scene (Eq. 5):

$$I(x, y) = L(x, y) R(x, y) \quad (5)$$

The first step in the homomorphic filtering consisted of the transformation of the multiplicative components into additive components by exchanging Cartesian coordinates into logarithmic coordinates (Eq. 6):

$$\log[I(x,y)] = \log[L(x,y)] + \log[R(x,y)] \tag{6}$$

where I is the image, L is scene illumination, and R is the scene reflectance.

The second step presents the analysis which was performed in the Fourier domain (Eq. 7):

$$I(u,v) = L(u,v) + R(u,v) \tag{7}$$

In the third step, a Gaussian high-pass filter was applied in the Fourier domain (Eq. 8):

$$M(u,v) = I(u,v)H(u,v) \tag{8}$$

where M is the filtered image and H is a Gaussian high-pass filter, respectively.

Finally, an inverse Fourier transform was applied to the filtered image [$M(u,v)$] in order to return from logarithmic coordinates into Cartesian coordinates. Thus, the output image $O(x,y)$ was rendered by Eq. 9:

$$O(x,y) = invF[M(u,v)] \tag{9}$$

Histogram. Histogram stretching increases the contrast of the image [9–12] using non-linear and non-monotonic transfer functions. Histogram represents the number of pixels afferent to any intensity in the image. In the OCT images the intensity varies from black to white. Histogram is stretching out the intensity range of the image. Thereafter, a better global contrast across the entire image results and the OCT image becomes more accurate when it renders on the computer's display.

Image Brightness. Brightness is a subjective property of an object rendered by a display. OCT images are faded, so that its fine details are not clearly distinguishable from the image background. Image brightening is useful in highlighting the OCT images details [9–12]. If the image brightness is the average of all the pixels within the image $f(x,y)$, then increasing the brightens means to map the pixels values of $f(x,y)$ into a new higher value via a function or a constants values applied to $f(x,y)$ as follows (Eq. 10):

$$g(x,y) = af(x,y) + b \tag{10}$$

where: $a > 1$ and $b > 0$ are constants.

3 Results of Simulations and Discussion

Original OCT images of periodontal tissue, acquired by *in vitro* scanning of a defrozen pig mandible, are blurred and faded. Thus, a four steps image processing algorithm [9–12] that filters the OCT images to improve the quality of the original images is employed (see Experimental Sect. 2.2).

OCT images were recorded in a cube as 3D volumetric images, both top (XY) and lateral (YZ and XZ) images being recorded. The optical image of the top view (XY) scanned area shows the gum in its upper part and the tooth in its downer part (Fig. 2c). The corresponding OCT images, original (Fig. 2a) and processed (Fig. 2b), show a partial detachment of the gum from the tooth surface in the marked areas 1–2 and 3–4, respectively, and some changes in the tooth surface at positions 5 and 6. The optical imaging allowed a correct identification of tooth (right part) and gum (left part) positioning in the YZ OCT images, based on the ratio of each corresponding length in both optical and OCT images (Fig. 2d, e). The proposed algorithm for OCT image quality improvement was also run on another lateral view (YZ), acquired from an intact periodontal tissue cross-section: original (Fig. 2f) and processed (Fig. 2g), respectively. In this case, a proper contact of the gum with the surface of the tooth is emphasized (Fig. 2f,g).

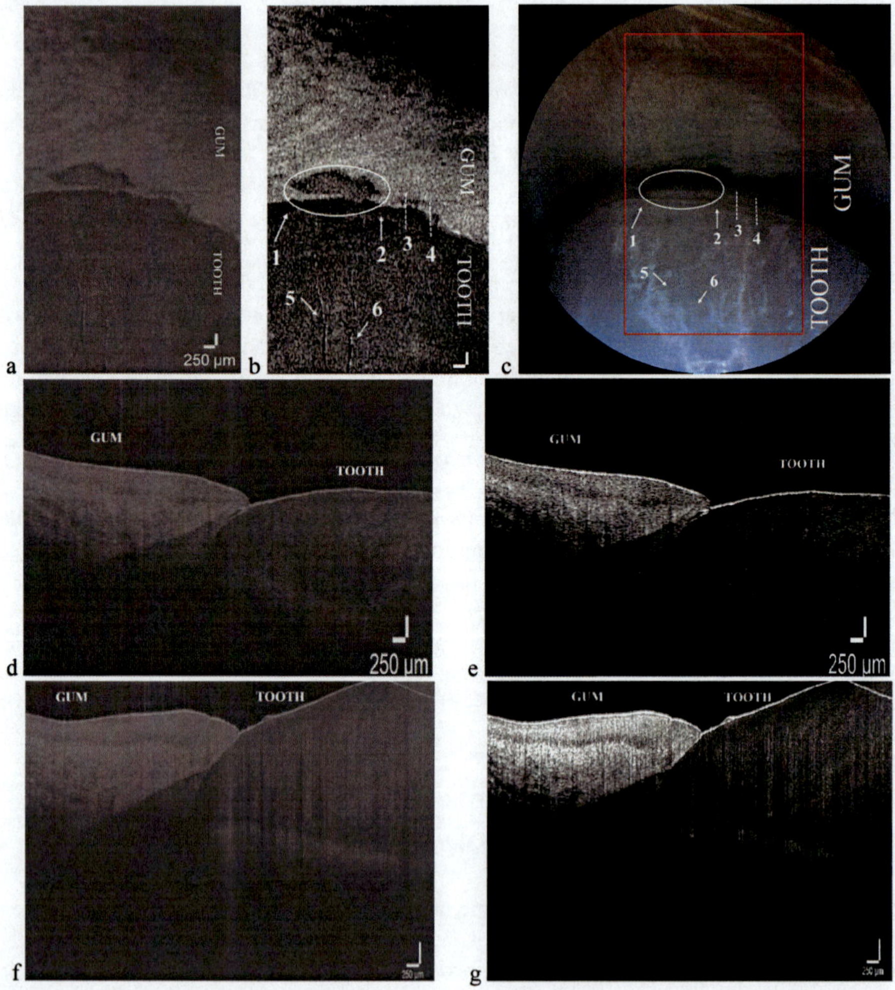

Fig. 2. OCT images of periodontal tissue (tooth and affected/intact gum) acquired by *in vitro* scanning of a defrozen pig mandible: top view (XY), a. original, b. processed; lateral view 1 (YZ), d. original, e. processed; lateral view 2 (YZ), f. original, g. processed; c. optical image of the OCT scanned area.

The periodontal tissue aspect differs from teeth to gum because they are optically inhomogeneous, their anatomical elements having variable refractive indexes [13]. The consequent scattering will limit the penetration depth and imaging resolution.

The OCT image presented in Fig. 2d was subjected to an image processing by colored manual segmentation (Fig. 3) to identify the periodontal tissues anatomical elements. Therefore, the gum is shaped with red, dental crown with white, root with yellow, alveolar bone with brown and gingival sulcus with gray. Considering the image scale of 250 microns, the size of the gingival sulcus (Fig. 3b) is between 0 and 0.5 mm. This dimension places the periodontal tissue in the category of healthy tissue, because only gingival sulcus with dimensions over 3 mm is considered in the pathological area, and thus, no proper periodontal pocket is assumed.

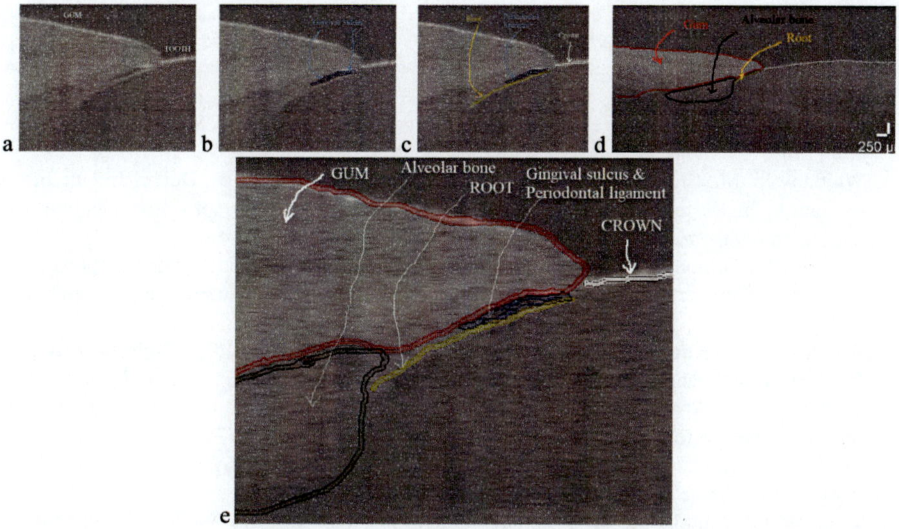

Fig. 3. Anatomical elements identification in original (YZ) OCT images of periodontal tissue: a. bare; b. gingival sulcus (gray); c. crown, root and periodontal ligament (white, yellow, gray); d. gum and alveolar bone (red and brown); e. all elements.

4 Conclusions

SD-OCT technique offers a high-resolution non-contact and non-invasive tissues imaging with a good contrast. The periodontal tissue aspect differs from teeth to gum because they are optically inhomogeneous, their anatomical elements having variable refractive indexes. The consequent scattering will limit the penetration depth and imaging resolution. The quality of the periodontal tissue images recorded *in vitro* on a defrozen pig mandible by an OCT handheld system is improved by image processing. The dedicated imaging processing algorithm uses a package of proper filters to remove the noise and the blur, but also to enhance the original images contrast. The clearer visualization of the OCT images allows identifying the anatomical elements of the periodontal tissue, necessary for diagnosis of periodontal disease by OCT imaging as a non-invasive alternative to periodontal probing. The proposed algorithm was run on different OCT images for testing the method and comparison with the optical method visualization of the OCT scanned area.

Acknowledgments. This work was funded by the Romanian Ministry of Education and Research, through the project number PN-III-P1-1.2-PCCDI-2017-0010/74PCCDI/2018, within PNCDI III.

Conflict of Interest. The authors declare that they have no conflict of interest.

References

1. Huang, D., et al.: Optical coherence tomography. Science **254**, 1178–1181 (1991)
2. Donald, I., MacVicar, J., Brown, T.G.: Investigation of abdominal masses by pulsed ultrasound. Lancet **1**(7032), 1188–1195 (1958)
3. Minsky, M.: Memoir on inventing the confocal scanning microscope. Scanning **10**(4), 128–138 (1988)
4. Wojtkowski, M., Srinivasan, V., Ko, T., Fujimoto, J., Kowalczyk, A., Duker, J.: Ultrahigh-resolution, high-speed, Fourier domain optical coherence tomography and methods for dispersion compensation. Opt. Express **12**(11), 2404–2422 (2004)
5. Drexler, W., Fujimoto, J.G.: Optical Coherent Tomography, Technology and Applications. 2nd edn. Springer International Publisher (2015). https://www.springer.com/gp/book/9783319064185
6. Podoleanu, A.: Advances in optical coherence tomography. In: 2018 20th International Conference on Transparent Optical Networks (ICTON), Bucharest, pp. 1–4 (2018)
7. Chin, C., et al.: Acousto-optic tunable filter for dispersion characterization of time-domain optical coherence tomography systems. Appl. Opt. **55**(21), 5707–5714 (2016)
8. Toadere, F., et al.: 1 MHz akinetic dispersive ring cavity swept source at 850 nm. IEEE Photonics Technol. Lett. **29**(11), 933–936 (2017)
9. Toadere, F.: Simulation of a digital camera pipeline. In: ISSCS 2007: International Symposium on Signals, Circuits and Systems, vol. 1&2, pp. 569–572 (2007)
10. Toadere, F.: Simulating the functionality of a digital camera pipeline. Opt. Eng. **52**(10), 102005 (2013)
11. Toadere, F., Tosa, N.: Enhancement of the raw OCT image quality. AIP Conf. Proc. **2206**(1), 040002, 1–5 (2020)
12. Toadere F., Tosa N.: Noise removal from raw OCT images achieved using an OCT system operating in the bandwidth 827 nm–873 nm. In: Proceedings SPIE 10977, Advanced Topics on Optoelectronics, Microelectronics and Nanotechnology IX, 109770N (2018)
13. Colston, B.W., et al.: In vivo optical coherent tomography of teeth and oral mucosa. Appl. Opt. **37**(16), 3582–3585 (1998)

Voting Based CAD Model for Breast Cancer Classification

S. R. Sannasi Chakravarthy(✉) and Harikumar Rajaguru

Department of ECE, Bannari Amman Institute of Technology, Sathyamangalam, India

Abstract. For the diagnosis of any disease, the design of Computer-Aided Diagnostic (CAD) model is a significant one that guides physicians to ensure their opinion on diagnosis and so it influences the decline in mortality of breast cancer. This paper intends to devise a CAD model for mammogram classification i.e. benign or malignant and the same is done in three steps. The preprocessing phase aims for the removal of noise, pectoral muscle, and other unwanted objects. The region of interest (ROI) is extracted from the preprocessing step. Then extraction of features is carried out by making use of two-dimensional discrete wavelet transform (DWT), and the significant features are selected by applying the discrimination power analysis (DPA) technique. Eventually, the last step involves the classification using the Logistic Regression Classifier (LRC), Nu-Support Vector Classifier (Nu-SVC), Complement Naïve Bayes classifier (CNBC). Finally, the above said classifiers are conceptually combined by making use of the average predicted probabilities to obtain robust performance. The CAD model is evaluated with the MIAS dataset that provides a classification performance of 97.39% accuracy with a soft voting technique.

Keywords: Mammogram · MIAS · Breast cancer · SVM · Wavelet · Voting

1 Introduction

Breast cancer is a highly crucial disease and remains a severe public health crisis in many nations. The incidence rate of this type of cancer is more in developed nations and it is progressively increasing in developing countries. It is anticipated that around two million women died globally due to breast cancer in the year 2019 [1]. Thus, there is a high demand for the robust design of computer-aided diagnostic (CAD) models to predict cancer earlier. The early prediction of cancer will surely reduce the mortality rate of breast cancer. For this, several imaging modalities are being popular at different stages of cancer. But for the earlier prediction of breast cancer, the mammography imaging technique remains a promising one in diagnosing the disease at its earlier stage. Conversely, the rapid increase in mammography procedure requires a need to build a robust CAD model that can support clinicians in the earlier prediction and classification of tumor severity as either malignant type or benign severity. Rapid improvements in both screening and diagnostic procedures will radically increase its rate of recovery since 1990 [1].

S. Vlad and N. M. Roman (Eds.): MEDITECH 2020, IFMBE Proceedings 88, pp. 151–160, 2022.
https://doi.org/10.1007/978-3-030-93564-1_17

A tumor in breast cancer may be benign (non-cancer) or malignant (cancer). The benign type will not be dangerous to health and so they are non-cancerous. Furthermore, benign tumors seem to be to normal in appearance, since they evolve gradually, and most importantly, they will not invade other parts of the breast [2]. In contrast, malignant types are dangerous to health i.e. cancerous. In this type of severity, the tumor cells eventually have the ability to spread to other areas of the breast. Thus, in general, "breast cancer" term makes mention of a malignant severity that has expanded from breast cells [2].

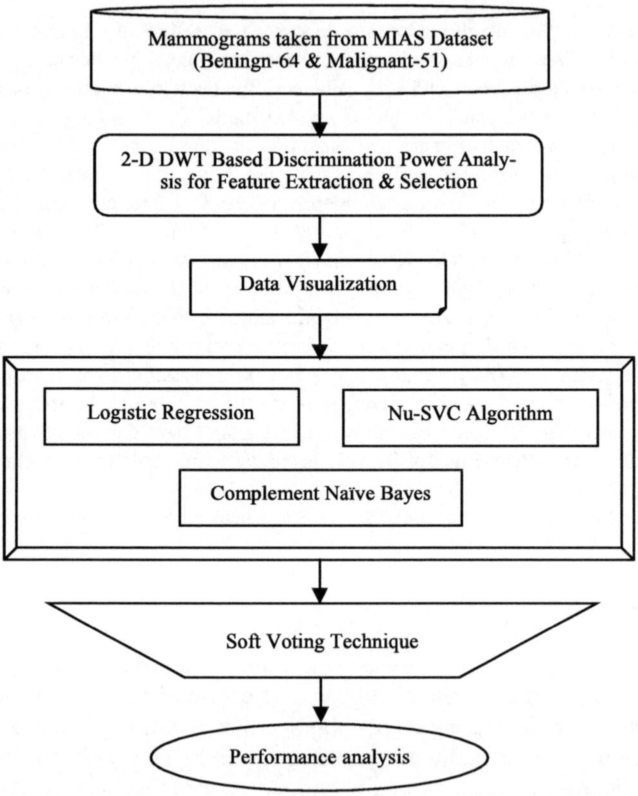

Fig. 1. Work flow of the CAD system

Figure 1 gives the design-flow and analysis of the CAD model. The input for the work is taken from the Mammogram Image Analysis Society (MIAS) dataset and the features are derived using Discrete Wavelet Transform (DWT) together with the discrimination power analysis (DPA) method. After extraction, the discriminative features are analyzed and visualized as plots. From this plot, the algorithms are chosen and then they are classified using three distinct classification models and finally combined using the soft-voting technique. The performance is analyzed and compared using standard metrics. The steps involved and shown in Fig. 1 will be discussed in the below subsequent sections.

2 Proposed CAD Model

2.1 Preprocessing

The pre-processing phase of digital mammogram images is being a vital role used for the ROI selection that contains the severities. The pre-processing is required to remove noise, unwanted background, and pectoral muscles. The adoption of effective preprocessing techniques guarantees a better performance for any CAD model. The details about the MIAS dataset [3] are given in Table 1. As shown in Table 1, the noise (salt-and-pepper) in input mammograms are removed by using an adaptive median filter [4]. After manual cropping, global thresholding [5] is used for the extraction of ROI.

Table 1. Information about MIAS dataset

Specification	Details	Sample Mammogram Image
Projection or View	Mediolateral oblique (MLO) view	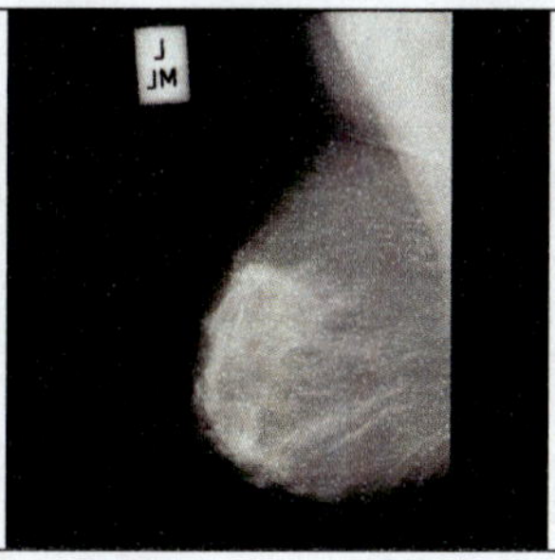
Spatial resolution	50 μm/pixel	
Gray-level quantization	8-bits	
Pixel resolution	1024 x 1024 pixels	
Digitizer	SCANDIG-3 Joyce-Loebl microdensitometer	
Database Size	322 Digital Mammograms	
Database Information	Normal – 207 Benign – 64 Malignant – 51	

2.2 Extraction and Selection of Features

The feature extraction process plays a vital role in the design of any CAD system used for the diagnosis of any medical problem and the choice of suitable technique is a challenging step for all classification tasks. Next to extraction, feature selection is an important phase because the extracted features may not have a similar capability to make a distinction among different output classes. The work makes use of DPA for selecting the most discriminating and significant coefficients of DWT. This way of feature selection provides larger values to the coefficients of DWT with a high range of discrimination power and is determined by the ratio of the within-class variance related to the between-class variance [6]. Hence, a higher range of discrimination power represents a higher discrimination ability of the respective coefficients.

A discrete wavelet transform (DWT) is a mathematical approach that provides a time-frequency representation of any image or signal. DWT has been used typically for the decomposition of image or signal utilizing a series of high-pass and low-pass filters to analyze the corresponding low and high-frequency components in the input [7]. For the decomposition of images, two filters are appealed across both columns and rows, in order to acquire the DWT coefficients. And this should provide an approximation sub-band and three detailed sub-bands on each decomposition level, and these coefficients of DWT provide the reconstruction of the input image.

The value of discrimination power (DP) corresponds to a coefficient that is dependent on both the smaller and larger variations between the output labels. Larger discrimination power values more discrimination power to the DWT coefficients. The coefficients of DWT are adaptably chosen based on their obtained DP values and the coefficients with larger DP values are taken for further process [8]. The so obtained DP matrix values are arranged as column vectors and then sorted in the descending order. The amount of feature selection is based on the threshold value that can be applied over the already sorted DP values. Thus, for choosing k feature vectors, a threshold that is same as the DP value is selected at position k. After that, the formation of a mask matrix is done by thresholding the DP matrix with the chosen threshold. The created mask matrix comprehends the position (= 1) of the most significant features in the wavelet domain. Soon after, in the process of classification, the mask created can be utilized for choosing the coefficients of wavelet and is given by,

$$Mask(x,y) = \begin{cases} 1, & DP(x,y) \geq threshold(k) \\ 0, & otherwise \end{cases} \tag{1}$$

The extracted and selected features are then plotted in Fig. 2. As plotted below in Fig. 2, the obtained features appear to be a non-linear one.

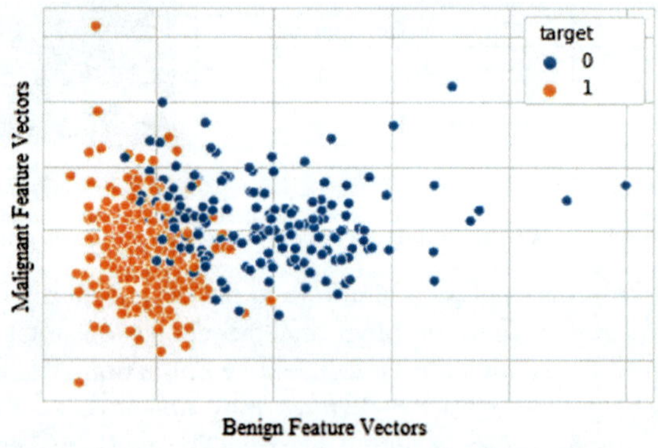

Fig. 2. Visualization of both Benign and Malignant output class features

3 Classification Models

3.1 Logistic Regression Classifier (LRC)

The logistic regression algorithm is a type of linear approach for the process of classification more willingly than for the regression process and so this approach is also termed as a log-linear classifier model. Logistic regression belongs to a statistical learning approach that comes under supervised techniques dedicated to the process of classification [9]. This model has attained remarkable popularity for the last two decades particularly in the image analysis applications because of its promising capability to detect defaulters. The hypothesis of this model makes it to limit the value of the cost function to be 0 *to* 1. In this classification problem, the probabilities relating to the possible findings of a unique trial are formed through a logistic function, as given by [9]:

$$f(x) = \frac{1}{1 + e^{-x}}. \quad (2)$$

3.2 Nu-Support Vector Classifier (Nu-SVC)

The Nu-SVC is similar to the support vector classifier, but here the *Nu* represents the advantage of controlling the number of support vectors [10]. As above said, a *Nu* parameter is included for controlling both the margin errors and the number of support vectors. In the Nu-SVC algorithm, a margin error occurs whenever a vector is placed on the incorrect side of its hyperplane boundary, i.e. it is either classified correctly (but not lies beyond the margin) or it is simply mis-classified [10]. Thus, $Nu \in [0, 1]$. represents a lower and upper bound on the fraction of support vectors and on the fraction of margin errors. The value of *Nu* is set as 0.6 for the classification of mammograms. In this paper, the classification is attained by maximizing the margin between the input vectors and the created hyperplanes. The hinge loss function is used in the work to maximize the margin [10].

3.3 Complement Naïve Bayes Classifier (CNBC)

Naive Bayes's techniques are a group of supervised learning approaches that follows the principles of Bayes mathematical theorem together with the "*naive*". assumption of conditional independence among all the feature vectors assigned with the value of output class [11]. Bayes theorem mathematically states that the given output class y. and the dependent feature x_1. through x_n [11],

$$P(y \mid x_1, \ldots, x_n) = \frac{P(y)P(x_1, \ldots, x_n \mid y)}{P(x_1, \ldots, x_n)} \quad (3)$$

The different variants of Naïve Bayes algorithms differ primarily in the way of assumptions they make in connection with the $P(x_i \mid y)$ distribution. CNB is an adaptation of the standard multinomial Naïve Bayes classifier (MNBC) that is particularly suited for imbalanced data sets. Based on the experimentation, the CNBC algorithm typically performed well for our classification problem than the MNBC algorithm. In this work, the CNBC algorithm makes use of statistics considered from the complement of every output class for calculating the model's weights.

3.4 Soft Voting Technique

The soft-voting technique is a type of voting classifier, and its idea is based on conceptually combining some machine learning classification models and employs the average predicted probabilities (soft voting) for predicting the output targets [12]. Such a classification strategy can be found to be helpful while employing a group of equally well-performing models to compensate for their individual weaknesses. Here, a weight parameter is used to assign specific weights to each classifier model. After assigning the weights, the computed class probabilities for each classification model are considered, multiplied by the assigned weights, and then finally averaged. The final output target is then derived from the output class having the highest value of average probability [12]. In this work, the uniform weights are assigned for the three classifier models.

4 Results and Discussion

As given in Fig. 1, all the steps involved in the work are carried out using MATLAB R2013a. There are so many wavelet families are available, but the paper utilizes the Bi-orthogonal wavelet (bior4.4) based on our previous experimentation done in [13]. The results are finally analysed and compared using the confusion matrix elements. There are four different confusion matrix elements: true positives & true negatives and false positives & false negatives. Based on these four elements, standard measures Sensitivity (Se), Specificity (Sp), Precision (Pr), F1 score (F-Score), Accuracy (Ac), Matthews Correlation Coefficient (MCC) are calculated and compared for finding the best one for the classification problem. The paper employs 64 benign and 51 malignant mammograms, thus a total of 115 ROIs are extracted and considered. These 115 ROIs are further divided into two sets: a training set containing 80 ROIs and a testing set containing 35 ROIs. The confusion matrix values are tabulated in Table 2 and are graphically visualized in Fig. 3.

Table 2. Confusion matrix values

Classifiers with bior4.4 wavelet	TP	FN	TN	FP
Logistic regression classifier	42	9	57	7
Nu-SVC classifier	46	5	58	6
Complementary NB classifier	47	4	60	4
Soft-Voting classifier	49	2	63	1

As shown in Table 2, the obtained number of false classifications (FN-9 and FP-7) is high for logistic regression classifier and the obtained number of true classifications (TP-49 and TN-63) is high for the soft-voting classifier. This indicates that the soft-voting method provides more correct classification than others. But the obtained true and false classification results of LRC, Nu-SVC, and CNBC algorithms are overlapping and appear to be overlaid with each other. That is the reason for this paper to employ the soft-voting technique in order to compensate for the individual weakness of LRC, Nu-SVC, and

CNBC algorithms for the classification of breast cancer. The same overlapping results are clearly illustrated in Fig. 3. Also, from Fig. 3, note that the hand of the soft-voting classifier is high in true classifications (TP & TN) and low in false classifications (FN & FP). For the results obtained using the confusion matrix, the standard performance metrics are computed and illustrated in Table 3 and are graphically visualized in Fig. 4.

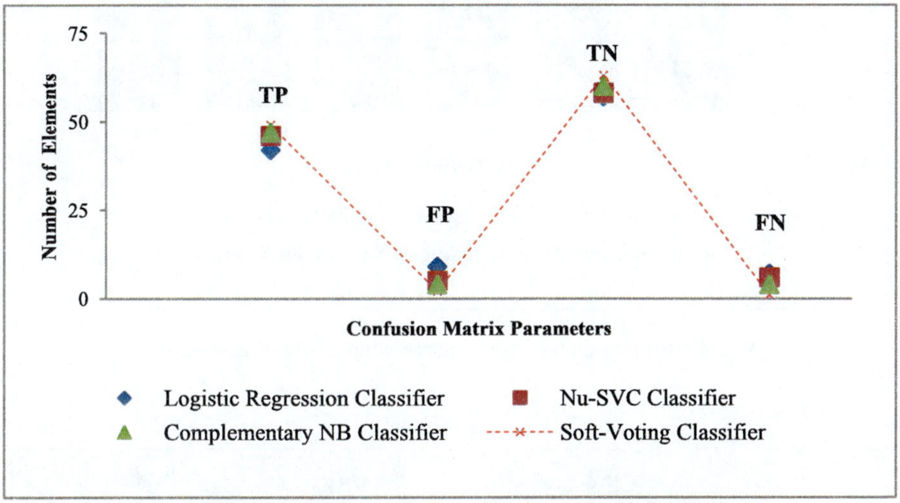

Fig. 3. Visualization of confusion matrix

Table 3. Performance analysis of classifier models

Classification models	Performance measures (%)					
	Se	Sp	Ac	Pr	F-Score	MCC
Logistic regression classifier	82.35	89.06	86.09	85.71	84.00	71.75
Nu-SVC classifier	90.20	90.63	90.43	88.46	89.32	80.67
Complementary NB classifier	92.16	93.75	93.04	92.16	92.16	85.91
Soft-Voting classifier	96.08	98.44	97.39	98	97.03	94.72

Table 3 gives the comparison of the performance of LRC, Nu-SVC, CNBC, and Soft-Voting algorithms used for the classification of breast cancer. For this, six different performance measures are considered and computed for each algorithm. As shown in Table 3, the classification accuracy is obtained as low for the LRC algorithm (86.09%) and high for the soft-voting classifier (97.39%). The lowest performance is obtained for the LRC algorithm with precision (85.71%), F1-Score (84%), and MCC (71.75%). This is due to their more misclassifications obtained for each severity, benign and malignant as shown in Table 2 and Fig. 3. The overall superior performance is obtained for the soft-voting classifier with the precision (98%), F1-Score (97.03%) and MCC (94.72%). Thus, it provides more true classifications rather than false classifications as shown in Table 2.

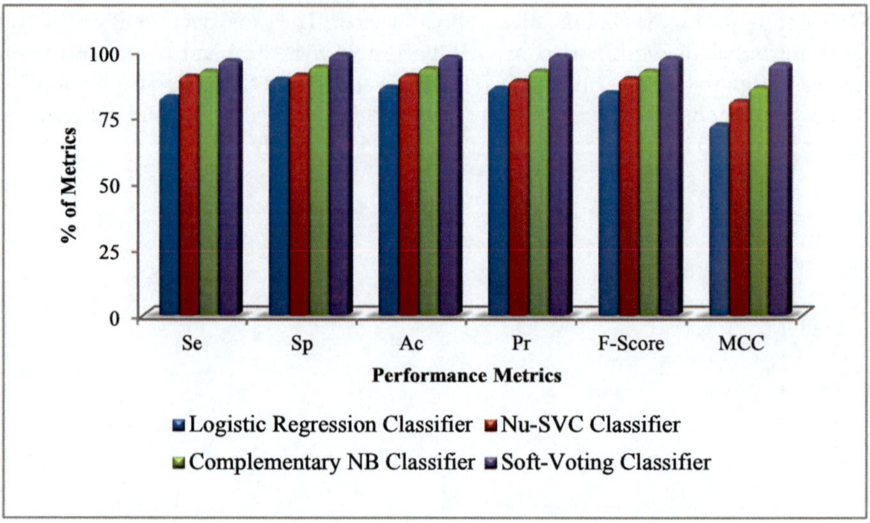

Fig. 4. Visualization of performance comparison of algorithms

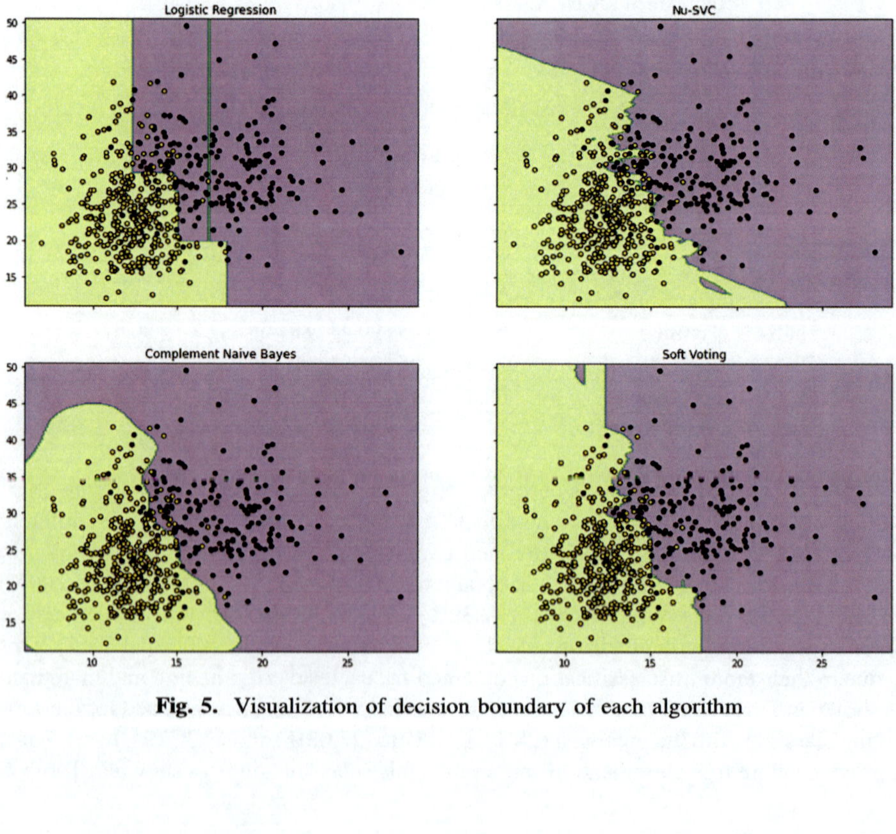

Fig. 5. Visualization of decision boundary of each algorithm

From Fig. 4, the individual classifiers had accuracy rates of (86–93)% on the input data, but employing the soft-voting technique moved up the classification accuracy rate to about 97%. This is not a massive improvement, but worth the reasonably little extra endeavor. This little extra classification improvement attained by the soft-voting classifier can be possible due to the classification computed based on the average probability of the decisions of individual algorithms and this decision boundary of each algorithm is visually plotted in Fig. 5.

5 Conclusion

The work in the paper conferred about the efficient CAD model design for the classification of breast cancer. The input mammograms are taken from the MIAS dataset and they are pre-processed to obtain the ROIs. The features are extracted and selected using a discrete wavelet transform with bior4.4 and discriminant power analysis techniques. The classification performance of individual classifiers, logistic regression, Nu-support vector classifier, and complementary Naïve Bayes is improved by utilizing the soft-voting technique. How the soft-voting technique improves the classification performance is illustrated by visualizing the decision boundary of all algorithms. The decision boundary of the soft-voting classifier is improved with the inputs from the individual LRC (86.09%), Nu-SVC (90.43%), and CNBC (93.04%) algorithms. Thus, the soft-voting technique (97.39%) compensates for the individual weakness of above said classification models. Future work involves the classification of a greater number of mammograms with different datasets and with different pre-processing techniques.

Conflict of Interest. The authors declare that there is no conflict of interest.

References

1. Siegel, R.L., Miller, K.D., Jemal, A.: Cancer statistics, 2019. CA Cancer J. Clin. **69**(1), 7–34 (2019)
2. Marmot, M.G., Altman, D.G., Cameron, D.A., Dewar, J.A., Thompson, S.G., Wilcox, M.: The benefits and harms of breast cancer screening: an independent review. Br. J. Cancer **108** (11), 2205–2240 (2013)
3. Matheus, B.R., Schiabel, H.: Online mammographic images database for development and comparison of CAD schemes. J. Digit. Imag. **24**(3), 500–506 (2011)
4. Zheng, D., Da, F., Kemao, Q., Seah, H.S.: Phase-shifting profilometry combined with Gray-code patterns projection: unwrapping error removal by an adaptive median filter. Opt. Express **25**(5), 4700–4713 (2017)
5. Mapayi, T., Viriri, S., Tapamo, J.R.: Comparative study of retinal vessel segmentation based on global thresholding techniques. Comput. Math. Methods Med. **2015**, 1–16 (2015)
6. Sannasi Chakravarthy, S.R., Rajaguru, H.: Lung cancer detection using probabilistic neural network with modified crow-search algorithm. Asian Pac. J. Cancer Prev. APJCP **20**(7), 2159 (2019)
7. Lai, C.C., Tsai, C.C.: Digital image watermarking using discrete wavelet transform and singular value decomposition. IEEE Trans. Instrum. Meas. **59**(11), 3060–3063 (2010)

8. Ly, N.H., Du, Q., Fowler, J.E.: Collaborative graph-based discriminant analysis for hyperspectral imagery. IEEE J. Sel. Top. Appl. Earth Observations Remote Sens. **7**(6), 2688–2696 (2014)
9. Chen, W., et al.: Novel hybrid artificial intelligence approach of bivariate statistical-methods-based kernel logistic regression classifier for landslide susceptibility modeling. Bull. Eng. Geol. Env. **78**(6), 4397–4419 (2018). https://doi.org/10.1007/s10064-018-1401-8
10. Ketabchi, S., Moosaei, H., Razzaghi, M., Pardalos, P.M.: An improvement on parametric Nusupport vector algorithm for classification. Ann. Oper. Res. **276**(1–2), 155–168 (2019)
11. Seref, B., Bostanci, E.: Sentiment analysis using naive Bayes and complement naive Bayes classifier algorithms on Hadoop framework. In: 2018 2nd International Symposium on Multidisciplinary Studies and Innovative Technologies (ISMSIT), pp. 1–7. IEEE, 19 October 2018
12. Cao, J., Kwong, S., Wang, R., Li, X., Li, K., Kong, X.: Class-specific soft voting based multiple extreme learning machines ensemble. Neurocomputing **3**(149), 275–284 (2015)
13. Sannasi Chakravarthy, S.R., Rajaguru, H.: Detection and classification of microclassification from digital mammograms with firefly algorithm, extreme learning machine and non-linear regression models: a comparison. Int. J. Imag. Syst. Technol. **30**(1), 126–146 (2020)

Detection of Liver Cirrhosis in Ultrasonic Images from GLCM Features and Classifiers

R. Karthikamani(✉) and Harikumar Rajaguru

Bannari Amman Institute of Technology, Sathyamangalam 638 401, Tamil Nadu, India
karthikamani@bitsathy.ac.in

Abstract. In this paper the detection of Liver Cirrhosis from Ultrasonic Images are investigated using Gray Level Co-Occurrence Matrix (GLCM) features and six classifiers. Liver Cirrhosis is the most affected human health problem. A Computer Aided Diagnostic system is proposed to categorize the liver images as Normal or Cirrhosis. Eighty four liver ultrasonic images are collected from standard liver ultrasonic image database and GLCM features are extracted. The liver cirrhosis is detected using Gaussian Mixer Model Classifier (GMM), Firefly Algorithm (FA), Expectation Maximization Algorithm (EM), Principal Component Analysis (PCA), Firefly GMM, and EM-PCA. The performances of classifiers are analyzed by the parameters like, Sensitivity, Specificity, Accuracy, Error Rate, Precision, F1 Score, MCC, Jacard Metric and BCR. The firefly with GMM hybrid classifier attained higher accuracy of 89.28% when compared with all other classifiers.

Keywords: Cirrhosis · Firefly · GMM · EM-PCA · GLCM · MCC

1 Introduction

The liver is considered the most important organ in the human body. The main purpose of a liver is to filter the blood flow from the digestive path. We have to give more attention to diagnose the liver diseases to live long life [1]. The data given by the world health organization, 27.7% of expiries in India is due to liver associated problems. Normally the liver diseases can be clustered into two main classes, the first one is focal diseases and the second one is disperse liver diseases. Focal liver diseases are cyst or abnormalities determined within a minor area, but the diffused liver diseases are blowout through an area or the whole liver. Diagnosis is very important for deadly diseases like liver fibrosis, cirrhosis, etc. So it is very important to detect such diseases at an earlier stage. Medical image processing acts as an important role in these applications [2]. The most commonly used techniques to diagnose the abnormalities in a human body are Computer Tomography (CT), Magnetic Resonance Imaging (MRI), X-ray, and Ultrasound. Ultrasonography is the widely used technique to detect the abnormalities in the liver because it is a non-invasive, cost-effective, generally available and non-radioactive in nature [3]. The organization of the paper is as follows subdivision 1 introduces the paper. The Sect. 2 explains the materials and methods. Section 3

S. Vlad and N. M. Roman (Eds.): MEDITECH 2020, IFMBE Proceedings 88, pp. 161–172, 2022.
https://doi.org/10.1007/978-3-030-93564-1_18

discusses about the classifiers. The results are explored in the Sect. 4. Section 5 concludes this article.

2 Materials and Methods

The overall process of this work contains Image Acquisition, Preprocessing, Extraction of features and classification. In preprocessing the filters are employed to remove the noises. GLCM features are extracted and classified using Six Classifiers. The classifier performance are analyzed and compared through the bench mark parameters.

2.1 Block Diagram

Figure 1 shows the block diagram of the proposed system to detect the Cirrhosis in the liver images. In this paper, a set of 84 liver ultrasonic images are collected from the SP lab Database for classification. In that 42 images are normal and other 42 images are Cirrhosis.

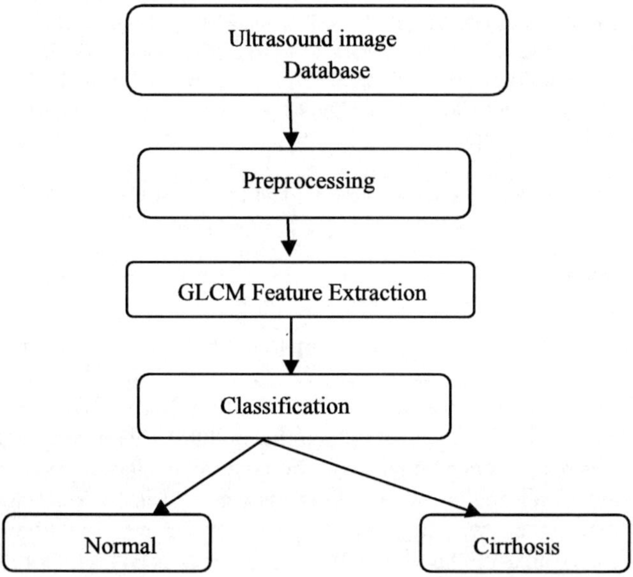

Fig. 1. Block diagram model for proposed system

2.2 Preprocessing

The preprocessing is required for classifying the images, because the effectiveness of ultrasonic imaging is reduced by the occurrence of unwanted signals called speckle noise. Due to this noise the significant information of the images like edge, shape, intensity values are degraded and the diagnostic value of the image modality is also affected. Ahmed S. Bafaraj [4] stated that the median filter is highly effective for noise removal and it preserves image edges and corners. Therefore, 12th order Median Filter of 5 * 5 window size is employed to remove the noise in this work [5].

2.3 Feature Extraction Using GLCM

Texture is one of the significant parameter used for detecting the objects or ROI in an image [6]. In 1970 Haralick and Shanmugam proposed this technique to extract the texture data from an image. The GLCM is useful to expose the texture features of an ultrasonic image. The co-occurrences of the pixels with the values m, n is calculated using the co-occurrence matrix $G(m,n)$ [7]. In this study Twenty Two features are extracted to describe the characteristics. These features and their equations are given in Table 1.

Table 1. GLCM features and its mathematical description

S. No	Features	Mathematical description	Description
1	Autocorrelation	$\sum_{m,n=0}^{N-1} (m.n)P_{m,n}$	Measure of linear dependency of a variable between two points in time
2	Contrast	$\sum_{m,n=0}^{N-1} P_{m,n}(m-n)^2$	Measure the local intensity variation
3	Correlation (corrm)	$\sum_{m,n=0}^{N-1} P_{m,n}\left[\frac{(m-\mu_m)(n-\mu_n)}{\sqrt{(\sigma_1)^2(\sigma_2)^2}}\right]$	The linear dependencies of the gray level values are calculated based on neighboring pixels
4	Correlation (corrp)	$\sum_{m,n} \frac{(m,n)P(m,n)-\mu_x\mu_y}{\sigma_x\sigma_y}$	
5	Cluster prominence	$\sum_{m,n=0}^{N-1} P_{m,n}(m+n-\mu_m-\mu_n)^4$	Measure the skewness
6	Cluster shade	$\sum_{m,n=0}^{N-1} P_{m,n}(m+n-\mu_m-\mu_n)^3$	Measure the skewness and uniformity
7	Dissimilarity	$\sum_{m,n=0}^{N-1} P_{m,n}\lvert m-n\rvert$	Measure comparisons between segmentations created by different algorithms
8	Energy	$\sum_{m,n=0}^{N-1} P^2_{m,n}$	Sum of squared elements
9	Entropy	$\sum_{m,n=0}^{N-1} P_{m,n}(-ln\ P_{m,n})$	Measures the statistical randomness
10	Homogeneity (homop)	$\sum_{m,n=0}^{N-1} \frac{P_{m,n}}{1+(m-n)^2}$	Closeness of distribution in diagonal
11	Homogeneity (homom)	$\sum_{m,n=0}^{N-1} \frac{P_{m,n}}{1+\lvert m-n\rvert}$	
12	Maximum probability	$\max_{m,n} P_{m,n}$	Occurrences of the most predominant pair of neighboring intensity values

(continued)

Table 1. (*continued*)

S. No	Features	Mathematical description	Description
13	Sum of squares: Variance [1]	$\sum_{m,n=0}^{N-1} (m - \mu)^2 P_{m,n}$	Measure of variation from the gray level intensities
14	Sum average	$\mu_m = \sum_{m,n=0}^{N-1} m\left(P_{m,n}\right),\ \mu_n = \sum_{m,n=0}^{N-1} n\left(P_{m,n}\right)$	Average of a set of two/more values
15	Sum variance	$\sigma_m^2 = \sum_{m,n=0}^{N-1} P_{m,n}(m - \mu_m)^2,\ \sigma_n{}^2 = \sum_{m,n=0}^{N-1} P_{m,n}(n - \mu_n)^2$	Measure of dispersion of a set of data calculated as the square root of variance by determining the difference between each pixel
16	Sum entropy	$-\sum_{m=2}^{2N} P_{x+y}(m)\ \log\left(P_{x+y}(m)\right)$	sum of neighborhood intensity value differences
17	Difference variance	$\sum_{m=2}^{2N}\left(m - \left[\sum_{m=2}^{2N} m\, P_{x-y}(m)\right]\right)^2$	Measure of heterogeneity that places higher weights ondiffering intensity level pairs that deviate more from the mean
18	Difference entropy	$-\sum_{m=2}^{2N} P_{x-y}(m)\ \log\left(P_{x-y}(m)\right)$	Measure of the randomness/variability
19	Information measure of correlation1	$\frac{H_{AB} - H_{AB1}}{Max\{H_A, H_B\}}$	Correlation between the probability distributions of 'i' and 'j' using mutual information I(x, y):
20	Information measure of correlation2	$\sqrt{1 - \exp[-2.0(H_{AB2} - H_{AB})]}$	Correlation between the probability distributions
21	Inverse difference normalized (INN)	$\sum_{m,n=1}^{G} P_{mn} \frac{C_{mn}}{1 + \lvert m - n\rvert^2 / G^2} \quad C_{m,n} = \frac{P_{mn}}{\sum_{m,n=1}^{G} P_{mn}}$	Measure of the local homogeneity
22	Inverse difference moment normalized	$\sum_{m,n\neq 1}^{G} P_{m,n} \frac{C_{m,n}}{1 + (m-n)^2 / G^2}$	Measures image homogeneity

The extracted GLCM features characteristics are analyzed in this section through the scatter plot and statistical parameters. Figure 2 shows the scatter plot GLCM features for Normal and Cirrhosis Cases for Liver Disease.

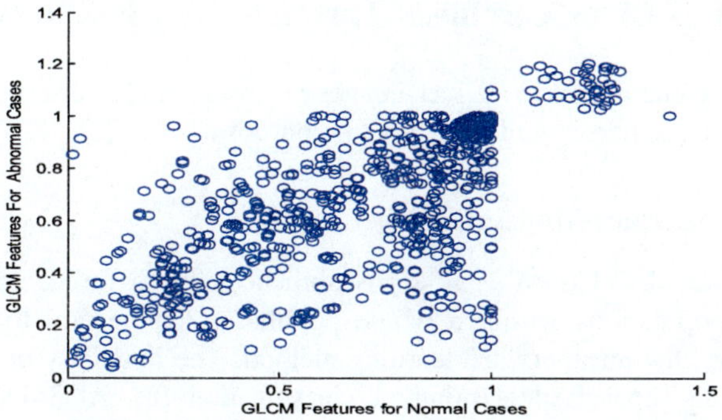

Fig. 2. Scatter plot for GLCM features for normal and abnormal cases for liver disease

As shown in the Fig. 2 that the overlapping nature of GLCM features among the two classes is indicating the inseparable quality of the class features.

Table 2 demonstrates the analysis of statistical features in the extracted GLCM features for liver images. As shown in the Table 2 that all the parameters are closely followed in the normal and cirrhosis classes. The parameter Higuchi Fractal dimension value indicates that the features are anti-persistence in the nature. The Canonical Correlation Analysis (CCA) exhibits the non-linearity and closely correlated of the features among the classes. Before classification the GLCM features are normalized using the formula

$$X_N = \frac{X - X_{\min}}{X_{\max} - X_{\min}} \tag{1}$$

Where, X_N is the normalized data and its range is between 0 and 1. The Target value for normal image is 0.1 and abnormal image 0.85.

Table 2. Statistical parameters of the GLCM features

S. No	Statistical parameters	GLCM features	
		Normal	Liver cirrhosis
1	Mean	0.721373	0.688417
2	Variance	0.078306	0.063751
3	Skewness	−0.15922	0.039307
4	Kurtosis	0.026416	−1.46914
5	Geometric mean	0.647788	0.64091
6	Harmonic mean	0.564972	0.592667
7	Permutation entropy	1.6638	1.6142
8	Sample entropy	4.2276	4.3664
9	Higuchi fractal dimension	1.9865	1.9703
10	Canonical correlation analysis (CCA)	0.641045	

3 Detection of Liver Cirrhosis Through the Classifiers

In this section the classification of liver images as normal and liver cirrhosis affected is carried out. Six classifiers are utilized for this purpose.

3.1 Gaussian Mixture Model (GMM)

Gaussian Mixture Model (GMM) is a probabilistic method for demonstrating usual distributed subpopulations within a whole population. In general to learn the sub-population it uses the unsupervised learning method. The boundary of subpopulations are learned using supervised learning or classification [8]. A GMM is a class of probabilistic model which defines that all bred data models are derived from mixture of finite Gaussian densities. A Gaussian density in a dimensional space is stated as

$$N(x/\mu, \Sigma) = \frac{1}{(2\pi)^{1/2} (|\Sigma|)^{1/2}} \exp\left(-\frac{1}{2}(x-\mu)^T \Sigma^{-1} (x-\mu)\right) \tag{2}$$

Where a d-component feature vector is x, the d-component vector comprising the mean of each feature is μ, the d-by-d covariance matrix is Σ.

3.2 Firefly Algorithm

Firefly algorithm is one of the bio – inspired algorithm used to solve the optimization problem in various fields. This procedure was developed on the idealized and combined behavior of the flashing features of fireflies [9].

1. All the fireflies are attract each other irrespective of the gender.
2. Attractiveness of the fireflies are based on their brightness of the fireflies.
3. The illumination is based on the objective function.

To enhance the optimization problem, the light Intensity I of a firefly at a particular distance x can be selected as

$$I(x) = f(x) \tag{3}$$

The brightness of the firefly I(r) differs with distance r monotonically and exponentially.

$$I = I_o e^{-\gamma r} \tag{4}$$

Where I_o is the brightness and γ is the light absorption coefficient. The attractiveness of the fireflies seen by adjacent firefly is given as

$$\beta = \beta_O e^{-\gamma r^2} \tag{5}$$

Where β_o represents the light intensity of firefly. The equation of 'r' is given by the formula

$$r = \|x_i - x_j\| = \sqrt{\sum_{k=1}^{d} (x_i^k - x_j^k)^2} \quad (6)$$

Where i & j represents the location of fireflies. A firefly 'i' move towards another brighter firefly 'j' and it is calculated using the formula

$$x_i = x_i + \beta_0 e^{-\gamma r_{ij}^2}(x_j - x_i) + \alpha \in_i \quad (7)$$

Where the x_i denotes the current location of firefly, $\beta_0 e^{-\gamma r_{ij}^2}(x_j - x_i)$ represents the firefly's attraction and $\alpha \in_i$ denotes the randomization with random variable $\in_i$. During the implementation of firefly algorithm the value of $\beta 0 = 1$, $\alpha \in [0, 1]$ and the variation of attractiveness $\gamma = 1$.

3.3 Expectation Maximization

The Expectation Maximization algorithm is used for calculating certain unfamiliar parameters θ, given a measurement data set D. To estimate the parameters of statistical model maximum likelihood of EM is used [10]. Consider that a set of X observed data is generated by statistical model, a set of mislaid values Z and the unidentified parameter θ, along with a likelihood function $L(\theta, X, Z) = p(X, Z|\theta)$, the maximum likelihood estimate (MLE) of the unidentified parameters is found by maximizing the marginal likelihood of the observed data.

$$L(\theta; X) = p(X|\,\theta) = \int p(X, Z|\theta)\, dZ \quad (8)$$

EM method consists of 2 main steps.
Step I-Expectation E-Step:

$$G = (H^T H)^{-1} H^T J \quad (9)$$

Step II-Maximization M-Step:

$$H^{new} = JG^T\,(GG^T)^{-1} \quad (10)$$

3.4 Principal Component Analysis (PCA)

A general statistical procedure used for data investigation and preprocessing is Principal Component Analysis (PCA). PCA is intended to convert the data in a compact system and retain maximum of the unique variance existing in the primary data. In scientific terms PCA is used to convert 'n' correlated variables into a d (d << n) uncorrelated variables named the principal components (PCs) [11].

The resulting SVD of matrix X is stated as

$$X = P\Delta Q^T \tag{11}$$

Where left singular vectors are $I = P \times L$ matrix, right singular vectors are $Q = J \times L$ and the diagonal matrix of singular values is Δ. In PCA the maximum significant components are attained by SVD [12].

3.5 EM PCA

It is a linear measurement reduction method and is measures as the finest in the mean square error. It is 2nd order technique depends on the covariance matrix of the particular variables utilized here. PCA discovers to diminish the data measurement by training a few linear groupings of the original variables which are orthogonal [13]. The first PC S_1 is stated as

$$S_1 = x^T W_1 \tag{12}$$

Where the p-dimensional coefficient vectors $w_1 = (w_{1,1}, \ldots..w_{1,p})^T$, Where w_1 is derived as

$$w_1 = \arg\max_{\|w\|} Var\{x^T w\} \tag{13}$$

The second principal component is the linear combination with the second highest variance and orthogonal to the previous principal component. If a consistent data is supposed with the experimental covariance matrix

$$\sum\nolimits_{pxp} = \frac{1}{n} XX^T, \quad \sum = U\Lambda U^T \tag{14}$$

Where $\Lambda = \text{diag}(\lambda 1, \ldots, \lambda p)$-Diagonal Matrix and $U = pxp$-Orthogonal Matrix. The Principal Component $'p'$ rows of the $'p \times n'$ matrix S is given as

$$S = U^T X \tag{15}$$

PCA yet has some limitations. With huge number of data sets, training the principal components is moderately tough, so EM algorithm is used in co-ordination with PCA. The classifiers are trained by 70% of the images and 30% of images are used for testing purpose. The Mean Square Error is the stopping criteria for the training of the classifiers.

The Mean Square Error (MSE) is represented as follows:

$$MSE = \frac{1}{N}\sum_{i=1}^{N} (O_i - T_j)^2 \tag{16}$$

Where O_i is the observed value at time, T_j is the target value at model $j; j = 1$ to 84, and N is the total number of images in our case, it is 84.

According to the binary classification, the prediction to a particular class can fall under one of four outcomes. Exact prediction that the instance belongs to positive class or negative class called as True Positive (TP) and True Negative (TN) respectively. The misclassifications are called False Positive (FP) and False Negative (FN) when the positive class member is mispredicted as negative class and vice versa for the other case.

Table 3 indicates the classifier outputs for GLCM Features based on the MSE values. All the classifiers achieved good MSE values except for the PCA Classifier.

Table 3. Classifiers outputs for GLCM features

S.No	Classifiers	TP	TN	FN	FP	MSE
1	GMM	36	36	6	6	5.3E-06
2	Firefly	29	35	13	7	2.31E-05
3	EM	24	33	18	9	6.23E-05
4	PCA	34	31	8	11	1.85E-05
5	Firefly GMM	38	37	4	5	2.75E-06
6	EM PCA	35	35	7	7	7.84E-06

4 Results and Discussion

The results are discussed in the following section of the paper.

4.1 Performance Parameters of Classifier

The various parameters considered while evaluating the performance of classifier like Sensitivity, Specificity, Accuracy, Error Rate, Precision, F1 Score, MCC, Jacard Metric and BCR. Accuracy being a primitive metric of measure of any classifier, the number of occurrences of exactprediction of class determines the percentage of accuracy of classifier. Hence on calculating the sum of the prediction of instances, the following equation is applied to calculate the accuracy of the classifier,

$$Accuracy = \frac{TP + TN}{TP + TN + FP + FN} \times 100\% \tag{17}$$

Comprehensive amount of wrong classification is Error Rate,

$$Error\ Rate = \frac{\mathrm{FN} + \mathrm{FP}}{\mathrm{TN} + \mathrm{TP} + \mathrm{FN} + \mathrm{FP}} \times 100\% \tag{18}$$

The Sensitivity and Specificity of the classifier is defined as,

$$Sensitivity = \frac{TP}{\mathrm{TP} + \mathrm{FN}} \times 100\% \tag{19}$$

$$Specificity = \frac{TN}{\mathrm{TN} + \mathrm{FP}} \times 100\% \tag{20}$$

$$Precision = \frac{TP}{\mathrm{TP} + \mathrm{FP}} \times 100\% \tag{21}$$

$$\mathrm{F1\ score} = \frac{2TP}{2\mathrm{TP} + \mathrm{FP} + \mathrm{FN}} \times 100\% \tag{22}$$

The correlation between observed and predicted classification is given by Mathews Correlation Coefficient (MCC),

$$\mathrm{MCC} = \frac{TP \times TN - FP \times FN}{\sqrt{(TP + FP)(TP + FN)(TN + FP)(TN + FN)}} \times 100\% \tag{23}$$

Jacard Metric (JM) explicitly ignores the perfect classification rate of true samples,

$$JM = \frac{TP}{TP + FN + FP} \times 100\% \tag{24}$$

Balanced Classification Rate (BCR) will be very much useful in imbalanced datasets and it will be given as,

$$BCR = \frac{1}{2}\left(\frac{TP}{TP + FN} + \frac{TN}{TN + FP}\right) \times 100\% \tag{25}$$

Table 4. Performance analysis of classifiers for GLCM features

S. No	Parameters (%)	Classifiers					
		GMM	Firefly	EM	PCA	Firefly GMM	EM PCA
1	Sensitivity	85.71	69.05	57.14	80.95	90.48	83.33
2	Specificity	85.71	83.33	78.57	73.81	88.10	83.33
3	Accuracy	**85.71**	76.19	67.86	77.38	**89.29**	83.33
4	Error rate	14.29	23.81	32.14	22.62	10.71	16.67
5	Precision	85.71	80.56	72.73	75.56	88.37	83.33
6	F1 score	85.71	74.36	64.00	78.16	89.41	83.33
7	Jacard metric	75.00	59.18	47.06	64.15	80.85	71.43
8	Balanced classifier rate	85.71	76.19	67.86	77.38	89.29	83.33
9	MCC	**0.71**	0.53	0.37	0.55	**0.79**	0.67

Table 4 denotes the performances of classifiers based on the bench mark parameters. The hybrid classifier Firefly GMM is achieved higher accuracy of 89.28% with MCC of 0.785 and low error rate of 10.71%. Then GMM classifier comes next in the performance with accuracy of 85.71% and MCC of 0.71428 with error rate of 14.28%. EM classifier is ebbed at low accuracy of 67.85%. However the Hybrid classifier EM PCA improves the accuracy of the classifier towards 83.33%.

5 Conclusion

Herein, we describe the comparative analysis of a general CAD framework for the classification of Cirrhosis and Normal Liver Images. The evaluation is performed using Ultrasonic Liver images from Signal Processing Database. Six classifiers like GMM, FA, EM, PCA, EM-PCA, Firefly-GMM classifiers are used in this analysis. The extracted GLCM features are inputted to the classifiers. The classifier performance is analyzed, and compared, which in turn discovered that the Firefly- GMM hybrid classifier attains an average accuracy of 89.28%, where as the average accuracy of GMM,FA, EM,PCA, EM PCA are 85.71%, 76.19%, 67.85%, 77.38%, 83.33% respectively. The firefly GMM classifier has improved performance over other classifiers. Future research will be in the direction of detecting abdominal abnormalities with different ultrasonic database.

Conflict of Interest. The authors declare that they have no conflict of interest.

References

1. Alivar, A., Daniali, H., Helfroush, M.S.: Classification of Liver Diseases Using Ultrasound Images Based on Feature Combination. IEEE Proceedings (2014)
2. Randhawa, S.R., Sunkaria, R.K., Bedi, A.K.: Prediction of Liver Cirrhosis Using Weighted Fisher Discriminant Ratio Algorithm. IEEE Proceedings (2018)
3. Xu, S.S-D., Chang, C.-C., Su, C.-T., Phu, P.Q.: Classification of liver diseases based on ultrasound image texture features. Appl. Sci. **9**(2), 342 (2019)
4. Bafaraj, A.S.: Performance analysis of best speckle filters for noise reduction in ultrasound medical images. Int. J. Appl. Eng. Res. **14**(6), 1340. ISSN 0973-4562 (2019)
5. Mahmoud, A.A., Rabaie, S.E.L., Taha, T.E., Zahran, O.: Comparative Study between Different Denoising Filters for Speckle Noise Reduction in Ultrasonic BMode Images. IEEE Proceedings, November 2012
6. Haralick, R.M., Shanmugam, K.: Textural features for image classification. IEEE Trans. Syst. Man Cybern. **3**(6), 610–621 (1973)
7. Park, B., Jang, W.S., Yoo, S.K.: Texture analysis of supraspinatus ultrasound image for computer aided diagnostic system. Healthc Inform. Res. **22**(4), 299–304 (2016)
8. Wan, H., Wang, H., Scotney, B., Liu, J.: A novel Gaussian mixture model for classification. In: 2019 IEEE International Conference on Systems, Man and Cybernetics (SMC), Bari, Italy, 6–9 October 2019
9. Sannasi Chakravarthy, S.R., Rajaguru, H.: Detection and classification of microcalcification from digital mammograms with firefly algorithm, extreme learning machine and nonlinear regression models: a comparison. Int. J. Imag. Syst. Technol. **30**, 126–146 (2020)

10. Arnaout, A., Esmael, B., Fruhwirth, R.K., Thonhauser, G.: Automatic threshold tracking of sensor data using Expectation Maximization algorithm. In: 11th International Conference on Hybrid Intelligent Systems (HIS), 5–8 December 2011
11. Hadri, A., Chougdali, K., Touahni, R.: Intrusion detection system using PCA and fuzzy PCA techniques. In: International Conference on Advanced Communication Systems and Information Security (ACOSIS), 09 February 2017
12. Prabhakar, S.K., Rajaguru, H.: PCA and K-Means clustering for classification of epilepsy risk levels from EEG signals – A comparitive study between them. In: ICIIBMS 2015, Track1: Signal Processing, Computer Networks and Telecommunications (2015)
13. Rajaguru, H., Prabhakar, S.K.: Multilayer Autoencoders and EM-PCA with genetic algorithm for epilepsy classification from EEG. In: Proceedings of the 2nd International conference on Electronics, Communication and Aerospace Technology (ICECA 2018), IEEE Xplore. ISBN 978-1-5386-0965-1 (2018)

Telemedicine and Health Care Information Systems

Audio Events Detection to Help TIAGo to Act as a Medical Robot

Lorena Muscar and Lacrimioara Grama(✉)

Signal Processing Group, Faculty of Electronics,
Telecommunications and Information Technology,
Technical University of Cluj-Napoca, Cluj-Napoca, Romania
{Lorena.Muscar,Lacrimioara.Grama}@bel.utcluj.ro

Abstract. Nowadays service robots are used extensively in and out hospitals to improve the level of patient care. In this paper we present how Moving Picture Experts Group-7 features together with the Hidden Markov Models can aid the TIAGo service robot to understand the context through audio signals, in such a way that can be used as a medical robot. Due to the help of medical robots, the person-to-person contact can be reduced, thus the physicians and medical staff workload can be narrow down, also improving the healthcare facilities. Through experimental results we prove that, in the test phase, the overall classification accuracy is 96.86%, at least for the given framework.

Keywords: TIAGo service robot · MPEG-7 features · HMM · Isolated audio events detection

1 Introduction

Nowadays more and more service robots are used as medical robots, in and/or out hospitals, in order to improve the overall level of patient care. The International Federation of Robotics stated that the request for medical robots will increase in the next years; there will be spent about $9 billion until 2022 [1]. Recently, the role of medical robots in global COVID-19 management was the subject of [2]. An investigation from Juniper Research revealed that in 2024 more than 74 million robots will be acquired, far from an estimation of 28 million in 2019 [3].

Due to the help of medical robots, the person-to-person contact can be reduced, thus the physicians and medical staff workload can be narrow down, also improving the healthcare facilities. Most of the robots that are used in the hospitals are alternatives of the service robots with a limited degree of freedom and with a high payload capacity [2]. The field of service robots is expended from robots which assist with learning language [4], to robots who assist elderly with dementia [5], or children with cognitive impairment [6].

Such a robot, that can be use as a medical one, is the TIAGo service robot from Pal Robotics [7]. TIAGo can carry out tasks which come as audio signals. Based on the received audio commands from patients or medical staff, TIAGo could perform certain tasks, such as: providing medicines to patients, warning medical staff if the patient is at risk (for example, has tripped and fell). Also, based on the audio information, TIAGo

S. Vlad and N. M. Roman (Eds.): MEDITECH 2020, IFMBE Proceedings 88, pp. 175–182, 2022.
https://doi.org/10.1007/978-3-030-93564-1_19

may signal to the medical staff that something inappropriate has occurred in the patient's condition (for example, when the patient coughs or sneezes).

The paper is organized as follows. In Sect. 2 we present the actual framework and some theoretical fundamentals. The results obtained experimentally are the subject of Sect. 3. Finally, we conclude the research in Sect. 4 and present some future developments.

2 Framework and Theoretical Fundamentals

Previously, we have tried to improve the audio capabilities of the TIAGo service robot. We wanted TIAGo to be capable to classify isolated audio signals which correspond to some indoor events [8, 9]. An improvement of the orinal proposed signal acquisition and processing diagram for the above mentioned service robot is presented in [10]. In [8] three types of features were used in the extraction phase for each audio signal: Mel Frequency Cepstral Coefficients, Linear Predictive Coding Coefficients, and Linear Predictive Cepstral Coefficients. In [7] only the Mel Frequency Cepstral Coefficients features were used.

Through this paper we will show how Moving Picture Experts Group-7 (MPEG-7) features can help TIAGo to understand audio commands. The isolated audio events that TIAGo is able to recognize should be consistent to the audio signals available in the database [8, 10].

Actually, the database consists in 46 classes. In each class there are 30 audio files (which means a total of 1380 isolated audio signals). The classes correspond to five different scenarios: appliances (5 classes – toaster alarm, microwave open, microwave close, microwave alarm, washing machine), room (8 classes – door knock, door key, door open, door close, page turn, Velcro, zip open, zip close), kitchen (8 classes – tap water, drop water, shower water, porcelain dish, cutlery, plastic-bag-rush, cardboard-drop, chair), female voice (20 classes – numbers from one to ten, hello, medicines, yes, no, right, left, stay, come, go, TIAGo) and non-verbal (5 classes – hand-clap, finger-clap, cough, laugh, whistle). The words from the voice classes are for Romanian language. For the classification phase we shall use the Hidden Markov Models (HMM).

2.1 Feature Extraction: Moving Picture Experts Group-7

The low level audio descriptors of MPEG-7 are of general importance for the representation of audio signals. In applications 17 descriptors can be used (spectral and temporal), but for this research, we have used only three of them: the Audio Spectrum Envelope (ASE), the Audio Spectrum Basis (ASB) and Audio Spectrum Projection (ASP).

The power spectrum of logarithmic frequencies which can be used for generating the original audio signal, but with reduced spectrogram, is called ASE. This is achieved by adding the energy of the initial power spectrum within a frequency band series. The frequency bands are distributed logarithmically (base 2) between the lower and the higher frequency edge [11]. To extract ASE a sliding window was implemented (Fast Fourier Transform (FFT) analysis; it was used resampling to logarithmic space bands).

After computing the window length, the FFT size is determined. Then, the Short-Time Fourier Transform (STFT) is performed, using a Hamming window of the same length. Finally, the squared magnitude of the FFT coefficients is computed to obtain the power spectrum coefficients.

The ASB descriptor is the reference container function used to transfer a spectrum to a lower-dimensional sub-space suitable for model classifiers probability. For each sound class, a basis is calculated which captures the statistically most regular sound features. The foundational functions are kept in a column of a matrix with row numbers corresponding to the length of the vector spectrum and the column numbers represents the number of basis functions. Using singular value decomposition (SVD) a basis is extracted. In many software packages it can be found as a built-in function.

The MPEG-7 specification adopted a generic structure for sound processing. The reduced dimension, decorrelated log-spectral features (ASP), are mostly used for training Hidden Markov Model (HMM) for specific audio signal classification [12]. The system's extraction feature is based on low-dimensional subspace spectrum projection through reduced spectral base functions, the ASB. To perform extraction for ASP, the spectrum matrix must be multiplied with the basis vector obtained. The output will be formed of the maximum value of L2-norm envelope, which is used for training of HMM and the matrix of the projection.

2.2 Classification: Hidden Markov Model

An HMM is a statistical tool which is frequently used in classification of patterns. HMM-based applications that are very popular include speech recognition, speaker verification and recognition of handwriting. They are used to model processes with characteristics that very in time [13].

The main idea, which was also used in this paper, is to train, for each audio class, the statistical model on the basis projection features. MPEG-7 handles HMM; it consist in several states. In the training phase, by evaluating the training sets of features, the parameters are calculated for each state of an audio model. The aim is to acquire a knowledge to recognize the audio classes delineated in the database.

Each state represents a part of similar behavior of a recognizable symbol sequence process. At each instant, for each sequence, the recognizable symbol stays at the same state or moves to another state; this depends on a set of probabilities for the state transition. Using the Baum-Welch algorithm one can obtain the HMM parameters. For each class of audio events one HMM is calculated; from the statistical point of view it encapsulates the most typical characteristics of the signal feature space.

3 Results Obtained Experimentally

In order to reproduce a real-life scenario, the data from each audio events classes was split into training data (70% - 21 audio events from each class) and test data (30% - 9 audio events from each class). All the steps were carried out in MATLAB 2020a environment.

For each class, the basis functions and the HMM were stored in a file, that was later loaded, to test the theories. After that, the Sound Model State Path presents the audio segments selected to every sound model. At the beginning, the code extracts and compute the ASE, and later project it in the lower dimension feature; for this the basis function of the model was used.

The Viterbi path and log likelihood of the sequence and instantaneous observation probabilities are calculated based on the Viterbi decoder. The sound in test is introduced to all 46 version; the HMM with the largest likelihood value is chosen as the emblematic class for the tested audio event.

We shall exemplify the extracted features for two types of audio signals that can be present for medical robots: "medicines" and "cough". When the medical staff pronounces the word "medicine", TIAGo should know that is the time to give medicines to the patient/patients. On the other hand, when the robot hears that the patient coughs, he should alarm the medical staff that something inappropriate has occurred in the patient's condition.

In Fig. 1, on the left part, we have the ASE of "medicines" sound and in the right one, the one of a sound that correspond to class "cough". We can observe that the "medicines" audio signal presents less energy at the higher frequencies and more energy at the lower frequencies. The "cough" has a different distribution of energy level and has a maximum value of 0.7 which is much greater than 0.16, the one of "medicines" sound.

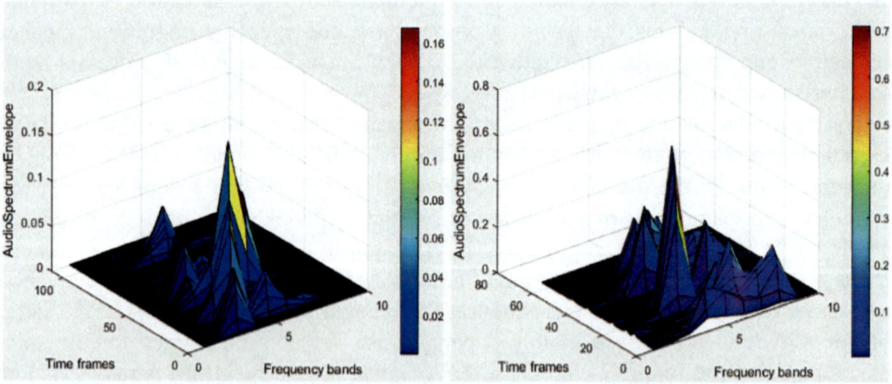

Fig. 1. Audio spectrum envelope of "medicines" (left) and "cough" (right).

In Fig. 2 we have the ASB for 10 basis functions extracted on each sound. We can observe that the basis are concatenated and a higher spectrum energy at the beginning of the sounds is present.

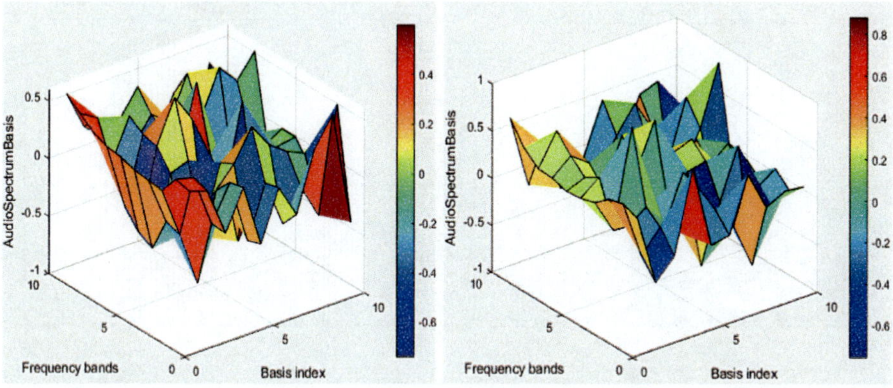

Fig. 2. Audio spectrum basis of "medicines" (left) and "cough" (right).

After we apply the steps presented in Sect. 2.1, and we obtain the projection matrix, we can say that our aim for minimizing data dimensionality while maintaining full information is achieved. Also, in Fig. 3 we observe the difference between the two sound signals that belongs to different classes.

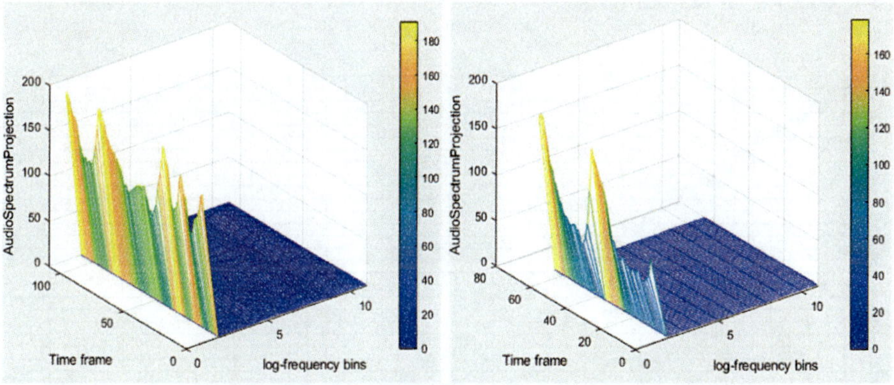

Fig. 3. Audio spectrum projection of "medicines" (left) and "cough" (right).

The audio signals are illustrated as paths in the feature space (see Fig. 4). Based on the noticeable evidence, the states are chosen to maximize the probability of the model. In the features space of the audio event the states are clusters.

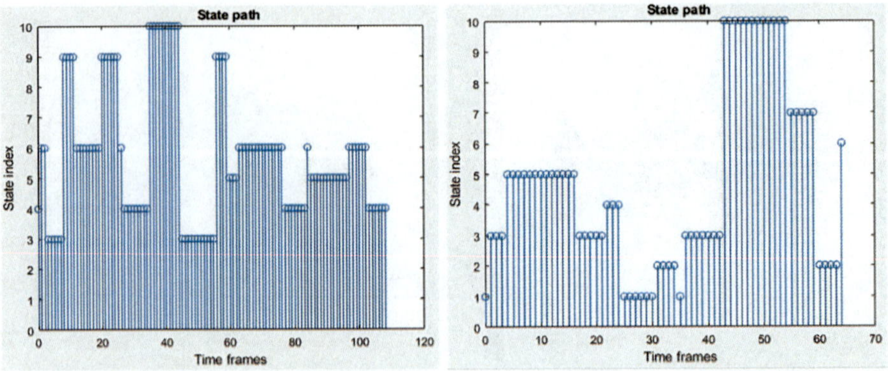

Fig. 4. Sound model state path of "medicines" (left) and "cough" (right).

The confusion matrix contains information on the actual and anticipated audio events. Figure 5 illustrates the confusion matrix for all classes, in the test phase. It presents the optimal case. The entries in the confusion matrix in the context of our study are all the sounds on the test list (9 sounds/class $\times$ 46 classes). At the output, we obtained the classified sounds. It is a matrix with 46 columns and 46 rows. We obtained 37 classes perfectly classified out of 46 classes, meaning that the overall classification rate is 96.86%, in the test phase.

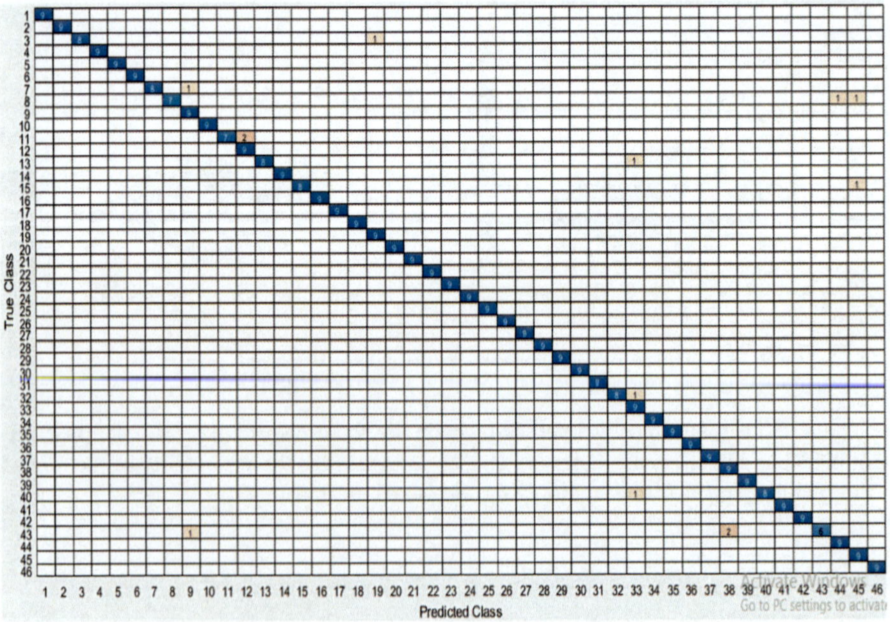

Fig. 5. Confusion matrix – test phase – 96.86% accuracy.

The misclassified audio signals are:

- One "drop_water" signal is misclassified as "microwave_close"
- One "plastic bag rush" signal is misclassified as "page turn"
- Two "cardboard drop" signals are misclassified as "cough" and as "laugh"
- Two "zip open" signals are misclassified as "zip close"
- One "door knock" signal is misclassified as "medicines"
- One "door open" signal is misclassified as "laugh"
- One "hello" signal is misclassified as "medicines"
- One "go" signal is misclassified as "medicines"
- One "finger clap" signal is misclassified as "page turn" and two "finger clap" signals are misclassified as "stop".

4 Conclusions and Future Development

Through this research paper we suggested an audio signal classification system for environmental sound events, which can be used for a service robot to get a clear understanding of the context, in order to help medical staff and patients in hospitals. The robot should be mindful of the indoor environmental quality. The sound groups selected for assessment pertain to indoor audio events. Based on the audio signals in the database we have demonstrated that TIAGo can decipher the environment context by the aid of MPEG-7 features. For the real-life scenario we have proposed in this paper, we have attained a correct classification rate of 96.86%, in the test phase.

In a future development we shall extend the database, such to be more reliable for medical robots. As all voice sounds are recorded with a female voice, more classes with a male voice could be implemented. A new scenario could be recorded containing the names of medicines with similar names. Because in a field hospital specially built for Covid-19 there are many more patients thus noises appear much easier than in a regular hospital, noise can be added to all sounds to see if TIAGo will be able to distinguish and process the basic sound. Another ability that TIAGo might have would be to analyze signals followed by prioritizing them according to their severity. This can be implemented through a signal overlap.

Acknowledgment. This work was supported by a grant of the Romanian Ministry of Research and Innovation, CCCDI – UEFISCDI, project number 52/2020, PN-III-P2-2.1-PTE-2019-0867, within PNCDI III.

Conflict of Interests. The authors declare that they have no conflict of interest.

References

1. Bieller, S.: International Federation of Robotics (IFR). Press Conference, Shanghai, China (2019)
2. Khan, Z.H., Siddiqueue, A., Lee, C.W.: Robotics utilization for healthcare digitization in global COVID-19 management. Int. J. Environ. Res. Public Health **17**(11), 3819 (2020)

3. Larner, M.: Why evolution is key to consumer robotics survival. Juniper Research (2019). https://www.juniperresearch.com/document-library/white-papers/why-evolution-is-key-to-consumer-robotics
4. van den Berghe, R., Verhagen, J., Oudgenoeg-Paz, O., van der Ven, S., Leseman, P.: Social robots for language learning: a review. Rev. Educ. Res. **89**(2), 259–295 (2019)
5. Leng, M., et al.: Pet robot intervention for people with dementia: a systematic review and meta-analysis of randomized controlled trials. Psychiatry Res. **271**, 516–525 (2019)
6. Ismail, L.I., Hanapiah, F.A., Belpaeme, T., Dambre, J., Wyffels, F.: Analysis of attention in child–robot interaction among children diagnosed with cognitive impairment. Int. J. Soc. Robot. 13(2), 141–152 (2020). https://doi.org/10.1007/s12369-020-00628-x
7. TIAGo: Pal Robotics: Barcelona, Spain (2018). http://wiki.ros.org/Robots/TIAGo
8. Grama, L., Rusu, C.: Adding audio capabilities to TIAGo service robot. In: 13th International Symposium on Electronics and Telecommunications, pp. 263–266, Timisoara, Romania (2018). https://doi.org/10.1109/ISETC.2018.8583897
9. Telembici, T., Grama, L.: Detecting indoor sound events. Acta Technica Napocensis Electron. Telecommunications **59**(2), 13–17 (2018)
10. Grama, L., Rusu, C.: Extended assisted audio capabilities of TIAGo service robot. In: International Conference on Speech Technology and Human Computer Dialogue (SpeD), pp. 1–8, Timisoara, Romania (2019). https://doi.org/10.1109/SPED.2019.8906635
11. Chiariglione, L.: Design and build the future of MPEG. MPEG Future (2020). https://www.mpegfuture.org/244-2/
12. Luque, A., Romero-Lemos, J., Carrasco, A., Barbancho, J.: Non-sequential automatic classification of anuran sounds for the estimation of climate-change indicators. Expert Syst. Appl. **95**, 248–260 (2018)
13. Akshay, S., Bazille, H., Fabre, E., Genest, B.: Classification among hidden Markov models. In: 39th IARCS Annual Conference on. Foundations of Software Technology and Theoretical Computer Science, pp. 1–14, Bombay, India (2019)

Horse Optimization Algorithm Based Recurrent Neural Network Method for Epileptic Seizures Classification

Dorin Moldovan(✉)

Department of Computer Science, Technical University of Cluj-Napoca, Cluj-Napoca, Romania
dorin.moldovan@cs.utcluj.ro

Abstract. The classification of the epileptic seizures is a challenging research problem as epilepsy is a central nervous system disorder that affects around one percent of the entire world population. Many medical professionals consider the application of the electroencephalography (EEG) in the monitoring of the electrical activity of the brain. In the method presented in this article we consider the classification of the epileptic seizures as a binary classification problem such that one class is represented by epileptic seizures and the other class is represented by non-epileptic seizures. The monitored data is classified using a two layers Recurrent Neural Network (RNN) such that the values of the dropout and of the recurrent dropout of the first layer, which is a Long Short-Term Memory (LSTM) layer, are determined using Horse Optimization Algorithm (HOA). The method is tested and validated on the Epileptic Seizure Recognition Data Set from the UCI Machine Learning Repository.

Keywords: Horse optimization algorithm · Classification · Recurrent neural network · Epileptic seizures · Deep learning

1 Introduction

Epilepsy is a chronic brain disorder which, according to the statistics of the World Health Organization (WHO) [1], is likely to affect around 50 million people globally. From a neurological perspective, in most cases the primary cause of the epilepsy is not known. However, in many cases epilepsy could be a result of genetic factors, brain injuries and tumors, and it results in altered behaviors such as temporary loss of respiration, consciousness or memory. The application of deep learning in the classification of the epileptic seizures is still not exploited enough as can be seen in [2] where the authors consider the application of a Feed-Forward Neural Network (FFNN) in the context of a more complex method.

The main contributions of this research article are: (1) the development of a method based on a Recurrent Neural Network (RNN) for the classification of the seizures in epileptic or non-epileptic seizures, (2) the application of the Horse Optimization Algorithm (HOA) [3] for the optimal selection of the dropout and of the recurrent dropout values of the Long Short-Term Memory (LSTM) layer of the RNN and (3) the

S. Vlad and N. M. Roman (Eds.): MEDITECH 2020, IFMBE Proceedings 88, pp. 183–190, 2022.
https://doi.org/10.1007/978-3-030-93564-1_20

comparison of the results obtained using the HOA based RNN method with representative results from the research literature.

2 Research Background

In [4] is presented a method in which the EEG signals are classified using a Particle Swarm Optimization (PSO) based Artificial Neural Network (ANN). The two EEG databases considered in that article are the Karunya Institute of Technology and Sciences (KITS) EEG database and the Temple University Hospital (TUH) database. Prior to the application of the classification approach, the EEG signals are decomposed in several sub-bands and for each sub-band several features are computed such as the Shannon entropy, the log energy entropy and the Stein's unbiased risk estimate (SURE) entropy. PSO is then applied in order to select the informative features and finally the ANN is used in the classification of the EEG signals.

In [5] is described an approach in which a Convolutional Neural Network (CNN) is used together with transfer learning in order to classify 8 types of epileptic seizures using as experimental support the TUH database. The best results were obtained when Googlenet and Inceptionv3 were applied. Moreover, the approach based on image features extraction outperformed the approach based on transfer learning.

Finally, the method presented in [6] considers a features extraction approach based on third order cumulant (ToC) prior to the application of the sparse autoencoder in the extraction of structural information from the ToC features. Then a neural network based on sparse autoencoder is applied in the classification of the data. The experiments were performed on the EEG database publicly available at the University of Bonn and the results demonstrate the effectiveness of that approach.

3 Experimental Dataset Description

The dataset used in experiments is the Epileptic Seizure Recognition Data Set and is public in the UCI Machine Learning Repository [7]. That dataset consists of 11500 samples characterized by 178 features as follows:

(1) 2300 samples labeled with 1 (seizure activity);
(2) 2300 samples labeled with 2 (tumor area);
(3) 2300 samples labeled with 3 (healthy brain area);
(4) 2300 samples labeled with 4 (eyes closed);
(5) 2300 samples labeled with 5 (eyes open);

Each sample corresponds to one second of monitored data. In the RNN based method presented in this research article the samples labeled with 2, 3, 4 or 5 are labeled with 0 thus the dataset is prepared for a two class epileptic seizures classification problem.

Therefore, in the experiments performed in this research article there are only two types of samples which are summarized in Table 1.

Table 1. Summary of the samples used in the experiments.

Sample type	Label value	Number of samples
Non-epileptic seizure	0	9200
Epileptic seizure	1	2300

4 Recurrent Neural Network Based on Horse Optimization Algorithm for the Classification of the Epileptic Seizures

The high-level view of the RNN based on HOA for the classification of the epileptic seizures is presented in Fig. 1.

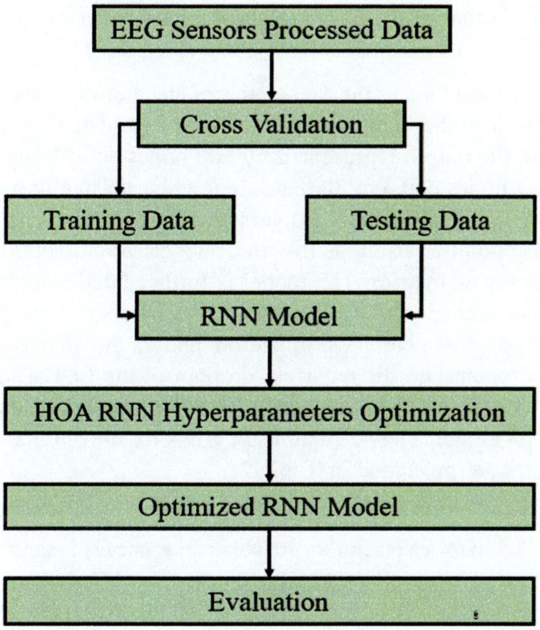

Fig. 1. HOA RNN method for epileptic seizures classification.

The input of the HOA RNN method for the classification of the epileptic seizures is the EEG Sensors Processed Data described in Sect. 3. In the Cross-Validation phase that data is split randomly in Training Data and Testing Data using a ratio equal to 80%:20%. The architecture of the RNN model for the classification of the epileptic seizures is presented in Fig. 2.

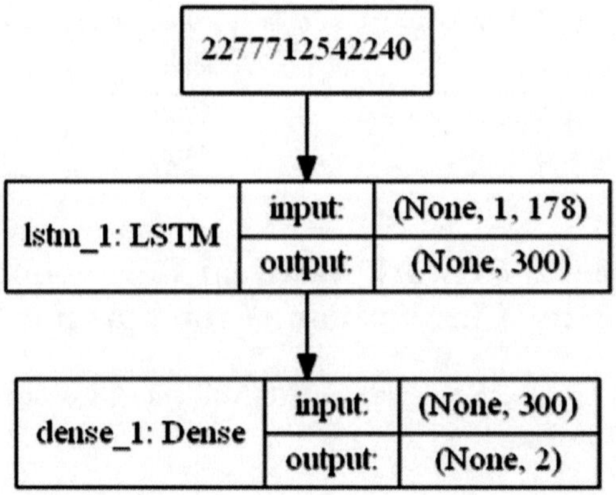

Fig. 2. Architecture of the RNN model for epileptic seizures classification.

The model was created using the Sequential model from keras and has two layers: (1) one LSTM layer with the input shape (1, 178) as each data sample is characterized by 178 features and the output represented by 300 nodes and (2) one dense layer with the number of input nodes equal to 300 and the number of output nodes equal to 2 as the problem is a binary classification problem;

The model was compiled using as loss the categorical crossentropy, the optimizer adam and the accuracy as metrics. The model is further fitted using the Training Data and the Testing Data, a batch size equal to 500 and a number of epochs equal to 10. In the HOA RNN Hyperparameters Optimization phase, the two parameters that are optimized are the dropout and the recurrent dropout of the LSTM layer. Their values are by default equal to 0 and they are known to influence significantly the overfitting behavior of the RNN model. The adaptations of HOA for the optimal selection of those two hyperparameters are presented in Table 2.

Table 2. HOA adaptation for RNN hyperparameters optimization.

HOA concept	Adaptation
Number of dimensions	2
First dimension of the search space	Dropout
Second dimension of the search space	Recurrent dropout
Search space	$[0, 0.5] \times [0, 0.5]$
Objective function	One minus the accuracy of the RNN model for the Training Data

The result of the HOA RNN Hyperparameters Optimization is represented by an Optimized RNN Model with the values of the dropout and of the recurrent dropout described by the position of the best horse. Finally, the results are evaluated in the

Evaluation phase. The confusion matrix considered in the evaluation of the model is presented in Table 3.

Table 3. Confusion matrix for epileptic seizures classification.

	Predicted epileptic seizure	Predicted non-epileptic seizure
Actual epileptic seizure	True Positive (TP)	False Negative (FN)
Actual non-epileptic seizure	False Positive (FP)	True Negative (TN)

5 Results and Discussions

The experiments were performed on a machine with the following configurations: Intel (R) Core(TM) i7–7500 CPU @ 2.70 GHz 2.90 GHz, 8.00 GB RAM, Windows 10 Pro N, 64-bit Operating System.

In Table 4 are summarized the values of the HOA used in experiments.

Table 4. HOA parameters values used in experiments.

Parameter	Value
Number of iterations ($N_{iterations}$)	10
Number of horses (N_{horses})	10
Horses hierarchy update frequency (M)	5
Dominant stallions percent (DSP)	10
Single stallions percent (SSP)	10
Horse memory pool (HMP)	2
Horses distribution rate (HDR)	10
Minimum velocity value (v_{min})	−0.001
Maximum velocity value (v_{max})	0.001
Minimum position value (p_{min})	0
Maximum position value (p_{max})	0.5
Standard deviation (sd)	1

After the application of HOA, the best horse has a fitness value equal to 0.039 and the values of the dropout and of the recurrent dropout are equal to 0 and 0.016, respectively. In Fig. 3 and in Fig. 4 are presented the confusion matrices for the case when default values of the dropout and of the recurrent dropout are considered and for the case when those values are determined using HOA, respectively.

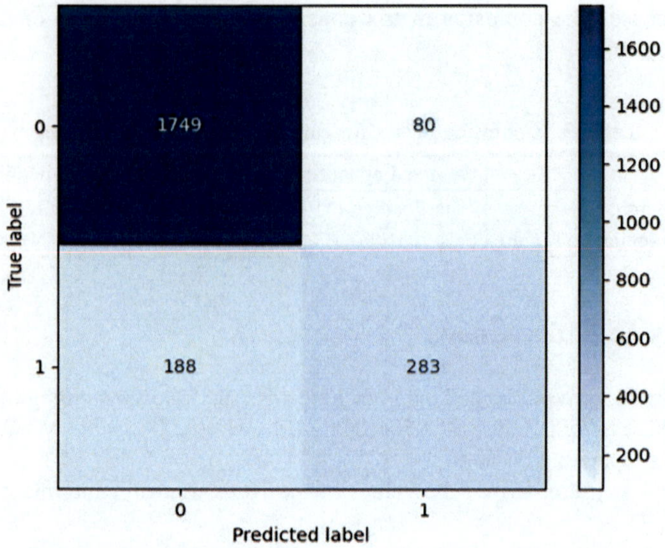

Fig. 3. Confusion matrix for the default dropout and recurrent dropout.

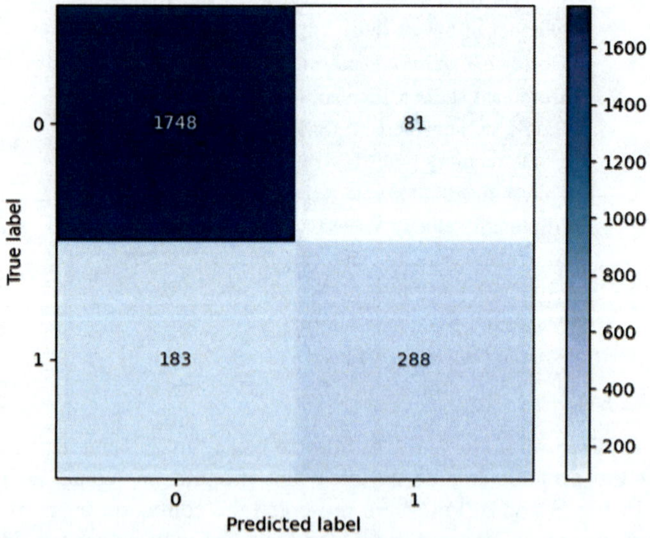

Fig. 4. Confusion matrix when the dropout and recurrent dropout are determined using HOA.

Figure 5 presents the evolution of the accuracy and of the training loss both for the Training Data and for the Testing Data.

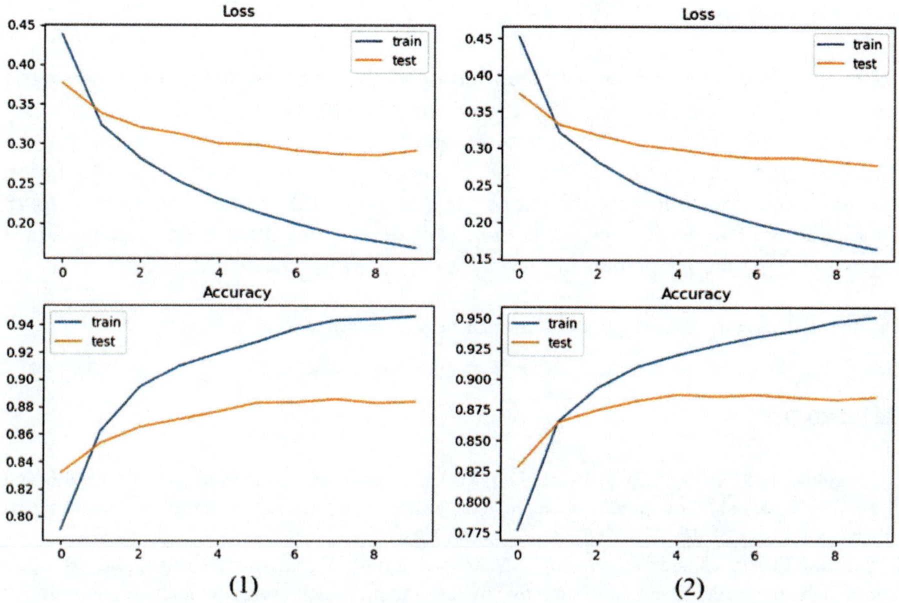

Fig. 5. Evolution of the loss and of the accuracy values for: (1) the case when the default values of the dropout and of the recurrent dropout are considered, (2) the case when the dropout and the recurrent dropout values are determined using HOA.

Finally, in Table 5 the values of the precision, recall, F1Score and accuracy obtained using the RNN method are compared with results from the research literature. The first row which corresponds to RNN (this approach) presents the results when the default values for dropout and recurrent dropout are considered and the second row which corresponds to HOA RNN (this approach) presents the results when those values are determined using HOA.

Table 5. Comparison of the obtained results with the literature results.

Approach	Precision	Recall	F1score	Accuracy
RNN (this approach)	0.877	0.883	0.877	0.883
HOA RNN (this approach)	**0.879**	**0.885**	**0.879**	0.885
Moldovan 2019 [8]	0.745	0.641	0.756	0.866
Rajendra Archarya et al. 2017 [9]	0.838	0.880	0.858	**0.903**

As can be seen in that table, the HOA RNN results are better than the RNN results and better than the results obtained by the method presented in [8] which is based on a features extraction phase. The results presented in [9] are not directly comparable because that method considers a CNN with 13 layers and 3 outputs, using the original EEG signals data from the University of Bonn as experimental support.

6 Conclusions

The results show an improvement compared to our previous research work even though the proposed method does not consider a separate features extraction step explicitly. Moreover, with further tuning and more samples, the classification results can be improved significantly like in the case of other deep learning methods. As future research work, the following directions are proposed: (1) the consideration of more combinations of hyperparameters, (2) the consideration of other deep learning architectures and (3) the adaptation of the method to other biomedical datasets.

Conflict of Interest. The authors declare that they have no conflict of interest.

References

1. Megiddo, I., Colson, A., Chisholm, D., Dua, T., Nandi, A., Laxminarayan, R.: Health and economic benefits of public financing of epilepsy treatment in India: an agent-based simulation model. Epilepsia **57**(3), 464–474 (2016)
2. Gómez-Gil, P., Juárez-Guerra, E., Alarcón-Aquino, V., Ramírez-Cortés, M., Rangel-Magdaleno, J.: Identification of epilepsy seizures using multi-resolution analysis and artificial neural networks. In: Castillo, O., Melin, P., Pedrycz, W., Kacprzyk, J. (eds.) Recent Advances on Hybrid Approaches for Designing Intelligent Systems. SCI, vol. 547, pp. 337–351. Springer, Cham (2014). https://doi.org/10.1007/978-3-319-05170-3_23
3. Moldovan, D.: Horse optimization algorithm: a novel bio-inspired algorithm for solving global optimization problems. In: Silhavy, R. (ed.) CSOC 2020. AISC, vol. 1225, pp. 195–209. Springer, Cham (2020). https://doi.org/10.1007/978-3-030-51971-1_16
4. Thomas George, S., Subathra, M.S.P., Sairamya, N.J., Susmitha, L., Joel Premkumar, M.: Classification of epileptic EEG signals using PSO based artificial neural network and tunable-Q wavelet transform. Biocybern. Biomed. Eng. **40**(2), 709–728 (2020)
5. Raghu, S., Sriraam, N., Temel, Y., Rao, S.V., Kubben, P.L.: EEG based multi-class seizure type classification using convolutional neural network and transfer learning. Neural Netw. **124**, 202–212 (2020)
6. Sharma, R., Pachori, R.B., Sircar, P.: Seizures classification based on higher order statistics and deep neural network. Biomed. Sig. Process. **59**, 101921 (2020)
7. Dua, D., Graff, C.: {UCI} Machine Learning Repository. University of California, Irvine, School of Information and Computer Sciences, http://archive.ics.uci.edu/ml. Accessed 28 Aug 2020
8. Moldovan, D.: Scalable hypothesis tests for detection of epileptic seizures. In: Silhavy, R., Silhavy, P., Prokopova, Z. (eds.) CoMeSySo 2019 2019. AISC, vol. 1047, pp. 157–166. Springer, Cham (2019). https://doi.org/10.1007/978-3-030-31362-3_16
9. Rajendra Acharya, U., Oh, S.L., Hagiwara, Y., Tan, J.H., Adeli, H.: Deep convolutional neural network for the automated detection and diagnosis of seizure using EEG Signals. Comput. Biol. Med. **100**, 270–278 (2018)

Binary Cat Swarm Optimization Feature Selection and Machine Learning Based Ensemble for the Classification of the Daily Living Activities of the Elders

Dorin Moldovan(✉), Ionut Anghel, Tudor Cioara, Cristina Pop, Viorica Chifu, and Ioan Salomie

Department of Computer Science, Technical University of Cluj-Napoca, Cluj-Napoca, Romania
dorin.moldovan@cs.utcluj.ro

Abstract. The classification of the daily living activities (DLAs) is a challenging research topic as the analysis of the DLAs data can help to improve the general health condition of the monitored subjects, to identify patterns that indicate various types of diseases such as the Alzheimer's disease and to prevent the falls caused by the poor health condition. The DLAs classification is approached in this article using a machine learning methodology as follows: (1) the standard Cat Swarm Optimization (CSO) algorithm is adapted to a binary version (BCSO) applicable for features selection, (2) the features are selected using the BCSO algorithm, (3) the cats are evaluated considering the sum of square errors of the DLAs and (4) the DLAs data is classified by an ensemble based on six machine learning classification algorithms. The data used in experiments is the PAMAP2 Physical Activity Monitoring dataset from the UCI Machine Learning Repository.

Keywords: Daily living activities · Classification · Features selection · Binary cat swarm optimization · Gradient boosted trees

1 Introduction

The classification of the daily living activities (DLAs) is an important research topic, considering that in the last years more and more people neglect the physical activities. As mentioned in [1], a decreasing of the physical activity is one of the major issues that affect the human health. The monitoring of the physical activities using wearable sensors such as accelerometer and gyroscope is attractive among the healthcare device manufacturers [2] and in the research literature there are machine learning algorithms that are designed for accelerometry data.

The in-home monitoring of the healthy independent elders in Ambient Assisted Living (AAL) systems [3] is a natural solution to the aging society demographic change, especially in the industrialized countries. Moreover, the context aware techniques can be used to process data related to the elderly home activity [4] and to analyze the context dynamics continuously.

S. Vlad and N. M. Roman (Eds.): MEDITECH 2020, IFMBE Proceedings 88, pp. 191–198, 2022.
https://doi.org/10.1007/978-3-030-93564-1_21

The main contributions of this research article are:

(1) the adaptation of the standard version of Cat Swarm Optimization (CSO) [5] to a binary version;
(2) the selection of the features of the DLAs data using an adapted version of the Binary Cat Swarm Optimization (BCSO);
(3) the development of an objective function that evaluates the cats considering the sum of square errors of the DLAs;
(4) the classification of the DLAs using an ensemble based on six machine learning classifiers;

The experiments were performed on the public PAMAP2 Physical Activity Monitoring dataset [6] from the UCI Machine Learning Repository [7]. The research article is organized as follows: Sect. 2 presents the research background, Sect. 3 describes the dataset used in experiments, Sect. 4 presents the machine learning methodology for the classification of the DLAs, Sect. 5 presents the results and Sect. 6 presents the conclusions.

2 Research Background

The approach presented in [8] which considers the PAMAP2 dataset among other DLAs datasets in the context of a machine learning methodology in which the features are selected using a bio-inspired algorithm, namely Kangaroo Mob Optimization (KMO), and the data is classified using the Random Forest (RF) algorithm shows that it is possible to obtain better classification results using a smaller number of features. However, the imbalance nature of the data samples was not exploited at maximum in that approach and the solution presented in that article was the consideration of an equal number of samples for each type of daily living activity.

Another approach is represented by the optimization of at least two objectives such as the accuracy of a machine learning algorithm and the number of selected features, as can be seen in [9]. However, since there is a trade-off in that case between the accuracy and the selected features, the value of the accuracy is often not maximized as that type of approach promotes the solutions with the lower number of features. In the approach presented in this article which has a single objective, the number of features is not penalized, and therefore the method searches the solution that returns the best result with respect to a classification performance metric.

CSO was used before in the research literature for features selection showing good results. As an example, in [10] the authors consider an improved version of CSO for big data classification which is evaluated using the accuracy of a Support Vector Machines (SVMs) algorithm. The drawback of that approach is that the imbalance nature of the data is not exploited at maximum by the CSO algorithm and is left as a challenge for the classifier. However, those results are promising and show that optimal modifications of standard bio-inspired algorithms can lead to better results.

3 Description of the Data Used in Experiments

The data used in the experiments was collected by three Colibri wireless inertial measurement units (IMUs) that have the sampling frequency equal to 100 Hz and that were placed as follows: (1) on the wrist of the dominant arm, (2) on the chest and (3) on the ankle of the dominant side. That data also contains information collected by a HR-monitor with a sampling frequency equal to 9 Hz, but in this article that column is ignored as the heart rate might be slightly different for the elders when they perform the same activities. Another column that is ignored is the one that describes the timestamp as in the experiments the data is shuffled when is split.

In the experiments described in this article was considered data collected from the first monitored subject of the PAMAP2 dataset. From the initial number of 19 activities, only 9 activities were considered as follows: lying (1), sitting (2), standing (3), walking (4), running (5), ascending stairs (12), descending stairs (13), vacuum cleaning (16) and ironing (17), as those activities are also representative for AAL systems and a part of the activities mentioned in the description of the dataset were not present in the data collected from the first monitored subject.

Moreover, for the last four activities, new labels were assigned as follows: ascending stairs (6), descending stairs (7), vacuum cleaning (8) and ironing (9), so that the data can be processed by the average ensemble of six machine learning classifiers. Therefore, the data used in experiments is characterized by one label from $\{1, 2, 3, 4, 5, 6, 7, 8, 9\}$ and 51 ($51 = 3 \times 17$) features that describe the data that is generated by the IMUs. The rows that contain NaN values were removed as their number was insignificant and they might not influence the classification results significantly.

In Table 1 are presented the initial and final values of the labels and the number of samples for each activity considered in the experiments.

Table 1. Summary of the data samples considered in the experiments.

Activity	Label		Number of samples	
	Original	Final	Original	Final
Lying	1	1	27187	27179
Sitting	2	2	23480	23465
Standing	3	3	21717	21593
Walking	4	4	22253	21383
Running	5	5	21265	20684
Ascending stairs	12	6	15890	15875
Descending stairs	13	7	14899	14892
Vacuum cleaning	16	8	22941	22875
Ironing	17	9	23573	23493

Therefore, out of the initial number of 193205 samples that describe the selected DLAs, the number of samples used in the experiments is equal to 191439.

4 Machine Learning Methodology for Daily Living Activities Classification

The high-level view of the machine learning (ML) methodology for the classification of the DLAs using BCSO is presented in Fig. 1.

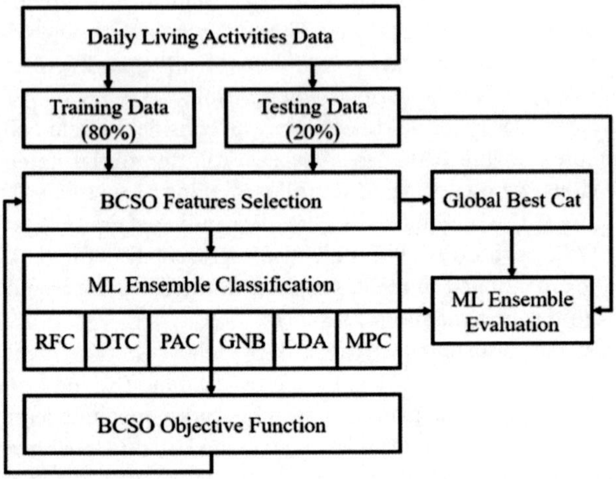

Fig. 1. High-level view of the machine learning methodology for the classification of the daily living activities.

The input of the machine learning methodology for the classification of the DLAs is represented by the *Daily Living Activities Data* presented in Table 1 after the processing of the original data. That data is split randomly in *Training Data (80%)* and *Testing Data (20%)* in the next phase of the methodology.

The *BCSO Features Selection* approach for the selection of the features is based on the original version of the CSO algorithm with the additional adaptation for binary optimization problems. In the approach presented in this article for each cat C_i with $i = \overline{1, N_{cats}}$ an additional vector bC_i of binary values is considered, which is derived from the current position of the cat as follows:

$$bC_{i,j} = \begin{cases} 1, & if\ rand < \frac{1}{1+e^{-C_{i,j}}} \\ 0, & otherwise \end{cases} \tag{1}$$

where *rand* is a random number from $[0, 1]$ and $j = \overline{1, D}$.

Then, according to the binary position of the cat, the *Training Data* and the *Testing Data* are filtered such that only the features which correspond to values equal to 1 are considered.

In the *ML Ensemble Classification* phase, 6 ML classifiers are considered as follows: Random Forest Classifier (RFC), Decision Tree Classifier (DTC), Passive

Aggressive Classifier (PAC), Gaussian Naïve Bayes (GNB), Linear Discriminant Analysis (LDA) and Multi-layer Perceptron Classifier (MPC);

The final prediction of the ML ensemble is the rounding of the average of the predictions of those ML classifiers to the nearest integer as follows:

$$y_{pred} = \text{Nint}\left(\frac{1}{6} \times \left(y_{RFC} + y_{DTC} + y_{PAC} + y_{GNB} + y_{LDA} + y_{MPC}\right)\right) \quad (2)$$

such that Nint represents the nearest integer function and y_{RFC}, y_{DTC}, y_{PAC}, y_{GNB}, y_{LDA}, y_{MPC} are the values predicted by RFC, DTC, PAC, GNB, LDA, MPC, respectively.

The BCSO Objective Function is represented by the sum of the square error of each DLA. In this article each DLA is considered separately as the data is imbalanced and the target is to obtain good classification results for each DLA not only a good overall classification result.

Therefore, considering as input the values y_{pred} predicted by the ML ensemble and the values y_{test} considered as testing data, the objective function OF is defined by the following formula:

$$OF\left(y_{pred}, y_{test}\right) = \sum_{k=1}^{9}\left(\sum_{i=1}^{\left|y_{test}^{(k)}\right|} \frac{1}{2 \times \left|y_{test}^{(k)}\right|} \times \left(y_{pred,i}^{(k)} - k\right)^2\right) \quad (3)$$

where $y_{test}^{(k)}$ are the labels equal to k and $y_{pred,i}^{(k)}$ is the value predicted by the ensemble for the ith sample that has the label k.

In the *ML Ensemble Evaluation* phase the ensemble is evaluated considering the values of the accuracy, precision, and recall such that the training data and the testing data are filtered according to the position G_{best} of the *Global Best Cat*.

5 Results

The experiments described in this article were written in Python and the standard scikit-learn implementation of the RFC, DTC, PAC, GNB, LDA, and MPC classifiers was considered. Some hyperparameter values were customized in order to get the prediction results faster as follows: (1) RFC: maximum depth = 20, number of estimators = 10, maximum number of features = 1, (2) DTC: maximum depth = 20 and (3) MPC: alpha = 1, maximum number of iterations = 1000. The values of the BCSO algorithm are presented in Table 2.

Figure 2 presents the confusion matrix when all features are considered by the ML average ensemble and Fig. 3 presents the confusion matrix when the features are selected according the value of G_{best}.

When all features were considered, the overall accuracy of the ML average ensemble was 0.93. When the features were selected according to the value of G_{best}, the accuracy was 0.96 using 20 out of the total 51 features as follows: f1, f2, f3, f5, f7, f14,

Table 2. BCSO parameters values used in experiments.

Parameter	Significance	Value
D	Number of dimensions of the search space	51
G	Number of generations	10
N_{cats}	Number of cats	10
MR	Mixture ratio	0.3
SMP	Seeking memory pool	5
CDC	Count of dimensions to change	34
SRD	Seeking range of the selected dimension	0.2
SPC	Self-position consideration	0
c_1	Constant used in seeking mode velocity update	0.5
C_{min}	Minimum possible value of the position of the cats	−5.0
C_{max}	Maximum possible value of the position of the cats	5.0
V_{min}	Minimum value of the velocity of the cats	−10.0
V_{max}	Maximum value of the velocity of the cats	10.0

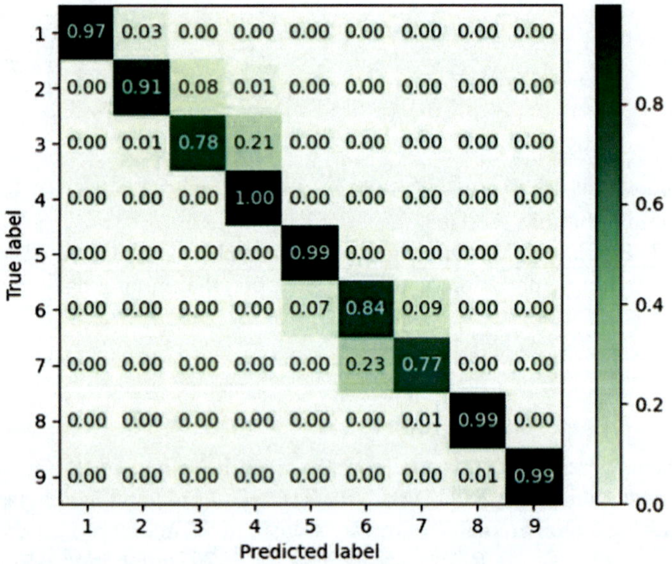

Fig. 2. Confusion matrix when all features are considered.

f15, f16, f18, f21, f23, f26, f28, f29, f30, f32, f33, f40, f41, f45. The value of G_{best} in the case of CSO after 10 iterations was equal to 0.193.

Finally, Table 3 presents the values of the precision and recall for each DLA when all features are considered and when the features are considered according to the value of the G_{best}.

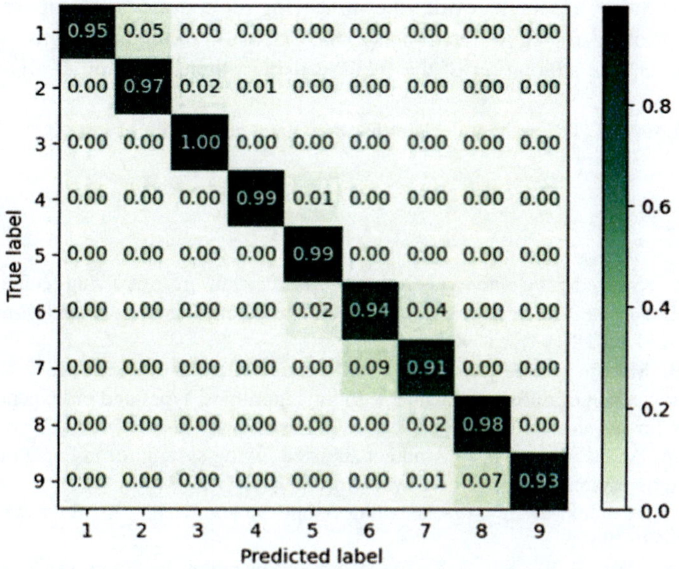

Fig. 3. Confusion matrix when the features are selected according to the global best.

Table 3. DLAs classification results summary.

Activity	All features		Features according to G_{best}	
	Precision	Recall	Precision	Recall
Lying	1.00	0.97	**1.00**	0.95
Sitting	0.96	0.91	0.94	**0.97**
Standing	0.90	0.78	**0.98**	**1.00**
Walking	0.81	1.00	**0.99**	0.99
Running	0.95	0.99	**0.97**	**0.99**
Ascending stairs	0.79	0.84	**0.91**	**0.94**
Descending stairs	0.88	0.77	**0.92**	**0.91**
Vacuum cleaning	0.99	0.99	0.94	0.98
Ironing	1.00	0.99	**1.00**	0.93

6 Conclusions

The novelty is justified by: (1) the DLAs classification using an ensemble of 6 classifiers, (2) the objective function which considers the mean square error for each different DLA and (3) the labelling of the DLAs such that similar DLAs have close labels. The precision values were equal or better for 7 out of 9 DLAs while the recall values were equal or better for 5 out of 9 DLAs when the CSO based method was applied. As future research work, the following directions are proposed: (1) the application of other objective functions, (2) the extension of the method for other DLAs datasets and (3) the adaptation of the method using other bio-inspired algorithms.

Conflict of Interest. The authors declare that they have no conflict of interest.

References

1. Lee, K., Kwan, M.-P.: Physical activity classification in free-living conditions using smartphone accelerometer data and exploration of predicted results. Comput. Environ. Urban **67**, 124–131 (2018)
2. Sheng, B., Moosman, O.M., Pozo-Cruz, B.D., Pozo-Cruz, J.D., Alfonso-Rosa, R.M., Zhang, Y.: A comparison of different machine learning algorithms, types and placements of activity monitors for physical activity classification. Measurement **154**, 107480 (2020)
3. Botia, J.A., Villa, A., Palma, J.: Ambient Assisted Living system for in-home monitoring of healthy independent elders. Expert Syst. Appl. **39**(9), 8136–8148 (2012)
4. Loke, S.W.: Context-aware artifacts: two development approaches. IEEE Pervasive Comput. **5**(2), 48–53 (2006)
5. Chu, S.-C., Tsai, P.-W., Pan, J.-S.: Cat swarm optimization. In: Yang, Q., Webb, G. (eds.) PRICAI 2006. LNCS (LNAI), vol. 4099, pp. 854–858. Springer, Heidelberg (2006). https://doi.org/10.1007/978-3-540-36668-3_94
6. Reiss, A., Stricker, D.: Creating and benchmarking a new dataset for physical activity monitoring. In: Proceedings of the 5th International Conference on PErvasive Technologies Related to Assistive Environments, pp. 1–8. Crete, Grece (2012)
7. Dua, D., Graff, C.: {UCI} Machine Learning Repository. University of California, Irvine, School of Information and Computer Sciences. http://archive.ics.uci.edu/ml. Accessed 27 Aug 2020
8. Moldovan, D., Anghel, I., Cioara, T., Salomie, I., Chifu, V., Pop, C.: Kangaroo mob heuristic for optimizing features selection in learning the daily living activities of people with Alzheimer's. In: 2019 22nd International Conference on Control Systems and Computer Science (CSCS), pp. 236–243. IEEE, Bucharest, Romania (2019)
9. Wei, B., Zhang, W., Xia, X., Zhang, Y., Yu, F., Zhu, Z.: Efficient feature selection algorithm based on particle swarm optimization with learning memory. IEEE Access **7**, 166066–166078 (2019)
10. Lin, K.-C., Zhang, K.-Y., Huang, Y.-H., Hung, J.C., Yen, N.: Feature selection based on an improved cat swarm optimization algorithm for big data classification. J. Supercomput. **72** (8), 3210–3221 (2016). https://doi.org/10.1007/s11227-016-1631-0

Quick Response in the SARS-COV-2 Context. How to Build up a Patient Video Monitoring System in Two days

Horia Hedesiu[1(✉)] and Dan Blendea[2]

[1] The Technical University of Cluj-Napoca, Cluj-Napoca, Romania
horia.hedesiu@emd.utcluj.ro
[2] University of Medicine and Pharmacy "Iuliu Hatieganu" Cluj-Napoca, Cluj-Napoca, Romania
Dan.Blendea@umfcluj.ro

Abstract. The spike in the number of cases of SARS-COV-2 in Cluj county in March 2020 brought up a set of challenges for the affected medical personnel due to this atypical context, tied to the rapid evolution of pandemic, rather old infrastructure requiring upgrade, limited expertise with a similar situation. To avoid unnecessary exposure of physicians and nurses to the SARS-COV-2 infected patients, a video surveillance system was deployed. This was not a trivial task because of the various technical issues related to infrastructure, short development time and other specific constraints. This paper presents how such solution got implemented, using a mixture of off-the-shelf equipment, combined with flexible graphical programming environment used in industry and academia.

Keywords: Remote monitoring · Graphical programming · Rapid prototyping

1 Context

1.1 SARS-COV2 Pandemic Situation, 2020

Given the circumstances of rapid evolution of SARS-COV-2 crisis in Romania and, specifically, Cluj county, the Cluj county management decided for a staged approach when distributing the newly diagnosed patients to the corresponding treatment facilities, just to cope with local resources in terms of available beds, specialized personnel, etc. As Fig. 1 shows, there was a rather slow evolution in the incipient phase of the crisis, followed by a burst at the end of March/early April [1, 2].

Therefore, considering also that Cluj-Napoca benefits from its privileged position of university town, where highly specialized faculty population is available, it was decided to have the Infectious Diseases Hospital as the front line defense unit, followed by other medical facilities, each with different levels of preparedness for dealing with such situation, unique for the vast majority of us.

The hospital unit considered here is by no means a facility specialized for treating SARS-COV2 patients, therefore it required a very quick accommodation to such operation environment, which implied the creation of two separate areas, the clean one

S. Vlad and N. M. Roman (Eds.): MEDITECH 2020, IFMBE Proceedings 88, pp. 199–206, 2022.
https://doi.org/10.1007/978-3-030-93564-1_22

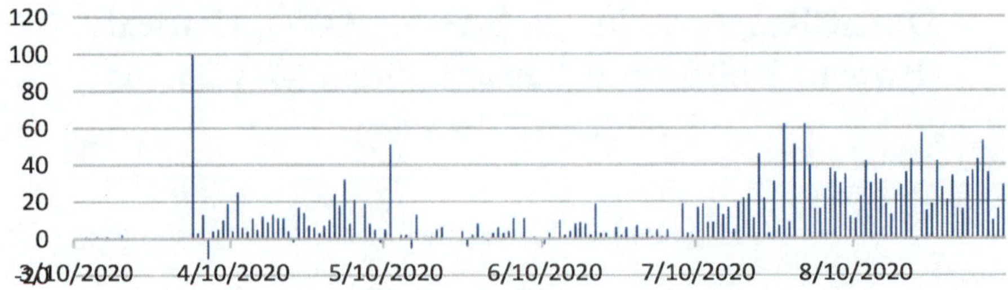

Fig. 1. Cluj county cases – March through September 2020. Source: [2]

and the infected one. These zones were clearly separated by a physical drywall, plastic made, just to avoid the spreading of the virus outside of the contaminated area. This retrofitting operation was needed to be complete in a really short time window, with very limited resources.

Part of this effort implied the deployment of a video surveillance system, with its main purpose to allowing specialized personnel to remotely view and observe infected patients in their respective dedicated rooms without direct interaction, minimizing thus the risk of contamination and also reducing the frequency of visits in that area, keeping also the number of such physical/direct contacts at the lowest as well.

2 Proposed Solution

2.1 Circumstances and Limitations

There are various factors to consider when designing and implementing such systems, which has a higher degree of particularity. These factors could be grouped in several categories:

- Topological: building floor plan, structure of walls affecting electromagnetic waves propagation, access ways, etc.;
- Equipment: Available equipment, operational features and characteristics, supply time, restrictions on use
- Ergonomics: Ease-of-use, trained personnel
- Engineering: Development and deployment time
- Budget: limitations and restrictions

Having all these limitations considered, our team has explored different options considering off-the-shelf equipment, mainly used in domestic security applications. All wired solutions were de facto disregarded since would have required intense deployment labor, generating debris and noise in an already operating medical facility. The other constraint, rather obvious, was the time pressure. A wired solution would have required a much different planning process, followed by intense work to laying the cables and accessories, etc.

As result, the next step was to consider the Wi-Fi based video monitoring commercial applications, delivered by a wide range of suppliers, with various operating results. Unfortunately, due to limited performances, none of these were able to meet reasonably decent operating performances, so the team had to drop on these as well.

The final step which finally proved to be the right one was to implement a dedicated system, involving specialized programmable hardware, able to:

- capture and process video images;
- remotely operable;
- easy to install and deploy.

2.2 Implementation

The proposed solution ended up being a custom one, tailored to meet the specifics of such application, having the high-level architecture as depicted in Fig. 2.

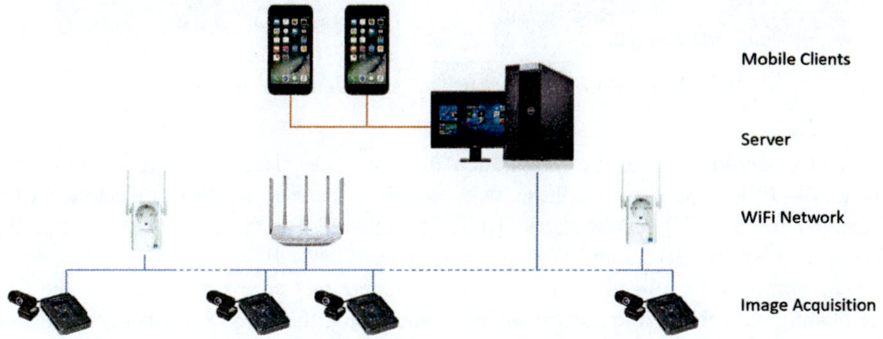

Fig. 2. System's high-level architecture

The Image Acquisition layer is using ten embedded dedicated systems developed by NI, www.ni.com, called myRIO. These devices are based on a Xilinx FPGA, and a dual-core ARM Cortex-A9 processor. As described in [3], NI MyRIO 1900, is designed for teaching the embedded design concepts for engineering students. Besides its capabilities, details here [4], there are some features that made it useful for this specific project, i.e.: (i) programmable; (ii) image acquisition capable; (iii) WiFi connection and remote operation.

According to [4], these devices can sense motion, take physical measurements, read barcodes and printed labels, inspect products for defects, and respond to colors. These features, easy to implement, finally convinced the team to choose MyRIO as preferred development hardware for the patients' video surveillance system.

Specifically, the image acquisition unit was provided with a Logitech C615 High Definition WebCam, supplied by Logitech, www.logitech.com, attached to its USB port, allowing image capturing from the patient rooms, as seen in Fig. 3.

This dedicated embedded system enabled the team to deploy the software application, developed in the NI LabVIEW™ environment. For programming MyRIO, besides the LabVIEW core module which supports the language among many other functions, a couple of other software modules are necessary for programming and

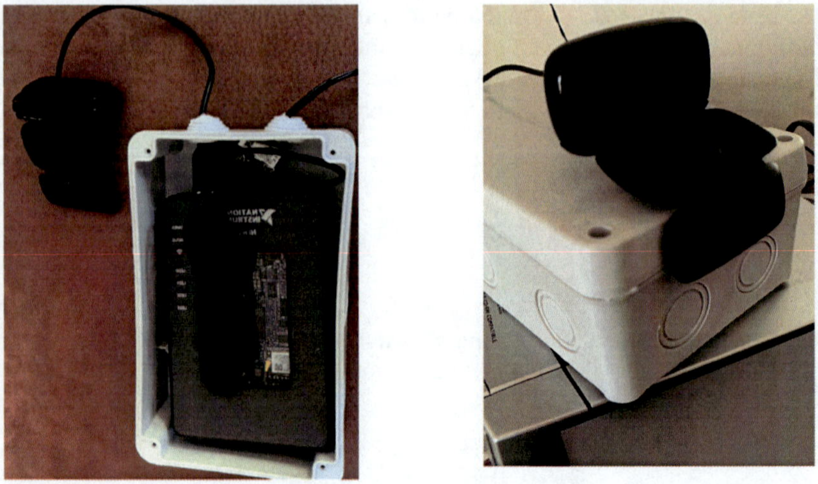

Fig. 3. MyRIO based image acquisition unit and its housing

eventually deploying to the embedded devices, i.e. LabVIEW Real-Time and LabVIEW FPGA. On top of these two modules, there is a third, specialized one, dedicated for MyRIO management itself. Particularly, for the described system, the IMAQ acquisition drivers and NI Vision were used as well.

The embedded application at MyRIO level contains two distinct sections, the first one dealing with the image acquisition and processing, the second one being dedicated to the bidirectional connection with the application' server.

The graphical nature of the code is seen below, being self-explanatory in some respect. The upper part of the application deals with the image acquisition part, followed by an image decimation, reducing thus the image size in order to optimizing the bandwidth usage [5, 6]. Once the image is acquired, it has to be prepared for transmitting to the server, therefore it is converted to a string, first, having an identifier attached in front of this string, allowing the server to recognize the source and manage it accordingly.

The lower part of this code diagram depicts the TCP/IP implementation of the image transfer to server. This is a simple solution that it opens the communication to the server, sending out the flattened image out after every single acquisition (Fig. 4).

All images generated by the *Image Acquisition* units are transmitted via the WiFi *Network*, which consists of commercial technology equipment, provided by TP-Link, www.tp-link.com. The router itself, an Archer AC1350 one, was not able to cover properly the floor area, so a couple of signal strengtheners/router extenders, TL-WA860RE model, same provider, were added for supporting data transmission at a decent rate. Deployment was easy as possible, requiring only power outlets, wall mounted.

On the *Server* side, there are several sections implemented as well:

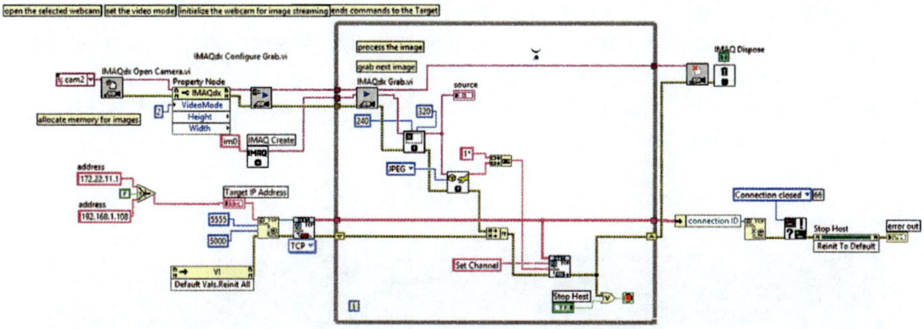

Fig. 4. Graphical code at the image acquisition unit level

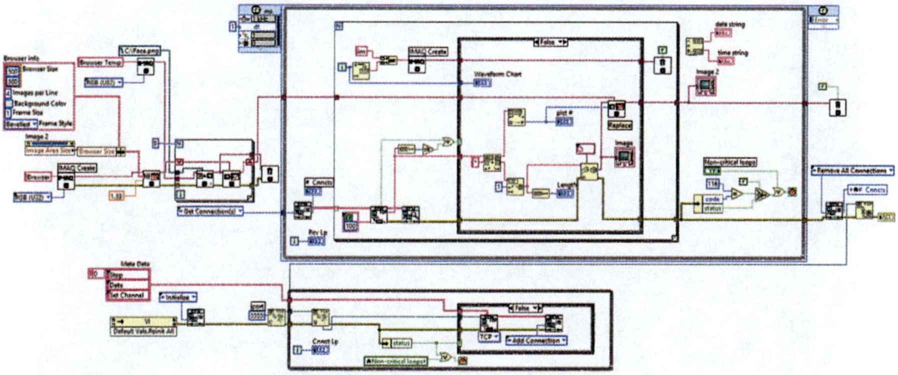

Fig. 5. The image server graphical code

1. Communication server, which collects the flattened images sent by the Image Acquisition units. There were different sever-client structures considered, but the team decided for a simple and robust implementation, with less chances for failure. Our solution took advantage of a freeware toolkit made available by NI, a LabVIEW API for Network Communication in Distributed Systems, called NI STM (Simple TCP Messaging), [7]. The main reasons for selecting this architecture are the easy-to-use data package manipulation along with data parsing, then the fact the transport layer (TCP/IP, UDP) is not specifically shown, making the developers' life easier, and, last but not least, STM minimizes network traffic by sending data only when it is needed. The lower loop of the diagram presented in Fig. 5 depicts the connection management (adding/deleting), listening incoming messages coming from the Image Acquisition units. The Main loop, placed above, receives the images, identifying the address they are coming from, then passing the information along to the next module. This simple server-client approach allows smooth clients addition/removal, with less interference for the other systems' components.
2. Decoding module, which retrieves the information collected as described above, unflattens it in the LabVIEW specific image format. The image arrives as a

dedicated string (packed and compressed), as it was sent by the Image acquisition units, so this module converts it back to the NI Vision image representation, allowing the appropriate processing.

3. Displaying module, which shows all these images on the UI (User Interface) and mobile *clients* (Fig. 6).

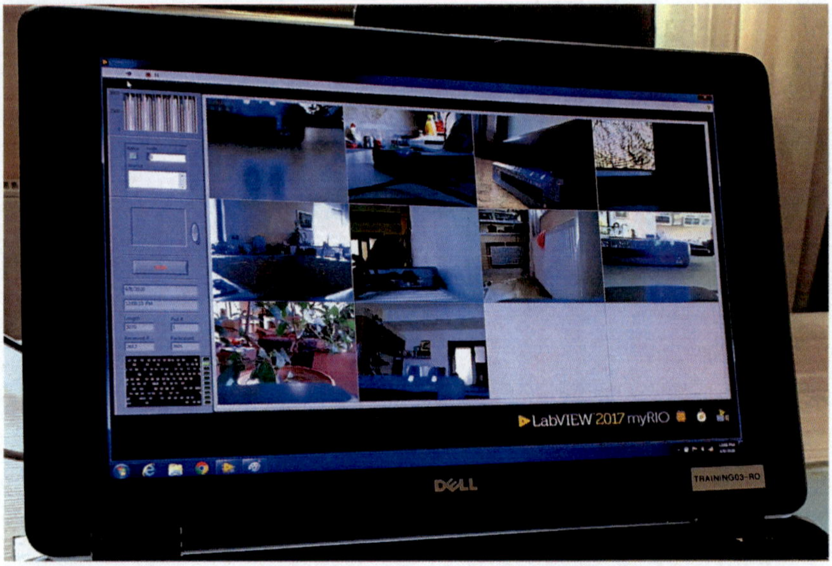

Fig. 6. UI of the video monitoring system. Proof-of-Concept phase

The UI (User Interface) consists of a collection of image snippets, as received from the Image Acquisition units. There are also other system info displaying elements, helping the user to monitor system's performance and letting him to debugging potential mis-functionalities. Such elements are the connection chart, (showing how the polling process goes, with potential breaks, if any), event log display, and user access info.

The implemented server is accessing the corresponding units per a polling protocol, designed to match the timing requirements for a continuous observation of the surveyed patients. Through its built-in capabilities, the LabVIEW application can forward the displayed information in a web suitable format, [8, 9], allowing remote access for a selected number of mobile clients.

3 Conclusions

This paper is presenting in a very succinct manner how a video surveillance system developed for COVID 19 patient's monitoring was put together and deployed. The implementation effort was reduced at minimum, only two working days/two specialists, taking advantage of off-the-shelf, graphically programmable embedded systems, with very versatile functional capabilities. This system had to comply with restrictive requirements related to the computer network connection and personnel access at medical facilities.

As future development, more functionality could be added with relatively small effort:

- Bidirectional audio communication patient-physician, since the embedded system displaced at patient level has such built-in capabilities. With minor software tweaks, such capability could be added with almost no effort, requiring minimal equipment investment.
- Alerting the supporting physician once the embedded system will be provided with dedicated alarming button or lever.
- Remote temperature monitoring of patients using non-contact sensors

The whole approach of this implementation was driven by time pressure, looking for and seeking an affordable approach in a crisis situation. It should be mentioned that there were severe limitations on urban travel/commute, with almost no access at equipment suppliers in lock-down circumstances.

The Image Acquisition units are designed and equipped for the engineering education, being flexible enough to match various teaching needs, so using them in this video surveillance system is just a testimonial of their flexibility and also team's ability to adapt and deliver as committed.

The software environment used for developing this quickly deployed solution, besides its easy-to-understand graphical nature, it has provided all necessary modules and elements to let the developers to assembly, prototype and commission the system with no major barriers and road-blocks. This was a major advantage, reducing the number of critical errors and bugs, saving valuable time, allowing the team to meet the strict deadlines they have committed for.

Conflict of Interest. The authors declare that they have no conflict of interest.

References

1. Casa Jurnalistului: Situația COVID-19 în România. https://docs.google.com/spreadsheets/d/1TafrOhj0AivTvTgNUGett6gEzn7m4mO2Urr-xYaIY4k/edit#gid=1963465719
2. Spitalul Clinic Judetean Cluj: Situatie Pacienti COVID-19. https://scjucluj.ro/index.php/covid-19/situatie-pacienti-covid-19
3. NI MyRIO User Manual Homepage. http://www.ni.com/pdf/manuals/376047c.pdf. Accessed 21 Sept 2020

4. Doering Ed: NI myRIO Vision Essentials Guide, National Technology and Science Press (2015)
5. Kwon, K.-S., Ready, S.: Practical Guide to Machine Vision Software: An Introduction with LabVIEW, 1st edn. Wiley-VCH, New York (2015)
6. Klinger, T.: Image Processing with LabVIEW and Image Vision (National Instruments Virtual Instrumentation Series), 1st edn. Prentice Hall, Hoboken (2003)
7. National Instruments: NI Simple Messaging (STM) Reference Library. http://sine.ni.com/nips/cds/view/p/lang/ro/nid/212055. Accessed 21 Sept 2020
8. Bress, T.: Effective LabVIEW Programming. National Technology and Science Press, Beijing (2013)
9. Jennings, R., De la Cueva, F.: LabVIEW Graphical Programming, 5th edn. McGraw-Hill Education, New York (2019)

Automatic Liver and Hepatic Tumors Segmentation in CT Images Using Convolutional Neural Networks

Alina S. Danciu[1], Simona Vlad[1(✉)], and Marius Leordeanu[2,3]

[1] Technical University of Cluj-Napoca, Cluj-Napoca, Romania
Simona.Vlad@ethm.utcluj.ro
[2] "Simion Stoilow" Institute of Mathematics of the Romanian Academy, Bucharest, Romania
[3] University Politehnica of Bucharest, Bucharest, Romania

Abstract. Automatic segmentation of liver and hepatic tumors is a task that has been widely studied in recent years but is still a demanding one. The largely demanded approach to patient care is a requirement for computer-aided diagnosis, treatment planning and liver cancer monitoring. In this paper is presented the performance of Convolutional Neural Networks (CNN) in liver and hepatic tumors segmentation, using a modified U-Net model. The architecture of the used CNN is composed by three down-sampling blocks, a central concatenated dilated bottleneck and three up-scaling blocks. The architecture was trained and tested on various datasets provided by the public medical dataset 3D-IRCADb-01. In liver segmentation the achieved Dice value was 0.9 with 99% accuracy and in the tumor case the accuracy obtained was about 98%. The aim of this paper is to explore the use of modified U-Net CNN for automatic segmentation of liver and hepatic tumors.

Keywords: Liver segmentation · Hepatic tumors segmentation · Convolutional neural networks

1 Introduction

The issue of liver and hepatic tumors segmentation has received a considerable attention in medical image computing community. Primary cancers such as breast, colon and pancreatic cancer also extend metastases to the liver over the course of the disease. As a result, the liver and its tumors are regularly analyzed in primary tumor stages. According to the World Health Organization (WHO) reports from 2018, cancer is a leading cause of death worldwide, accounting for an estimated 9.6 million deaths. The fourth most common cause of cancer death is liver cancer [1].

Computed tomography (CT) is currently one of the most commonly used imaging techniques for the identification and diagnosis of liver tumors, distinguished by its high spatial resolution and quick scanning speed. Traditionally, the liver and liver lesions are identified by radiologists on a slice-by-slice basis, manual or semi-manual, which is time-consuming, subjective and operator-dependent, this is why different operators often produce variable results. There is an urgent need to read, detect, and interpret CT

S. Vlad and N. M. Roman (Eds.): MEDITECH 2020, IFMBE Proceedings 88, pp. 207–216, 2022.
https://doi.org/10.1007/978-3-030-93564-1_23

scans automatically and rapidly. Liver and its tumors segmentation remains a challenging task due to the high variability of shapes, sizes and mostly because of challenges as a low-contrast between liver and lesion, different types of contrast levels and varying number or lesions. Figure 1 shows an example of the slight contrast difference from one organ/region to another.

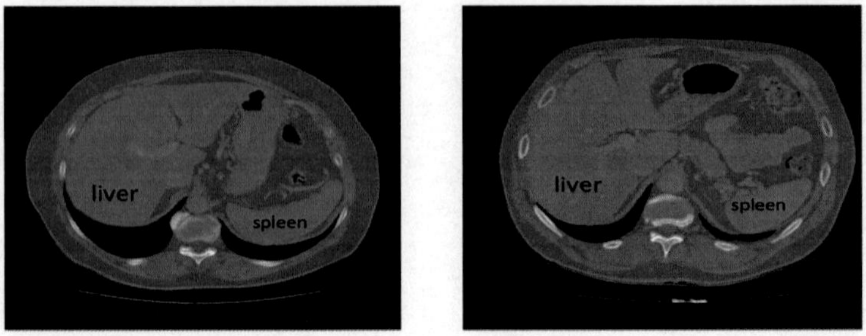

Fig. 1. Example of the similarity in gray levels between the liver and the spleen in computer tomography (CT) images, imported from 3DIRCADb-01.

Several interactive and automated methods for the segmentation of liver and liver lesions in CT volumes have been proposed in literature. In 2007 the Grand Challenge Conference on liver and liver lesions segmentation, in collaboration with MICCAI (Medical Image Computing and Computer-Assisted Intervention Society) conference took place. One of the papers [2] presents a comparative study of 10 automatic and six interactive methods for the segmentation of the liver from contrast-enhanced CT images. Methods proposed to the challenges were often focused on representations with structural forms and texture analysis.

More recently, Deep CNNs have received considerable interest in the science community to solve computer vision tasks such as target recognition, classification and segmentation [3, 4]. Additionally, new segmentation approaches based on CNN have been developed for medical image processing, with extremely successful outcomes relative to state-of-the-art ones [5].

Segmentation is a very popular topic for papers applying deep learning on medical imaging [6]. A well-known architecture for medical imaging processing is U-Net, published by Ronneberger et al. (2015) [7]. Through this work, the CNN approach was taken one step further and the new architecture combines learned upsampling paths with skip-connections to directly connect opposing contracting and expanding convolutional layers. The U-Net architecture was firstly used for segmentation of neuronal structures in electron microscopic stacks.

2 Method

2.1 Dataset

The database used in this research belongs to the French Research Institute against Digestive Cancer [8] and is publicly available. This is a database that includes several sets of anonymized medical images of patients and manual segmentation of various structures of interest performed by clinical experts.

The database contains 20 CT scans of patients, half of them for women and half for men, with hepatic tumors in 75% of the cases. All the scans are available in DICOM format, with the resolution of 512 $\times$ 512 pixels. For each case, we are presented with the original CT scan, used as input image to the CNN and the corresponding manual label portraying the liver region and also tumors. For the liver segmentation we considered all of the 20 cases, with a total of 2234 images. Because the hepatic tumor is absent in some of the investigations, for the tumor segmentation only the scans where the tumor is present have been used. The models have been trained and evaluated using the 13 scans containing hepatic tumors in the liver, the other two scans had very small tumors and were not conclusive enough. 3D-IRCADb-01 dataset is often used for both training and testing in different scenarios [9, 10].

2.2 Network Architecture

For the segmentation, an updated version of the U-Net has been used [11]. The approach [11] obtained satisfactory results in semantic segmentation for landing areas.

The proposed architecture in this paper is a U-Net-like network, as shown in the Fig. 2, composed by three down-sampling blocks, a central concatenated dilated convolutions bottleneck and three up-scaling blocks, with summed features. The sum has the role to reduce the computational cost of applying many convolutional operations over a large set of filters. For the down-sampling part are used two convolutional layers, one has stride 1 and the other one has stride 2 and in this way the input resolution is divided into equal parts. The up-sampling part of the architecture has a single convolutional layer for each block and also a feature map concatenation with its corresponding map. The dilation rates are increasing progressively, so the number of feature maps for the down-sampling blocks are raising from 16 to 64. For the up-sampling block, numbers decrease gradually, from 64 to 16.

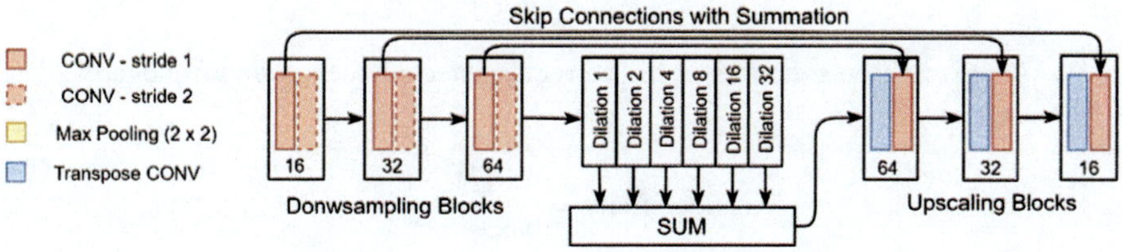

Fig. 2. The adopted CNN architecture [11]

2.3 Evaluation Metrics

The performance is measured using well-known quantitative evaluation metrics for segmentation such as precision, recall, specificity, accuracy and Dice coefficient. After obtaining the predictions from the model, the pictures were converted in PNG images for a faster and easier analysis. Using MATLAB software [12], evaluation is done over each prediction map with reference to the ground truth label.

The confusion matrix analyzes the performance of the classifier and measures how well the algorithm works. It provides four types of rates, as shown in Table 1. In image segmentation true positive (TP) means that the model's prediction was correctly assigned to the positive class and true negative (TN) express the number of predictions correctly assigned to the negative class. Also, false positive (FP) means that the neural network model predicted that some pixels belong to the positive class, which in reality belong to the negative class and false negative (FN) shows the number of pixels assigned by the model to the negative class, which in reality belong to the positive class. Using the metrics from Table 1, are obtained more indicators related to the capacity of classification of the proposed model.

Table 1. Confusion matrix

		Predicted	
		Positive	Negative
Actual	Positive	TP	FN
	Negative	FP	TN

Precision measures the trust in the predicted positive class, as formulated in the Eq. (1).

$$\text{Precision} = \frac{\text{TP}}{\text{TP} + \text{FP}} \tag{1}$$

Sensitivity, also known as recall, is used to quantify the ability of the algorithm to detect positive results. Equation (2) shows the way to determine recall value.

$$\text{Recall} = \frac{\text{TP}}{\text{TP} + \text{FN}} \tag{2}$$

Specificity represents the rate of the correctly detected background or negative data as formulated in Eq. (3).

$$\text{Specificity} = \frac{\text{TN}}{\text{TN} + \text{FP}} \tag{3}$$

Accuracy can be calculated with the formula in the Eq. (4) and is a ratio of correctly predicted cases to the total cases.

$$\text{Accuracy} = \frac{\text{TP} + \text{TN}}{\text{TP} + \text{TN} + \text{FP} + \text{FN}} \tag{4}$$

Dice score or *F1 measure* is a harmonic mean of recall and precision, as formulated in Eq. (5). It checks whether or not a point on the predicted boundary fits the ground truth boundary. The Dice score value is between 0 and 1, where 1 shows a perfect segmentation.

$$\text{DICE} = \frac{2\text{TP}}{2\text{TP} + \text{FP} + \text{FN}} = 2 \bullet \frac{\textit{Precision} \bullet \textit{Recall}}{\textit{Precision} + \textit{Recall}} \tag{5}$$

3 Experimental Results and Discussions

3.1 Liver Segmentation

Technical training details are presented as follow. 16 CT scans from the database were used for training and 4 CT scans were used for testing and evaluation. In order to achieve the best outcome from a very small database, the liver segmentation has been studied using two different instances of training. Training was carried out over 50 epochs; the batch was set at 32 and the learning rate was set at 0.001. The training time of the epoch lasted, on average of 140 s ($\sim$2 s/step).

In the first instance, the network was trained with all parts of each CT scan, having a total number of 2234 training images. The quantitative outcomes of the segmentation show a value for Dice of 0.87, while the precision, recall and accuracy achieved values of 0.89, 0.85 respectively 0.8. The qualitative results are presented in Fig. 3.

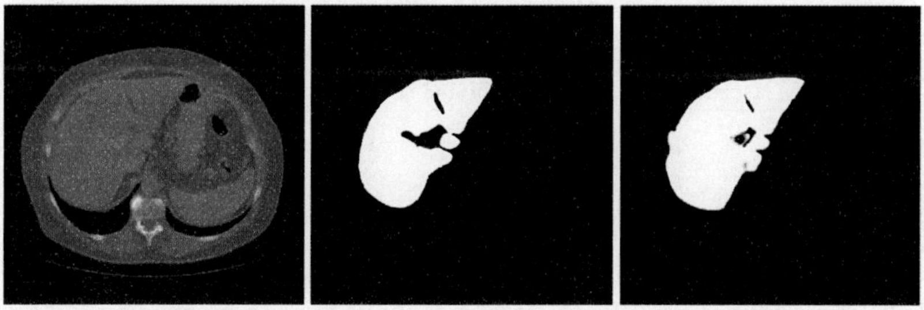

(a) (b) (c)

Fig. 3. Segmentation result using the proposed method: (a) – test image, (b) – ground truth (c) – predicted segmentation in the first instance.

The satisfactory results which were achieved in the first case, due to the value of Dice coefficient, can be improved. As a second instance, it was analyzed the effect of images that do not include an area of concern, on the learning process. There are a large number of images that do not provide detail on the position, shape and scale of the liver, so these images do not offer valuable information. In this situation we used a number of only 1678 images for training. The results are presented in Fig. 4.

(a) (b) (c)

Fig. 4. Segmentation result using the proposed method: (a) – test image, (b) – ground truth, (c) – predicted segmentation in the second case.

The basic training parameters for the neural network remained the same for each scenario and the testing set remained unchanged. In terms of quantitative results, improved values are observed for all metrics. We obtained a very satisfactory Dice score of 0.9 and an accuracy percentage of 99%. A comparison of other methods performing liver segmentation is presented in Table 2 and the results included are listed from the original papers.

Table 2. Comparison of liver segmentation results on different methods

Method	Dice	Precision	Recall
FCN-8s 3 slices [12]	0.89	0.95	0.86
Modified U-Net model	0.90	0.91	0.90
Cascaded U-Net [13]	0.93	-	-
Cascaded U-Net + 3D CRF [13]	0.94	-	-

3.2 Tumor Segmentation

Only 75% of the database used in this study includes liver tumors, contributing to a substantial decrease in data valuable for the study of tumor segmentation. As a result, with only 13 CT scans available for testing and evaluation, it becomes very difficult for the network to learn certain features. For this cause, a variety of pre-processing tasks were performed; augmentation techniques and noise reduction methods have been used to increase the number of data, to obtain more pixels with positive value and therefore better prediction.

The proportion of areas of concern with a hepatic tumor is very small relative to the total amount of images. The compromise approach for this situation is to eliminate CT slices without details, as in the case of liver segmentation. Thus, the tests to be mentioned have been validated and conducted on groups of images containing only slices in which the location of the tumor has been identified. In a much larger scale, with an expanded amount of details, it will be possible to locate tumors in the whole CT scan, without the need to eliminate images that do not represent any region of interest.

The dataset has been split in both training and testing, using 10 CT scans for the first set and 3 CT scans for the second phase. Training and testing parameters have been maintained: 100 epochs, 16 batch and the value of the learning rate stays 0.001.

First, it was applied normalization and then several techniques of augmentation, like rotation (90°, 180° and 270°) and flip (vertical and horizontal). Image augmentation is a very effective method used to artificially generate images related to the original version and extend the existing image data collection. This produces unique and distinct images from the current image data collection, which represents a comprehensive range of alternative pictures. Augmentation is very often used in the segmentation of medical images due to the small number of annotated images that exist and are available to use. It helped us to obtain 50 more images for training phase. The results of this stage and also of the next pre-processing methods are listed in Table 3.

Table 3. Quantitative evaluation for liver tumor segmentation

Method	Dice score
Normalization	0.08
Augmentation	0.14
Scaling	0.10
Scaling and addition	0.11
Factor 10 dilation	0.15
Factor 50 dilation	0.08
Contrast enhancement	0.12
Contrast enhancement and addition	0.18

Furthermore, the images were analyzed from a different perspective. By applying some bounding boxes to both CT images and related masks, the process acted as a magnification.

Another way that the images can be handled for segmentation is by adding dilations to the reference masks, leaving the CT image unchanged. In this method, the associated images can learn not only to recognize the location of the tumor but also the background in which it is located. The effect of dilation with a two-dimensional disk-type structural feature has been studied, with a value of 10 and 50. This experiment reveals that the existence of the background is only helpful to a certain degree, and that the over-extended region of the mask helps to recognize the presence of the tumor, but does not allow the contouring of its edges.

One more step was related to the small contrast between the regions that are viewable on the CT slices. The contrast control feature is used to minimize this interference (Fig. 5).

Fig. 5. Segmentation result using the proposed method: (a) – test image, (b) – ground truth, (c) – predicted segmentation.

Table 4 contains a comparison between other approaches and our results. It is necessary to note that the largest difference is the size of the database, which is why the Dice coefficient has a very low value. Our main aim, as a future work, is to increase the value of all measurement criteria in order to achieve a more precise segmentation. It requires access to a wider collection of images for testing and we expect more consistent outcomes.

Table 4. Comparison of tumor segmentation results on different methods

Method	Database size	Dice
Christ et al. [12]	100	0.56
U-Net [14]	179	0.65
DCNN: U-Net + ResNet [15]	200	0.67
Modified U-Net model	13	0.18

4 Conclusions

The experiments show how promising is this architecture in order to perform liver segmentation, although the task is not free of difficulties. Firstly, segmentation d struggling were related to the very slight variations in the size of the pixels belonging to the organ and those belonging to the neighboring organs. Second, the liver is not defined in each segment of the CT. Considering the Dice coefficient value as the most generic prediction parameter, there is an improvement from 0.8711 (first case) to 0.9063 (second case).

Hepatic tumors segmentation faces problems related to the limited amount of available data and also the low contrast between tumor tissue and healthy tissue. We showed an improvement when we applied different augmentation techniques. This experiment reveals the strong connection between the amount of training data used and model ability to produce the best outcomes.

It is known that the segmentation of CT images using neural networks is a procedure that can be automated, AI techniques providing a wide range of opportunities to develop algorithms.

As a future work, the main purpose is to obtain better results in both liver and hepatic tumors segmentation, using a wide database.

Acknowledgment. We would like to express our gratitude to Alina Marcu, Phd student at "Simion Stoilow" Institute of Mathematics of the Romanian Academy, for the all support and encouragement offered during this research.

Conflict of Interests. The authors declare that they have no conflict of interest.

References

1. World Health Organization: Global Status Report on Alcohol And Health 2018. World Health Organization (2019)
2. Heimann, T., et al.: Comparison and evaluation of methods for liver segmentation from CT datasets. IEEE Trans. Med. Imaging **28**(8), 1251–1265 (2009). https://doi.org/10.1109/TMI.2009.2013851
3. Krizhevsky, A., Sutskever, I., Hinton, G.E.: Imagenet classification with deep convolutional neural networks. In: Advances in Neural Information Processing Systems, pp. 1097–1105 (2012)
4. Long, J., Shelhamer, E., Darrell, T.: Fully convolutional networks for semantic segmentation. In: Proceedings of the IEEE Conference on Computer Vision and Pattern Recognition, pp. 3431–3440 (2015)
5. Havaei, M., et al.: Brain tumor segmentation with deep neural networks. Med. Image Anal. **35**, 18–31 (2017)
6. Litjens, G., et al.: A survey on deep learning in medical image analysis. Med. Image Anal. **42**, 60–88 (2017)
7. Ronneberger, O., Fischer, P., Brox, T.: U-net: Convolutional networks for biomedical image segmentation. In: Navab, N., Hornegger, J., Wells, W., Frangi, A. (eds.) MICCAI 2015. LNCS, vol. 9351, pp. 234–241. Springer, Cham (2015). https://doi.org/10.1007/978-3-319-24574-4_28
8. IRCAD. Hôpitaux Universitaires. 1, place de l'Hôpital, 67091 Strasbourg Cedex, France. https://www.ircad.fr/contact/. Accessed 2 Feb 2020
9. Sun, C., et al.: Automatic segmentation of liver tumors from multiphase contrast-enhanced CT images based on FCNs. Artif. Intell. Med. **83**, 58–66 (2017)
10. Badura, P., Wieclawek, W.: Calibrating level set approach by granular computing in computed tomography abdominal organs segmentation. Appl. Soft Comput. **49**, 887–900 (2016)
11. Marcu, A., Costea, D., Licaret, V., Pîrvu, M., Slusanschi, E., Leordeanu, M.: SafeUAV: learning to estimate depth and safe landing areas for UAVs from synthetic data. In: Leal-Taixé, L., Roth, S. (eds.) ECCV 2018. LNCS, vol. 11130, pp. 43–58. Springer, Cham (2019). https://doi.org/10.1007/978-3-030-11012-3_4

12. Ben-Cohen, A., Diamant, I., Klang, E., Amitai, M., Greenspan, H.: Fully convolutional network for liver segmentation and lesions detection. In: Carneiro, G. (ed.) LABELS/DLMIA -2016. LNCS, vol. 10008, pp. 77–85. Springer, Cham (2016). https://doi.org/10.1007/978-3-319-46976-8_9
13. Christ, P.F., et al.: Automatic liver and tumor segmentation of CT and MRI volumes using cascaded fully convolutional neural networks. arXiv preprint arXiv:1702.05970 (2017)
14. Chlebus, G., Meine, H., Moltz, J.H., Schenk, A.: Neural network-based automatic liver tumor segmentation with random forest-based candidate filtering. arXiv:1706.00842 (2017)
15. Han, X.: Automatic liver lesion segmentation using a deep convolutional neural network method. arXiv:1704.07239 (2017)

Brain Tumors Detection Using Convolutional Neural Networks

Catalina Maghiar, Viorica Rozina Chifu(✉), Cristina Bianca Pop, and Emil Ștefan Chifu

Computer Science Department, Technical University of Cluj-Napoca, Cluj-Napoca, Romania
{viorica.chifu, cristina.pop, emil.chifu}@cs.utcluj.ro

Abstract. This paper presents a method based on Convolutional Neural Networks (CNN) for detecting brain tumors. The proposed method consists of the following steps: dataset preparation, neural network preparation and training and performance evaluation. To improve the performance of the CNN in terms of results accuracy, ResNet50, a pre-trained model is used. The method has been validated on a Brain Magnetic Resonance Imaging (MRI) dataset and evaluated using appropriate metrics.

Keywords: Brain tumor diagnosis · Convolutional Neural Networks · ResNet50 model · Magnetic resonance images

1 Introduction and Related Work

According to [1], in 2017 in the United States there were approximately 168,494 people living with brain and other nervous system cancer. In 2020, it is estimated that the number of new cases will be 23,890 while the number of deaths will be around 18,020.

A delicate task for radiologists is detecting the brain tumors at an early stage. This is crucial because brain tumors grow fast and their average size can almost double in only 25 days. When proper treatment is not received in time, the survival rate of the patient drops to just half a year. As every medical research area, tumor detection requires a high accuracy and precision of the results since it involves the life of a patient. Computer aided diagnosis (CAD) is becoming more popular nowadays because it provides accurate results and is less time consuming. In particular, deep learning offers accurate results in detecting tumors in medical images. It is seen as a crucial tool for future medical applications that will lead to impressive changes in the medical world.

There is a lot of work in this field of application of the ML techniques in tumor diagnosis. Some of them are based on Support Vector Machine (SVM) classifiers [1, 2] or Artificial Neural Networks [3–9]. For example, in [2] is presented a SVM based method that combines multi model texture features with kernel based support vector machine to detect the presents of brain tumors in Magnetic Resonance Imaging (MRI) images. The performance of the proposed method was evaluated on 80 T1

S. Vlad and N. M. Roman (Eds.): MEDITECH 2020, IFMBE Proceedings 88, pp. 217–225, 2022.
https://doi.org/10.1007/978-3-030-93564-1_24

weighted brain MRI image sequences by using sensitivity, specificity and accuracy evaluation metrics. [5] proposes a CNN based method for classifying brain tumors from T1-weighted contrast-enhanced MRI images which consists of two steps: image preprocessing, and image classification using CNN. The method has been tested on a data set of 3064 images and evaluated using precision, recall and accuracy. [6] also proposes a CNN based method for multi-class classification of brain tumors and evaluates the method on the Contrast-Enhanced Magnetic Resonance Imaging (CE-MRI) data set. Due to the small dimension of the data set used in the evaluation, in the training phase, the weights of the CNN layers are initialized from the pre-trained VGG19 model and then fine-tuned on CE-MRI data set. The method has been comparatively evaluated with others CNN based methods and the results demonstrated that it outperforms them. Similar with [6, 7] uses a pre-trained CNN model to classify brain tumors in three main classes: glioma, meningioma and pituitary tumors. The data set used in experiments contains 3064 brain MRI images collected from 233 patients, diagnosed with different types of brain tumors and the accuracy of the obtained results was of 98%.

This paper proposes a method based on CNN for detecting brain tumors. To improve the performance of the CNN in terms of the accuracy of the results, ResNet50, a pre-trained model has been used. The method has been integrated into an experimental prototype, validated on Brain MRI Images data set [10] and evaluated using appropriate metrics.

The paper is organized as follows. Section 2 describes the CNN based method, while Sect. 3 presents the experimental results. We end our paper with conclusions.

2 CNN Based Method for Brain Tumor Detection

This section presents the main steps of our CNN based method for brain tumor detection.

Data Set Preparation and Image Augmentation. Resources for medical images are usually low when it comes to online availability because the people to whom they belong have to give their consent for researchers to use those images in their work. Since, we have used in our experiments a data set [10] containing a small number of magnetic resonance images (MRI) of the brain (our data set consists of 160 images out of which 80 represent normal brain MRIs and the other 80 represent brains with tumorous tissue), to enrich the data set, we have applied different types of image augmentation techniques. Those *augmentation techniques* are provided by Keras deep learning library and consists of applying a series of operations to slightly alter/modify the available images in order to create similar ones such as: brightening the images or making them darker in order to illustrate different compositions for the tumorous tissue, zooming in and out to vary the size of the tumors and rotating the images at different angles to diversify the shapes of the tumors. In this way the data set become ten times larger.

An example of image augmentation obtained for some images from our data set is presented in Fig. 1.

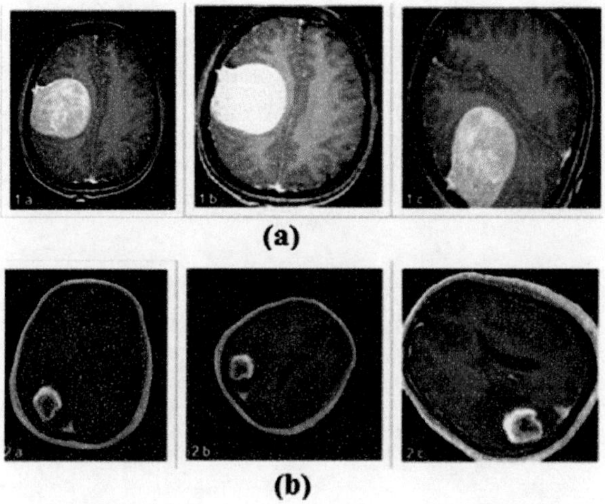

Fig. 1. Image augmentation example - (a) the original images, (b) the augmented images

For our data set, in the case of rotation transformation, the range of rotation was set to 90 degrees for diversity of angles because the shape of a tumor is irregular. Horizontal or vertical flips were not considered necessary since most of the images are taken in the same way. Also, the brightness range is varying between 0.4 and 1.4. Anything smaller than 1 will make the image darker and any value greater than 1 will brighten the image. The values must be floats in the range of 0 to 2. For the zoom range, the same interval is valid for the values. A zoom value smaller than 1 will zoom in the image, while a value greater than 1 will zoom out. The shear range was also used for the process of image augmentation with a value of 0.1. Shearing is an image altering technique where a part of the image is pulled in one direction and the other part is pulled in the opposite direction. This can create different forms and sizes of the tumors from the images already available, to increase the diversity. We have applied these types of transformations only on the images reserved for training since the validation data set does not need them.

We have also, generated batches of a given size of augmented data. A batch represents how many images should be loaded at once by the network. If the batch size is too large, the model might lack the ability to generalize. The number selected for a batch size is usually a power of two such as 16, 32, 64 and so on. The default value for batch size is 32 in most cases. Setting a suitable batch size will also prevent overfitting of the model. Since the number of images available in our dataset is reduced, even with the augmentation techniques are applied, the best performance seemed to be achieved when the batch size was set to 16 images.

Network Preparation and Training Using Transfer Learning. In our approach we have used the ResNet50, a pre-trained model trained on ImageNet data set [11], which is a CNN that consists of 50 layers. The architecture of ResNet50, as a residual network, consists of a sequence of residual blocks that are made of more convolutional layers. ResNet50 has 16 residual blocks, according to the number of skip connections. In order to be able to use the ResNet50 model with good results in our case, the output layer is cut off and custom layers will be added to the model. To preserve the knowledge obtained so far by the model, the weights will be frozen for the remaining layers. By doing this process, those layers will be excluded from training because they contain valuable information that should not be altered or lost by retraining them. The following step was to add the new custom layers to the ResNet50 model (see Fig. 2).

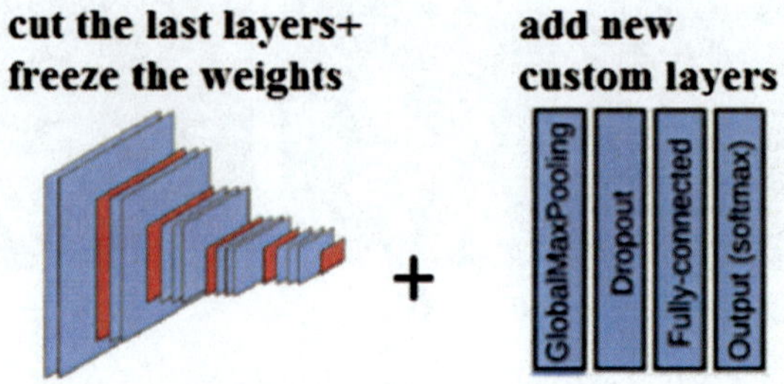

Fig. 2. Custom layers added to RestNet50

First, a pooling layer is needed. It has the role of reducing the feature maps obtained from the convolutional layers and making them more open to changes in the position of the features in the image. The chosen layer is a global maximum pooling layer for spatial data, since the problem works with images. Maximum pooling will take a region from the image and get the largest value from that region, the result being a single value. The global pooling will take the whole feature map, not just a region and output a single value. This type of pooling was used because it is the last pooling layer in the model and the main feature can be summarized more roughly.

After the pooling layer, a dropout layer was added. The purpose of it is to avoid the overfitting of the model. A rate for dropout must be set carefully. The rate represents the percentage on which the outputs from the activation function at the previous layer will be dropped (set to zero) before feeding them to the next layer after the dropout layer. For our network, a rate of 0.2 has been set.

The next layer added is a fully connected one with 128 units. The output space is set by this parameter representing its dimensionality. For this layer, the activation function used is hyperbolic tangent function (tanh function). An activation function is needed in order to determine which outputs of the neurons are relevant and should be passed forward to the following layers. The activation function takes the sum of the weighted inputs and transforms that number to fit between a lower and a higher limit, according to an interval. For tanh function, that interval is between -1 and 1.

At the end, an output layer will be the one that makes the predictions. It should consist of a number of neurons equal to the number of classes for the classification, two in our case. This layer will be a fully connected layer. The activation function is desired to be able to give the result in percentage values for each possible class. This is exactly what softmax activation function does, hence the reason it was chosen. The output will be given as an array of predictions for the two classes, namely tumor and healthy brain.

After the network is built, it is ready for the training process. As loss function chosen for our model we have used categorical cross-entropy and as optimization algorithm we have used Adam optimization algorithm because it allows parameters to have a unique learning rate each one and to be adapted individually. The setting value for the learning rate is very important because when the learning rate has a value too high, the model overshoots. Overshooting means that the steps taken in minimizing the loss value are too big and they step over the optimal value, without reaching it. The opposite happens if the value set for the learning rate is too small. The learning process will be too slow and will probably require a larger number of epochs. In our case we have set a learning rate of 0.0001 and for this learning rate a number of epochs of around 45 has provided good results.

Performance Evaluation. The final step is evaluating the performance of the obtained model. The history of the training process provides valuable information that can be used for performance evaluation of the model. The accuracy and loss values are the main indices when evaluating the model in general. The history of the training provides information about those two parameters at each stage, after each epoch. If the training loss begins to differ too much from the validation loss, it could imply that overfitting occurred. In this case, some changes should be made such as reducing the number of epochs, lowering the learning rate. This situation happened during training and it led to the optimal result obtained in the end.

3 Experimental Results

This section presents the experimental results obtained by applying our method on a data set provided by Kaggle [10]. The considered data set consists of 160 images out of which 80 represent normal brain MRIs and the other 80 represent brains with tumorous tissue. A sample of the data set can be seen in Fig. 3.

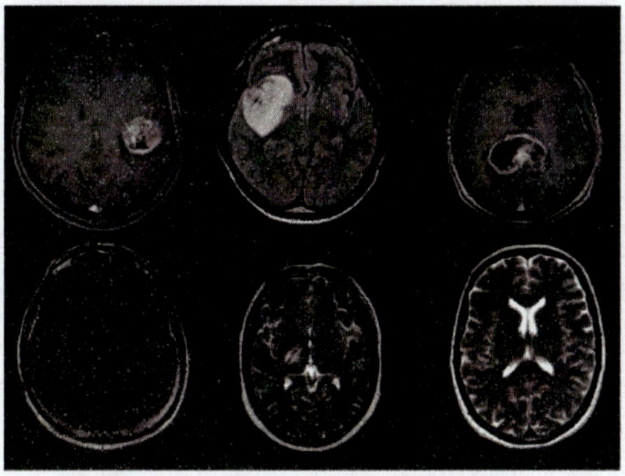

Fig. 3. Sample from the set of data

The images on the first row are from the class containing tumors, while the images on the second row represent healthy brain MRIs. The data set was divided into training and validation sets, following the 80% to 20% rule.

By applying the CNN based method, our goal was to obtain a model with a high accuracy that gives reliable results. Figures 4 and 5 illustrate some examples of brain tumor detections that was obtained with our method applied on images from the considered data set. The resulted class is the one with the percentage value above 50%.

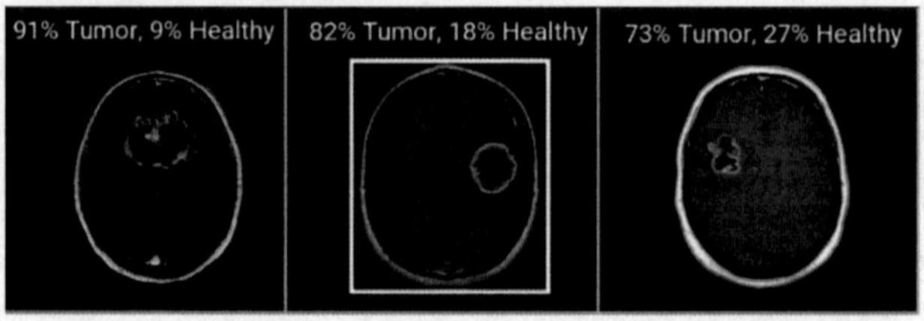

Fig. 4. Brain MRIs with tumors

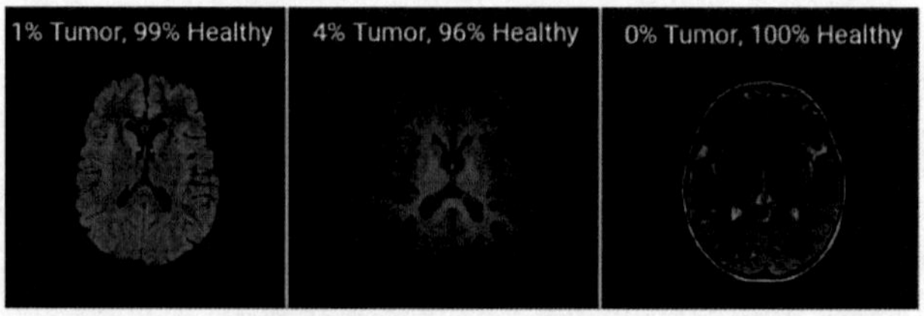

Fig. 5. Healthy brain MRIs

The above detection was made using the best trained model which has the highest accuracy (i.e. training accuracy of 88%) and the lowest loss (i.e. loss value of 0.27) value in comparison with all the other models created through a process of trial and error. The validation accuracy for this model was of 81% with a loss of 0.32. Other metrics such as *precision, recall (i.e. sensitivity) and F1 score* was computed for the best model. They are all calculated depending on the number of false positive, false negative, true positive and true negative samples and the values has. *The obtained values for these metrics are: 1 for precision, 0.75 for recall and 0.85 for F1 score.*

A comparative analysis between the best 5 trained obtained models is presented in Table 1. As it can be seen, for the obtained models we have varied the batch size, the number of epochs, the used optimizer, the learning rate, the loss function, the dropout, the FC layer and the output layer. From Table 1, it can be noted that better results were obtained when the learning rate was lowered but the epochs number was at the same time increased to compensate the slower learning process. Also, the best results were obtained for Adam optimization algorithm with a learning rate of 0.0001. The dropout rate was also lowered which helped the learning process.

Table 1. Comparative analysis between the best trained obtained models

	Model1	Model2	Model3	Model4	Model5
Accuracy	88%	72%	69%	71%	73%
Loss	0.27	0.46	0.62	0.53	1.04
Val_acc	81%	78%	71%	62%	71%
Val_loss	0.32	0.37	0.46	0.54	1.10
Batch size	16	16	16	16	32
Epochs	45	15	20	10	3
Optimizer	Adam	AdaDelta	Adam	AdaDelta	AdaMax
Learning rate	0.0001	0.001	0.0001	0.001	0.001
Loss function	Categorical cross-entropy	Categorical cross-entropy	Categorical cross-entropy	Categorical cross-entropy	Squared hinge
Dropout	0.2	0.5	0.5	0.5	0.5
FC layer	128 units + tanh	128 units + tanh	512 units + tanh	512 units + tanh	128 units + tanh
Output layer	softmax	softmax	softmax	softmax	softmax

Figure 6 presents the values of accuracy over epochs, while Fig. 7 presents the loss values over epochs obtained during the training process for the best trained model. In the case of Fig. 6, it can be seen that after 20 epochs the validation accuracy started to decrease while the training accuracy increased but they corrected and intersected again after 35 epochs.

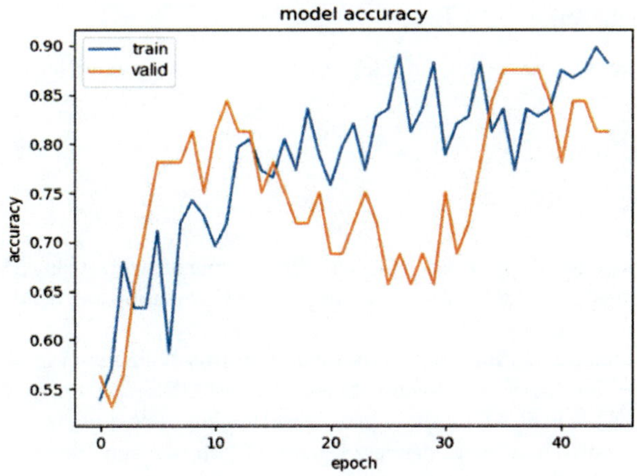

Fig. 6. Accuracy over epochs diagram for the best model

In Fig. 7, it can be observed that similar with the case of the accuracy, at approximately 25 epochs the validation loss increased dramatically in comparison with the training loss, but it regularized again after 30th epoch.

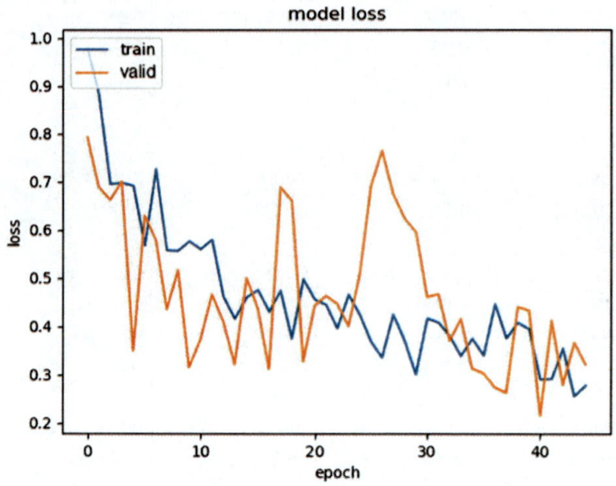

Fig. 7. Loss values over epochs diagram for the best model

4 Conclusions

In this paper we have proposed a method for brain tumors detection using CNN. Keras Deep Learning library was used for implementing the neural network. In order to solve the problem of insufficient data, a number of image augmentation techniques implemented in Keras were applied. The method has been validated on a magnetic resonance images (MRI) data set.

Conflict of Interest. The authors declare that they have no conflict of interest.

References

1. Kumar, T.S., Rashmi, K., Ramadoss, S., et al.: Brain tumor detection using SVM classifier. In: 3rd IEEE International Conference on Sensing, Signal Processing and Security, pp. 318–323 (2017)
2. Prabin, A., Veerappan, J.: Robust classification of primary brain tumor in MRI images based on multi model textures features and kernel based SVM. Biomed. Pharmacol. J. (2015)
3. Abdalla, H.E.M., Esmail, M.Y.: Brain tumor detection by using artificial neural network. In: International Conference on Computer, Control, Electrical, and Electronics Engineering, pp. 1–6 (2018)

4. Virupakshappa, A.B.: An automated approach for brain tumor identification using ANN classifier. In: International Conference on Current Trends in Computer, Electrical, Electronics and Communication, pp. 1011–1016 (2017)
5. Das, S., Aranya, O.F.M.R.R., Labiba, N.N.: Brain tumor classification using convolutional neural networks. In: 1st International Conference on Advances in Science, Engineering and Robotics Technology (2019)
6. Khan, S., Zhao, Z.N., et al.: Brain tumor classification for MR images using transfer learning and fine-tuning. Comput. Med. Imag. Graph. **75**, 34–46 (2019)
7. Deepak, S., Ameer, P.M.: Brain tumor classification using deep CNN features via transfer learning. Comput. Biol. Med. **111** (2019)
8. Guan, S., Loew, M.: Breast cancer detection using transfer learning in convolutional neural networks.In: IEEE Applied Imagery Pattern Recognition Workshop (2017)
9. Talo, M., Baloglu, U.B., et al.: Application of deep transfer learning for automated brain abnormality. Cogn. Syst. Res. **54**, 176–188 (2019)
10. BMI. https://www.kaggle.com/navoneel/brain-mri-images-for-brain-tumor-detection
11. ImageNet. http://www.image-net.org/

Machine Learning for Fatigue Estimation and Prediction "An Introduction Study"

Lilia Aljihmani[1(✉)], Doru Ursutiu[2], Samoila Cornel[3], and Khalid Qaraqe[1]

[1] Department of Electrical and Computer Engineering, Texas A&M University at Qatar, 23874 Doha, Qatar
{lilia.aljihmani,khalid.qaraqe}@qatar.tamu.edu

[2] Department of Electronics and Computers, Transylvania University of Brasov, 500036 Brașov, Romania
udoru@unitbv.ro

[3] Department of Materials Science, Transylvania University of Brasov, 500036 Brașov, Romania
csam@unitbv.ro

Abstract. Fatigue is considered as reduced workability and motivation that affects physical, emotional, and mental activeness. It is a critical concern that influences the precision and accurate implementation of some tasks or the emotional condition. Early detection of fatigue onset is crucial, such that preventative or corrective controls may be presented to minimize work-related traumas, the inexact performance of a task that require high-level accuracy, as well as to avoid making a wrong decision as a result of tiredness.

Our goal is to create a non-invasive, proactive model for real-time fatigue estimation based on typical features as tremor, heart rate, and blood oxygen saturation. We expect to set up a relation among the handshaking, heart rate, and oxygen level on one side and the weariness onset on the other. We will use a compact high-precision accelerometer to capture the low-frequency physiological tremor and an optical sensor to detect the heart rate and blood oxygen saturation. Intelligent learning algorithm will be used to personalize user characteristics, such as baselines of the tremor, heart rate, and oxygen level.

Keywords: Fatigue · Tremor · Heart rate · Blood oxygen saturation · Machine learning

1 Introduction

Fatigue is a common non-specific condition that identifies with reduced ability to work and motivation to implement voluntary tasks or emotional disturbance [1]. Tiredness accumulation could cause a chronic fatigue syndrome, overtraining syndrome, disorders, dysfunction, or diseases on human health. Fatigue is classified as physical when it affects the muscles' motion performance, mental in case it is related to overworking or working under pressure and stress, and emotional that is known as long-term or chronic fatigue with incapability to retain mental focus [2].

S. Vlad and N. M. Roman (Eds.): MEDITECH 2020, IFMBE Proceedings 88, pp. 226–231, 2022.
https://doi.org/10.1007/978-3-030-93564-1_25

Fatigue is influenced by multiple factors, which is the main reason for the lack of a tool that estimates the exhaustion level. The main techniques that are used for tiredness evaluation are motion detection and estimation, electromyography, questionaries, and fatigue scales, perceived exertion and discomfort scales, visual attention tests, analysis of the work quality, productivity and number of errors, and some physiological parameters as heartbeat/heart rate, blood pressure, blood oxygen saturation (SpO_2), biomarkers, force measurement, etc. Typical signs of physical load and fatigue are increased heartbeat/heart rate and diminished blood oxygen level. It is reported that reduced SpO_2 in muscles has significant effect on muscle tiredness [1].

A significant indication of the tiredness or exhaustion is tremor [4], defined as a slightly involuntary shaking of a body part [5]. It can be detected in healthy individuals (physiological tremor). Physiological tremor is rarely visible and is not considered as a disease. It is a normal human phenomenon that is a result of physiological properties in the body (for example, rhythmical activities such as heartbeat and muscle activation). It is amplified by anxiety or panic, stress, some medicine, etc. It can also be caused by fatigue as a natural event after making a long-time effort or short-time efforts with high-intensity. Physiological tremor with a frequency between 8 to 12 Hz and an amplitude of the oscillations in the range 50–100 μm range [6], is usually reversible once the underlying cause is addressed. The quantitative measurements of tremor frequency and amplitude cannot give a detailed information regarding the tremor type. To determine the characteristics that are independent of the body part motion, features in time and frequency domain as amplitude, power spectra density distribution, frequency dispersion, frequency peak, etc., should be calculated [7].

Machine learning (ML) techniques are applied to recognize daily human activities and tremor severity evaluation [8, 9], as well as to interpret the tremor data and predict fatigue [10, 11]. Recently in microsurgery, handled instruments with active tremor compensation are used to avoid the influence of the oscillations of the operator's hand based on the physiological tremor or handshaking affected by tiredness. Alongside the improvement of clinical methods, the robotic systems with the highest significance of the precision, stability, and controllability are used. For tremor compensation, distinctive and accurate recognition and separation of the unwanted involuntary hand movements in real-time are required.

2 Objectives

We aim to create a non-invasive, proactive model for real-time fatigue estimation based on typical signs of tiredness as tremor, heart rate, and SpO_2. The framework aims are to distinguish the early signs of tiredness and exhaustion in workers to prevent them from unfavorable outcomes of fatigue and predict fatigue onset using the motion and physiological data. We will apply advanced ML techniques to create a mobile phone application to alert the user when the first signs of fatigue appear.

Accomplishments in ML and wearable sensor innovation have made the potential for creating frameworks that use the information gathered from the sensors and predict fatigue onset in real-time. The suggested approach will give an answer to address a fundamental and pragmatic requirement for care for the user's physical, mental, and emotional well-being. It is an advanced methodology for distinguishing the first manifestations of the tiredness.

3 Methodology

A group of 20 volunteers older than 40 years will be invited to participate in the study. Three target subgroups will be recruited: subjects working in unfavorable conditions such as high level of contamination, heat, humidity, noise, etc. (field or outdoor workers); subjects implementing long-time effort or short-time effort with high-intensity (machine operators, loader, etc.); subjects working under stress (emergency room workers, etc.). They will be asked to wear the watch for one month, implementing the daily activities.

A compact high-precision accelerometer integrated into smartwatch or in a commercially available device will be used to capture the low-frequency physiological tremor. Smartwatch's optical sensor will collect the heart rate data. The blood oxygen saturation will be examined by a sensor integrated into a ring available in the market. Collected data will be Bluetooth transferred to a smartphone where signal analysis software will estimate the data and will create a model that predicts fatigue based on the tremor and heartbeat quantification (Fig. 1).

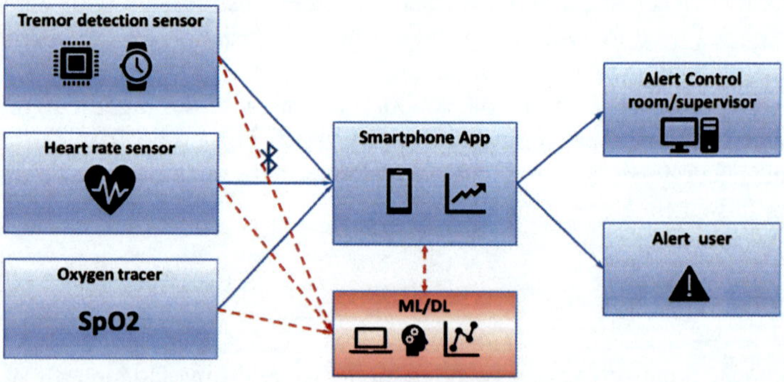

Fig. 1. The principle of operation of the system for detection and prediction of fatigue onset.

To collect the accelerometer and heart rate data continuously in real-time, we could use smartwatch as TicWatch E or Polar M600 (Fig. 2) that combine both sensors. Smartwatch application (Android or iOS) will record and plot the accelerometer data. A smartphone application will receive collected by the smartwatch data and will analyze it.

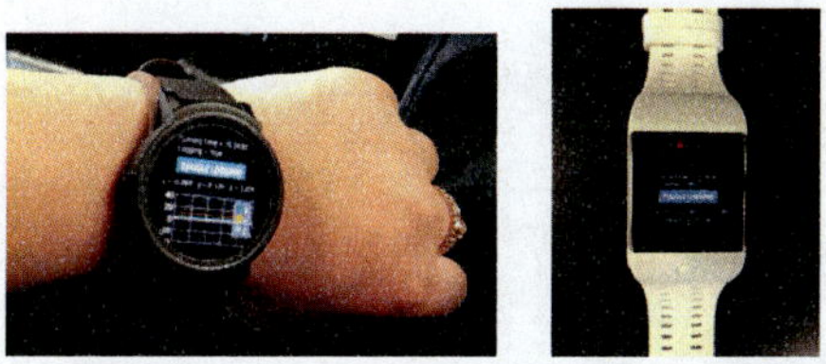

Fig. 2. Smartwatches that will be used in the study: TicWatch E and Polar M600.

Motion data could also be collected using a MetaMotionR sensor by MbientLab (Fig. 3). It is a circular sensor with a diameter of 20 mm and a lifetime upwards of 24 h that uses a Li-Po rechargeable battery. A personal wearable sleep tracker GO2SLEEP HST by Sleepon, will be used to monitor the heart rate and blood oxygen level (Fig. 4). It consists of 660 nm and 940 nm dual-wavelength reflective photoplethysmogram sensor to detect the finger arterial capillaries. Data will be recorded with a sampling frequency of 50 Hz; the calibration range is from 70% to 100% for SpO_2. Data from both devices will be transferred to a smartphone for analysis.

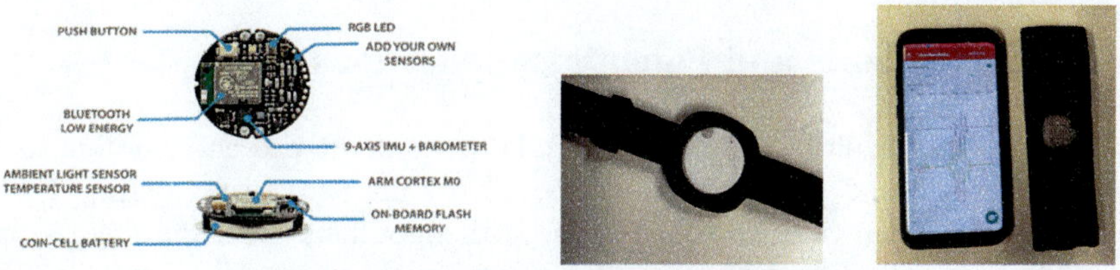

Fig. 3. MetaMotionR sensor by MbientLab.

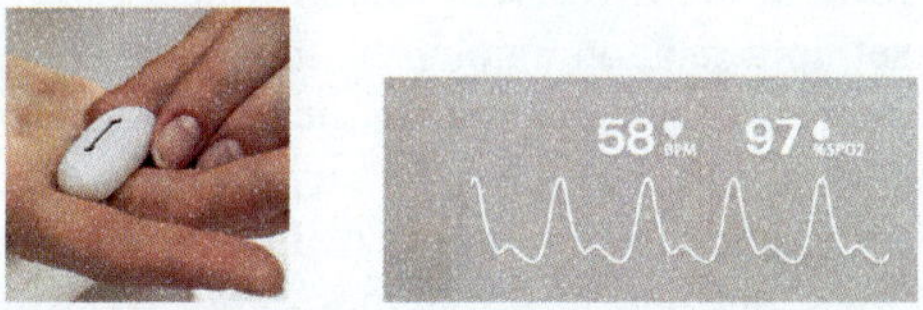

Fig. 4. Rings measuring heart rate: GO2SLEEP HST by Sleepon.

Intelligent learning algorithms will be used to personalize user characteristics, such as the baselines of the tremor, heart rate, and oxygen level. ML techniques as Support Vector Machine, Ensemble classifiers, and Neural Network will be applied to discover arbitrary models or structures in the recorded data. This approach will be based on the innovative extraction of the features in time and frequency domain from the raw accelerometer data, heart rate, and SpO_2 that are relevant to fatigue detection (Fig. 5). Auditory and vibrotactile alerts will trigger when tremor characteristics changes correspond to the fatigue onset.

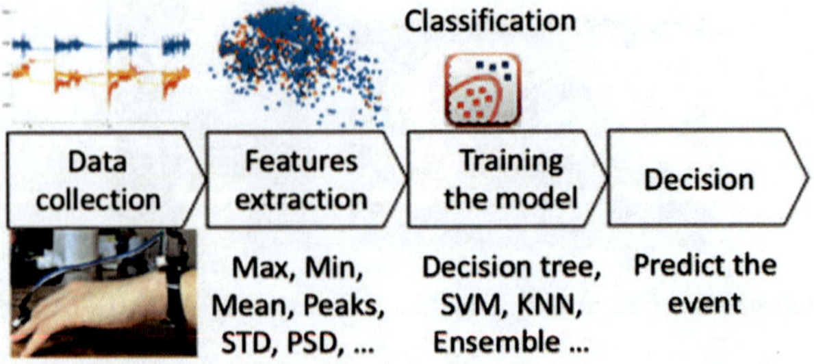

Fig. 5. Overview of the machine learning schema.

To accomplish this, we will design and develop a model for fatigue estimation and detection of the early onset of fatigue based on the recorded tremor, heart rate, and SpO_2 data. The data enable tracking of the health status of risk groups. We will create a proactive model that will identify the user's first signs of tiredness in real-time and will alert him/her and the control center/supervisor to provide the rest spots of the vulnerable area and ensure the relaxation to continue work and make the right decisions.

4 Expected Results and Potential Applications

In a previous study with results published in [12], we developed an algorithm for the recognition of rest and effort tasks implemented by the participants during the performance of maximum voluntary contractions. ML algorithms (Decision Tree, Support Vector Machine with different kernels, k-Nearest Neighbor algorithm, and Ensemble Bagging Tree classifier) were applied to classify and identify the rest and the effort tasks. The achieved performance of the models was 96%.

We are target to create a non-invasive, proactive model for real-time fatigue estimation. We expect to set up a relation among the tremor, heart rate, and SpO_2 on one side and the onset of weariness and fatigue on the other. After the tremor detection, accurate recognition and separation of the voluntary and undesirable involuntary hand movements in real-time should be implemented. In the developed ML model, age, gender, co-morbidities that can impact the tremor, and exercises being performed as well, will be taken into consideration to create a personalized weariness level prediction for the user.

We expect to define a relevant admonition framework for the presence of early indications of fatigue. It will keep up the physical, emotional, and mental health of the user by distinguishing the early symptoms of tiredness and cautioning that break is required.

The framework could be utilized in microsurgery or in the fields where hand/body exactness, steadiness, and controllability are critical. It can help to recognize the tremor of the operator and develop models for tremor indemnity to isolate undesirable hand motions in real-time. The fatigue detection device will be suitable for usage from the people working in unfavorable conditions such as high level of contamination, heat, humidity, noise, etc., to alert the supervisors for the fatigue level of the workers and to take preventative or corrective controls to minimize work-related traumas, as well as in other cases where the making the right decision quickly is with high importance (pilots, public transport drivers, etc.). Since the exhaustion is additionally a weakening result of long work hours, lacking recovery, and a helpless errand plan (e.g., motion repetition, intense efforts, etc.), it is crucial, as it can influence the exactness and precise execution of particular assignments.

Conflict of Interest Statement: The authors declare that they have no conflict of interest.

References

1. Wan, J.-J., Qin, Z., Wang, P.-Y, Sun, Y., Liu, X.: Muscle fatigue: general understanding and treatment. Exp. Mol. Med. **49**(10), e384–e384 (2017)
2. Gonzalez, K., Sasangohar, F., Mehta, R.K., Lawley, M., Erraguntla, M.: Measuring fatigue through heart rate variability and activity recognition: a scoping literature review of machine learning techniques. In: Proceedings of the Human Factors and Ergonomics Society Annual Meeting, pp. 1748–1752. Sage, Los Angeles (2017)
3. Elble, R.J., Higgins, C., Moody, C.J.: Stretch reflex oscillations and essential tremor. J. Neurol. Neurosurg. Psychiatry **50**(6), 691–698 (1987)
4. Ebenbichler, G.R., et al.: Bochdansky T:Load-dependence of fatigue related changes in tremor around 10 Hz. Clin. Neurophysiol. **111**(1), 106–111 (2000)
5. Elble, R.J.: Tremor Neuro-geriatrics, pp. 311–326. Springer (2017)
6. Tatinati, S., Veluvolu, K.C., Hong, S.-M., Latt, W.T., Ang, W.T.: Physiological tremor estimation with autoregressive (AR) model and Kalman filter for robotics applications. IEEE Sens. J. **13**(12), 4977–4985 (2013)
7. Darnall, N.D., et al.: Application of machine learning and numerical analysis to classify tremor in patients affected with essential tremor or Parkinson's disease. Gerontechnology **10** (4), 208–219 (2012)
8. Jeon, H., et al.: High-accuracy automatic classification of Parkinsonian tremor severity using machine learning method. Physiol. Meas. **38**(11), 1980 (2017)
9. Zheng, X., Vieira, A., Marcos, S.L., Aladro, Y., Ordieres-Meré, J.: Activity-aware essential tremor evaluation using deep learning method based on acceleration data. Parkins. Relat. Disord. **58**, 17–22 (2019)
10. Tatinati, S., Veluvolu, K.C., Ang, W.T.: Multistep prediction of physiological tremor based on machine learning for robotics assisted microsurgery. IEEE Trans. Cybernet. **45**(2), 328–339 (2014)
11. Veluvolu, K.C., Ang, W.T.: Estimation of physiological tremor from accelerometers for real-time applications. Sensors **11**(3), 3020–3036 (2011)
12. Dandan, G., Kerdjidj, O., Aljihmani, L., Zhu, Y., Mehta, R.K., Qaraqe, K.: Rest and effort tremor detection using machine learning. Int. J. Adv. Electr. Comput. Sci. (IJAECS) **7**(7), 7–11 (2020)

Wireless Sensor Networks for Healthcare Monitoring

Ioana Borz, Tudor Palade, Emanuel Puschita, Paul Dolea, and Andra Pastrav(✉)

Technical University of Cluj-Napoca, 400114 Cluj-Napoca, Romania
Andra.Pastrav@com.utcluj.ro

Abstract. Monitoring systems are of great importance in patient healthcare. They provide real-time data about a patient's wellbeing and allow disease prevention. Employing wireless sensor networks for this task provides the advantages of remote monitoring, reduced hospitalization and increased mobility of the patient by reducing the amount of wiring. As such, this paper aims to design a healthcare monitoring system (HMS) based on the IEEE802.15.4 standard. Several simulation scenarios are implemented in QualNet Developer 6.1 to analyze the behavior of the HMS in a variety of cases: single static/mobile user, multiple users, data forwarding by means of cellular mobile networks, or periodical data collection by a drone. The performance of the system is evaluated in terms of QoS parameters. Energy consumption is also considered.

Keywords: Wireless sensor networks · Remote patient monitoring · IEEE 802.15.4 · ZigBee

1 Introduction

Careful monitoring of a patient's vital parameters ensures the early discovery of illness aggravation signs and rapid intervention, prevents hospitalization and leads to an increased quality of life of the patient. Moreover, to lower the costs of healthcare and to offer better patient care, remote patient monitoring has been widely employed [1].

Depending on the patient illness, several sensors might be needed to assess the bio-physiological parameters of interest. All the information collected by these specialized sensors is then sent to the healthcare provider to be processed and analyzed. To achieve a complete and complex patient monitoring system that enables continuous observation of patients, Wireless Sensor Networks (WSNs) can be used.

As such, this paper aims to design a wireless sensor network based on IEEE 802.15.4 for health monitoring, to implement the WSN in QualNet Developer 6.1, and evaluate its performance through simulation in terms of QoS parameters.

The remainder of this paper is organized as follows. Section 2 presents related works on WSNs for health monitoring, Sect. 3 describes the proposed WSN solution. Section 4 outlines the simulation scenarios while Sect. 5 shows the experimental results. Section 6 concludes the paper.

S. Vlad and N. M. Roman (Eds.): MEDITECH 2020, IFMBE Proceedings 88, pp. 232–239, 2022.
https://doi.org/10.1007/978-3-030-93564-1_26

2 Related Works

Taking into account the considerable advantages the WSNs are showing, especially in the health domain, a particular study about patient monitoring using Wireless Mesh Networks (WMN) is analyzed in [2]. The cabling needed for the connection to the backbone network is significantly reduced when WSNs are used. The monitoring of the patients is made respecting the following scenario: a bio-medical packet that is associated to the patient is used to automatically supervise the heartbeat, pressure and temperature of the patient.

Because WSNs are very suitable to deal with emergency situations and with abnormal conditions of the patients, in [3] is proposed an ECG and blood pressure supervision based on the wireless network technology and continuous information analysis using the mobile phone. In case of mobility, [4] provides an idea about a portable monitoring system where all the sensors are connected to a microcontroller, containing a wireless communication module. A prototype system has been designed with recommended medical standards, the obtained results indicating that the proposed healthcare system is effective and can be used to supervise outdoor patients [4].

The system proposed in [5] presumes a coordinator node attached on the patient's body to collect all the signals from the wireless sensors and send them to the base station. This system can detect the abnormal conditions, issue an alarm to the patient and send a SMS/E-mail to the physician [6, 7]. Also, the proposed system consists of several wireless relay nodes which are responsible for relaying the data sent by the coordinator node and forwarding them to the base station.

In [8], the recently proposed Body Sensor Networks (BSNs) combine context-aware sensing for increased sensitivity and precision, in addition to providing continuous assessment and control of physiological parameters. The BSN hardware development structure is developed to promote research and development in BSN and multi-sensor data fusion. The BSN nodes provide a scalable setting for wireless sensing technology development with their low power, versatile and compact design.

As most devices and their applications are wireless in nature, security and privacy concerns are among major areas of concern. Whether the data gathered from patients or individuals are obtained with the consent of the person or without it, misuse or privacy concerns may restrict people from taking advantage of the full benefits of the system. These issues are discussed and analyzed in detail in [9], indicating some possible solutions.

3 Proposed WSN Solution

The key goal of this paper is to design and test through simulation a WSN for health monitoring. The proposed WSN relies on the ZigBee IEEE 802.15.4 specifications and is implemented in QualNet Developer 6.1 network simulator.

The WSN consists of several sensors (Reduced Function Devices, RFDs) that collect data regarding the patient health and transmit it to the Personal Area Network (PAN) coordinator (a Full Function Device, FFD, or data sink).

The most common set of medical sensors used in patient supervision were considered, being important variables for disease prevention and treatment especially in the case of elder patients. The type and number of bio-physiological factors to evaluate in a mobile patient monitoring system can differ based on the implementation situation and the patient's health condition [1]. Table 1 summarizes the general characteristics and the performance requirements of some reliable, low power and small-size medical sensors [1].

Table 1. Transmission requirements for bio-physiological variables [1].

Bio-physiological variable	Delay sensitivity	Data rate (bps)
Electrocardiography (ECG)	$\sim$1 s	7,200–9,600
Motion detection	NO	7,200
Glucose monitoring	NO	640–3,200
Blood pressure	NO	1,920
Blood oxygen (SpO_2)	NO	1,920
Cardiac output	NO	640
Respiration	NO	300

For accurate simulation results, the features in Table 1 were considered when configuring the simulation scenarios. As such, Constant Bitrate (CBR) applications were configured to simulate the traffic generated by each wireless sensor (or acquisition front-end in case of multiple leads). Given the multiple leads in the ECG, the corresponding CBR resulted in a data rate of 9600 bps. A high sampling frequency for glucose monitoring resulted in a CBR of 3200 bps.

The simulation scenarios underwent a calibration phase to determine the set of configuration parameters that yields the best performance results. If not efficiently selected, parameters like MAC channel polling, routing protocol, or buffer size may result in decreased performance of the network. The main physical and MAC layer parameters are synthesized in Table 2.

After several studies performed on different scenarios, it was noticed that Ad-hoc On-demand Distance Vector (AODV) protocol is suitable only in case of monitoring less than 4 bio-physiological variables per patient. When the number of sensors is increased, the Dynamic Source Routing (DSR) protocol needs to be employed, offering better performances. In case of needing to forward the data to a collecting center through LTE or WLAN technologies, a challenge occurs taking into consideration that DSR protocol implementation in QualNet cannot support more than one interface. To solve this inconvenience another routing protocol was tested, the most satisfactory one for this situation being Dynamic MANET On-demand (DYMO) protocol.

Table 2. Main PHY and MAC settings in QualNet 6.1.

PHY parameter		MAC parameter (RFD)		MAC parameter (FFD)	
Radio type	802.15.4 Radio	MAC protocol		802.15.4	
Tx power (dBm)	3.0	Device type	RFD	Device type	FFD
Packet Rx	PHY802.15.4 Rx Model	Poll interval	1	FFD mode	PAN coord
Modulation scheme	O-QPSK	Beacon order	3	Enable GTS	Yes
CCA mode	Carrier sense	SF order	3	GTS trigger precedence	1
Antenna model	Omnidirectional			Data ack	No
				Beacon order	6
				SF order	6

4 Simulation Scenarios

4.1 Monitoring One Static Patient

The first simulation scenario considers the health monitoring of a hospitalized patient as presented in Fig. 1. In this case, the data collected by the wireless sensors are transmitted to the sink, which, in a real case, would allow the doctor in charge to verify the data. Device 2 is the PAN coordinator while the other devices are RFDs, all of them being connected to a subnet. The scenario dimension is set to be 4 × 4 m.

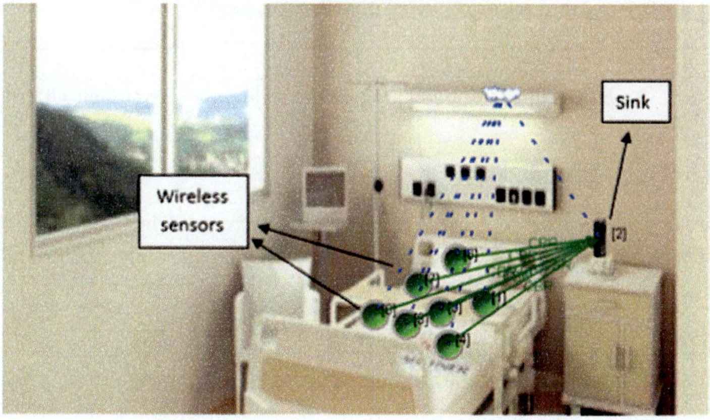

Fig. 1. Monitoring of one static patient.

4.2 Remote Patient Monitoring – Data Forwarding Through LTE

This scenario takes into account the case where the patient is at home but still requires health monitoring. As in the previous scenario, the sensors collect health data from the patient and transmit it to the sink. Then, the data collected by the sink are transmitted to the healthcare provider for analysis by means of an LTE network. This means that the sink must be equipped with an LTE radio interface as well. Making use of cellular mobile technologies to transmit the data to the care provider ensures the quality of the transmission, distance and delay not being an issue. The terrain dimension is set to be 10 × 15 m to account for the patient's home (Fig. 2).

This scenario also considers a mobile patient, Fig. 3 showing that the patient has a random path inside the house (marked with red flags), while the sink remains in the same position. As the patient moves, so do the sensors, thus varying the propagation conditions during data transmission. Device 9 represents a data collection center in the healthcare provider facility.

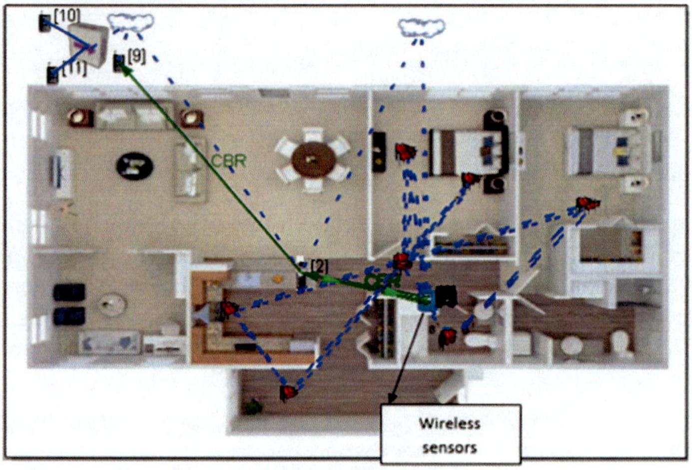

Fig. 2. Remote patient monitoring and data forwarding through LTE.

4.3 Monitoring Multiple Patients

In a similar approach, this scenario accounts for the monitoring of group of patients from an elderly center. There is a sink for each patient, gathering health data from the sensors and forwarding them to the medical staff in charge of patient monitoring. Given that data forwarding is done by means of an LTE network, the data collection center can be in the same building as the patients, or in a remote area. Figure 3 illustrates this scenario, where 5 patients are being monitored.

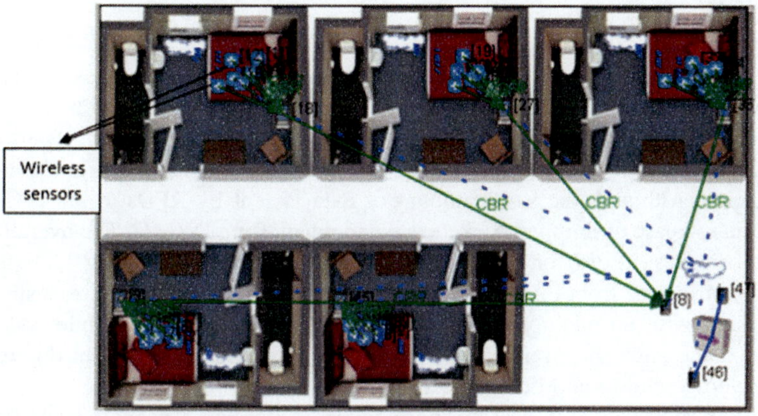

Fig. 3. Monitoring of multiple patients.

4.4 Remote Monitoring of Multiple Patients in an Isolated Area

While the previous case made use of LTE to ensure proper data transmission to the monitoring center, this scenario accounts for the monitoring of patients living in an isolated area, where the infrastructure for cellular mobile communications (e.g. LTE) is not available. In this case, the data can be collected using a drone which will fly by each house and wait to download the data from the sink (1 min in this case). After the data is downloaded from the first patient, the drone will fly to the next one and so on until all the information is collected. In this scenario, illustrated in Fig. 4, the sink continuously collects the sensor data but forwards it only when the drone is in the vicinity. A temporary ad-hoc WLAN network is set up between the PAN coordinator and the drone for data collection. The terrain dimension is set to 400×300 m.

Fig. 4. Monitoring of multiple patients in an isolated area.

5 Results

For the first simulation scenario, the DSR protocol was used. The packet loss rate is low, being 0.87% overall. The overall average end-to-end delay is 0.42s.

For the second simulation scenario, where the data has to be forwarded to a monitoring center, the routing protocol used is DYMO, which changes the network performance. Although the same amount of data is sent by RFDs to the PAN coordinator, an increase of dropped packets is noted when using DYMO. The overall packet loss rate reaches 1.26% for a static patient. As expected, employing LTE for data forwarding from the from PAN coordinator to the collecting center ensures a successful transmission with no additional packet loss. When the patient is mobile and moves away from the sink, the overall packet loss rate increases to 1.28%. In this case the overall average end-to-end delay is 0.16 s.

In the third scenario, increasing the number of monitored patients leads to worse network performance. The packet delivery is affected by node placement, application start time, DYMO protocol settings and node opportunity to access resources. The overall packet loss is 1.92% and the overall average end-to-end delay is 1.04 s.

For the last simulation scenario, the overall packet loss rate reaches 2.3%, but the average end-to-end delay is 0.6 s.

Table 3 summarizes the obtained results which show that the performance of the network is affected by various factors such as number of sensors per patient, number of supervised patients, or interfacing with other networks for data forwarding. The packet loss rate increases with the complexity of the network, while the delay is highly dependent on the existence of other networks and possible interferences. The jitter is insignificant, given the application type.

Table 3. Overall QoS performance.

Scenario	Packet loss rate (%)	Average delay (s)	Average jitter (ms)
1 – one static patient	0.87	0.42	39
2 – one remote mobile patient	1.28	0.16	15
3 – multiple patients	1.92	1.04	70
4 – multiple patients in an isolated area	2.3	0.6	49

In order to estimate the battery lifespan, the simulation time was set to 6 h. The power consumption rate during this period is approximately 0.3%. Therefore, the battery will be empty after 2000 h, or about 83 days when sensors are sending data continuously. The battery can be changed for all sensors at once when the battery runs out at the most requested sensor, or in turn if necessary.

6 Conclusions

The effective implementation of a WSN for patient monitoring relies on several key aspects. First, the bio-physiological variables of interest and the associated traffic profiles have to be evaluated in order to determine the specific QoS requirements for each type of sensor. Next, the requirements in terms of number of devices, placement, propagation conditions and mobility have to be taken into account when deciding what topology is suited for the use-case scenario. Then, the WSN candidate is selected, determining the communication protocols and standard specifications (in this case ZigBee based on IEEE 802.15.4) to meet the previously determined requirements. Interference with existing networks must be considered when selecting the operating frequency. Device interfacing with other wireless communication technologies may be required if data needs to be forwarded to a remote monitoring center.

Due to their performance, small size and low power consumption, WSNs are suitable for patient health monitoring and transmitting the collected results to the healthcare center, leading to decreased healthcare costs and an increased quality of life of the monitored patients.

Compliance with ethical standards

Conflict of Interest. The authors declare that they have no conflict of interest.

References

1. Casillas, M., Villarreal-Reyes, S., González, A.L., Martinez, E., Perez-Ramos, A.: Design guidelines for wireless sensor network architectures in mHealth mobile patient monitoring scenarios. In: Mobile Health: A Technology Road Map, pp. 401–428 (2015)
2. Vijayalakshmi, B., Ram, K.C.: Patient monitoring system using wireless sensor based mesh network. In: Third International Conference on Computing, Communication and Networking Technologies (2012)
3. Chung, W.-Y., Lee, S.-C. Toh, S.-H.: WSN based mobile u-healthcare system With ECG, blood pressure measurement function. In: 30th IEEE Annual International Conference on Engineering in Medicine and Biology (2008)
4. Kasim, A.-A., Mutairişi, A.A., Ahmad Mohammad, D.: Real-time healthcare monitoring system using wireless sensor network (2017)
5. Media, A., Hamid, R.N.: A hospital healthcare monitoring system using wireless sensor networks. J. Health Med. Inform. **4**, 2 (2013)
6. Standford, V.: Using pervasive computing to deliver elder care. IEEE Pervas. Comput. **1**, 10–13 (2002)
7. Anliker, U., Ward, J., Lukowicz, P., Gerhard, T., Françoisşi, D., Michel, B.: AMON: a wearable multiparameter medical monitoring and alert system. IEEE Trans. Inf. Technol. Biomed. pp. 415–427 (2004)
8. Benny, P.L., Surapa, T., Kingşi, R., Yang, G.-Z.: Body sensor network – a wireless sensor platform for pervasive healthcare monitoring (2005)
9. Moshaddique, A.A., Jingwei şi, L., Kyungsup, K.: Security and privacy issues in wireless sensor networks for healthcare applications. J. Med. Syst. **36**, 93–101 (2012)

Smart House Control Using Hand Gestures Recognition LabVIEW Applications

Rodica Holonec(✉), Simona Vlad, Nicolae M. Roman, and Laszlo Rapolti

Department of Electrotechnics and Measurements, Technical University of Cluj-Napoca, Cluj-Napoca, Romania
rodica.holonec@ethm.utcluj.ro

Abstract. Smart homes have a positive impact on people with physical disabilities by improving their quality of life. Thus, depending on the health and physical condition of the person, a certain type of architecture can be chosen. This paper presents a system designed to command a smart house low-cost prototype being based on two modalities of hand gestures recognition. The first approach is referring to the usage of an envelope signal taken from an electromyography (EMG) sensor. The gestures sequence of tightening and releasing an elastic object in the hand was coded in such a way that through the digital output signals, different appliances in a smart home can be controlled. The second application uses hand gestures images and with the help of a recognition algorithm the same smart home prototype can be commanded. A particle classifier was used in order to recognize nine hand gestures. The beneficiary of this system could be mainly a person with motor disabilities that can manage in a simple and relatively inexpensive way any household devices such as lights, doors, fans, TV sets, heaters and others. The applications are implemented in LabVIEW environment from National Instruments (NI) and they offer a friendly interface and also a simple and robust way to communicate with the relays module that controls the intelligent house.

Keywords: Smart house · Hand gestures recognition · Electromyography · Image processing · LabVIEW

1 Introduction and Related Work

Home automation refers to the integration of systems like lighting, security, heating, air conditioning, video, audio and other household appliances in a single central interface. Thus, the smart homes can be controlled in different implementation approaches starting from the simple button press to voice or gestures recognition.

There are many assistive domotics that are specially designed for elderly people with disabilities [1]. In this case the the interaction with appliances based on the hand gesture recognition is considered often more natural and simpler than that of using different computer peripherals like keyboard, mouse or remote-operated devices [2]. The hand gesture systems were also designed for medical assistance like in [3] where a wheelchair is driven intelligently by using a human-computer interaction based on a

S. Vlad and N. M. Roman (Eds.): MEDITECH 2020, IFMBE Proceedings 88, pp. 240–249, 2022.
https://doi.org/10.1007/978-3-030-93564-1_27

gesture vocabulary. In the same area of interest, hand gesture recognition was used to switch different electrical equipment (lights, fan) or to communicate with the hospital staff [4].

The hand gesture can be recognized in real time from electrical signals or images. In the first case, the information can be acquired from diverse wearable devices like accelerometers or gyroscopes. Thus, such systems like that presented in [5] uses ten gestures for controlling a smart house. Similarly, in [6] a hand gesture recognition system was developed and based on two deep models representing features from several mobile sensors.

Electromyography (EMG) signal is the basis of many applications in the field of hand gesture recognition. The signal captured by the electrodes can be processed by specialized circuits so that after digitization it can be processed and interpreted for directing different prosthetic devices, machines or robots [7, 8]. The acquisition of the EMG signal can be done non-invasively using surface EMG or invasively using intramuscular EMG. Surface EMG involves the sensors placing on the skin in order to measure the electrical activity of the muscles. The second variant, intramuscular EMG, consists in inserting the needle into the muscles. The real time gesture recognition based on EMG signal uses often machine learning models [9]. Here new techniques were developed in preprocessing, feature extraction, classification and postprocessing stages. In [10] a Myo Sensor is used for designing a device model that helps people with disabilities to have a normal live.

The gestures recognition using hand images was a permanent preoccupation regarding the field of human-computer interaction. There are studies, like in [11], where diverse machine learning algorithms are presented and compared. In [12] the authors propose an accurate RGB hand gesture recognition model using a two-stage deep convolutional neural network (CNN) architecture. Another approach uses a wavelet-based feature extraction and supervised Artificial Neural Networks (ANNs) based classification [13]. In [14] the gestures can be recognized with high accuracy from low resolution images.

The concept of smart home is often correlated with people with special needs, and consequently there are a large variety of smart home models. The main objective in designing a smart home for people with mobility disabilities is the installation of devices to assist in mobility. The selection of devices varies from one category to another depending on the needs of the person. The right choice of devices must give to the inhabitant the confidence regarding the mobility and possibility to communicate with domotic environment [15].

Our paper presents a smart house prototype that is controlled by using two different hand gestures recognition LabVIEW applications. The first one uses the rectified envelope signal from a surface EMG (MyoWare) sensor. This signal is used to command the household appliances. In the second application, different hand gesture images are acquired, processed and classified in order to operate the the same house devices.

2 System Design

Figure 1 shows the schematics of our hand gesture recognition system and its interaction with the smart house. The system contains two main parts: the first one contains the prototype where the appliances are switched using a module of relays and the second one is referring to one of the hand gesture recognition LabVIEW applications: the one based on an EMG signal and the one based on hand gesture image recognition.

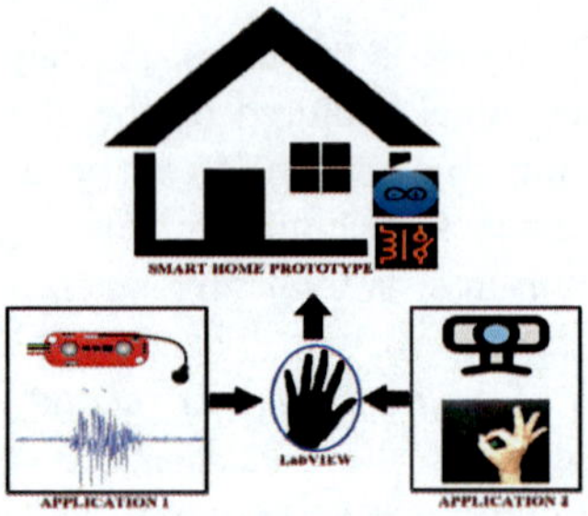

Fig. 1. The schematics of the hand gesture recognition system

2.1 The Smart House Prototype

The idea of designing and implementing of smart house prototypes was and continues to remain the main preoccupation of many researchers. Thus, in [16] an integrated system representing a smart house was presented in detail. This approach is useful when an evaluation of the system compliances with its requirements is needed.

The goal of our research was to implement and test two types of applications regarding a smart house control by using the hand gesture recognition. In order to achieve these objectives, we developed a low-cost smart house prototype where the opening and closing of the doors, the switching of lights, TV station and fan are possible due to the user hand gestures. The proposed system contains the home environment, the Arduino Uno unit and the a module with 12 relays. The digital outputs of Arduino board and the relays inputs were used to test a software simulation with all the commands necessary to switch the appliances in the house. Each ON/OFF state will represent one of the recognized hand gesture. The prototype components were installed on a wooden house provided with fours rooms and two doors like in Fig. 2a. The electrical diagram of the house prototype and its main parts: Arduino board, relays module, Light Emitting Diode (LED), dc motors, fan, power supplies are presented in Fig. 2b.

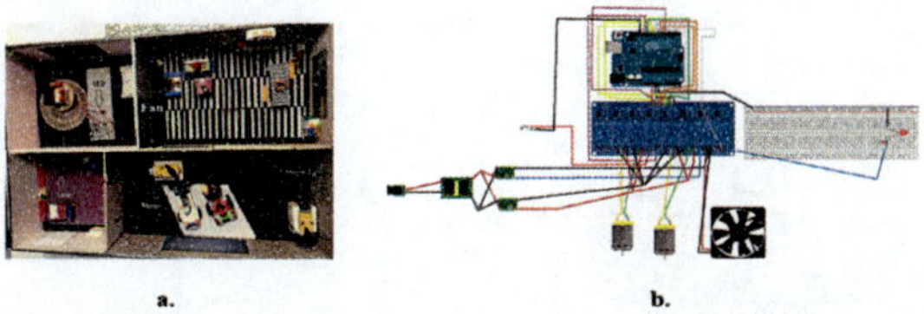

Fig. 2. The smart house prototype: a. Top view, b. Electrical diagram

3 Application 1. Smart House Control Using EMG Signal

In the first LabVIEW application an EMG sensor is used to detect the gestures and thus different devices in the smart house can be operated. An elastic ball is tightened and released several times by the user and the number of these actions will be coded. The opening or closing of the appliances in the house will be decided, simply, only by playing with the ball.

3.1 Hardware Components

The hardware part for the application based on EMG signal consists of a MyoWare surface sensor placed on the users hand and a NI-myDAQ data acquisition board connected like in Fig. 3.

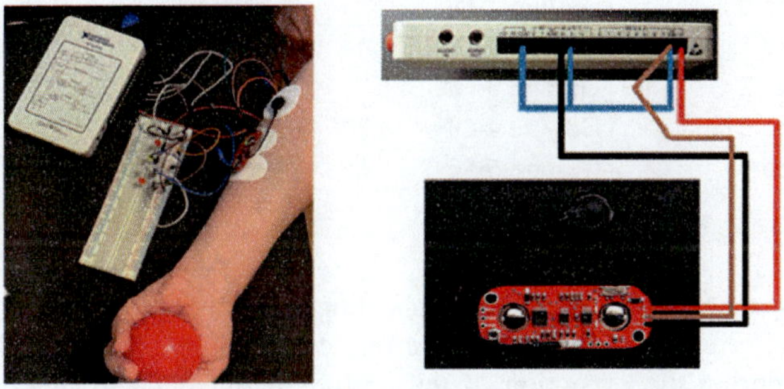

Fig. 3. Applications 1-the hardware components

The electrodes placement was chosen in order to have the maximum response for the user's hand gestures. The envelope signal from the SIG output on MyoWave sensor board was used to codify the gestures. The application was tested for recognition of five gestures and for this purpose five LEDs were used as seen in Fig. 4.

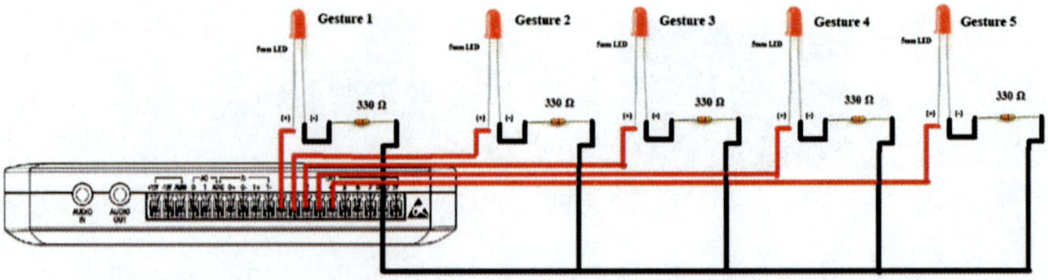

Fig. 4. The schematics for testing the application 1

3.2 Methodology and LabVIEW Application

The main objective of the application was to use the signal from the MyoWave EMG sensor in order to drive the digital outputs of the NI-myDAQ acquisition board. The user has the possibility to tighten and release the object in the hand 1, 2, 3, 4 or 5 times. Two successions of repeated movements performed by the user were considered as one gesture. The identified code can thus be used to drive the various functions of a smart home. These functions were tested on the five LEDs placed at the digital outputs of the acquisition board (Table 1).

Table 1. Hand gesture codification

NI myDAQ DO	Gesture LED-ON	Code	Gesture LED-OFF	Code
Line 0	Gesture 1-ON	55	Gesture 1-OFF	15
Line 1	Gesture 2-ON	44	Gesture 2-OFF	14
Line 2	Gesture 3-ON	33	Gesture 3-OFF	13
Line 3	Gesture 4-ON	22	Gesture 4-OFF	12
Line 4	Gesture 5-ON	66	Gesture 5-OFF	16

The front panel of the LabVIEW application is shown in Fig. 5 and it contains a graphical waveform chart indicator where one can view the signal received from the EMG sensor during the execution of the gesture. This signal is also displayed in a waveform graph indicator for 6 s, on a number of 30,000 points, with a sampling frequency of 5k points/second. The gesture code (message) can be viewed on a string indicator. This code can be obtained if the tightening maneuvers of the elastic object are repeated twice in the signal analysis window. The coding is based on determining the number of peaks in the signal values that exceed a predetermined threshold. Thus, before running the application, a calibration stage as well as an initialization stage is required. Once the gesture is identified, for example if the code is 55, the LED (virtual and physical) related to it will switch to ON state. For the LED-OFF code status the user has to make a new gesture: for example, 5-followed by a single tightening maneuver, i.e. 1. Consequently, the resulting gesture code will be 15 and this means that the LED for gesture 5 will be turned OFF. The front panel has a logical indicator

(Boolean) that will announce the user of the application when he can start executing the next gesture. This is signaled simultaneously by a beep sound.

The front panel also has the possibility of displaying in real time the entire experiment in an “Image display” type window.

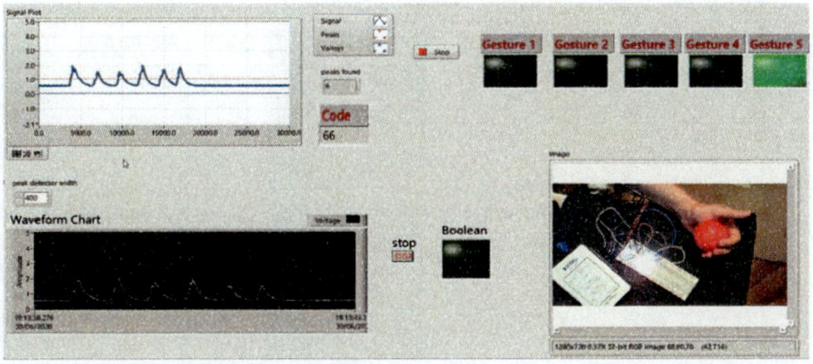

Fig. 5. The LabVIEW front panel of the application 1

An important element in the block diagram was the Peak Detection.vi, where the threshold and the width were chosen in order to have a stable and strong detection of signal peaks. The number of gestures can be increased, and the codes can be expressed in more than two digits, thus there is the possibility to extend the application if the number of outputs of the acquisition board permits this.

4 Application 2. Smart House Control Using Hand Gestures Images

4.1 Hardware Components

The application 2 (Fig. 6) is based on hand gestures images and is implemented using the following hardware components: the smart home prototype (1), the webcam placed above the user's hand (2), the computer (3) and the NI-myDAQ acquisition board (4).

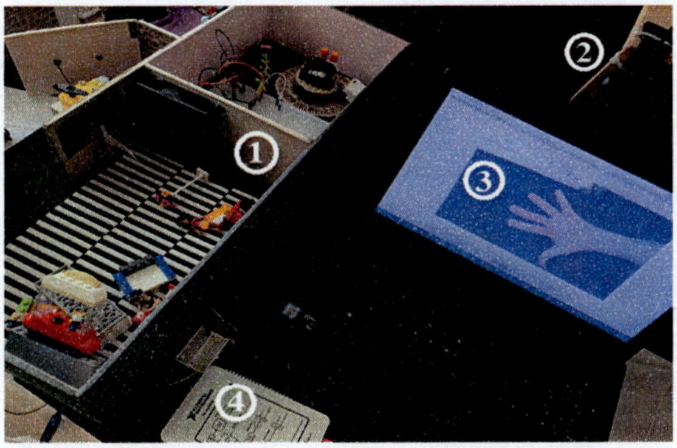

Fig. 6. Application 2 - the hardware components

The acquired images are processed and classified in order to recognize hand gestures. Thus, nine such gestures were learned and then associated with the ON/OFF commands for the relay's inputs according to Table 2.

Table 2. Hand gestures images associated with the appliances commands

Relay	NI-myDAQDO	Gesture - ON		Gesture - OFF	
1-Light	Line 0	Gesture 1 True		Gesture 2 False	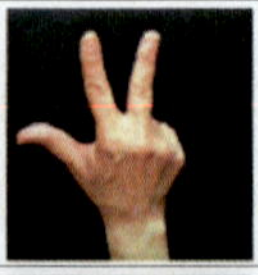
2-fan	Line 1	Gesture 3 False		Gesture 4 True	
4,7-Doors 1, 2 Closing	Line 2	Gesture 5 False	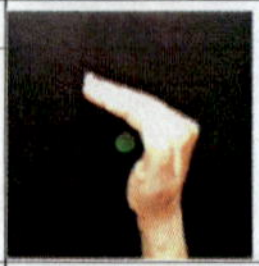	Gesture 6 True	
3,6-Doors 1, 2 Opening	Line 3	Gesture 7 False	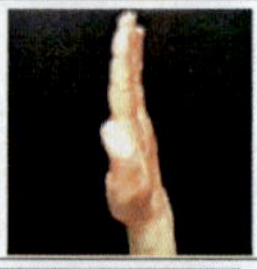	Gesture 6 True	
7-TV Set	Line 4	Gesture 8 True	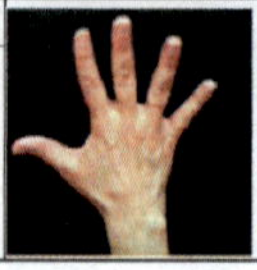	Gesture 9 False	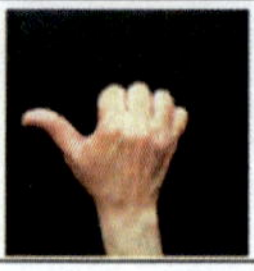

4.2 Methodology and LabVIEW Application

In order to recognize hand gestures, the NI Vison Assistant program was used in the classification learning stage. For this, the images were pre-processed by applying a mask (Fig. 7), extracting the color plane and respectively applying a smoothing filter. Then the gestures were learned by generating the classification file (Fig. 8).

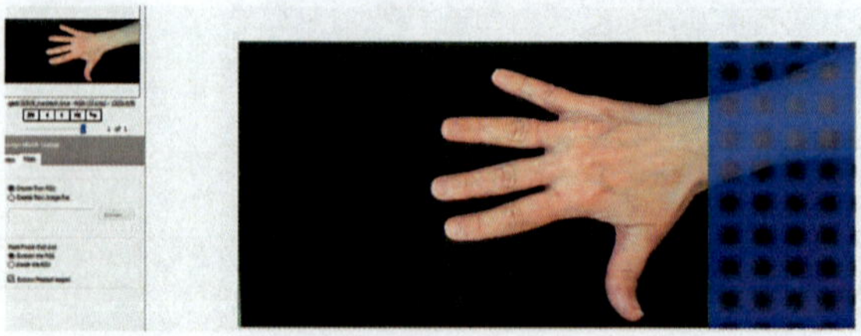

Fig. 7. Hand image preprocessing: appling the mask

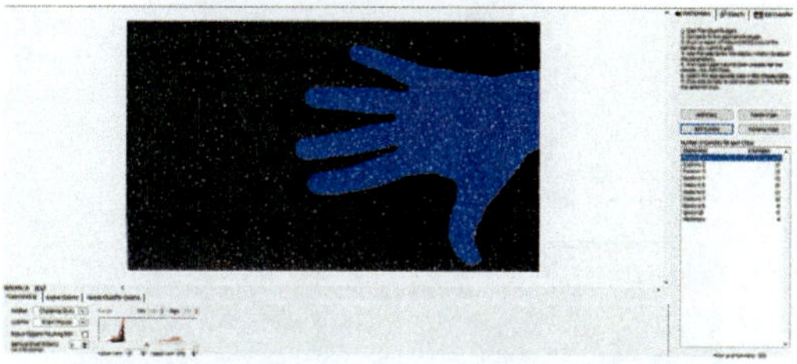

Fig. 8. The generation of the classification file.

The front panel of the application (Fig. 9) contains two image displays in which images with hand gesture before and after preprocessing are presented. If the gesture identification score is maintained above a 900 threshold for more than 2 s, meaning a static and stable image, the command will be considered valid and consequently the relay associated with the gesture will change its state. The recognition of the gesture is signaled by the Boolean indicators. To align the camera and to maintain the state of the relays on the chosen position, a reference image was used. This is represented by a green round landmark disposed on the work surface.

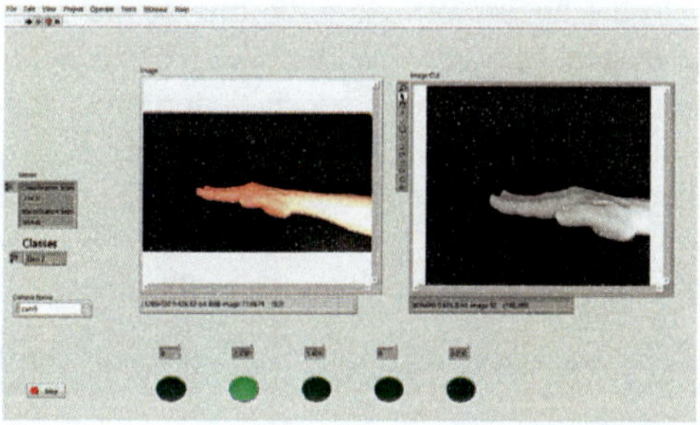

Fig. 9. The LabVIEW front panel of the application 2

The application was tested for all the possible situations. Thus, for example, with the Gesture 7 followed by Gesture 6 the doors 1 and 2 will open and with the Gesture 1 the light in the room will turn on (Fig. 10).

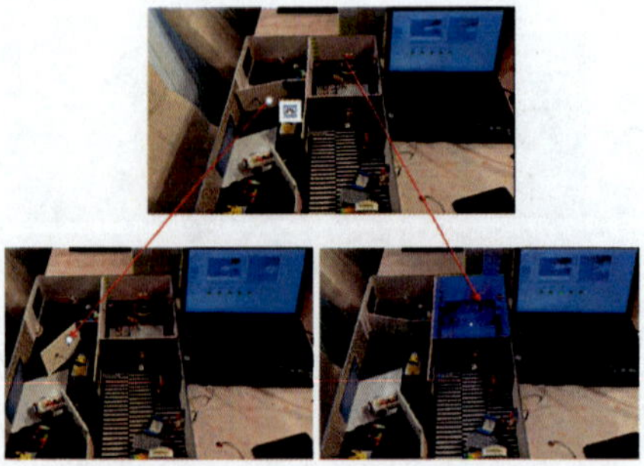

Fig. 10. Testing the application 2

5 Conclusions

The design of a smart home is currently oriented on software development strategies, when through dedicated algorithms the devices in the house can be controlled by human gestures.

The applications presented in this paper permit the user to interact with the smart house environment. The first application offers the advantage of using EMG sensor for gesture recognition by counting, in a simple way, the numbers of hand movements. The drawbacks are related to the uncomfortable mounting of the sensor on the user's hand and to the necessity of system periodical calibration. The recognition system based on hand gesture images does not imply any special conditions for interaction with the appliances. The lighting conditions can be kept relatively constant if a local illumination system is installed. In an improved solution for this application a camera or more could be mounted on the ceiling of the house and the user, based on his hand gestures, could interact with the different devices disposed in the house.

Conflict of Interest. The authors declare that they have no conflict of interest.

References

1. Kshirsagar, S, Sachdev, S., Singh, N., Tiwari, A., Sahu, S.: IoT enabled gesture-controlled home automation for disabled and elderly. In: Proceedings of the Fourth International Conference on Computing Methodologies and Communication (ICCMC), pp. 821–826. IEEE, Erode (2020)
2. Dinh, D.-L., Kim, J.T., Kim, T.-S.: Hand gesture recognition and interface via a depth imaging sensor for smart home appliances. Ener. Procedia **62**, 576–582 (2014)
3. Zeng, J., Sun, Y., Wang, F.: A natural hand gesture system for intelligent human-computer interaction and medical assistance: In: Proceedings of the 3rd Global Congress on Intelligent Systems (GCIS 2012), pp. 382–385, IEEE, Wuhan (2012)
4. Bhat, A., Jeevith, H.M., Chirag, B.: Hospital power automation system using live gesture. In: International Conference on Wireless Communications, Signal Processing and Networking, pp. 2279–2282, IEEE, India (2016)

5. Alemuda, F., Lin, F.J.: Gesture-based control in a smart home environment. In: Proceedings of IEEE International Conference on Internet of Things, IEEE Green Computing and Communications, IEEE Cyber, Physical and Social Computing, IEEE Smart Data, iThings-GreenCom-CPSCom-SmartData, pp. 784–791, IEEE, Exeter (2017)
6. Trong, K.N., Bui, H., Pham, C.: Recognizing hand gestures for controlling home appliances with mobile sensors. In: Proceedings of the 11th International Conference on Knowledge and Systems Engineering (KSE), pp. 1–7. IEEE, Da Nang (2019)
7. Patil, N.M., Patil, S.R.: Review on real-time EMG acquisition and hand gesture recognition system. In: Proceedings of the International Conference of Electronics, Communication and Aerospace Technology (ICECA), pp. 694–696, IEEE, Coimbatore (2017)
8. Ding, I. Jr., LinA, Z.-Y.: Service robot design with an advanced HCI scheme for the person with disabilities. In: Proceedings of the International Conference on Information, Communication and Engineering (ICICE), pp. 531–534, IEEE (2017)
9. Jaramillo, A.G., Benalcazar, M.E.: Real-time hand gesture recognition with EMG using machine learning. In: Proceedings of the IEEE Second Ecuador Technical Chapters Meeting (ETCM), pp. 1–5, IEEE, Salinas (2017)
10. Boonlar, T., Prasatkaew, W.: The Development of hand controller for persons with disabilities by Myo sensor, In: Proceedings of the 14th International Conference on Signal-Image Technology & Internet-Based Systems (SITIS), pp. 375–379, IEEE, Las Palmas de Gran Canaria (2018)
11. Trigueiros, P., Ribeiro, F., Reis, L.P.: A comparison of machine learning algorithms applied to hand gesture recognition. In: Proceedings of Iberian Conference on Information Systems and Technologies (CISTI), pp. 1–6, IEEE (2017)
12. Fu, X., Zhang, T., Bonair, C.H., Coats, M.L.: Wavelet enhanced image preprocessing and neural networks for hand gesture recognition. In: Proceedings of IEEE International Conference on Smart City/SocialCom/SustainCom together with DataCom 2015 and SC2 2015, pp. 838–843, IEEE, Chengdu (2015)
13. Liu, J., Furusawa, K., Tateyama, T., Iwamoto, Y., Chen, Y.-W.: An Improved hand gesture recognition with two-stage convolution neural networks using a hand color image and its pseudo-depth image. In: Proceedings of IEEE International Conference on Image Processing, pp. 375–379, IEEE, Taipei (2019)
14. Vishwakarma, D.K, Maheshwari, R., Kapoor, R.: An efficient approach for the recognition of hand gestures from very low resolution images. In: Proceedings of Fifth International Conference on Communication Systems and Network Technologies, pp. 467–471, IEEE (2015)
15. Stefanov, D.S., Bien, Z., Bang, W.-C.: The smart house for older persons and persons with physical disabilities: structure, technology arrangements, and perspectives. IEEE Trans. Neural Syst. Rehabil. Eng. **12**(2), 228–250 (2004)
16. Howedi, A., Jwaid, A.: Design and implementation prototype of a smart house system at low cost and multi-functional. In: Future Technologies Conference (FTC), pp.1–9, IEEE, San Francisco (2016)

Using Probabilistic Functional Dependencies for Determining Correlations in Skewed Data

Luminita Ioana Loga(Iancu)[1,2], Gabriel Cristian Dragomir-Loga[3(✉)], and Lucia Dican[4]

[1] Clinical Institute of Urology and Renal Transplantation, Cluj-Napoca, Romania

[2] Department of Cellular and Molecular Biology, "Iuliu Hatieganu" University of Medicine and Pharmacy, Cluj-Napoca, Romania

[3] Computer Science Department, Technical University of Cluj-Napoca, Baritiu St. 26-28, Cluj-Napoca, Romania
gabriel.dragomir@cs.utcluj.ro

[4] Biochemistry Department, "Iuliu Hatieganu" University of Medicine and Pharmacy, Cluj-Napoca, Romania

Abstract. In this paper we present a method of analysis of skew data. Skew data exists in many areas including medical data. Our contributions consist in the use of probabilistic functional dependencies together with statistical evaluators and elements of information theory to find correlations in skewed data. The experimental part uses real medical data that shows the skew property. The results highlighted the most representative classes in the practical field analyzed which was determining HLA-CMV correlation in renal transplantation.

Keywords: Probabilistic databases · Probabilistic functional dependencies · Skewed data · HLA · CMV

1 Introduction

1.1 Context and Motivation

The context that generated the research presented in this paper is the following:

- There is a collection of patients suffering from a malady X.
- Every person has intrinsic properties.
- In the treatment of the disease D_1, there is susceptibility to some patients acquiring another disease D_2 (viral).

The question that is asked and to which this article tries to present a possible solution is whether some of the patient's properties facilitate illness D_2.

Data skew is common in various areas, including medical data. One of the difficulties we have over which to pass in data analysis is positive skewness of the data.

Distributions of this kind have been extensively studied in the literature - Pareto [1], Zipf [2], power-laws.

S. Vlad and N. M. Roman (Eds.): MEDITECH 2020, IFMBE Proceedings 88, pp. 250–258, 2022.
https://doi.org/10.1007/978-3-030-93564-1_28

Our experiments were performed for $\mathbf{D_1}$ **= kidney dysfunction** and $\mathbf{D_2}$ **= cytomegalovirus (CMV) disease**, and the characteristics studied were human leukocyte antigens (HLA). In kidney transplantation, CMV disease leads to rejection of the graft. The rationale behind our analysis is that by demonstrating the causality "the existence of an HLA locus leads to the appearance of CMV after transplant", one can predict and treat early.

1.2 Probabilistic Databases

Relational data model remains a model widely used although it took almost half a century after the publication of E. F. Codd's works [3, 4]. The most important aspect related to the relational model is normalization of relations. About the normalization functional dependency (FD) between the attribute single or compound X and the attribute single or compound Y ($X \rightarrow Y$), means that when there is a tuple which contains the combination of x and y, where $x \in X$ and $y \in Y$ in the case that another tuple contains $x \in X$, then the Y attribute contains the value y.

In the field of medical database, functional dependencies cannot be defined with 100% certainty. In the medical area for events already occurred the relationship between an $x \in X$ and $y \in Y$ can be specified but we can make only an approximate statement, with some confidence, related $X \rightarrow Y$ for future events. Therefore, in medical field probabilistic approach is more appropriate.

In the last decade, researchers have extensively dealt with database type probabilistic. The field of applications include data integration systems, multimedia databases with image recognition, stock market prediction. Probabilistic database prototype systems include: MayBMS - A Database Management System for Uncertain and Probabilistic Data from Cornell University [5], The MystiQ project at the University of Washington [6], Trio - A System for Integrated Management of Data, Uncertainty, and Lineage from Stanford University [7].

Related work regarded functional dependencies assuming uncertainty include among others [8, 9].

2 Proposed Method for Determining Correlations in Skewed Data

Definition 1

We define probabilistic-functional-dependency (pFD) on relation R between attribute X and attribute Y, notated $X \rightarrow {}_pY$, a FD $X \rightarrow Y$ which holds with probability p.

Definition 2

We define conditional-probabilistic-functional-dependency (cpFD) on relation R between attribute X and attribute Y, notated $X \rightarrow p,cY$, if $X \rightarrow pY$ holds with respect of condition c ($X \rightarrow pY$ holds only for a subset of tuples of relation R).

Input

Some relation, noted R; the set of subject attributes, noted X; consequence attribute, noted Y; target class of Y.

Output
The set of cpFDs on relation R.

Step 1
For the first step of the proposed method we cleanse the data in the input relation R. The data must adhere to some standards.

Because of the assumed property of skewness some normalization of data may be necessary.

Step 2
This step involves data preparation. We count the cardinality of each class in the set of attributes of relation R. The probabilities (or support of classes) are simple obtained (see [10]). In general case, the number of columns in probability relation is determined by the cardinality of Y domain. In our experiments the cardinality was two.

Step 3
We test for each combination of interest X and Y the truth of the joint entropy H(X, Y) [11].

We continue with step 4 only with that $X_g \in X$ for which joint entropy of X_g and Y is less than the sum of individual ones.

Step 4
We determine statistical evaluators [12–14] that will help us finding the pFDs and cpFDs.

We calculate confidence and all-confidence for each X_g.

We continue with the calculation of lift, conviction, j-measure and z-score.

We shall use z-score to exclude the edge interval values (long tail) and the other evaluators in validating the excluded classes and in finding the most relevant classes.

Step 5
We determine pFDs. We initiate some voting system. For each class of X_g we determine the vote for one of the classes of Y supplemented with a dummy class 'Unknown' which corresponds to equal probabilities. Next, we maximize the vote. This way we found orientation of pFD. We consider average of individual probabilities for elected candidate as p of pFD.

Step 6
We suppose that one class of Y attribute is considered "the most important citizen" (target class). Conditional probabilistic FD will help us in determining the most relevant classes of X_g i.e. with a probability that encounter a certain confidence threshold (the condition).

3 Experimental Results

Cytomegalovirus (CMV) is an important pathogen that leads to significant morbidity and mortality. With more patients undergoing transplants, along with the expanding indications for immune-modulating agents, the number of patients at risk for developing CMV disease is increasing.

Major Histocompatibility Complex (MHC) is a group of genes on a single chromosome that mediate rejection of grafts between two genetically different individuals. HLA (human leukocyte antigens) are the MHC antigens of humans and are called so because they were first detected on leukocytes.

We evaluated the association of CMV disease and HLA genotype's in recipients after transplantation, out of 670 kidney transplant recipients.

Typing was performed by complement-dependent cytotoxicity (CDC) method using magnetic beads for cell isolation and monoclonal typing trays (One Lambda, USA), while HLA-DR alleles were determined by PCR-SSP technique (Olerup, Sweden).

Input relation for our experiments was the one from Table 1.

Table 1. Patient relation

Patient No	Gender	Age	HLA_A		HLA_B		HLA_DR		CMV
			1	2	1	2	1	2	

As shown in Fig. 1, 2 and 3 skewness is expressed approximately as a straight segment with a certain slope on the log-log representation of rank and frequency. That is, the number of patients holding a certain property (HLA - A, B and DR) decreases orderly by the rank.

Step1:
Firstly, we have assured that HLA data are consistent with standard names of alleles. Each patient has pairs of alleles HLA -A, -B, -DR thereby we ordered alleles inside pairs.

Secondly, from the age of patient at transplant date we derived Age_Group so as to increase the frequency and to overcome the problem of skewness.

Step 2:
We counted the total number of patients (the cardinality n of the relation showed in Table 1).

Because the position of allele in the pair does not have any importance we have calculated probabilities for each domain (A, B, DR) in the following way: for each distinct value in HLA_A1 we count the number of appearances, for each distinct value in HLA_A2 we count the number of appearances and we sum the partial results by distinct HLA-A allele. The probabilities (or support of alleles) are obtained as in [10].

The process is the same for HLA-B and HLA-DR. Our study has the characteristic that Y from a $X \rightarrow Y$ FD test is categorical data (Y domain has cardinality two: "yes" or "no" for events already occurred).

The probabilities for Age_Group and Gender are calculated much simpler by counting the number of appearances by distinct value in domain.

Step 3:
For Patient relation, X varies with {A, B, DR} and Y is CMV.

In Table 2 we can see individual results for HLA-A, -B, -DR but in Table 3 the results are for pairs (A1-A2, B1-B2, DR1-DR2). It is obvious that the entropy grows when pair is considered (Table 4).

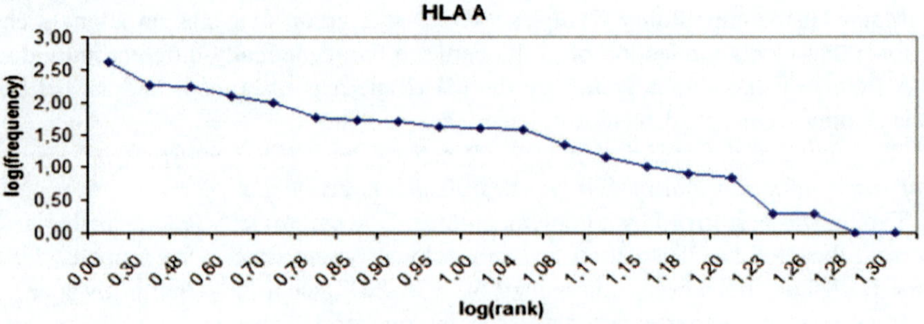

Fig. 1. HLA A skew = 2.6

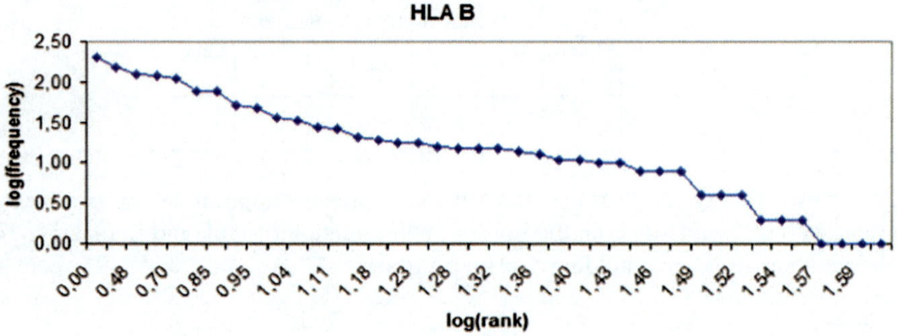

Fig. 2. HLA B skew = 2.1

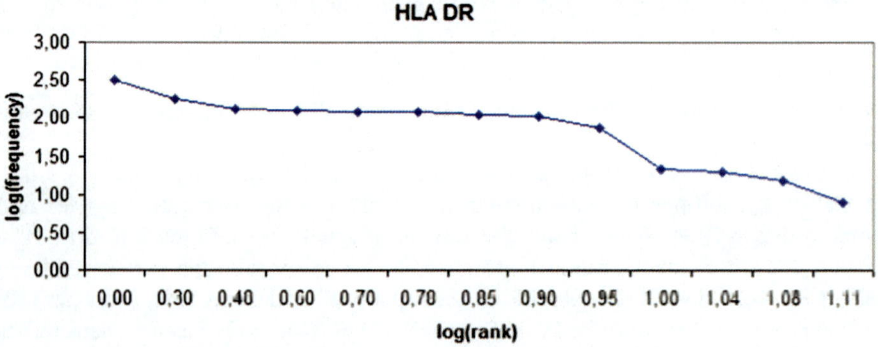

Fig. 3. HLA DR skew = 1.2

Table 2. The contribution of information H(x) results

Patient number	Gender	Age group	HLA A	HLA B	HLA DR	CMV
6.5132	0.6654	2.2729	2.2491	2.9792	2.2654	0.2042

It follows when considering joint entropy which are the attributes of interest to continue the test for pFD respectively cpFD. In our case are all three HLA (A, B and DR). Gender and Age Group does not present interest.

Table 3. H(X) results

HLA		
A1, A2	B!, B2	DR1, DR2
3.8092	4.9462	3.8799

Step 4:
We calculated statistical evaluators for each allele (class) of HLA -A, B and DR (X_g). The classes selected for exclusion by considering a threshold of −0.5 for z-score were as follows:

- HLA – A alleles: 69(28), 30(19), 2403, 10, 66(10), 34(10) with z-score = −0.5709, confidence = 0, lift = 0, conviction = 1, and with corresponding j-measure = 0.0001, 0.0004, 0, 0.0003, 0.0003, 0;
- HLA – B alleles: 56(22), 46, 73, 75(15), 45(12), 64(14), 15, 14, 62(40), 17, 37, 52 (5), 62(15), 70, 42, 16 with z-score = −0.8353, confidence = 0, conviction = 1 and with corresponding j-measure = 0.0004, 0.0001, 0, 0, 0.0002, 0.0001, 0.0004, 0.0005, 0.0014, 0.0001, 0.0006, 0.0007, 0.0003, 0, 0, 0.0002;
- HLA – DR allele/z-score: *09/−2.2775, *01/−1.1106, *03/−0.9737, *04/−0.5228 with corresponding confidence = 0, 0.025, 0.0279, 0.0376, lift = 0, 0.4814, 0.5379, 0.724, conviction = 1, 1.0234, 1.0249, 1.0352, j-measure = 0.0003, 0.0008, 0.0009, 0.0002.

Step 5:
For each HLA (-A, -B, -DR) we determined the vote for one of categorical value {'CMV-Yes', 'CMV-No', 'CMV-Unknown'} by comparing ('>', '<', '=') the number of 'CMV-Yes' occurrences with the number of 'CMV-No' occurrences for each allele. After that we maximized the vote and considered the average of individual probabilities for elected candidate as p of pFD. For example, if we take pFD between HLA-A and CMV we calculated one vote for 'CMV-Unknown' and 19 votes for 'CMV-No' with P_{avg} = 0.9377. We can say therefore that HLA_A $\rightarrow$ 0.9377CMV in direction 'CMV-No'.

Step 6:
For clinical adequate treatment of transplanted patients, the purpose we are interested in is the correlation between some values of HLA domains and certain value from domain CMV namely 'CMV-Yes'. Conditional probabilistic FD will help us in this goal. Using statistical evaluators presented in section above we can determine the most relevant alleles i.e. with a probability that meet a confidence threshold to specify disease risk.

In searching for cpFD, for each person to whom the status of CMV is known, we use pFDs previously determined and calculate max(P_i) where i $\in$ {'HLA-A1',

Table 4. H(X, Y) results

Patient number, CMV	Gender, CMV	Age group, CMV	HLA A, CMV	HLA B, CMV	HLA DR, CMV
6.5132	0.8684	2.4731	0.1971	0.1931	0.2008

'HLA-A2', 'HLA-B1', 'HLA-B2', 'HLA-DR1', 'HLA-DR2'} for both conditions "CMV = 'Y'" and "CMV ='N'". Afterwards we test the definition of FD namely for "CMV ='Y'" condition what are the alleles which do not appear in combination with 'CMV-No' and for "CMV ='N'" condition what are the alleles which do not appear in combination with 'CMV-Yes'. The results must contain only the values that respect the rule. An example is HLA-A → 0.9545; c = HLA-A IN {32(19); 68(28)}CMV. We found that none of HLA-A,-B,-DR alleles determine positive CMV.

4 Usage of the Proposed Method

For a new transplanted person for which we don't know the CMV value we will try to estimate {'CMV-Yes', 'CMV-No', 'CMV-Unknown'} based on cpFD.

Results of proposed method for CMV positive (patients) are presented in Table 5 and Table 6, where for a patient number it can be interpreted as how useful the most relevant allele is in predicting CMV in both directions.

What we can see for example at patient number 3 in Table 5 is that if we did not exclude edge positive interval values by using z-score, HLA_A 36 was elected for direction "CMV-Yes". The confidence measure is 0.5, the lift measure is 9.6286, the conviction measure is 1.9985, the j-measure is 0.0012 and z-score is 4.0077. The explanation is the noise induced by small counting (two transplanted, from which one patient and one control).

Table 5. Excerpt of patient CMV positive

Patient number	Probability CMV "No"	Probability CMV "Yes"
3	HLA_DR *03 p = 0.972	**HLA_A 36 p = 0.500**
1779	HLA_B 18 p = 0.948	**HLA_B 48 p = 0.250**
1450	HLA_A 24(9) p = 0.942	**HLA_B 57(17) p = 0.200**
3713	HLA_A 1 p = 0.978	**HLA_B 57(17) p = 0.200**
2488	HLA_DR *04 p = 0.962	**HLA_B 57(17) p = 0.200**
3457	HLA_B 35 p = 0.965	HLA_A 33(19) p = 0.136
3122	HLA_A 11 p = 0.980	HLA_A 33(19) p = 0.136
2925	HLA_A 3 p = 0.945	HLA_A 33(19) p = 0.136
3040	HLA_B 35 p = 0.965	**HLA_B 63(15) p = 0.125**
3332	HLA_DR *03 p = 0.972	**HLA_B 65(14) p = 0.125**

In the same situation with HLA_A 36 are HLA_B 65(14), 53, 63(15), 57(17), 48 and HLA_DR *09, *01 (left side of z-score), *16, *15.

Table 6. Excerpt of patient CMV positive after z-score correction

Patient number	Probability CMV "No"	Probability CMV "Yes"
3	HLA_DR *03 p = 0.972	HLA_B 44(12) p = 0.082
1779	HLA_B 18 p = 0.948	HLA_A 2 p = 0.065
1450	HLA_A 24(9) p = 0.942	HLA_B 47 p = 0.100
3713	HLA_A 1 p = 0.978	HLA_B 38(16) p = 0.075
2488	HLA_DR *04 p = 0.962	HLA_A 2 p = 0.065
3457	HLA_B 35 p = 0.965	HLA_A 33(19) p = 0.136
3122	HLA_A 11 p = 0.980	HLA_A 33(19) p = 0.136
2925	HLA_A 3 p = 0.945	HLA_A 33(19) p = 0.136
3040	HLA_B 35 p = 0.965	HLA_DR *11 p = 0.061
3332	HLA_DR *03 p = 0.972	HLA_B 44(12) p = 0.082

In step 6 of our method if we consider z-score in the interval (−1, +1) and we determine conditional probabilistic FD again we found the results presented in Table 6. For patient number 3 cpFD has for "CMV-Yes" the most representative allele = "HLA_B 44(12)" with p = 0.082.

However, as already mentioned, probabilistic values in column '**Probability CMV "Yes"**' from Table 6 indicate "possible" but cpFD does not applies. Therefore, for people with CMV value unknown column '**Probability CMV "No"**' must be considered conditioned by the presence of alleles in cpFDs already determined and the chance of illness is obtained by 1-p.

5 Conclusion

The conclusion is that conditional probabilistic functional dependencies could highlight the most representative alleles correlated with CMV disease (negative): {HLA-A: 32 (19), p = 0.962; 68(28), p = 0.947}, {HLA-B: 7, p = 0.936}, {HLA-DR: *14, p = 0.947; *07, p = 0.937}.

Conflict of Interest. The authors declare that they have no conflict of interest.

References

1. Pareto, V.: La Courbe de la Repartition de la Richesse. In: Busino, G. (ed.) Oevres Completes de Vilfredo Pareto, pp. 1–5. Librairie Droz, Geneva (1965) (Originally published in 1896)
2. Zipf, G.K.: Selective Principle Studies and the of Relative Frequency in Language. Harvard University Press, Cambridge (1932)
3. Codd, E.F.: A relational model of data for large shared data banks. Commun. ACM **13**(6), 377 (1970). https://doi.org/10.1145/362384.362685

4. Codd, E.F.: Further normalization of the data base relational model. In: Courant Computer Science Symposia Series 6, "Data Base Systems", May 24–25, New York City (1971)
5. Antova, L., Jansen, T., Koch, C., Olteanu. D.: Fast and simple relational processing of uncertain Data. In: Proceedings of the. 24th International Conference on Data Engineering, ICDE 2008, Cancun, Mexico, 7–12 April 2008, pp. 983–992 (2008)
6. Boulos, J., Dalvi, N., Mandhani, B., Mathur, S., Re, C., Suciu, D.: MYSTIQ: a system for finding more answers by using probabilities. In: SIGMOD (2005)
7. Benjelloun, O., Das Sarma, A., Halevy, A., Widom, A.: ULDBs: databases with uncertainty and lineage. In: Proceedings of the 32nd International Conference on Very Large Data Bases, pp. 953–964, Seoul, Korea, September 2006
8. Das Sarma, A., Ullman, J.D., Widom, J.: Schema design for uncertain databases. In: Proceedings of the 3rd Alberto Mendelzon Workshop on Foundations of Data Management, Arequipa, Peru, May 2009
9. Wang, D.Z., Dong, L., Sarma, A.D., Franklin, M.J., Halevy, A.: Functional dependency generation and applications in pay-as-you-go data integration systems. In: Proceedings of SIGMOD WebDB 2009, June 28, Providence (2009)
10. Agrawal, R., Imielinski, T., Swami, A.: Mining associations between sets of items in large databases. In: Proceedings of the ACM SIGMOD International Conference on Management of Data, pp. 207–216, Washington D.C, May 1993
11. Shannon, C.E.: A mathematical theory of communication. Bell Syst. Tech. J. **27**(3), 379–423 (1948). https://doi.org/10.1002/j.1538-7305
12. Omiecinski, E.R.: Alternative interest measures for mining associations in databases. IEEE Trans. Knowl. Data Eng. **15**(1), 57–69, January/February 2003
13. Brin, S., Motwani, R., Ullman, J.D., Tsur. S.: Dynamic itemself counting and implication rules for market basket data. In: Proceedings of the ACM SIGMOD International Conference on Management of Data (ACM SIGMOD 1997), pp. 265–276 (1997)
14. Smyth, P., Goodman, R.M.: Rule induction using information theory. In: Piatetsky-Shapiro, G., Frawley, W. (eds.) Knowledge Discovery in Databases, pp. 159–176, The MIT Press, Cambridge (1991)

Towards Detecting Fake Medical Content on the Web with Machine Learning

Radu Razvan Slavescu[1(✉)], Florina-Ionela Pop[1], and Kinga Cristina Slavescu[2]

[1] Technical University of Cluj-Napoca, Cluj-Napoca, Romania
Radu.Razvan.Slavescu@cs.utcluj.ro,
Florina.Nagy@student.utcluj.ro
[2] "Iuliu Hațieganu" University of Medicine and Pharmacy, Cluj-Napoca, Romania
Cristina.Slavescu@umfcluj.ro

Abstract. High speed information dissemination in the online environment is useful, but could also lead to fast spread deceptive content. Among the areas affected by online misinformation is the medical field, hence the need of tools able to help the users filter inaccurate information. This paper is focused on the detection of false medical information distributed online. The proposed model is a sequential neural network with three layers which employs word embeddings. When tested on a dataset concerned with vaccine hesitancy, its accuracy reached 83%. We show that effective detection of potentially deceptive articles on vaccination can be achieved by using this neural network.

Keywords: Neural networks · Misinformation detection · Fake news · Anti-vaccination · Natural language processing

1 Introduction

At the beginning of 2020, SARS-CoV-2 virus started spreading across the world and so did COVID-19 news, at an unprecedented level [1]. This has been described by World Health Organization (WHO) also as a "massive infodemic" [2]. When Oliver Véran, the French Health Minister, suggested that anti-inflammatory painkillers could worsen the effect of coronavirus [3] on the following days it has been retweeted over 40 000 times. This is just an example of how misinformation can cause harm by influencing attitudes and beliefs. Individuals are more engaged to understand and make health-related decisions on treatments and test [4], instead of having personalized recommendations from specialized personnel [5]. Most people tend to trust all the health information available on the Web without attempting to validate the sources [6].

The vaccine misinformation has been recognized as a global public-health threat. In 2019 World Health Organization included the anti-vaccination movement as a top-10 global health hazards [2]. The overload of conflicting information, misinformation or manipulated information on social media started to cause serious public health consequences, as a decrease in vaccination rates and increasing the risks of disease outbreaks [7]. Thus, online anti-vaccination articles may lead parents to question the safety

S. Vlad and N. M. Roman (Eds.): MEDITECH 2020, IFMBE Proceedings 88, pp. 259–266, 2022.
https://doi.org/10.1007/978-3-030-93564-1_29

of vaccine, distrust health professionals, and seek non-medical vaccine exemptions, targeting people with low confidence in vaccination [8].

Social media and online information growth since 2010, started to have a considerable impact on society opinion and since then many tools have been developed to combat fake news and satire articles shared online [9]. In the early days, manual methods were preferred over automatic ones in fake news detection, denoting the incipient phase of Natural Language Processing [9]. Automatic methods used on fake news detection are based on linguistic analysis such as lexical, syntactical or semantical criteria [10]. Rubin et al. [10] work defines three type of fake news after a research conducted on deceptive news as follows: fabrication, hoaxing and satire detection. Other works such as Al Asaad et al. [11] have addressed different on fake news detection using a machine learning algorithm.

The problem that has been studied on this paper is focused on the detection of false information distributed online on medical field, on vaccines topic. In our research, we used a neural network model for Natural Language Processing meant to identify whether an article has a misleading content or provides legit information. This paper presents our results obtained in this respect.

The paper is organized as it follows. Section 2 describes the related work on fake-news detections and anti-vaccines articles. Section 3 provides the knowledge for the proposed model, including the pre-processing, modelling and training phase. Section 4 describes the experimental results obtained. Section 5 provides conclusions for the proposed architecture and future work related to the goal of this study and how it can be improved.

2 Related Work

The paper studies two current topics: one of them being misinformation from online content on vaccination topic and the second one is referring to automatic methods using neuronal networks to filter content shared online.

Due to the acceptability of online content sharing, people are prone to bias [11] and they do not intend to validate the source of information as long as it follows their preconceived ideas [12]. This has determined a growth in discerning truth and deceptive articles shared online due to the impact on emotional and personal beliefs [13], including medicine misinformation. Since 2012 the interest in developing automated tools to identify satire and fake articles has helped increasing the automated detection using Natural Language Processing and machine learning techniques on misleading information and legitimate content [13]. Online misinformation is a well-known topic in researchers field [11]. Public perception of vaccines misinformation is an exemplar problem of online misinformation spread through the web [15]. Shah et al. [14] studied the credibility of online content shared on Twitter using a recurrent neural networks (RNN) classifier indicating that 12% of websites with vaccine content have low credibility. Meanwhile, another approach proposed by Guibon et al. [9] classifies the textual content of the articles into three classes: Fake News, Trusted and Satire. With an optimized LGBM classifier they predicted better the cases of satire, while neural networks outperformed for fake and trusted classifications [9]. In their study, better reults were obtained with machine learing algorithms compared with the results obtained with the convolutional neuronal network model. The work proposed by Rubin et al. [13] towards a news verification system using Rhetorical Structure Theory (RST) on content data from NPR achieved a 63% prediction accuracy over a 56% baseline. The study [13] aims to identify differences in narrative between different stories, topic which is well suited on vaccines article content detection. However, all the above work is focused to develop computational tools used for detection of false information distributed online. Therefore, when implementing solutions for binary text classification algorithms they are focused on the laxical use of the words, evaluating data on the word level. For the stated problem, the developed solution classifies medical articles evaluating data at a paragraph level.

When comparing conventional machine learning algorithms with neural networks, we find that neural networks represent a more complex subset of machine learning algorithms which have a better performance and are scalable for complex applications [16]. In [17] Masino et al. proposed a convolutional neural network used to identify drug reaction from Twitter data. Du et al. [18] adopted CNN models to analyze Twitter data regarding the public perception of a measles outbreak. In [19] the proposed solution is a 3-layer neural network to characterize the discussion about antibiotics on Twitter. Looking at the related work, in the last years the interest in public perceptions of diseases and medical knowledge has increased significantly.

3 Problem Aproach

The solution we propose relies on a sequential neural network model. The model was implemented in Python. NLTK was used for preprocessing, while Tensorflow and Keras were employed for building the model. The architecture of the neural network model is displayed in Fig. 1. The layers are described in detail in Sect. 3.2.

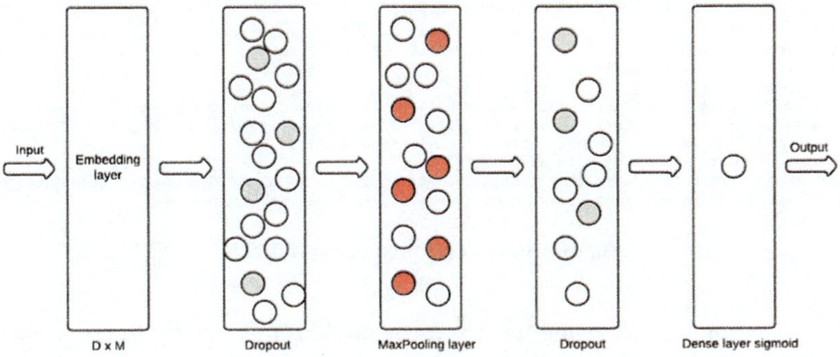

Fig. 1. Model architecture

3.1 Dataset

The dataset used is the Facebook Anti-Vaccination Dataset [20]. This set consists of over 50,000 data containing medical articles, scientific papers and videos which deals with the subject of vaccination and anti-vaccination movement. For model development were extracted 3000 articles from the dataset, where 2400 were used for the training the model and 600 for testing. Data distribution according to the "anti-vaxx" label is presented in Fig. 2.

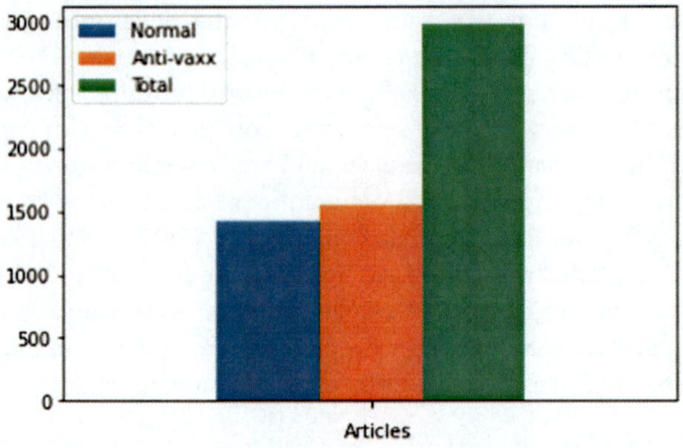

Fig. 2. Data distribution sorted by label value

A set of processing steps were conducted in order to enable the data for feature extraction. Multiple steps of normalization, substitution and tokenization processes, including the removal of stopwords and tokenize the articles, as displayed in Fig. 3.

```
>>> print(posts_data['text'][0])
Propaganda piece from 2002. What I would like to know is: Have they done ANY kind of follow up with
the soldiers that actually took this vaccine?? What is their health like now, 12 years later?? Woul
d they recommend this vaccine to anyone else?? Side effects? I have so many question while reading
this & wish I was a real journalist: .
>>> print(posts_data['clean_text'][0])
['propaganda', 'piec', 'would', 'like', 'know', 'kind', 'follow', 'soldier', 'actual', 'take', 'vac
cin', 'health', 'like', 'year', 'later', 'would', 'recommend', 'vaccin', 'anyon', 'els', 'side', 'e
ffect', 'mani', 'question', 'read', 'wish', 'real', 'journalist']
>>> print(posts_data['clean_text_to_string'][0])
propaganda piec would like know kind follow soldier actual take vaccin health like year later would
recommend vaccin anyon els side effect mani question read wish real journalist
```

Fig. 3. Stages of a preprocessed article

Data encoding is made using the Tokenization API provided by Keras, where each word is represented by a unique integer, so it can be feed into the model. Each sequence that has been vectorized, having the same defined length of 150. An example of representation is given in Fig. 4.

```
[377 443 211  13   1   1 200   1   1   1  72 264  46 240   1   1   1   1
   1   1   1   1 241   1   1 240   1   1  17  72  37   1  69   1   1   1
 393  10  83 264   2  46   1   1 241   1   1 393  38   1  10 264   2 261
 444  46 161 335   1 393  70  89   1   1   1   1   1 363 264   2  46 116
 464   0   0   0   0   0   0   0   0   0   0   0   0   0   0   0   0   0
   0   0   0   0   0   0   0   0   0   0   0   0   0   0   0   0   0   0
   0   0   0   0   0   0   0   0   0   0   0   0   0   0   0   0   0   0
   0   0   0   0   0   0   0   0   0   0   0   0   0   0   0   0   0   0
   0   0   0   0   0   0]
```

Fig. 4. Padded sequence with defined length

The goal was to label paragraphs like the one in Fig. 3 as either "normal" or "anti-vaxx". We assumed the "pro-vaccination" opinions possess enough scientific support to be considered "normal". Once a text is automatically classified into one of these two classes, this information could further be used for assessing user/site credibility.

The corpus vocabulary we used the articles from the dataset, instead of the whole English vocabulary. Due to the diversity introduce by such a huge amount of data, the embedding layer will be to generic, and it will cause the underfitting of the model. In order to avoid unbalanced data distribution, we crated a domain related vocabulary for the stated problem.

Also, another advantage provided by the dataset was that all the data was labeld, and a lot of time was saved. Each article had a label of 0 and 1 using "antivaxx" label, indicating if the article is a deceptive one or it contains valid information.

3.2 Classification Model

The developed solution is based on a neural network model with 3 layers. The model consists of an embeddig layer with a droput applied, a maximum pooling layer aplied with droput and the output layer with sigmond activation function applied.

The embedding layer is used as the first layer of the neural network and it is fit in a supervised way using the Backpropagation algorithm. It is the first hidden layer of the network and it is defined by the size of the vocabulary or the total number of unique words in the corpus, the number of the dimensions for each word vector and the length of input sequences. Using word embedding, the words are represented by dense vectors where a vector represents the projection of the word into a continuous vector space.

The Embedding layer has a vocabulary of 500 with an input length of 150 and the embedding space of 24 dimensions. The output from the Embedding layer will be 150 vectors of 24 dimensions each, one for each word, creating a dense matrix. We apply dropout to the output of the embedding layer by selectively dropping certain ids to prevent overfitting. The second layer consists of a one-dimensional global maximum pooling layer, used to compute the maximum vector of a from the vector representation of an article. The third layer is a dense layer with one neuron and a sigmoid activation function. Since the classification problem is a binary one, the output layer uses sigmoid as activation function for probabilistic approach in terms of decision making and variation between 0 and 1 for predictions.

The model was compiled using binary cross-entropy for the loss function and Adam optimizer with the learning rate equal to 0.001. The training data and testing data is fit into the model in 150 epochs and the verbosity is set to 1 to observe the model evolution.

4 Experimental Results

During our research, another experimental model investigated was the one in [9]. A convolutional neural network which has an embedding layer, the second one is a convolutional layer were the kernel slides along one dimension with a dropout activation function, the next layer is a global maximum pooling one for one dimension and the output layer is a dense layer with 3 neurons with softmax activation function, because the proposed model uses three labels to solve the classification problem. The proposed architecture is suggested in articles who discuss about Natural Language Processing problems and it was adapted for the stated problem. We added an extra dropout function on [9the second layer and adjusted the hyperparameters to maximize performance. This new model obtained an accuracy 80% as shown in Fig. 5.

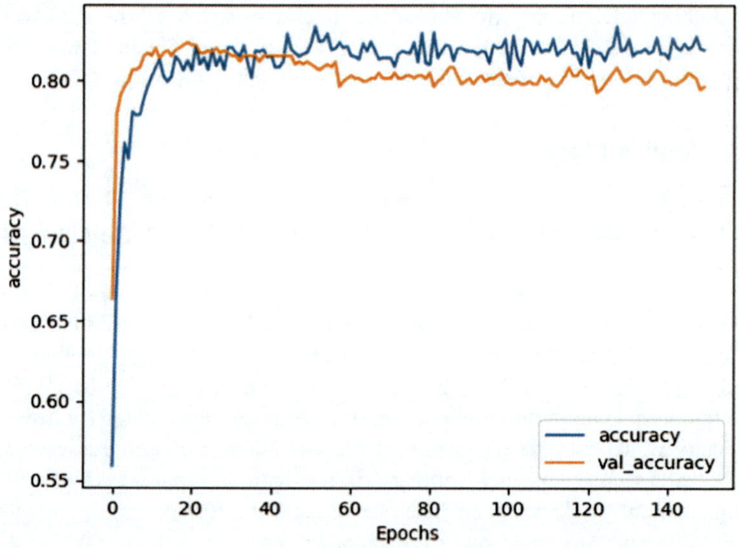

Fig. 5. Model accuracy

To estimate the expected level of fit of the proposed model to a data set that is independent of the data that were used to train the model, we performed a k-fold cross-validation, where we chose k = 5. The mean of the model accuracy scores obtained was 83% and validates sufficiently the model for the proposed classification problem. In both situations, it outperformed the baseline model.

5 Conclusions

The online environment allows the spread of deceptive content, which can lead to misinformation of users and contouring wrong society opinions. Figuring out sources in social media or forums where content with no factual support originates or is spread is part of a bigger effort of assessing the credibility of websites and publishers. To achieve this, the first step is to detect fake content in raw text. We explored this by focusing on the anti-vaccination case, i.e., classifying paragraphs of texts into "normal" or "anti-vaxx".

The problem has been addressed by a sequential neural network with 3 layers. The solution employs Natural Language Processing techniques for modeling the text content in a manner which allows classifying it as either pro- or anti-vaccination. The model classifies medical texts by evaluating data at a paragraph level.

The proposed model obtained an accuracy of 80.95% on the dataset. Also, the model obtained a mean score of 83% accuracy of expected level of fit when performing a 5-fold cross-validation. Thus, we can conclude that effective classification o deceptive articles vaccination topic could be achieved using the neural network model developed. Information about the presence of non-factual content could then be used, for example, for automatically displaying warning for readers.

As future work, we intend to employ pre-trained models such as ELMo or FastText for the NLP part. Experimenting with some other datasets will also be pursuit.

Acknowledgments. This work was supported by the Computer Science Department of the Technical University of Cluj-Napoca, Romania.

Conflict of Interest. The authors declare they have no conflict of interest.

References

1. Orso, D., Federici, N., Copetti, R., Vetrugno, L., Bove, T.: Infodemic and the spread of fake news in the COVID-19-era. Eur. J. Emerg. Med. **27**(5), 327 (2020)
2. WHO: World Health Organization (2019). https://www.who.int/news-room/feature-stories/ten-threats-to-global-health-in-2019. Accessed 2020
3. Véran, O.: Twitter, 14 May 2020. https://twitter.com/olivierveran/status/1238776545398923264. Accessed 18 June 2020
4. Schwitzer, G.: A review of features in internet consumer health decision-support tools. J. Med. Internet Res. **4**(2), 53 (2002)
5. Pluye, P., et al.: Health outcomes of online consumer health information: a systematic mixed studies review with framework synthesis. J. Am. Soc. Inf. Sci. **70**(7), 643 (2019)
6. Weaver, J.B., Thompson, N.J., Weaver, S.S., Hopkins, G.L.: Healthcare non-adherence decisions and internet health information. Comput. Hum. Behav. **25**(6), 1373 (2009)
7. Steffens, M.S., Dunn, A.G., Wiley, K.E., Leask, J.: How organisations promoting vaccination respond to misinformation on social media: a qualitative investigation. BMC Public Health **19**(1), 1–12 (2019)

8. Jones, A.M., Omer, S.B., Bednarczyk, R.A., Halsey, N.A., Moulton, L.H., Salmon, D.A.: Parents' source of vaccine information and impact on vaccine attitudes, beliefs, and nonmedical exemptions. Adv. Prev. Med. 2012, Article ID 932741 (2012)
9. Guibon, G., Ermakova, L., Seffih, H., Firsov, A., Le Noé-Bienvenu, G.: Multilingual fake news detection with satire. In: Proceedings of the CICLing: International Conference on Computational Linguistics and Intelligent Text Processing, La Rochelle, France (2019)
10. Rubin, V., Conroy, N., Chen, Y.: Deception detection for news: three types of fakes. In: ASIS&T (2015)
11. Al Asaad, B., Erascu, M.: A tool for fake news detection. In: 20th International Symposium on Symbolic and Numeric Algorithms for Scientific Computing (SYNASC) (2018)
12. Lazer, D.M.J., et al.: The science of fake news. Science **359**(6380), 1094 (2018)
13. Rubin, V., Conroy, N., Chen, Y.: Towards news verification: deception detection methods for news discourse. In: Hawaii International Conference on System Sciences (HICSS48) Symposium on Rapid Screening Technologies, Deception Detection and Credibility Assessment Symposium (2015)
14. Shah, Z., Surian, D., Dyda, A., Coiera, E., Mandl, K.D., Dunn, A.G.: Automatically appraising the credibility of vaccine-related web pages shared on social media: a Twitter surveillance study. J. Med. Internet Res. **21**(11), e14007 (2019)
15. Larson, H.J.: The biggest pandemic risk? Viral misinformation. Nature **562**(7726), 309–310 (2018)
16. Singh, S.: Towards Data Science, Deep learning. https://towardsdatascience.com/cousins-of-artificial-intelligence-dda4edc27b55. Accessed 2020
17. Masino, A.J., Forsyth, D., Fiks, A.G.: Detecting adverse drug reactions on Twitter with convolutional neural networks and word embedding features. J. Healthc. Inform. Res. **2**(1–2), 25–43 (2018). https://doi.org/10.1007/s41666-018-0018-9
18. Du, J., Tang, L., Xiang, Y., Zhi, D., Xu, J., Song, H.-Y., Tao, C.: Public perception analysis of tweets during the 2015 measles outbreak using convolutional neural network models. J. Med. Internet Res. **20**(7), e236 (2018)
19. Kendra, R.L., Karki, S., Eickholt, J.L., Gandy, L.: Characterizing the discussion of antibiotics in the twittersphere: what is the bigger picture? J. Med. Internet Res. **17**(6), e4220 (2015)
20. Heylar, A.: Kaggle Facebook Antivaccination Dataset (2019). https://www.kaggle.com/alechelyar/facebook-antivaccination-dataset. Accessed 2020

Prototype Music Recommendation System – Preliminary Results in Using Music as Therapy

Cristina Simina Ciubăncan, Laura-Nicoleta Ivanciu(✉), and Emilia Șipoș

Bases of Electronics Department, Technical University of Cluj-Napoca, 26-28 G. Barițiu street, 400027 Cluj-Napoca, Romania
laura.ivanciu@bel.utcluj.ro

Abstract. Music is currently used as a method of therapy in different areas of medicine, as it can reduce stress and anxiety, increase the level of relaxation and improve the patients' mood. This paper presents a prototype music recommendation system, which provides a result based on the correlation of the user's emotional states. The prototype was implemented and developed using Python and the Google Colab environment. Preliminary results were obtained using a free music samples data set, but the algorithm can be scaled and optimized for larger, more complex data sets. Mobile versions of the algorithm can be included in apps designed to improve the user's mood.

Keywords: Music therapy · Music recommendation · Python · Correlation

1 Introduction

Music is a source of inspiration, healing and expression of feelings. It evolved gradually over time and it plays an important role in all cultures. Nowadays, music is not just a symbol of art and culture, it is used also as a method of therapy in different areas of medicine.

It was proven that music has effects on physiology and symptoms. If carefully selected, music can reduce stress and anxiety, offer distraction from pain, increase relaxation and comfort level, boost the energy, improve the mood of the patients and enhance clinical performance. For example, music can improve mood in oncology patients, reduce anxiety and blood pressure for surgical patients, reduce depression, decrease patients' perception of pain and classical music increase weight gain for premature infants [1].

Music supported therapy is used for patients who suffered a stroke because it helps them in cognitive rehabilitation [2]. In addition, music therapy can help people who have gone through traumatic events to move on easily [3].

Listening to music is a complex process, which involves many motor, cognitive, perceptive and emotional components that work together to create our subjective experience of music [2]. Our mood and emotional state are greatly influenced by the music we listen to, and we can steer away from negative emotions by means of appropriate music.

S. Vlad and N. M. Roman (Eds.): MEDITECH 2020, IFMBE Proceedings 88, pp. 267–273, 2022.
https://doi.org/10.1007/978-3-030-93564-1_30

The user's emotions are the key to developing a music recommendation system. Currently, recommendation systems are among the most popular applications, in various fields, such as engineering, research, finance, sales and to industries such as applications that recommend music or movies. The main purpose of a recommendation systems is to predict what users might be interested in [4]. The recommendation can be made according to the user's preferences, recommending a ranking of the most suitable elements [5].

Recommendation systems can be classified based on three main models: the collaborative filtering model, the content-based model and the hybrid model, which is a combination of the two [6]. There is also the knowledge-based model and the demographic recommender system [5].

This paper presents the implementation and development of a prototype music recommendation system, based on the correlation of the user's emotional states. The system uses tags or keywords assigned to songs and the correlations between them. An approach similar to collaborative filtering is suitable for a recommendation algorithm that uses correlations between emotions, because this model allows making recommendations for new users. This kind of recommendations are based on similarities between users and between recommended elements. The implementation uses the Python programming language and the Google Colab environment.

The remainder of the paper is organized as follows: Sect. 2 describes the implementation of the prototype, Sect. 3 shows user scenarios and preliminary results, while Sect. 4 concludes the paper and states sets future work objectives.

2 Work Description

2.1 Tools and Objectives

The Python programming language was used to develop the prototype music recommendation system, in the Google Colab environment used.

The objective is to develop a system that recommends music based on the correlations of the nine emotions presented in the data set. The database is created through annotations, collected using the *GEMS* scale (*Geneva Emotional Music Scale*). This scale contains nine tags of emotions: *amazement, solemnity, tenderness, nostalgia, calmness, power, joyful activation, tension* and *sadness*. Each of the 400 songs from the data set is labeled according to these emotions.

The system computes the similarities between emotions, which are interpreted as tags, for making recommendations. The first goal of the prototype is to make recommendations of emotions, similar to the one introduced by the user. In the second phase, music recommendations are made based on the similarities between these emotions.

2.2 Implementation

The input data is taken from the data set and processed in the first phase, within the correlation algorithm.

The correlation algorithm receives the emotions from the data set as input, which are interpreted as tags, and computes the correlations.

These correlations are inputs to the feelings (emotions) recommendation algorithm. This algorithm receives a feeling as input, and it generates two other similar feelings as a result. The obtained results are implemented in the final algorithm that will recommend songs based on the similarities between emotions, given by the calculated correlations (Fig. 1).

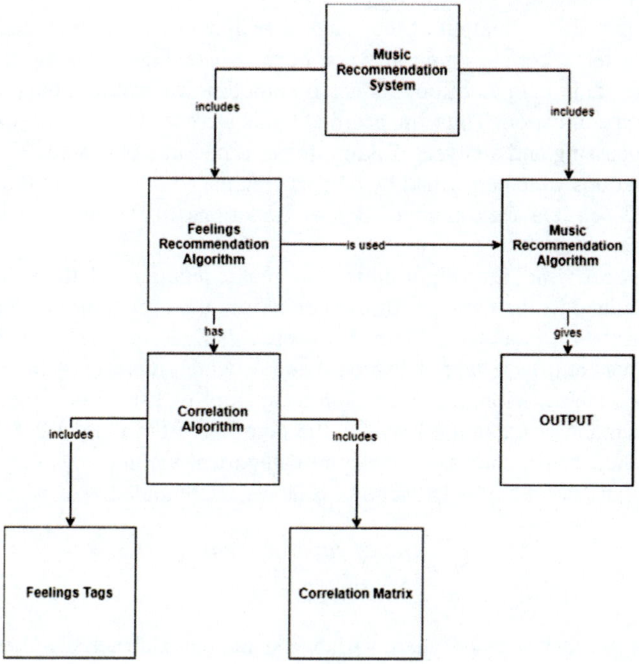

Fig. 1. Conceptual diagram of the prototype music recommendation system

The data set in [8] was used to implement the recommendation algorithm. This system is based on 400 excerpts from songs, lasting one minute, divided into four musical genres (rock, pop, classical and electronic). Sorting or recommending songs by genre was not a priority in achieving the proposed goal. The developed algorithm focuses on the idea of making recommendations based on emotions.

During the implementation of the system, emphasis was placed on the emotions and the evaluations of the songs, not on the popularity of the songs, their level of appreciation or on the demographic information of the participants.

The data was divided into two subsets. The dataset used to develop the system contained information about the nine tags of emotions, the musical genre, the id of each song and its name, added later. The file with the 400 songs that were evaluated was also attached to the data. This file with songs is the second reason why this dataset was chosen, because it is desirable that once the songs have been recommended according to the mood, the user can listen to them.

In the dataset, the evaluation of the songs is realized by a unary evaluation system. If a participant felt a certain emotion when he was listening to a song, that state was assigned to the value 1. In addition, if a certain emotion was not felt, the participant did not make any specification. Thus, unspecified fields were initialized with the value 0, to allow data processing and analysis. Taking this to account, a binary matrix is obtained.

The adnotations were performed by 64 participants, who listened to the songs. Each of them could select a maximum of 3 tags (emotions that were experienced when listening to the song).

The purpose of the first algorithm is to make recommendations based on the emotions contained in the data set. In the beginning, a correlation was made between tags. This correlation is an association between tags that are similar. The correlation between two tags returns a value between 0 and 1, which represents the probability of finding the two labels belonging to the same category or list. This suggests the similarity or the semantic correlation between the emotions from the dataset. This type of correlation is necessary, since the labels are categorical variables, as opposed to continuous ones. The correlations between two labels is computed as a probability:

$$p = \frac{No.\ of\ favorable\ cases}{No.\ of\ possible\ cases} \quad (1)$$

Where p is the number of cases where variables a and b are labeled together, against the number of cases where just a or just b are labeled in that category. There are three possible cases: (1) both variables are labeled in a song (*a_and_b*); (2) only a is labeled (*a_not_b*); (3) only b is labeled (*b_not_a*). Variables a and b are initialized with the rows of the dataset where a or b is equal to 1 (meaning that the emotion was labeled). The dataset is parsed three times, to count all three possible cases. Finally, the number of favorable and possible cases are computed and the correlation between a and b is computed, using Eq. (1).

With these values, the correlation matrix was formed (Fig. 2). The matrix shows a correlation of 1 (maximum) on the main diagonal and is symmetrical with respect to it.

The algorithm explained above was integrated into another algorithm that recommends music based on the correlation between emotions from the data set. The reason for choosing this kind of approach is that a certain song can be defined by several tags or can convey more emotions. This is more relevant if a data set with more tags is used. However, the associations are also suitable for the data set used and the functionality of the algorithm is highlighted.

index	amazement	solemnity	tenderness	nostalgia	calmness	power	joyful_activation	tension	sadness
amazement	1.000000	0.072907	0.092105	0.050427	0.074290	0.106764	0.176310	0.042590	0.015121
solemnity	0.072907	1.000000	0.071429	0.098221	0.138386	0.116854	0.083158	0.109244	0.091894
tenderness	0.092105	0.071429	1.000000	0.158890	0.201553	0.025653	0.080645	0.024687	0.084388
nostalgia	0.050427	0.098221	0.158890	1.000000	0.183062	0.037412	0.064927	0.052973	0.181420
calmness	0.074290	0.138386	0.201553	0.183062	1.000000	0.036477	0.076218	0.029980	0.098174
power	0.106764	0.116854	0.025653	0.037412	0.036477	1.000000	0.180008	0.133217	0.037418
joyful_activation	0.176310	0.083158	0.080645	0.064927	0.076218	0.180008	1.000000	0.090136	0.018418
tension	0.042590	0.109244	0.024687	0.052973	0.029980	0.133217	0.090136	1.000000	0.115482
sadness	0.015121	0.091894	0.084388	0.181420	0.098174	0.037418	0.018418	0.115482	1.000000

Fig. 2. The correlation matrix between the emotions from the data set

3 User Scenario and Preliminary Results

The operating principle of the correlation system is illustrated in Fig. 3: when the user chooses an emotion, the system will recommend two similar emotions, which have highest correlation with the emotion given as input. This suggests that the three previously mentioned emotions can be integrated into a certain category and they frequently occur together as tags for the same songs in the data set.

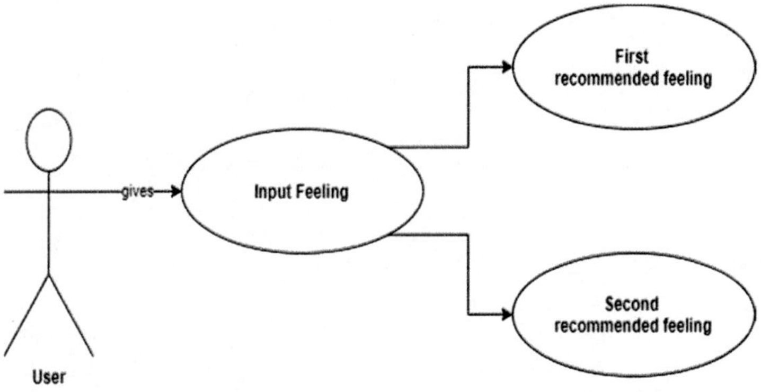

Fig. 3. Principle of operation for the correlation system

For example, if the user chooses to give as input the emotion "*calmness*", the returned emotions will be "*tenderness*" and "*nostalgia*". The semantic bond between these three emotions is clearly visible. This type of recommendation can help the user define and better understand his mood, and then choose a song according to his current emotional state or a song compatible to a new emotional state he wants to experience.

In another possible user scenario, the input emotion is "*amazement*" and the received emotions are "*joyful activation*" and "*power*". These three emotions are also connected and can be placed in the same category.

Next, the music recommendation algorithm receives as input data, the output data of the algorithm that generates correlations.

The user has the possibility to choose a certain emotion from the list of emotions in the data set used and two similar emotions will be returned to him, which have the greatest correlations with the emotion given as input. After the user received these recommendations for similar emotions, he will be given song recommendations based on the first emotion received as output. The recommendations are made this way because there are certain similarities between the emotion given by the user and the first recommended emotion, as they are often tagged together in the dataset. In the end, the user can listen to the chosen song.

The principle of operation for the music recommendation system is visible in the usage scenario illustrated in the Fig. 4.

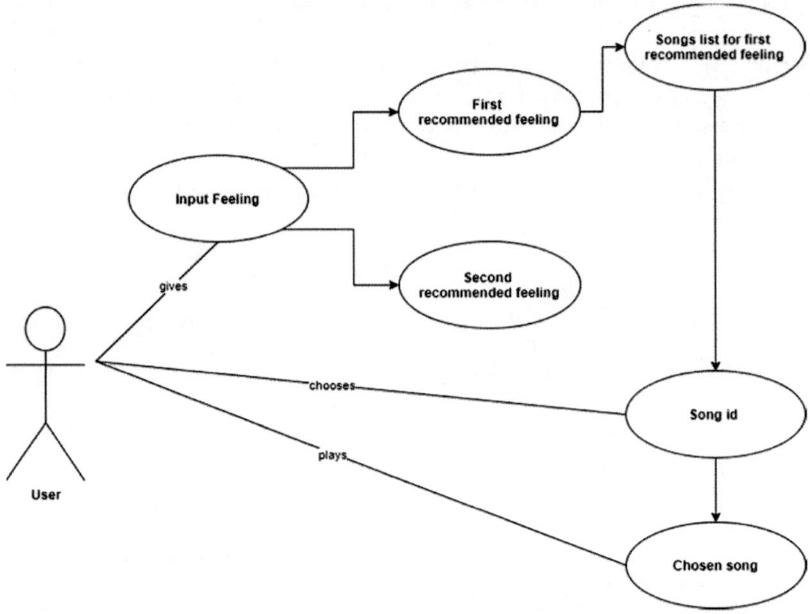

Fig. 4. Usage scenario for the music recommendation system

4 Conclusions

Emotions govern our daily lives, whether we want to accept this or not. The emotional state we are in influences the quality of our work and relationships, as well as our mental and physical state. One of the handiest tools in quickly changing and improving our emotional state is music.

The proposed system can make music recommendations based on the correlations of the emotional states from the data set. This approach allows the users to define and understand better their mood and to listen to music from a certain category of songs.

Also, the users can choose the emotion they want to have and listen to music according to the desired mood.

This system is a prototype and was implemented for small scale integration, but large scale optimization can be achieved, provided the access to larger data sets. Also, testing the system and collecting user feedback is crucial to further development. The long-term goal is to include the system into a mobile app designed for mood improvement, with music and visual effects as the main instruments of the mood changing process. A wider range of emotions and tags would significantly increase the precision of the system.

Conflict of Interest. The authors declare that they have no conflict of interest.

References

1. Kemper, K.J., Danhauer, S.C.: Music as therapy. Sothern Med. J. **98**(3), 282–288 (2005)
2. Dalla Bella, S., et al.: The Neurosciences and Music III. Blackwell Publishing on Behalf of New York Academy of Sciences, Boston (2009)
3. Sutton, J.P.: Music, Music Therapy and Trauma. Jessica Kingsley Publishers, London
4. Advani, G., Soni, N.: A novel method for music recommendation using social media tags. Int. J. Comput. Appl. **122**(2) (2015)
5. Aggarwal, C.C.: Recommender Systems: The Textbook. Springer, Cham (2016). https://doi.org/10.1007/978-3-319-29659-3
6. Medium. https://medium.com/@briansrebrenik/introduction-to-music-recommendation-and-machine-learning-310c4841b01d. Accessed 04 Sept 2020
7. Displayer blog. https://www.displayr.com/what-is-correlation/. Accessed 04 Sept 2020
8. Projects.science. http://www2.projects.science.uu.nl/memotion/emotifydata/

A Game-Theoretic Analysis of the Recurrent Patient-Doctor Interaction Using Genetic Algorithms

Paul Faragó(✉), Mihaela Cîrlugea, Sorin Hintea, and Călin Fărcaş

Technical University of Cluj-Napoca, Cluj-Napoca, Romania
paul.farago@bel.utcluj.ro

Abstract. Medical service relies on patient-doctor interaction, which is modeled in this work with the Stag Hunt game. As prescribed by game theory, the Stag Hunt game has two pure-strategy Nash equilibria in which either both players cooperate or both players defect. This paper investigates means of achieving cooperation in a recurrent patient-doctor interaction scenario. For this purpose, genetic algorithms (GA) are employed to solve the Stag Hunt game. As such, evolutionism models learning along the generations, whereas randomness in the evolutionary process models trembles in the choice of strategy. Four test scenarios are simulated: variable number of GA iterations, variable initial choice of strategy, variable number of recurrent interactions and variable fraction of population partitioning. Simulation results provide valuable insights into the means of achieving patient-doctor cooperation.

Keywords: Game theory · Stag Hunt game · Genetic Algorithms

1 Introduction

Medical service relies on patient-doctor interaction which, in a modern ubiquitous healthcare environment, is carried out through a variety of communication channels. The quality of the medical service relies very much on patient-doctor cooperation, which requires openness and honesty from both sides.

Such a relationship however is not always bilaterally supportive. On one hand, the patient may (purposely) omit symptoms because of fear of diagnosis, fear of treatment, fear of long term commitment with sacrifices regarding change of lifestyle, etc. Also, in case of a long-term medical treatment for example, the patient may neglect to follow the prescribed medication, give up after minor improvement in the state of health, return to former lifestyle, etc. On the other hand, from the doctor's point of view, establishing a diagnosis when facing an improperly drawn case history becomes problematic, follow-ups also become inefficient when the patient doesn't keep to the prescription, changes of lifestyle, etc. Also, the doctor could be overwhelmed by all the legal situations and stipulations he needs to take care of. All these situations that can appear, can affect the patient-doctor relationship.

S. Vlad and N. M. Roman (Eds.): MEDITECH 2020, IFMBE Proceedings 88, pp. 274–281, 2022.
https://doi.org/10.1007/978-3-030-93564-1_31

A question which stands in the field targets the means to acquire patient-doctor cooperation, which becomes a problem of rational decision making in the context of inter-human interaction.

Game theory is a very strong mathematical tool to analyze interaction among rational self-interested individuals. Although initially employed as a rational decision-making tool in economics, game theory found applications in a variety of domains, ranging from political science to biology and medicine.

To the best of our knowledge, application of game theory to analyze and model interactions in healthcare is incipient, with relatively little literature available in the field. Some examples are provided as follows. Game theory attempts to formalize a quantitative and prescriptive means for optimal decision-making in [1]. Game theory also provides valuable models of interaction in-between medical staff. For example, Saposnik et al. present a scenario of reporting an error in medication administration which involved the physician and the pharmacist, modelled by the Prisoner's Dilemma [2]. Patient-doctor interaction is also modeled with game theory. For example, an employment of the Prisoner's Dilemma was reported by Djulbegovic et al. to model opioid prescription to potentially drug-seeking patients [3]. Alternatively, Saposnik et al. reported an observation where patient tracking in a stroke prevention clinic could be modeled by the Stag Hunt and Assurance game [2].

In this work, we analyze the means of achieving cooperation in the patient-doctor relationship. We employ the Stag Hunt game for modeling the interaction as described in [2]. The story behind the Stag Hunt has two hunters, unable to communicate, choose whether they go hunting for stag or hare. In the absence of communication, each hunter makes his choice independently. Both hunters prefer stag to hare, but hunting for stag requires both hunters to cooperate whereas hunting for hare can be done individually.

The Stag Hunt game is illustrated generically in the normal form from Fig. 1, with parameters a, b, c and d following

	Stag (*cooperate*)	Hare (*defect*)
Stag (*cooperate*)	a, a	c, b
Hare (*defect*)	b, c	d, d

Fig. 1. The Stag Hunt game in normal form

$$a > b \geq d > c \tag{1}$$

and where (u_{row}, u_{column}) from each cell of the normal form representation specifies the utilities of the row and column player respectively.

Modeling patient-doctor interaction with the Stag Hunt game requires the following assumptions. The Stag strategy assumes *cooperation* from the player's part, and only brings utility to the players if all cooperate. The Hare strategy on the other hand assumes *defection* from the player's part, e.g. untruthful report of symptomatology to get medication, medical documentation etc., and only brings some utility, smaller in magnitude, to the player who defects.

Following the reasoning to determine pairs of mutually dominant strategies, it comes to the result that Stag Hunt exhibits two pure-strategy Nash equilibria: s_1 = (Stag, Stag) and s_2 = (Hare, Hare). Indeed, if the column player chooses Stag, the row player will be best off by also choosing Stag. On the other hand, if the column player chooses Hare, the row player will be best off by also choosing Hare. Thus, Stag Hunt is an example of a pure coordination game.

In addition to determining the optimal play in the stage game, we assume repeated interaction which enables learning along the recurrent medical appointments. Furthermore, we allow trembles to model mistakes in choosing the optimal strategy.

Given the nature of the repeated interaction, learning can be modeled by evolution and trembles can be modeled by randomness. As such, Genetic Algorithms (GA) are suitable to solve the recurrent Stag Hunt game in an evolutionary fashion and provide valuable insights into the means of achieving cooperation.

2 Mathematical Formulation

Given the availability of multiple Nash equilibria, players face the challenge of coordinating onto the same pair of actions. One approach is to search for the payoff dominant equilibrium, which Pareto dominates all other equilibria. One will notice that the (Stag, Stag) equilibrium is strictly Pareto dominant in the sense that no other solution will make both players better off [4]. Indeed, as prescribed by the formal definition of Pareto dominance [5], a strategy profile s Pareto dominates a strategy profile s' if and only if

$$\begin{cases} u_i(s) \geq u_i(s'), i \in N \\ \exists j \in N \text{ such that } u_i(s) \geq u_i(s') \end{cases} \tag{2}$$

where N is the set of players. Relations (2) are satisfied only by s = (Stag, Stag).

Although the theory of Pareto dominance prescribes (Stag, Stag) as the payoff dominant solution, practical experimentation on situations of pure coordination show that players prefer the "safe" equilibrium solution (Hare, Hare) [4]. The motivation is that players prefer a sure utility b or d from Hare, which is larger than c from the unsuccessful coordination on Stag. Indeed, failure of coordination on Stag results in a deviation loss

$$b - a \tag{3}$$

Similarly, failure of coordination on Hare results in a deviation loss

$$c - d \tag{4}$$

Taking both players into account, if the deviation loss on Hare is larger than the deviation loss in Stag, namely

$$(c - d)^2 > (b - a)^2 \tag{5}$$

it is the case that solution (Hare, Hare) risk dominates (Stag, Stag). The power of 2 in Eq. (5) stands for having the multiplicative product of the deviation losses for both players in the given equilibrium.

To handle uncertainty regarding the actions of the other player, let p be the probability of the other player choosing Stag. The risk factor p_{Stag} of equilibrium (Stag, Stag) is the value of p which equals the expected utility gained via both actions

$$\begin{aligned} & p \cdot u_i(Stag, Stag) + (1 - p) \cdot u_i(Stag, Hare) \\ = & p \cdot u_i(Hare, Stag) + (1 - p) \cdot u_i(Hare, Hare), i \in N \end{aligned} \tag{6}$$

The risk factor q_{Hare} of equilibrium (Hare, Hare) is calculated similarly, letting q be the probability of the other player choosing Hare. Then, an alternative definition of risk domination assumes the equilibrium with the smallest risk factor, i.e. (Hare, Hare) is risk dominant provided $q_{Hare} < p_{Stag}$.

Risk domination also allows an evolutionary interpretation. Consider a population of individuals and let p be the fraction of the population choosing Stag. Resistance of equilibrium (Hare, Hare) against equilibrium (Stag, Stag) is defined as the maximal value of p such that individuals playing Stag have no reproductive advantage over individuals playing Hare [4],

$$\begin{aligned} & p \cdot u_i(Stag, Stag) + (1 - p) \cdot u_i(Stag, Hare) \\ < & p \cdot u_i(Hare, Stag) + (1 - p) \cdot u_i(Hare, Hare), i \in N \end{aligned} \tag{7}$$

Then, equilibrium (Hare, Hare) is risk dominant if its resistance against (Stag, Stag) is larger than the other way around.

To be noticed is that Eq. (7) constitutes the definition of evolutionarily stable strategies [6]. This yields the result that the risk dominant equilibria will be the outcome of the evolutionary process. Accordingly, the preference for the safe solution is evolutionary (i.e. acquired by former experience, learning, etc.) rather than purely mathematical.

3 The Genetic Algorithm

As presented in Sect. 2, game theory provides several solution concepts to solve the Stag Hunt game. In this paper, we aim to analyze the recurrent Stag Hunt game modeling patient-doctor interaction. Some particularities had to be considered for this scenario as follows. First, we have considered two classes of players: patients and doctors, thus a play will always account for a patient-doctor interaction. Second, we have assumed repeated interaction with recall in the sense that players learn which strategy to play based on former interactions. Third, we have also allowed mistakes. Additionally, we have assumed randomness in establishing the patient-doctor pair.

Given the recurrent Stag Hunt scenario modeling patient-doctor interaction, we have employed Genetic Algorithms (GA) to solve the game, aiming to maximize the payoff gained within the population. The block diagram of the GA implemented in Matlab is illustrated in Fig. 2, and its particularization to the given game is explained as follows.

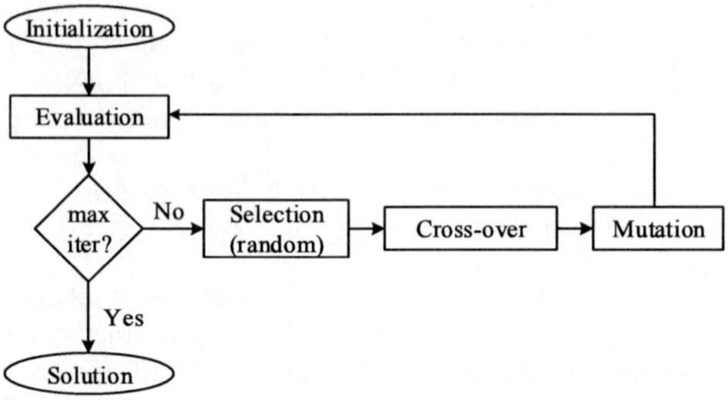

Fig. 2. The Genetic Algorithm implemented for evolving a population playing Stag Hunt, under a recurrent patient-doctor interaction scenario

The GA was initialized with a population size of 50 individuals partitioned into two classes: doctors and patients, with an arbitrarily chosen doctor-patient ratio $1 : k$. Binary representation of the individuals was considered for strategy encoding. Depending on the simulation scenarios, we have implemented two creation functions for the generation of the initial population. One creation function implements a purely random generation of the initial population, i.e. the choice for either strategy is random. The second creation function imposes the initial strategy for each player, having a population distribution with fraction p playing Stag, leaving $1 - p$ to play Hare.

For evaluation, the Stag Hunt game is played by randomly generated patient-doctor pairs, and the fitness of each individual is computed as the sum of the payoffs gained in each individual play.

In this application, the GA was designed to evolve for a fixed number of generations, with no additional stopping criteria. Computation of the next generation is performed as follows. Recombination is implemented with single-point crossover among randomly selected parents. Mutation is applied to 10% of the population by randomly complementing loci of the binary string.

As described, the GA implemented to evolve the population playing Stag Hunt resembles an evolutionary scenario with trembles. Evolutionism models learning, having knowledge, i.e. which strategy to play, transferred along generations via crossover. Trembles, i.e. mistakes in which strategy to play, are modeled with randomness during mutation.

4 Simulation Results

In order to determine the evolutionary outcome of the Stag Hunt game in a recurrent patient-doctor interaction scenario, the game parameters were arbitrarily set to $a = b = 3$, $b = d = 2$ and $c = 0$. Accordingly, the game exhibits both risk and payoff dominant solutions respectively with a risk factor $p = 1/3$ for Hare.

Four test scenarios are presented as follows to study the effects of evolution length, population distribution regarding the initial strategy choice, population distribution regarding the doctor – patient ratio and number of interactions for the same doctor – patient pair. The pairs of individuals to play the game are chosen randomly in either test. As such, 10 GA runs were carried out for each test setup and we have considered the test result to be the average of the evolutionary outcomes of each run.

The simulation results are plotted in Fig. 3 and are explained as follows. The nominal test conditions are an evolution length of 100 generations, random distribution p of the strategy choice, random partitioning of the population $1 : k$ into patient and doctor classes and random generation of the patient-doctor pairs (with no account for repeated interaction). The parameters of interest are varied according to each scenario.

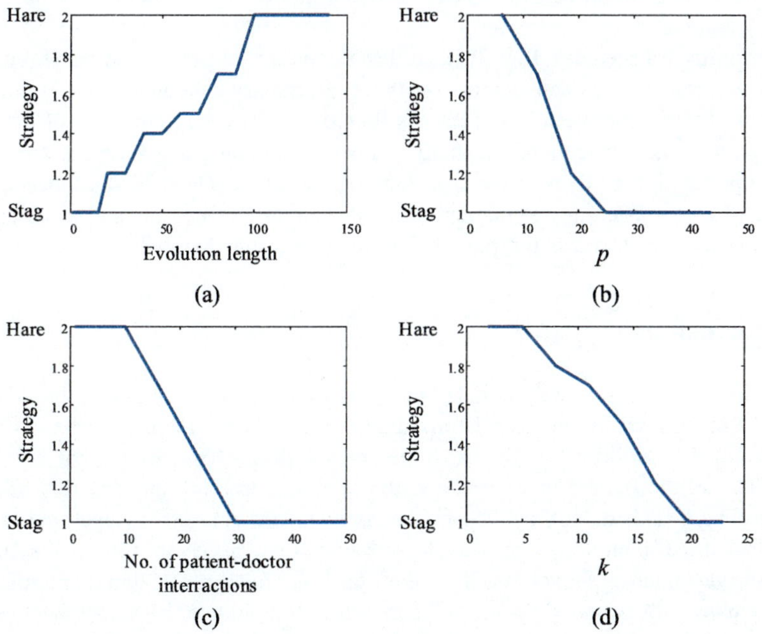

Fig. 3. The GA outcome – the Stag Hunt strategy under various test scenarios: (a) length of the evolution, (b) distribution of the strategy within the population, (c) number of recurrent patient-doctor interactions and (d) partitioning of the population into doctors and patients

The first test scenario, Fig. 3(a), studies in what respect the evolutionary strategy for Stag Hunt is influenced by the evolution length. The evolutionary outcome illustrates that the payoff dominant solution (Stag, Stag) prevails for short evolution lengths up to 15, while the risk dominant solution (Hare, Hare) prevails for long evolution lengths above 100. For evolution lengths in between, the outcome varies in-between the payoff and risk dominant solutions respectively, depending of the random generation of pairs to compete.

The second test scenario, Fig. 3(b), studies the effect of the initial choice of strategy on the evolutionary outcome. For this purpose, the fraction *p* of individuals playing Stag was varied between 1/8 to 7/8. As expected, the risk dominant solution (Hare, Hare) prevails for values of *p* below the risk factor, while the payoff dominant solution prevails for values of *p* above the risk factor. However, as illustrated, the evolutionary outcome varies between the payoff and risk dominant solutions respectively for values of p around the risk factor, determined by the random generation of pairs to compete.

The third test scenario, Fig. 3(c), studies the effect of the number of recurrent patient-doctor interactions on the evolutionary outcome. Thus, the number of interactions was varied from 1 to 50. As illustrated, the risk dominant solution (Hare, Hare) prevails for a small number of interactions below 15, whereas the payoff dominant solution (Stag, Stag) prevails for a large number of interactions above 25. One will notice a critical number of interactions around 20 where the evolutionary outcome depends on the random distribution of the population and the random generation of pairs to compete.

The fourth test scenario, Fig. 3(d), studies the effects of population partitioning, i.e. number of patients for each doctor, on the evolutionary outcome. For this purpose, factor *k* was varied between 2 and 23. As illustrated, for a large number of doctors in the population and consequently a small number of patients per doctor, i.e. small value of *k*, the population evolves to the risk dominant solution (Hare, Hare). Once with the increase of factor *k*, i.e. fewer doctors and consequently more patients per doctor, the population evolves towards the payoff dominant solution (Stag, Stag).

5 Conclusions

Patient-doctor interaction was modeled in this work using a recurrence of the Stag Hunt game. As prescribed by the Stag Hunt game, patient-doctor interaction has two pure-strategy Nash equilibria: the payoff dominant solution where the patient and doctor cooperate, i.e. the patient is sincere and the doctor is trustful, and the risk dominant solution where both defect, i.e. the patient is insincere and/or uncooperative and the doctor is distrustful and/or suspicious. It is obvious that the payoff dominant solution is preferred, yet practice shows that it is often the case for the risk dominant solution.

This paper investigated means of achieving cooperation in a recurrent patient-doctor interaction. For this purpose, the Stag Hunt game modeling the patient-doctor interaction scenario was solved using GAs. As such, evolutionism models learning along the generations, whereas randomness in the evolutionary process models trembles in the choice of strategy. The conclusions of the proposed evolutionary test scenarios, transcribed to the medical field, are summarized as follows.

- The players, both patients and doctors, become uncooperative with the length of the learning process. This can be translated from the patient's point of view into a generalized distrust into the medical staff, and from the doctor's point of view into suspicion regarding the patients.
- The recurrent patient-doctor interaction evolves towards cooperation provided the interaction starts off by cooperating for a rather large fraction of the population. This result is straightforward, as cooperation acquires a larger utility to both players, and this knowledge is transmitted along the evolutionary process.
- The recurrent patient-doctor interaction evolves towards cooperation provided the number of recurrences increases. This result illustrates that a larger number of interactions within the same patient-doctor pair grows mutual trust which leads to cooperation.
- Recurrent patient-doctor interaction evolves to cooperation for a smaller number of doctors within the population, i.e. a larger number of patients per doctor. This result illustrates that, provided the medical staff is outnumbered, both patients and doctors are bound to cooperate in order to attain medical care and reduce overcrowding of the medical institution. Adversely, a large number of doctors will rather enable defection, such as competition for patients, lying with regard to symptomatology, mistrust, suspicion, etc.

Conflict of Interest. The authors declare that they have no conflict of interest.

References

1. Diamond, G.A., Rozanski, A., Steuer, M.: Playing doctor: application of game theory to medical decision-making. J. Chronic Dis. **39**(9), 669–682 (1986)
2. Saposnik, G., Johnston, S.C.: Applying principles from the game theory to acute stroke care: learning from the prisoner's dilemma, stag-hunt, and other strategies. Int. J. Stroke **11**(3), 274–286 (2016)
3. Djulbegovic, B., Hozo, I., Ioannidis, J.: Modern health care as a game theory problem. Eur. J. Clin. Invest. **45**, 1–12 (2014)
4. Myerson R.B.: Game Theory: Analysis of Conflict. Harvard University Press, Cambridge (1991)
5. Cvetkovic, D., Parmee, I.C.: Genetic Algorithm–based Multi–objective Optimisation and Conceptual Engineering Design, Congress on Evolutionary Computation. Washington D.C., USA (1999)
6. Leyton-Brown, K., Shoham, Y.: Essentials of Game Theory: A Concise Multidisci-plinaryIntroduction. Morgan and Claypool, San Rafael (2008)

A System Proposal for Collecting Medical Data from Heterogeneous Sensors Using Software-Defined Networks

Sorin Buzura(✉), Bogdan Iancu, and Vasile Teodor Dadarlat

Computer Science Department, Technical University of Cluj-Napoca, Cluj-Napoca, Romania
Sorin.Buzura@cs.utcluj.ro

Abstract. Software-defined networking (SDN) is an emerging network paradigm which allows networks to become more customizable through programmability. In the current healthcare context, services and modern standards (e.g. FHIR) can be integrated in the networking layer for faster content delivery in the desired format. The purpose of the current work is to define a system proposal where heterogeneous sensor data can be collected from different network sensors and delivered in a standardized way to the applications plane. Compared to related studies which try to achieve interoperability in the applications layer, the current work focuses on defining a communication protocol which allows the SDN controller to interpret the sensor data and to format it accordingly. This enables the lower networking layer to process and format the data to match the needs of the applications. The chosen context is that of Wireless Body Area Networks (WBANs) which can contain different types of sensors, but the data must reach the applications layer in a unified way.

Keywords: Software-defined networking · Medical data · Heterogeneous · Sensors

1 Introduction

The term e-health covers a broad variety of domains, ranging from medicine, public health, economics to technology. It can encompass terms such as telemedicine, telecare, telehealth, health information systems/technology, or interactive health. WHO defines eHealth as "the use of information and communication technologies (ICT) for health" in [1]. The main goals of e-health systems are to reduce costs (both for patients and health professionals), improve health care, built trust between all involved parties, encourage innovation and also support various educational and research activities (by using anonymized medical data) [2].

Given the current Internet of Things (IoT) context, where the number of devices connected to the internet is rising, it is a normal phenomenon that medical devices connect to the internet as well via wireless sensor networks (WSN). A WSN is composed of several types of wireless sensors (called nodes) that communicate with each other or with a central station (called a gateway or a sink) using radio frequencies, in a wide range of application areas [3]. Wireless Body Area Networks (WBAN) represent

S. Vlad and N. M. Roman (Eds.): MEDITECH 2020, IFMBE Proceedings 88, pp. 282–289, 2022.
https://doi.org/10.1007/978-3-030-93564-1_32

a specialized type of WSN designed for medical monitoring and other applications. WBANs are autonomous and mobile devices that can be placed inside or outside the human body [4] with the aim of improving the quality of life of their users (Fig. 1).

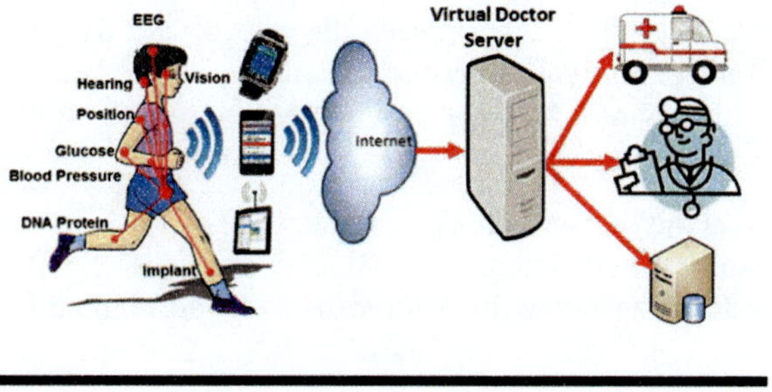

Fig. 1. WBAN architecture [5]

Software-defined network (SDN) is a novel network paradigm which allows networks to become more customizable through programmability. A SDN architectures, compared to a classical network architecture, separates the network responsibilities between different planes (or layers): data plane, control plane and the applications plane. The control plane acts as the intelligence of the SDN managing the data plane and offering services to applications, through the application plane. The data plane represents the physical devices that forwards data using the rules specified by the control plane.

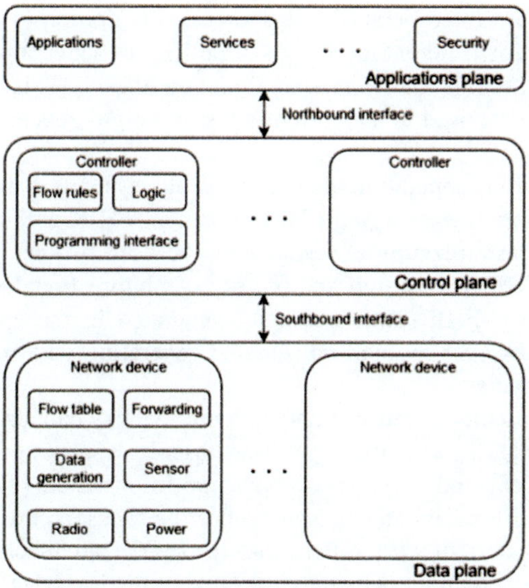

Fig. 2. SDWSN architecture [6]

Using the SDN paradigm, it is possible to bring an additional benefit to the current work through the fact that it allows the WSN to be programmable and customized to meet the applications' expectations and therefore ensure high quality medical services being delivered to the end users. This leads to the SDWSN architecture, where a WSN is controlled by an SDN. Figure 2 shows the main architecture of a SDWSN. As seen in the figure, there is a separation between the control and data planes. The control component is aware of the entire network topology in the data plane and oversees defining rules and behavior which the lower layer devices will follow.

The current work is important because:

- It allows interoperability with modern standards
- It allows sustainable innovations in health care practices
- It combines eHealth technologies with state-of-the-art communication networking technology.

The paper is structured as follows: Section 2 provides background information related to the current research together with a summary of related work. Section 3 describes a system proposal which achieves the sensor data collection from a heterogenous environment. Section 4 concludes the paper and adds some proposals to extend the current work.

2 Related Work

The standards used for medical data exchange are in continuous transformation and together with the increasing number of wearable devices which generate new medical data, modular standards need to be used for better healthcare interoperability between entities. Various data formats can be used to transmit data over the internet, such as FHIR (Fast Healthcare Interoperability Resources). In addition to data representation, the systems can employ different topologies or designs to better support the application layer above, such as WSNs, SDNs or a combination of both. This chapter presents the key theoretical concepts used in the current work and highlights several related studies in this domain area.

In order to ensure interoperability between eHealth systems, a unified standard must be used. This is where the emerging FHIR comes in. FHIR uses a modular structure which allows easy customization of resources to be used by specific applications. [7] explores the use of FHIR to design and prototype a bidirectional system to exchange patient health records. FHIR data is usually generated in the applications layer by medical software, however the current work is generating data in this format in the lower networking layers.

The same functionality to retrieve data in a unified way can also be achieved using protocols such as MQTT, which is a publish/subscribe application layer transfer protocol allowing a client and a server to exchange information [8]. In the context of sensor networks, each sensor can choose to subscribe to a server to report changes and receive updates. The current work differs with the MQTT architectural approach as the data processing and formatting is being performed in the lower networking layer, with the advantage of reducing overhead in the applications layer.

An important aspect of next-generation eHealth systems is represented by their ability to integrate and extract knowledge not only from medical sensor data, but also from home, hospital or other health care environments. Peng and Goswami [9] proposed a web ontology language (OWL)-based integration ontology for data from different health services/systems and home environment. A combination between SDN technology, to relieve data and sensors heterogeneity and existing data representation could further benefit the creation and adoption of next-generation eHealth systems.

WBANs are characterized by their heterogeneity (both in terms of their types and communication protocol used) and limited power supply. In order to interconnect heterogeneous WBANs, SDN can be used as a unifying technology. The future of eHealth should provide medical systems that interoperate securely, reliably and in a scalable fashion, thus allowing heterogeneous devices to work together. The ICE++ architecture [10] combines mobile edge computing paradigm with SDN and virtualization techniques to automatically manage all the components of an eHealth system. The SDN approach provides flexibility to the proposed architecture, by enabling real-time and on-demand communication between the systems' components.

Cicioğlu and Çalhan [11] proposed a flexible SDN-WBAN architecture with a communication interface protocol and the results demonstrated that an SDN-enabled solution outperforms traditional WBAN architectures. The proposed SDN-based solution adheres to ISO/IEEE 11073 Health informatics - Medical/health device communication standards [12] in terms of real-time and efficient data exchange and plug-and-play capabilities.

The current work offers a system proposal that uses the SDN paradigm to collect heterogeneous medical sensor data from different network sensors and delivered it, in a standardized manner, to the applications plane.

3 System Proposal

As previously stated, the current research proposes a system where data can be collected from a network of WBANs composed of heterogeneous sensors. Each WBAN is associated with a patient. This system focuses mostly on a communication protocol which allows the different WBANs to collect sensor data in any way and using any technology, but they must all deliver the data in the same format. This means that each WBAN will forward the data to the central SDN controller in a predefined way.

The key feature of this communication protocol is that it supports a variable number of sensors. It is important to note that each WBAN can act as an emitting station or a relay station as depicted in Fig. 3, therefore the communication protocol must contain relevant information so that the SDN controller can correctly match each incoming data packet to the corresponding WBAN.

With the above system constraints, a common sensor object must be defined (Fig. 4), to make it possible for a network administrator to identify a sensor in the WBAN. The generic sensor data object contains a tuple consisting of the sensor ID and the sensor serial number or the sensor's physical address (can be the MAC address). This is also useful to allow keeping the human in the loop and allow further customizations.

Next, after having defined the generic sensor data object, it can be integrated in the payload delivered by each WBAN (defined in Fig. 5). This payload is parsed by the SDN controller and the operations it performs are done in the following order:

1. Extract the first four bytes which represent the patient ID. This ID is mapped directly on the WBAN making it unique.
2. Extract the last four bytes representing the checksum. Verify the checksum.
3. Verify that the remaining bytes are a multiple of 15, since this middle portion represents the array of sensor data values collected from each WBAN. After this verification, extract every sensor data value and convert it to the format acceptable in the applications layer (e.g. FHIR).
4. Forward the data in the desired format towards the applications layer.

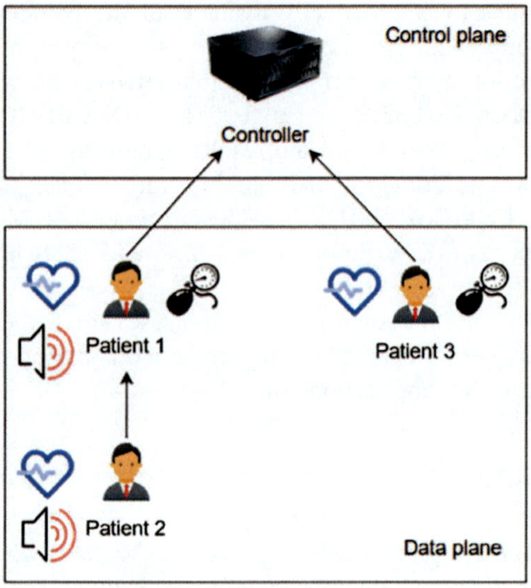

Fig. 3. System proposal

Sensor ID	Sensor Serial Number	Value
1 byte	6 bytes	8 bytes

Fig. 4. Individual sensor data object

Patient ID	Sensor Data	...	Sensor Data	Checksum
4 bytes	15 bytes		15 bytes	4 bytes

Fig. 5. Payload format delivered by each WBAN

Figure 3 shows that Patient 1 generates data from three sensors, whereas Patients 2 and 3 generate data only from two sensors. This means that the traffic volume generated by Patient 1 is greater than the traffic generated by the other two patients. Therefore, another desirable factor is that the network of WBANs configures itself (if possible) to place the high traffic WBANs closer to the SDN controller. Figure 6 demonstrates how the placement of the WBANs in the network affects the total traffic volume. In Scenario 1 the patient closer to the SDN controller generates data from three sensors, this results in a data payload of 53 bytes, whereas Patient 2 only generates data from two sensors, resulting in a data payload of 38 bytes. The number of sensors for the patients is reversed in Scenario 2. Since the patient closer to the Controller must also forward the payload generated by the other patient, the total traffic volume will differ. This results in a total of 129 bytes of relevant application data in the first scenario compared to a total of 144 bytes in the second scenario. Therefore, this demonstrates that it is better to place the WBANs generating multiple data as close as possible to the SDN controller.

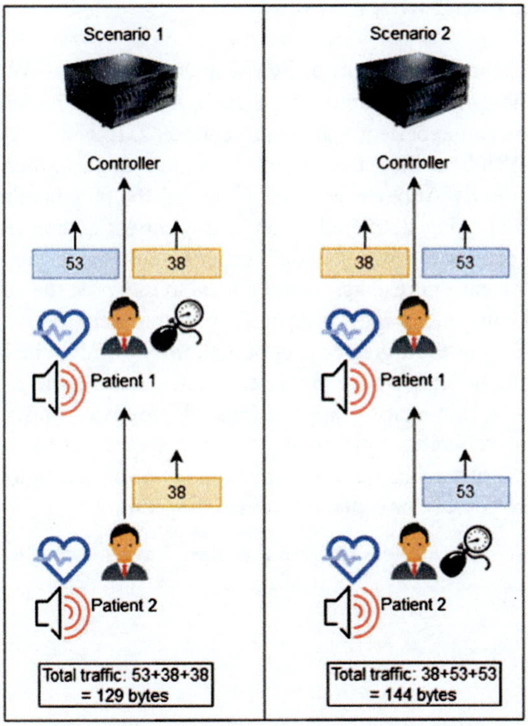

Fig. 6. Scenarios for WBAN placement with different number of sensors

Using an SDN controller brings many other benefits to the approach. The SDN controller can define caching policies or implement optimizations to increase the network lifetime or sensors battery life [6]. It also allows storing information related to the network traffic in order to detect patterns or perform an in-depth analysis on the patient data to identify any possible risks.

Another benefit of the programmable SDN-based approach is the fact that the network is easily scalable by adding multiple WBANs or by adding multiple sensors to the same WBAN. Adding a new WBAN to the network requires that the WBAN implement the needed communication protocol (described above in Fig. 5) and that it connects to the same network as all the other devices. After that, the network will take care of transmitting the traffic and the controller will take care of processing the WBAN information. Adding multiple sensors to an already existing WBAN requires that information about these sensors is configured on both the WBAN and on the controller. This is because the controller is the entity that must be allowed to process and format the data received by each sensor. After that, the new sensors are connected to the WBAN and the WBAN will take care of formatting the sensor data and deliver it upstream using the predefined communication protocol.

4 Conclusions and Future Work

The current work proposed a system design where a network of WBANs is managed using the SDN paradigm. Interoperability in the healthcare domain is a very important topic so that patients can experience the best treatments. Exchanging data between different systems can best be achieved using middleware systems which adhere to modern communication standards. Another desirable feature is the fact that the data is processed (or formatted) as it is being delivered through the network, thus reducing some time delays during data processing which would otherwise occur in the applications layer. A future direction the current research could take is to integrate the current proposal in a system where applications can register specific requirements (e.g. data formats) in the SDN controller, allowing even greater programmability through the SDN paradigm.

The main benefits brought by the proposed SDN system to the e-Health community are: reducing costs by automating applications' deployment and resource allocation processes, increasing reliability by allowing network specialists to control WBANs in a centralized manner, optimizing nodes power consumption and allowing flexibility in adding new WBAN devices or replacing an existing one.

Conflict of Interest. The authors declare that they have no conflict of interest.

References

1. eHealth at WHO. https://www.who.int/ehealth/about/en/. Accessed 21 Sept 2020
2. Eysenbach, G.: What is e-health? J. Med. Internet Res. **3**(2), e20 (2001)
3. Iancu, B., Kovacs, R., Dadarlat, V., Peculea, A.: Interconnecting heterogeneous non-smart medical devices using a wireless sensor networks (WSN) infrastructure. In: Vlad, S., Roman, N. (eds.) International Conference on Advancements of Medicine and Health Care through Technology; 12th–15th October 2016, Cluj-Napoca, Romania. IFMBE Proceedings, vol. 59, pp. 207–212. Springer, Cham (2017). https://doi.org/10.1007/978-3-319-52875-5_45

4. Khan, J.Y., Yuce, M.: Wireless body area network (WBAN) for medical applications. New Develop. Biomed. Eng. **1**, 591–628 (2010)
5. Barakah, D.M., Ammad-uddin M.: A survey of challenges and applications of wireless body area network (WBAN) and role of a virtual doctor server in existing architecture. In: Third International Conference on Intelligent Systems Modelling and Simulation, Kota Kinabalu, pp. 214–219 (2012)
6. Buzura, S., Iancu, B., Dadarlat, V., Peculea, A., Cebuc, E.: Optimizations for energy efficiency in software-defined wireless sensor networks. Sensors **20**, 4779 (2020)
7. Saripalle, R., Runyan, C., Russell, M.: Using HL7 FHIR to achieve interoperability in patient health record. J. Biomed. Inform. **94**, 103188 (2019)
8. Larmo, A., Ratilainen, A., Saarinen, J.: Impact of CoAP and MQTT on NB-IoT system performance. Sensors **19**, 7 (2019)
9. Peng, C., Goswami, P.: Meaningful integration of data from heterogeneous health services and home environment based on ontology. Sensors **19**(8), 1747 (2019)
10. Celdrán, A.H., García Clemente, F.J., Weimer, J., Lee I.: ICE++: improving security, QoS, and high availability of medical cyber-physical systems through mobile edge computing. In: IEEE 20th International Conference on e-Health Networking, Applications and Services (Healthcom), Ostrava, pp. 1–8 (2018)
11. Murtaza, C., Ali, Ç.: HUBsFLOW: a novel interface protocol for SDN-enabled WBANs. Comput. Netw. **160**, 105–117 (2019)
12. ISO/IEE 11073. https://ec.europa.eu/eip/ageing/standards/healthcare/e-health/isoiee-11073_en. Accessed 11 Sept 2020

The Building of Telemedicine Platforms

Cornelia Melenti[1(✉)], Bianca-Maria Dobrescu[1], and Ovidiu Florin Coza[2]

[1] Computer Science Department, Technical University of Cluj-Napoca, Cluj-Napoca, Romania
cornelia.melenti@cs.utcluj.ro
[2] "Iuliu Hatieganu" University of Medicine and Pharmacy Cluj-Napoca, Cluj-Napoca, Romania

Abstract. The following paper gives general insight regarding the use and creation of medical platforms.

Keywords: Telemedicine · SDLC · Functionalities · Standards

1 Introduction

In today's age, health in a digital form is in a continuous progress and demand. New technologies can help better treat patients and monitor their illnesses. This monitoring offers doctors up-to-date data about their patients. This type of health, meaning the remote exchange of medical information/services using electronic communication, is also called telemedicine. Telemedicine was practiced for the first time in 1950, when NASA used it to monitor the astronauts' health status. Today, along with major progress in the medical field, the telemedicine grew along with it.

In more developed countries, medical field are using technology and the IT field more and more. This usage has a positive effect on the efficiency of medical services, on improving the quality offered, on the communication between patient and medical staff and on reducing human error in the medical processes. The health field has a great impact on the well-being of the society, respectively on the people that live in that society.

A medical platform is designed as helpful tool for the institutions that use it for a better management of confidential data. Keeping the information digitally, in databases, results in a positive impact towards patients care. Even though problems can be identified in the process of saving the information digitally, the benefits outweigh them. For hospitals or very big institutions, in which there are a lot of patients coming in daily, there will be a big influx of information as well. In such a context, managing the medical files physically will generate a need of a big storage space, space that can be difficult to maintain and the information hard to access.

Together with the growth in the percentage of digitalized patient information, the problem of data confidentiality and protection arouse. Patients want their personal data to be available only to the right persons, so it's imperative to design a system that doesn't allow unauthorized persons to access it. It's necessary for the patients' data to be well secured in order to guarantee the quality of medical services. With the right

S. Vlad and N. M. Roman (Eds.): MEDITECH 2020, IFMBE Proceedings 88, pp. 290–298, 2022.
https://doi.org/10.1007/978-3-030-93564-1_33

information, the medical staff can take precise decisions related to the diagnosis and treatment of patients.

Who can benefit from such a platform? The main advantage of such a system is allowing its users to socially distance each other in a context where human face-to-face interaction is not desired. So, in the context the world is in right now, a telemedicine platform represents an answer to the distancing measures and hurries the process of using technologies in the field of health. Anyone with a device and internet connection can use such a system and can connect with their doctor for a consultation. In this day and age, digital health is becoming a necessary tool in people's lives. The more people get used to telemedicine, this can become a viable alternative to the traditional medical practice.

2 Why Use It?

There are several reasons as to why someone should use a medical platform [1]:

- Helps you save time and money used for the hospital journey,
- Helps you book an appointment online, based on your preferences and availability,
- Reminds you to take the drug prescription when needed ~~be~~,
- Allows you to check the lab results in a couple of clicks only,
- Shows you the route to the hospital/clinic where the appointment is booked at,
- Allows you to access the medical records 24/7,
- Permits you to conduct more convenient patient-doctor communication and the other way around,
- Reminds you of an upcoming appointment,
- Allows you to find information regarding the hospital/clinic, the medical staff and their qualification and the services offered in a quick way.

3 The Process of Creating a Medical Platform

When trying to build a digital system, we will face the usual software development lifecycle, which can be divided in the following phases [2]:

3.1 Business Analysis Phase

This is one of the most important steps in the software development lifecycle. In order to make sure that the system we build in the later phases is usable by the clients, we need to have clear specifications from the business. The BAs (Business Analysts) take the role of translating the requirements from the business into software requirements, which will be the foundation for the future application.

3.2 Design Phase

In this phase, the elements of the system such as the architecture, the modules and components and the data and different interfaces for the components are designed. In order to create these elements, the software requirements gathered in the previous phase are used as an input.

3.3 Software Development Phase

Here, the coding of the medical platform starts. The development team receives the design document which then will be translated into source code by them. The team will involve a variety of software developers such as front-end developers, back-end developers, mobile developers and full-stack developers. All the components that are a part of the system are implemented and tested in this stage.

3.4 Maintenance Phase

After the telemedicine system is implemented and tested, it can be deployed to the client. After the deployment, the system should be updated regularly and supported by the team so it stays reliable and productive regardless of the moment in time and the circumstances.

4 Basic Functionalities of a Medical Platform

A medical platform should be a reliable piece of technology that comes with diverse functionalities that helps both the doctor and the patient. It should help the doctor in doing their job, respectively the patient in receiving the medical services from the staff. Any telemedicine application is created to minimize the amount of time the patients spend in medical establishments. Such a system can come in many different sizes, such as: WEB applications, desktop applications, mobile applications and so on. Regardless of that, they all have some features in common. So, a good medical system should have the following functionalities [2]:

- Appointment Booking: the platform should allow its users to book appointments and manage their schedule at the clinic/hospital.
- Monitoring overall health status and offering decision support for patients: telemonitoring is extremely useful in monitoring persons that have chronic cardiac illnesses and diabetes. Patients that suffer from these diseases need permanent monitoring due to the nature of the illnesses and the drugs they take. Using the data introduced or synchronized in the platform, the system can support the decisions that the patients take. For example, if the glucose level of a diabetes patient is too low, the system will suggest for him/her to consume some sugar or glucose. If the glucose level is too high, the system will suggest for him/her to administrate insulin.
- Chatting and video meetings: this is at the core of any telemedicine system. A patient should have the possibility to connect with the doctor in order to be diagnosed

and/or to get consulted. Depending on the complexity of the consultation, the doctor could request a video consultation with the patient.

- Medical records visualization: There should be a medical records storage, which is updated as soon as a new medical file is uploaded and that contains the entire patient's medical history.
- Billing: the patient should be able to pay for a virtual visit or any of the other services offered through the platform.
- Treatment: the doctors should be able to prescribe drugs using the system and the patients should receive further information regarding the drugs they were prescribed.
- Decision support for the medical staff: the system, using machine learning, should be able to help the doctors in diagnosing the patients and offering drug prescriptions.
- Offer urgent care information: in case of an urgent issue that needs to be attended, the patient should be able to really easily find information that redirects them to their local emergency room.

5 Database Structure

One of the most important things in an application is the possibility to save/load saved data. The data in a medical platform should be persisted in a well secured and efficient database. An example of a database is in Fig. 1.

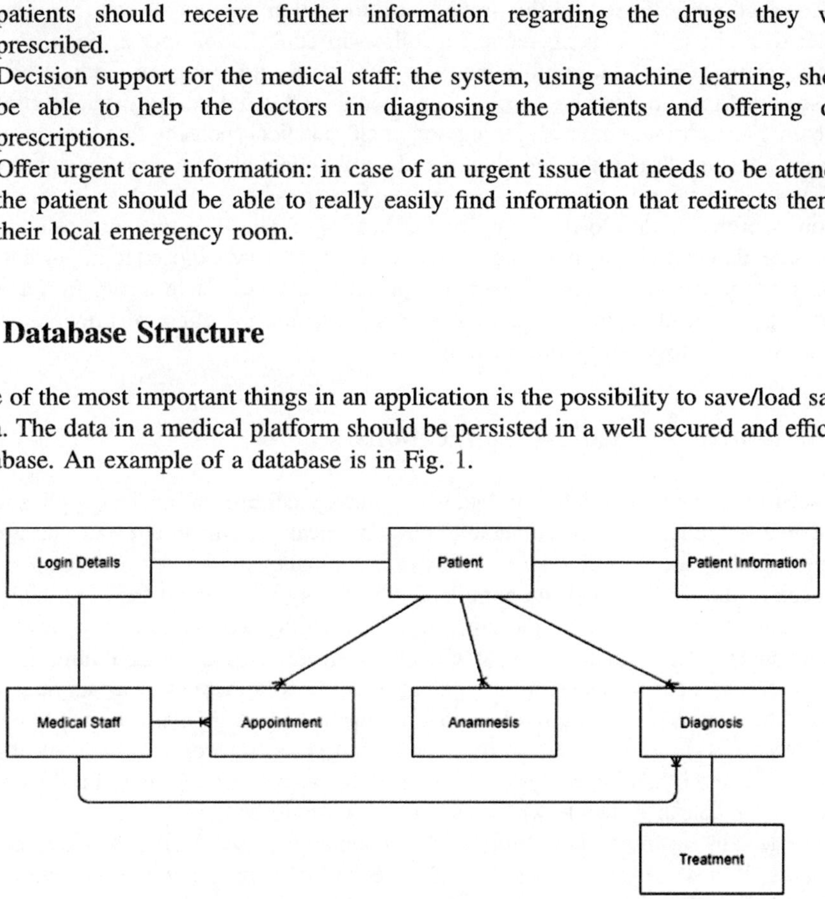

Fig. 1. Example of a simple medical database structure

We must save the information used by the patient and the medical staff to login into the system. Also, we must save patient data such as address, email, telephone and others. When creating an appointment we need as a foreign key the primary key of the doctor for which the patient makes the appointment at, as well as the primary key of the patient as a foreign key. In the tables Anamnesis and Diagnosis we have as foreign key the primary key of the patient. Here we save the anamnesis and diagnosis for each patient in particular. Related to a diagnosis, we can also have a treatment for it and the diagnosis must have the information of the doctor that put the diagnosis.

Some of the most crucial information for the doctors in order to make a diagnosis should be the anamnesis. The anamnesis, also called medical history of a patient, is the information about the patient that contains the current symptoms that he/she has, the history of present illnesses, previous treatments, current medications that he/she is taking, history of present illnesses, family history, whether he has been abroad or not and so much more. Based on this important information, the doctor can then request further data about the patient, request a follow-up consultation, put a diagnosis and offer a treatment plan. More often than not, patients have a complex medical history and some information received from them should be filtered before inserted into the database. For example, in the otolaryngology field (medical specialty that is focused on the ears, nose and throat) the age, the time of employment and the actual occupation is the key data to save. For instance, a significant progress of hearing loss in a retiring person is probably due to the aging of the hearing organ. The type of noise in the workplace also plays a role in the hearing loss, people that are exposed to impulse noise being more prone to it than the people exposed to a steady-state noise. In the ophthalmology field, other information like the risk factors, the effect of previous treatments, ocular history is key information.

6 Standards in Medical Applications

The security and confidentiality of data is the primary concern of medical applications. All network applications are vulnerable, but a medical system should be as secure as possible as they are desirable targets for cybercriminals.

Besides the security and the confidentiality, the system should focus on patients. The system needs to increase the convenience, usability and access for patients and doctors/nurses. The patient users of the platform are used to other complex retail systems and as a consequence, have high expectations for other customer services, like health. The best resource of input information for the healthcare platform would be the medical staff. Such a system would function at its best when involving the medical staff to decide how the healthcare system can be used. Based on their answers, the development team can then decide which technologies would be best to use.

The Health Insurance Portability and Accountability Act (HIPAA) is one of the standards all healthcare systems should abide by. HIPAA regulated the protection and confidentiality of individual medical information. In order to keep the systems secure, medical platform team should [3]:

- Encrypt all data,
- Back up the data,
- Train employees,
- Use multifactor authentication,
- Monitor the usage of the system.

7 Relevant Technologies

There are many technologies out there already being used in the medical industry. Some of the most relevant ones are:

7.1 IoMT

IoMT is a combination of medical devices with software applications that can connect to health care IT systems using networking technologies. At a time when a lot of IoT devices are being used in multiple industries, they fulfill a crucial role in delivering medical services in some remote places or in times when people need to distance each other. The rise of Internet of Medical Things came about due to an increase in the number of medical devices that can collect, can analyze or can transmit data from patients to a health care provider. Potentially, using IoT in the medical industry can lead to a better workflow management in the clinics and overall improvement in the process of health care of the patients.

The IoMT approach allows the medical personnel to use the secure network of services to better offer medical services to their patients while also enabling a more transparent and fast access to information for the patients (Fig. 2).

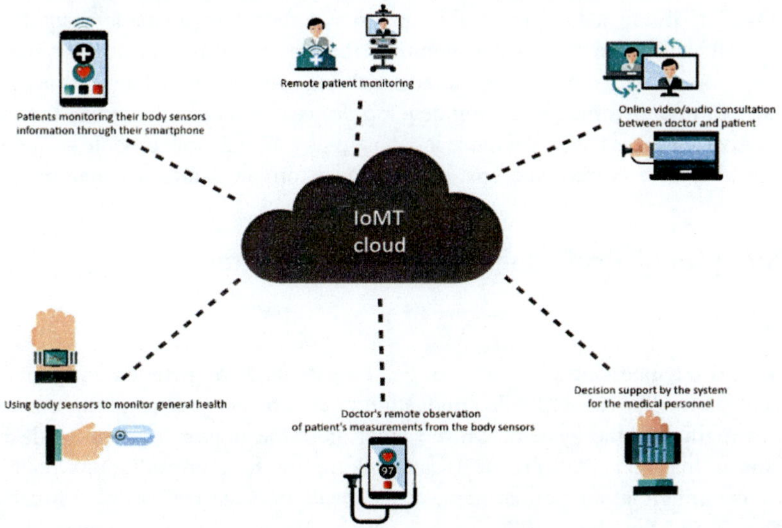

Fig. 2. Example of IoMT facilities

There are several facilities that IoMT can bring to the table. Using some monitoring devices patients and doctors alike can monitor their health. Examples of such devices: sensors that monitor the insulin intake and/or sugar levels in the blood, devices that are embedded with respiratory rate sensors, heart rate sensors, temperature sensors etc., These devices are, most generally, adviced by the doctors and can improve some chronic health conditions by the quick information they provide. Consultations can take place online now and they only require a connection to internet, a microphone and a video camera. In this way it will also benefit rural communities by providing good medical services remotely for a lesser cost. Also, to keep track of the state of the discharged patients that need a continuous observation of their health, medical personnel can use remote patient monitoring devices. This can prevent patient readmission into the hospital and can accelerate their recovery process.

7.2 Artificial Intelligence (AI)

Artificial intelligence describes the use of computers and technology to simulate an intelligent behavior and critical thinking that is comparable to that of a human being. AI in medical applications uses computer techniques to give diagnoses and suggest treatment for the patients. In a similar way in which the medical personnel is educated through years of medical school by doing lab work, assignments and projects, taking practical exams and learning from mistakes, the AI can also learn to do its job [5].

A very important branch of AI is machine learning. One of the challenges for this technology is the ability to obtain patient sets of data which are needed to be used in order to train the models, data which is protected by very strict privacy rules and laws. Fortunately, machine learning and data science is starting to be used more and more in the medical industry. Not only that its ability to process the data is way beyond human capability, but it can also convert the analysis of data into valuable insight for the medical staff which can take better planning, diagnosis and treatment decisions.

Even though AI is currently already helping doctors to more efficiently give diagnoses, to give treatments or even develop drugs, the more we digitize the medical data and give AI access to it, the more it can help us find valuable models – models that can be used to give correct and cost-efficient decisions in complex situations [6].

8 Examples of Professional Medical Platforms

8.1 MDLive

MDLive is a telemedicine platform that can be used both by patients and the medical staff. The patient can use a mobile application or a website to get a appointment with a doctor to login into the system. On the other side, the doctor logs into a dashboard where the online consultations are located. Some of these consultations happen on-demand, meaning that the patient requests a consultations and is left in a virtual lobby. The algorithm of the platform determines which patient is next in the queue based on the relevant information. The doctors have the ability to create a schedule for when they are available. One of the difference between this telemedicine application and others is also having a call-center. The patient that call in for a video consultation through this application are redirected and helped by the operator to the website/mobile application to make an appointment.

The system is said to be secure by the CTO by having a HIPAA compliant security and privacy rules and using end-to-end DTLS encryption [7].

8.2 Doclandia

Doclandia is the biggest telemedicine platform in Romania. It can be used by both the patients and doctors. They can have video consultations, patients having the option of searching for the doctor wanted in whatever speciality he/she wants, while the doctors can also give pro-bono consultations at their choice. The users can create medical profile for each of their family members, the medical history beign accessible 24/7 from the smartphone. There is also an option for the patient to save the video consultation in order for it to be replayed whenever needed.

The technology behind this platform is compliant with HIPAA, which means that it offers high standars of security, being specially dedicated to the medical industry as opposed to other video sharing/communication platforms that doesn't cover the needs of its users such as the security of data which is extremely important in this field. All the traffic on Doclandia.ro is secured and encrypted. The platform has a capacity of holding over 100.000 simultaneous consultations. More so, over the emergency state here in Romania at the beginning of 2020, the platform offered free access to its services to all doctors and patients [8].

9 Conclusions

The medical field progressed a lot in the past century and continues to do so with a fast pace. Until the discovery of antibiotics, people used to die because of a simple infection produced by a small cut. Today we are capable even to program robots that execute an operation instead of a person with a lot more precision than the human hand. Each day we advance more and more from a scientific point of view and we build smart tools and applications that take our daily life to a higher level. A medical platform is only a small piece from the big set of functionalities that the IT field can bring to the medical field, but even so, brings a lot of improvement to doctors and patients compared to traditional medical practice.

All in all, telemedicine platforms and medical applications are the steps for a better tomorrow, the future being more safe with them around. They improve the quality of life for patients with chronic diseases and of elderly people that need continuous observation, reducing so the frequency of visits at the doctor's office and the number of hospital admissions. Using them efficiently and in a correct way they can be an important resource for the health care system after the crisis produced by the global pandemic COVID-19.

Conflict of Interest. There are no conflicts of interest.

References

1. Bessarabova, E.: Healthcare App Development: Types of Medical Apps, Basic Features and Tips for Smooth Development. https://themindstudios.com/blog/healthcare-app-development/

2. Bogdan, M.: Creating a Telemedicine Platform. https://kindgeek.com/blog/post/cre-ating-a-telemedicine-platform
3. Brook, C.: What is a Health Information System? https://digitalguardian.com/blog/what-health-information-system
4. Jangir, M.: IoMT Devices Will Revolutionize HealthTech in 2020. https://www.rtinsights.com/iomt-devices-will-revolutionize-healthtech-in-2020/
5. Greenfield, D.: Artificial Intelligence in Medicine: Applications, implications, and limitations. http://sitn.hms.harvard.edu/flash/2019/artificial-intelligence-in-medicine-applications-implications-and-limitations/
6. Schmitt, M.: Artificial Intelligence in Medicine. https://www.datarevenue.com/en-blog/artificial-intelligence-in-medicine
7. Siwicki, B.: Comparing 11 top telehealth platforms: Company execs tout quality, safety, EHR integrations. https://www.healthcareitnews.com/news/comparing-11-top-telehealth-platforms-company-execs-tout-quality-safety-ehr-integrations
8. Radescu, A.: Platforma de telemedicină Doclandia oferă acces gratuit pentru toți medicii din România. https://www.revistabiz.ro/platforma-de-telemedicina-doclandia-ofera-acces-gratuit-pentru-toti-medicii-din-romania/

Biomechanics, Robotics and Rehabilitation

Design of a Pet-Like Assistive Robot for Elderly People

Alexandru Ianoşi-Andreeva-Dimitrova(✉), Alis-Maria Maier, and Dan-Silviu Mândru

Technical University of Cluj-Napoca, Cluj-Napoca, Romania
Alexandru.Ianosi@mdm.utcluj.ro

Abstract. This paper presents a summary of the state of the art in regard to pet-like robots and the design process of one specially focused on elderly people and their emotional as well as physical needs; our designed pet-like robot aims to provide companionship to people that are no longer able to adequately support a real pet, therefore the mechanical design took into consideration modelling of a general shape that resembles a dog and design several mechanisms that bestow some real life interaction. In light of the fact that there might be some health issues that have to be monitored, the electronical and control subsystems were designed to sense the heart rate and alert via SMS a predetermined contact person if there are any problems detected.

Keywords: Pet-like robot · Assistive robot · Social robot · Design

1 Introduction

World population is slowly aging, which is not a negative thing in itself as this only means that medical services are successful in providing a longer and healthier life to an ever-growing number of citizens. An important issue for elderly people is that some of them are unable to take proper care of an actual pet, yet the desire to have one is still strong. A pet robot solves both issues, as it is usually low maintenance and highly tolerant to neglect.

Pet robots is a natural development of electronic toys; far from providing simple amusement, this type of robots require special abilities not normally encountered or associated with electronic devices: perceiving (sensing the environment, assessing owner desire), acting (responding to stimuli), communication (learning from the interactions with the owner) and surviving (preserving own integrity, e.g. not bumping into furniture, self-charging, etc.) [1]. These robots belong to a class called "personal service robots" which includes companion, assistive and entertainment robots, being characterized by their permanent interaction with humans. This is the reason why for their acceptance are important some features, like: adaptability, ease of use, appearance, emotive expressions, humanness, and facial dimensions [2]. Pet robots potential as tools for increasing quality of life for elderly people was early recognized, as a 2002 case report [3] found that even after 20 sessions there were significant improvements observed in speech, emotional words and satisfaction index; other mental health related

S. Vlad and N. M. Roman (Eds.): MEDITECH 2020, IFMBE Proceedings 88, pp. 301–307, 2022.
https://doi.org/10.1007/978-3-030-93564-1_34

parameters, like loneliness and mental stress, showed improvement. Subsequent randomized controlled clinical trials in a residential care facility [4] confirmed that those who had interaction with the robot had significant decrease in loneliness in comparison with the control group; also, interaction with the robot was far more frequent than with the resident dog and conversation about it were more frequent between residents.

Our proposed pet-like robot resembles a dog, our focus being on the functional aspects of the head that are related to human-dog communication/interaction. The paper describes the design process of the mechanical as well as the control subsystem. A prototype for testing purposes was not ready at the time of this writing.

2 State of the Art

Currently, the research in the area of pet like robots is mature enough that there are successful examples of such devices on the market. This section will look closely to the research that surround this type of robots in order to infer the strengths and weaknesses of systems that attempts to replicate a pet. Two complimentary approaches are evident in the literature, one that deals with healthy adults and another one that deals with people that suffer from cognitive impairment, the latter being much extensively researched.

An important issue with dealing with individuals in various stages of cognitive impairment is the development of behavioral problems, which increase the burden of care. To solve this matter, Petersen et al. [5] proposed a randomized study to see how well a robotic pet is able to reduce the patients stress and anxiety related to their treatment; their result are consistent with similar study and it reports that throughout the trial psychoactive and pain medication could be safely reduced. Another study [6] states that despite the continuous confirmation of multiple benefits of interaction between a pet robot and a patient with cognitive impairments, there are some important issues that might not be easily overlooked: the cost of such systems (expressed not only as the price, but also the necessary maintenance), concern of possible infectious vector (especially in hospital settings), and ethical concern related to social stigma, as the robot might be seen as a toy and the usage of it by cognitive impaired adults might be perceived as infantilizing. These concerns are valid, but applied with care in a patient centered approach, the chance of acceptance is high, as elderly patients with various stages of dementia, despite the fact that they saw the pet robot as a toy, treated it like a technological companion, helpful for supporting their psychosocial needs [7].

A literature review [8] found that social robots are useful for elderly people regardless of their cognitive health, as it provide emotional support in terms of entertainment, stimulation and calm; a troubling finding is that despite high satisfaction levels, the intention to actually use it was lowest in healthy individual, as they felt that they do not necessary need it. Given the fact that living alone is prone to slowly erode mental sanity, further efforts must be made to improve acceptability. In this regard, Whelan et al. [9] found that acceptability improve if the robot use humanlike communication and meet the emotional, psychological, social and environmental users' needs; another important factor is the attitude of the significant others from the person's social circle regarding the pet-like robot. A robust predictor for post-interaction positive affect is the engagement frequency [10], despite personal beliefs as the robot is more useful for others and not for him-/herself, which is an interesting finding that shows that if an elderly person is persuaded to use such a device positive outcomes are not dependent on personal assumptions. A critical view is expressed by Abdi et al. [11] which found many methodological issues in studies related to pet-like robots, and recommends a more rigorous delimitation about the role a robot serves in the study: affective therapy, cognitive training, social facilitation, companionship and/or physiological therapy.

3 Design Process

The design process took two main directions: designing of a mechanical subsystem, with a general shape easy relatable to the final user and with enough biomimetism to be accepted as a pet robot, and designing of the electronics and control subsystem to animate the robot and to provide basic health sign monitoring.

3.1 Mechanical Subsystem

The first step was to determine which animal to consider as a model; it was chosen a dog (*Canis familiaris*), as this is the kind of pet the authors are more familiarized with. Next, the function related to movement were identified: locomotion, tail-wagging, head orientation, jaw open/close movement, tongue movement, eyes and ears orientation. Early in the design process it was determined that the best strategy is to concentrate the effort on the head, therefore, locomotion, tail-wagging and head orientation were not yet studied; furthermore, given the fact that the objective of the implemented motions was to give an emotional anchor, only the jaw open/close movement, tongue motion and eye orientation are considered.

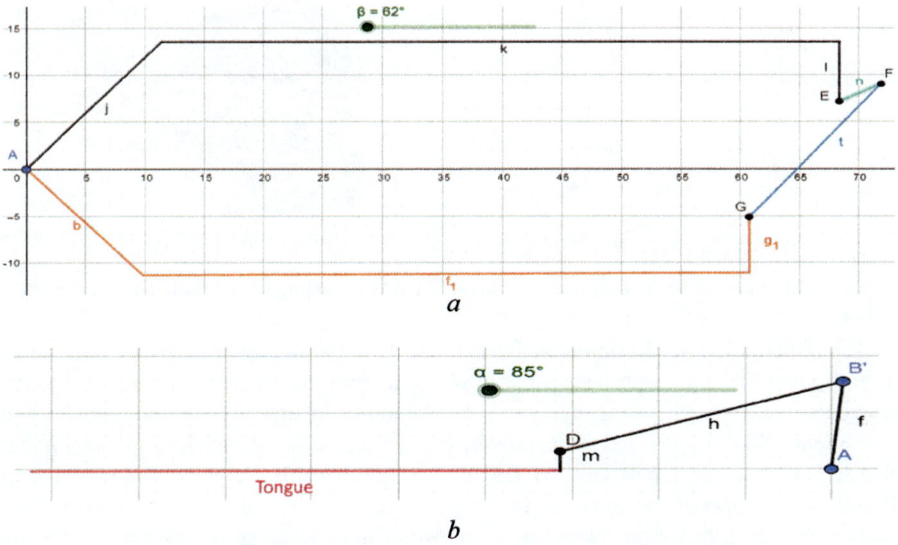

Fig. 1. Mechanism for (a) jaw and (b) tongue movement

The jaw is actuated by a MiniServo RS-3 servomotor, capable of producing a peak torque of 250 Nmm, maintaining a maximum angular velocity of 6.5 rad^{-1}. Its mass is 16 g and takes a volume of only 29 mm × 13 mm × 32 mm. The servomotor is placed on the upper half of the muzzle, its center of rotation at 64 mm removed from the jaw joint, and forms with the lower half (to movable part) of the muzzle a crank-rocker four-bar linkage (Fig. 1a); in this case, the crank moves 140° (out of the 180° allowed by the servomotor internals), which translates in a 5° rotation for the rocker (the jaw); this limited motion doesn't allow for yawning, and this is by design, as a large opening of the snout might expose the internal mechanisms, but it is enough to realistically emulate jaws motion during barking.

The tongue is also actuated by the same type of servomotor; this motion was deemed important for the capacity of the robot emotional expression: tongue hanging out is a universal sign of satisfaction and relaxation for dogs. For that purpose, the employed mechanism is the slider-crank linkage (Fig. 1b), where the crank executes a 90° motion, and the tongue exits the snout for a distance of 19 mm. The tongue rests on multiple rollers build into the lower half of the snout that reduce the friction forces and preclude scratching sounds. The range of movement might be further extended with minimal design modification.

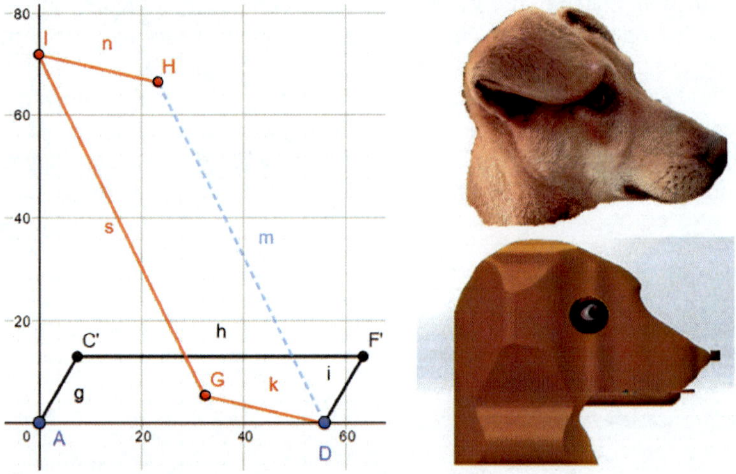

a *b*

Fig. 2. Mechanism for eyes movement (a) and CAD modeling after real-life reference (b)

Eye orientation is accomplished by using two parallelogram linkages in series (Fig. 2a); the first linkage transfers the movement from the servomotor (identical to the previous ones) to the right eye, and the second linkage ensure the coordinated movement of both eyes; range of movement is −32°… + 32° relative to the sagittal plane. The first linkage is needed because is not feasible to place the servomotor directly under one of the eyes, as is not possible to simultaneously maintain biomimetism and stack this motor with the other two. The external shell was designed using the dog owned by one of the authors as photographic reference (Fig. 2b).

3.2 Electronical and Control Subsystem

The electronical and control system is built around an Arduino Uno board; Fig. 3 shows the system architecture: as inputs, the robot features a microphone, a miniature electrocardiograph (ECG), a distance measurement unit, and a tactile sensor. The board controls 3 servomotors (powered separately), an MP3 shield and a GSM module.

At this stage, the microphone, an ADMP401, is used for simple background noise monitoring; the robot reacts by turning its eyes in response to the sound pressure level, a randomizing function making sure that this reaction does not leave the impression of predetermined response. The miniature ECG, AD8323, is activated only if the IR proximity sensor, GP2Y0A21YK, detects that the hands are closing on the electrodes, strategically placed on both sides of the robots neck; the proximity sensor has useful range of 10…80 mm and its sole purpose is to turn the ECG on/off to prevent needlessly power drain. The ECG signal is read by one of the analog-to-digital converter (ADC) enabled pin, the peaks are extracted and timestamped, after which simple statistics is employed to determine the mean heart rate and the standard deviation of it. If certain thresholds (currently for tachycardia, bradycardia, and arrhythmia) are exceeded an alert is send to a predetermined contact person using the SIM800L GSM module.

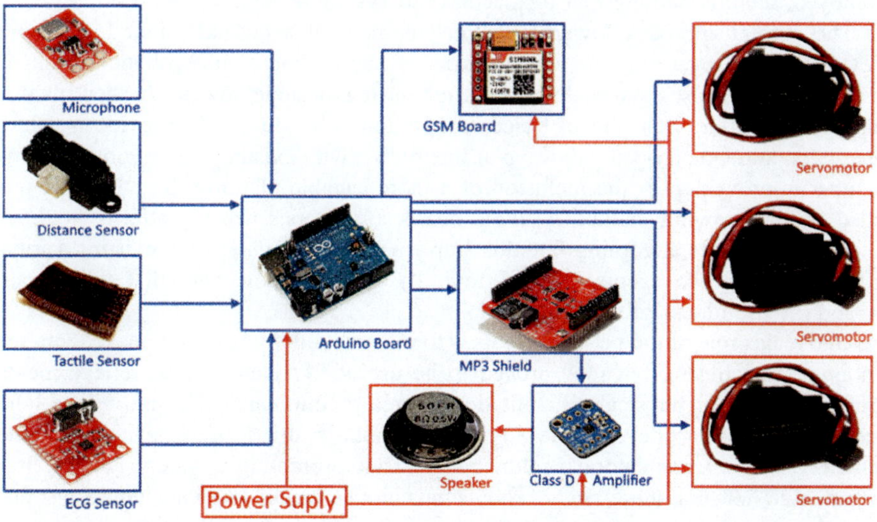

Fig. 3. System architecture

The tactile sensor is custom made using a Velostat™ sheet sandwiched between two conductive textile materials (the textile materials are insulated from each other); this sensor is integrated into the synthetic fur of the pet robot and, as force is applied on it, its resistance decreases, which, together with a fixed resistor connected in series determines a proportional voltage change that is easily read by an ADC pin. This signal is used to control the MP3 shield that access (currently 15) audio files that contain various dog sounds, from whining to aggressive barking; the assertiveness is primary determined by the intensity of the signal received from the tactile sensor, but there is also an element of randomness introduced to give a more lifelike interaction. The MP3 shield drives an class D amplifier MAX98300 that works at a voltage level of 5 V, and, provided adequate power supply, has a power output of up to 2.6 W; the sound volume is adjusted by a trimmer, and is subject to individual fine-tuning.

The sounds are synchronized with the movement of the jaw; the position of the servomotor and implicitly of the jaw are dependent of the assertiveness score of the played sound and a random element. Also, the tongue is sticked out based on the same assertiveness score and another random function. All randomizing functions mentioned here are initialized with a random seed that is determined by taking an ADC read of an unconnected ADC pin: the random environmental electromagnetic fluctuation and the 10-bit ADC resolution provide enough randomness for this application.

4 Conclusions

The main limitation of the presented system is that, due to the current pandemic, it could not be adequately build and tested, therefore one of the very first task is to experimentally validate the design; although appropriate design procedures were employed, there is nothing that tops actual full system testing.

The current approach towards the development of a pet-like robot follows our previous work regarding the development of anthropomorphic robotic heads, as assistive robots, also. As such, this design is a milestone toward further planned improvements: the inclusion of voice recognition software would provide a natural interaction and would open pathways of integration with existing smart home systems. Another improvement is the inclusion of a more capable platform, like the ones provided by the growing market of single board computers: such a platform would be essential to provide a learning function implemented with one of the existing various artificial intelligence agents. Adding mobility is an improvement that will greatly expand the usefulness of the robot.

One major role of the pet-like robots is to interact with humans and make them feel emotional attachment, but they proved to be useful for relaxation, to relieve mental stress, exercise for physical rehabilitation, increase communication and reduce loneliness. The efforts to develop these types of robotic systems aim to encourage independence and maintain good health, but without promoting a lonely elderly. It is important to note that these kind of robots must not be treated like innocuous toys; they are fully fledged medical devices, their usage being subject to medical prescription and responsibility that carefully considers individual related variables like how psychoactive medication might change the human-robot dynamic or how preexisting conditions might preclude their usage.

Conflict of Interest. The authors declare that there is no conflict of interest regarding any commercial products (equipment, components and/or software packages) described here, nor does they endorse any product mentioned; names of equipment, components and software packages were given solely for reference purposes.

References

1. Kubota, N., Nojma, N., Baba, N., Kojma, F., Fukuda, T.: Evolving pet robot with emotional model. In: Proceedings of the 2000 Congress on Evolutionary Computation (2000). https://doi.org/10.1109/CEC.2000.870791
2. Broadbent, E., Stafford, R., MacDonald, B.: Acceptance of healthcare robots for the older population: review and future directions. Int. J. Soc. Robot. **1**, 319–330 (2009). https://doi.org/10.1007/s12369-009-0030-6
3. Kanamori, M., Suzuki, M., Tanaka, M.: Maintenance and improvement of quality of life among elderly patients using a pet-type robot. Jpn. J. Geriatr. **33**(2), 214–218 (2002). https://doi.org/10.3143/geriatrics.39.214
4. Robinson, H., Macdonald, B., Kerse, N., Broadbent, E.: The psychosocial effects of a companion robot: a randomized controlled trial. J. Am. Med. Dir. Assoc. **14**(9), 661–667 (2013). https://doi.org/10.1016/j.jamda.2013.02.007
5. Petersen, S., Houston, S., Qin, H., Tague, C., Studley, J.: The utilization of robotic pets in dementia care. J. Alzheimer's Dis. **55**(2), 569–574 (2017). https://doi.org/10.3233/JAD-160703
6. Hung, L., Liu, C., Woldum, E., et al.: The benefits of and barriers to using a social robot PARO in care settings: a scoping review. BMC Geriatr. **19**, 232 (2019). https://doi.org/10.1186/s12877-019-1244-6
7. Hung, L., Gregorio, M., Mann, J., et al.: Exploring the perceptions of people with dementia about the social robot PARO in a hospital setting. Dementia (2019). https://doi.org/10.1177/1471301219894141
8. Alonso, S.G., Hamrioui, S., de la Torre Diez, et al.: Social robots for people with aging and dementia: a systematic review of literature, Telemed. J. E Health **25**(7), 533–540 (2019). https://doi.org/10.1089/tmj.2018.0051
9. Whelan, S., Murphy, K., Barrett, E., Krusche, C., Santorelli, A., Casey, D.: Factors affecting the acceptability of social robots by older adults including people with dementia or cognitive impairment: a literature review. Int. J. Soc. Robot. **10**(5), 643–668 (2018). https://doi.org/10.1007/s12369-018-0471-x
10. McGlynn, S.A., Kemple, S., Mitzner, T.L., King, C.H.A., Rogers, W.A.: Understanding the potential of PARO for healthy older adults. Int. J. Hum. Comput. Stud. **100**, 33–47 (2017). https://doi.org/10.1016/j.ijhcs.2016.12.004
11. Abdi, J., Al-Hindawi, A., Ng, T., Vizcaychipi, M.P.: Scoping review on the use of socially assistive robot technology in elderly care. BMJ Open **8**, e018815 (2018). https://doi.org/10.1136/bmjopen-2017-018815

Concept of Robotic System for Assistance/Rehabilitation of Persons with Motor Disabilities

Claudiu Schonstein(✉) and Aurora Felicia Cristea

Technical University of Cluj-Napoca, 103-105 Muncii Bld., Cluj-Napoca, Romania
{claudiu.schonstein, felicia.cristea}@mep.utcluj.ro

Abstract. The paper explores an intensively researched field, linked to the role of robots in assisting and rehabilitating people with motor disabilities. Thus, the paper is focused on presenting a concept of a mechanical system, able to provide solutions for interaction between the human, with locomotors problems, and a mobile assistance/recovery robot, in order to improve the quality of life. The mechanical system proposed for development is customizable, based on human needs and physical capabilities. The paper presents the constructive characteristics and the kinematic control functions which are the base of the automation of the robot's functionality.

Keywords: Robot · Medicine · Locomotors disabilities · Assistance

1 Introduction

Nowadays, robots and the robotization of certain mechanisms, represent a major challenge, hence they are being part of all fields of activity as food & beverage service, management, storage purposes, industrial and medical.

Regarding the medical robots used in the assistance and rehabilitation of people with disabilities, they can be classified as:

- robots that help with functional readjustment - helping to correct disability, when illness or accidents have altered body abilities, practicing physical training or remedying organic deficiencies with an exoskeleton, as in the case of amputated arms or tetraplegia robotic replaces the defective part/lack of the body);
- robots that help with social rehabilitation - which can be useful to provide physical and social support to people with disabilities or the elderly. This class includes robots which can act to supervise, or help physically people, when human aid is lacking or is rejected for pathological reasons.

Human-robot interaction is a multidisciplinary domain, with contributions in the field of human-computer interaction, artificial intelligence, robotics, and social sciences. The term assistive robotics, in the past, referred to robots that assisted people with disabilities through physical interaction. Currently, the term is used in a broader sense, generally referring to all robots that help or support human users. People often

S. Vlad and N. M. Roman (Eds.): MEDITECH 2020, IFMBE Proceedings 88, pp. 308–320, 2022.
https://doi.org/10.1007/978-3-030-93564-1_35

need assistants to help them with their daily activities. Thanks to the personal robots that have been developed in recent years, this has become easier, but research in this area is still ongoing and there are many shortcomings until a robot can help humans in a safe, efficient way and fully autonomous. These types of robots are generally called assistance robots and are intended to help a person live better, by sharing the environment with their human operators [1].

2 Achievements in Assisting People with Motor Deficiencies

The development of medical robots is part of an avant-garde field and perspective of our century, so many research centers have developed various rehabilitation or assisting mechanical systems laying the foundations of a new direction of research, the positive results obtained are encouraging for development of new systems.

Among the realizations, in Fig. 1, a mechanical structure is presented, named PAM-AID walker. It is a mobility aid for people with disabilities, which can provide both physical support for walking, but also navigational intelligence. Its objective is to allow users to retain their personal autonomy and take independent exercise [2].

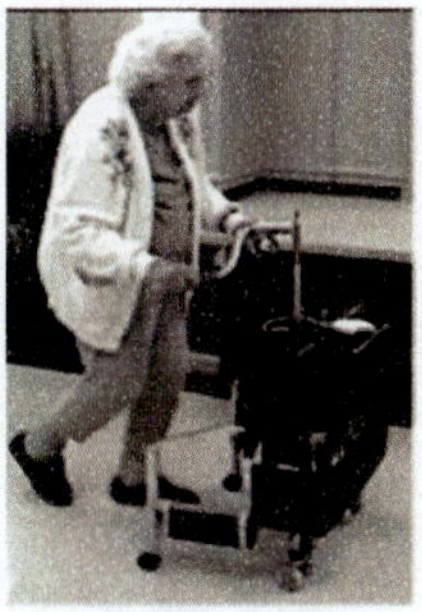

Fig. 1. The PAM-AID walker.

Another achievement that should be mentioned is a helping robotic walker for people with locomotors disabilities (see Fig. 2), Walkmate [3] that has two force sensing resistors embedded into the handlebar. These sensors are fixed with the handlebar directly so it can detect both pull and push pressure of the handlebar. Moreover, the force sensors embedded into the handlebars offer a natural negative feedback loop of the motion control. When the user wants to keep a constant velocity, the handle will be pulled/pushed to decelerate/accelerate the walker if the walker is faster/slower than the user.

Fig. 2. The prototype Walkmate.

Another robot worth mentioning, used in helping people with locomotor problems, is Lokomat 3D, presented in Fig. 3, by Hocoma [4]. The Lokomat is a driven gait orthosis that automates locomotion therapy on a treadmill and improves the efficiency of treadmill training [4]. The Lokomat improves the therapy outcome by providing highly intensive, individualized training in a motivational environment of constant feedback.

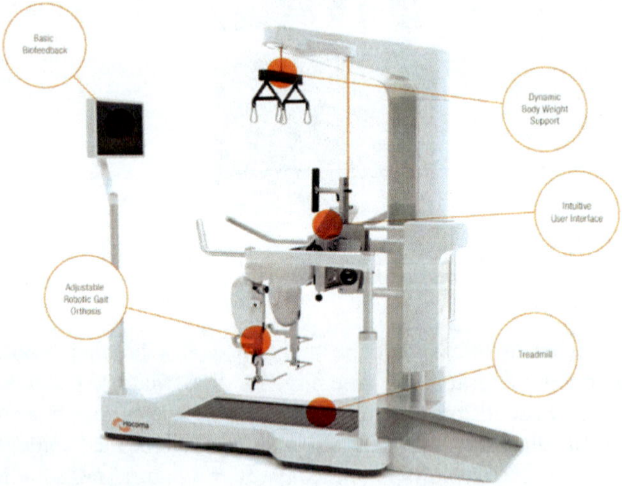

Fig. 3. The robotic system Lokomat 3D.

Further, considering the efforts made in improving the quality of life for people with musculoskeletal problems, by various research institutes or companies, the paper proposes a concept for prototyping of hybrid mobile structure.

3 The Design of ARPMD System

The structure proposed for development, abbreviated ARPMD (Assistant Robot for People with Motor Disabilities), is designed to be used both in assisting people with certain motor disabilities and with their recovery therapy, based on the movements required from the robot to perform locomotor rehabilitation operations [5–7].

As resulting from Fig. 4, the ARPMD system is based on a mobile structure marked with (1), on which a frame with two movable arms is fixed (2), which can perform a series of up-down or front-rear movements, which would allow easy customization for the user.

The movement of the mobile structure (1) is performed by means of four wheels, two of which are driving wheels, and the other two driven omnidirectional. The driving wheels are actuated by two DC motors, one for each wheel [8], the controlling of motors being performed with an L298 IC board from Arduino, [9].

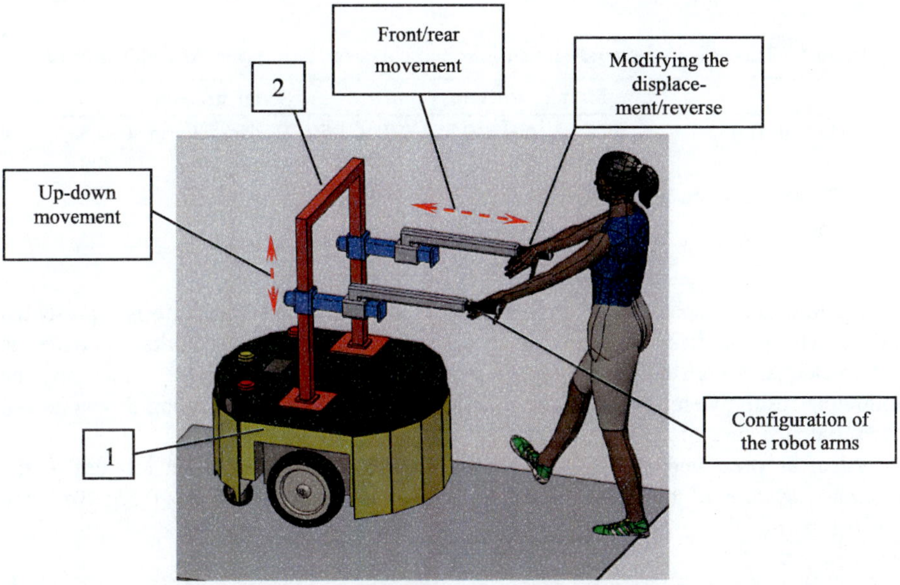

Fig. 4. The proposed prototype ARPMD.

The serial structure, (the arms of the ARPMD system), are driven by four DC motors [8], as can be seen in Fig. 5, they can be customized according to the characteristics of the operator, which can also change the moving speed of the robotic system. Given the average human body size [10], the ARPMD prototype has the following overall dimensions.

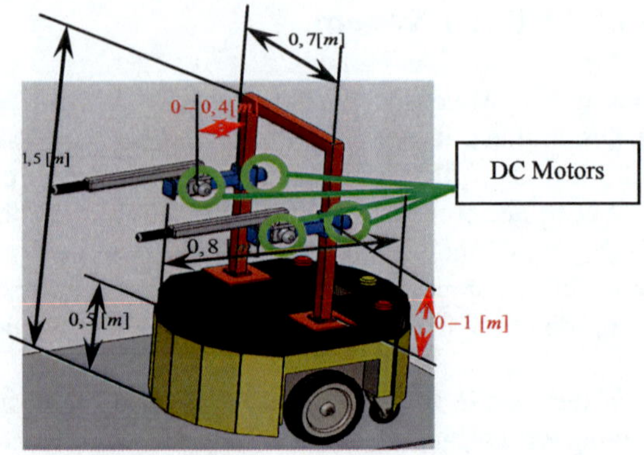

Fig. 5. Overall dimensions of the ARPMD structure.

In keeping with Fig. 5, the characteristics of ARPMD are concentrated in Table 1.

Table 1. The proposed overall dimensions and characteristics of the ARPMD structure.

	Mobile structure	Serial structure	
Length [m]	$L \times l \times H$ = 0,8 × 0,7 × 1,50	Horizontal arm	0–0,30
		Vertical arm	0–0.8
Travelling speed [m/s]	0,05–0,8	–	
Mass [kg]	85	10	

An important aspect in the construction of the mechanical structure is represented by the mass properties. Based on the analysis performed in Solid Works, to create the 3D model, and considering the high density materials used for the structures, the following results were obtained: the mobile structure of 85 [kg], adding the weight of the serial structure of 10 [kg], adding up together 95 [kg].

Also, as presented in Fig. 6, the position of the mass center in relation to the reference system of the robot, it is situated at x = 0.215 [m], y = 0.355 [m] and z = 0.115 [m].

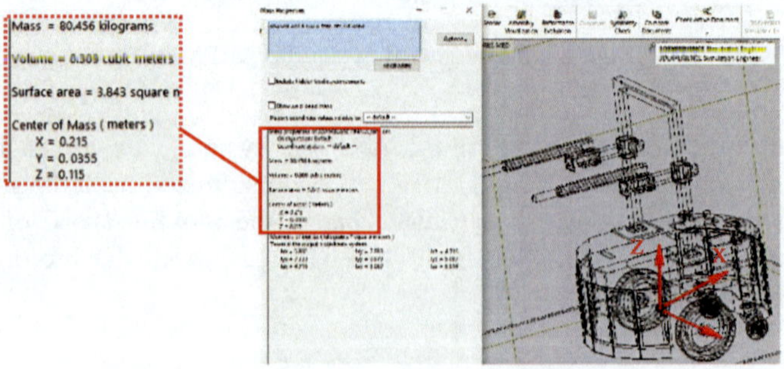

Fig. 6. Mass center position of the ARPMD structure.

Based on the Solid Works data, the robust structure, but also the mass center of the robot which is close to the floor, make it impossible for the robot to flip over and cause injuries to human operator, for which a weight of 90 [kg] was considered.

The structure incorporates inclinometers on its system that can detect if the surface is inclined or not. This information enables the robot to increase or decrease its velocity depending if it is an up or down slope. The goal is to adjust the robot velocity while climbing. This will help the user to walk in slopes, a type of terrain that usually leads to the fall of the user.

4 The Operating of the ARPMD Structure

As presented in the paragraph above, the ARPMD structure can be custom made, based on the needs of the human operator.

The right arm contains touch sensors, with the role of detecting the presence of the human operator, therefore, any functionality is linked to the grip of the handle. The operator can set the movement speed by spinning the right handle clockwise or anti-clockwise and, at the same time, holding down the safety button with the thumb of the right hand. If the handle is rotated clockwise and the safety button is held down, the robot will increase its speed by 0,005 [m/s]. Otherwise, if the handle is rotated anti-clockwise and the safety button is held down, the robot will reduce its speed by 0,005 [m/s] (see Fig. 7). To maintain a constant speed, the operator needs to release the safety button.

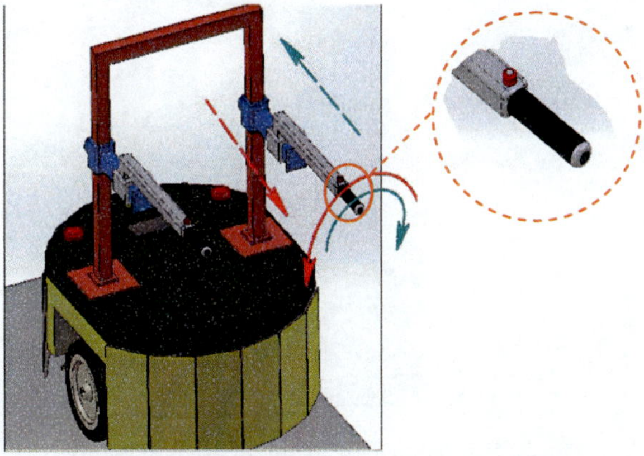

Fig. 7. Increasing and decreasing the speed of ARPMD structure.

As presented in Fig. 8, the robot can turn if the right handle is moved horizontally to the left or right side, based on the direction needed by the operator. Also, based on the same figure, if the operator wants to move the system backwards, he needs to raise the right handle slightly, without holding down the safety button.

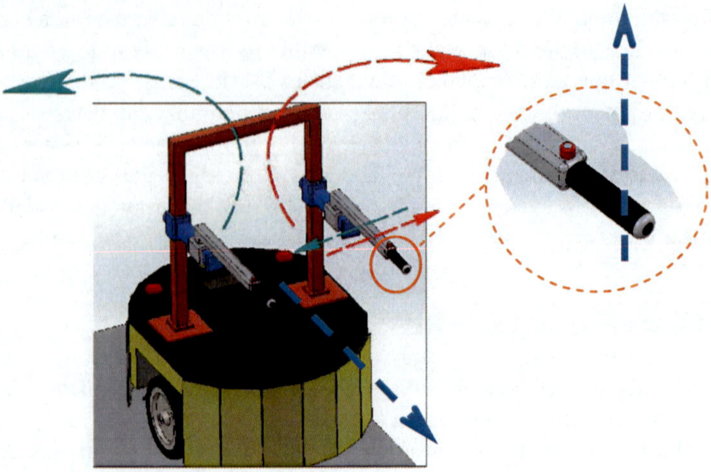

Fig. 8. Orientation and reversing of ARPMD structure.

In order to obtain movements of the robot's arms, the operator needs to hold down the safety button on the left handle and, for an ascending move, rotate the left handle clockwise or, for a descending move, rotate the left handle anti-clockwise, as presented in Fig. 9. The speed for this type of movement will be of 0,005 [m/s].

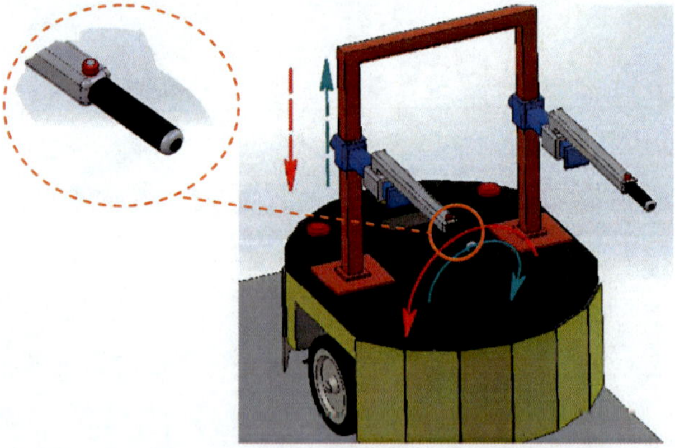

Fig. 9. Ascending/descending of ARPMD arms.

It is important to emphasize the fact that, as explained in the previous paragraphs, the robot is equipped with safety systems, to prevent any accidents.

For the elongation of the robot arms, the operator needs to hold down the safety button of the left handle and move it horizontally to the right or left, as shown in Fig. 10. Through this action the robot arms will become shorter or longer, with a speed of 0,001 [m/s].

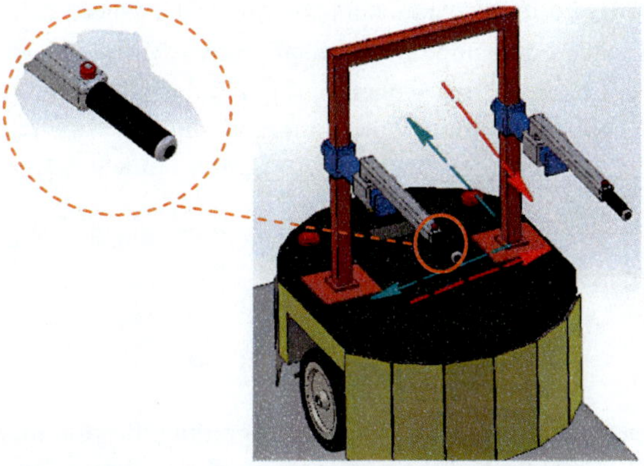

Fig. 10. Shorting/extending of ARPMD arms.

5 Establishing the Kinematic Control Functions for Velocities of the ARPMD Structure

Notions are presented below regarding the motion control of each component of the ARPMD structure. The aim is to obtain the expressions for velocities control, of each of the two structures, analyzed individually.

5.1 Mathematical Modeling for Mobile Robot (1)

As seen in Fig. 11, structure (1) performs a parallel- plane movement (translational and rotational movement in the same plane), which gives the disabled person the opportunity to move easily, in the home and/or outside of it (traveling to a store or meeting with other people), effectively replacing the use of a forerunner. Also, the robot base can be used to transport objects, facilitating the everyday activities of disabled persons (shopping, moving personal items in the home, performing basic operations). As

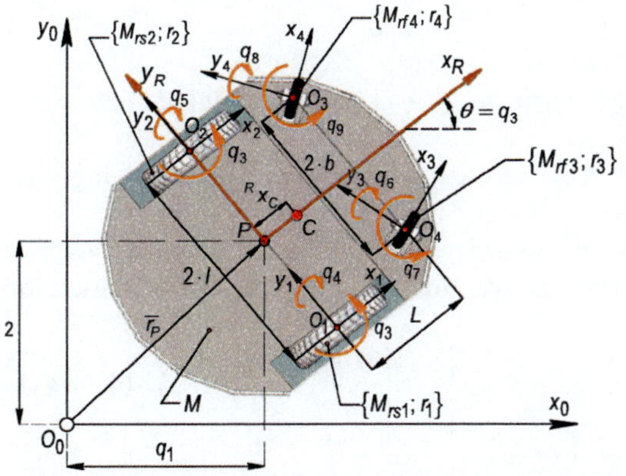

Fig. 11. Input parameters of mathematical modeling for mobile robot (1)-bottom view.

presented in Fig. 11, the input parameters for robot (1), where $q_j, j = 1 \rightarrow 9$, are the generalized coordinates corresponding to displacements; $M_{r(f,s)i}, i = 1 \rightarrow 4$ are representing the masses of back or front wheels and $r_i, i = 1 \rightarrow 4$ are the radii of the wheels.

In accordance with [11], the robot motion is studied with respect to reference frames: a fixed one, denoted $\{O\}$, having the origin in point $\{O_0\}$, and a mobile frame $\{R\}$, fixed on robot.

The column vector, of the operational velocities, which expresses the absolute motion of the robot is [12]:

$$ {}^{0}\dot{\overline{X}} = \begin{bmatrix} \dot{x}_p & \dot{y}_p & \dot{\theta} \end{bmatrix}^T = [\dot{q}_1 \quad \dot{q}_2 \quad \dot{q}_3]^T; \tag{1} $$

Following some mechanical considerations regarding the structure (1), according to [12], the kinematic parameters of the points in relation to the system $\{O_i,\ i = 1 \rightarrow 4\}$ attached to the centers of the wheels of the mobile robot, projected on the axes of the fixed reference system, are expressed with the following relations:

$$ \overline{v}_{O_1} = \begin{bmatrix} \dot{q}_1 + l \cdot \dot{q}_3 \cdot \cos q_3 \\ \dot{q}_2 + l \cdot \dot{q}_3 \cdot \sin q_3 \\ 0 \end{bmatrix} \qquad \omega_1 = \begin{bmatrix} 0 \\ \dot{q}_4 \\ \dot{q}_3 \end{bmatrix} \tag{2} $$

$$ \overline{v}_{O2} = \begin{bmatrix} \dot{q}_1 - l \cdot \dot{q}_3 \cdot \cos q_3 \\ \dot{q}_2 - l \cdot \dot{q}_3 \cdot \sin q_3 \\ 0 \end{bmatrix} \qquad \omega_2 = \begin{bmatrix} 0 \\ \dot{q}_5 \\ \dot{q}_3 \end{bmatrix} \tag{3} $$

$$ \overline{v}_{O_3} = \begin{bmatrix} \dot{q}_1 - L \cdot \dot{q}_3 \cdot \sin q_3 + b \cdot \dot{q}_3 \cdot \cos q_3 \\ \dot{q}_2 + L \cdot \dot{q}_3 \cdot \cos q_3 + b \cdot \dot{q}_3 \cdot \sin q_3 \\ 0 \end{bmatrix} \qquad \omega_3 = \begin{bmatrix} 0 \\ \dot{q}_6 \\ \dot{q}_7 \end{bmatrix} \tag{4} $$

$$ \overline{v}_{O_4} = \begin{bmatrix} \dot{q}_1 - L \cdot \dot{q}_3 \cdot \sin q_3 - b \cdot \dot{q}_3 \cdot \cos q_3 \\ \dot{q}_2 + L \cdot \dot{q}_3 \cdot \cos q_3 - b \cdot \dot{q}_3 \cdot \sin q_3 \\ 0 \end{bmatrix} \qquad \omega_4 = \begin{bmatrix} 0 \\ \dot{q}_8 \\ \dot{q}_9 \end{bmatrix} \tag{5} $$

and representing the linear $\overline{v}_{O_i},\ i = 1 \rightarrow 4$, and angular $\omega_i,\ i = 1 \rightarrow 4$ velocities of the wheels centers.

The previously obtained expressions are characterizing the kinematic control functions in velocities, with the help of the L298 IC board.

Based on the above considerations, and the kinematic parameters contained in the expressions determined above, the following relation is written in matrix form:

$$ \dot{\overline{X}} = \begin{bmatrix} \dot{q}_1 \\ \dot{q}_2 \\ \omega = \dot{q}_3 \end{bmatrix} = \begin{bmatrix} \cos q_3 & 0 \\ \sin q_3 & 0 \\ 0 & 1 \end{bmatrix} \cdot \begin{bmatrix} \frac{r}{2} \cdot (\dot{q}_4 + \dot{q}_5) \\ \frac{r}{2 \cdot l} \cdot (\dot{q}_4 - \dot{q}_5) \end{bmatrix}; \tag{6} $$

which expresses the velocity of the point belonging to the mobile robot projected on the fixed system and are representing the kinematics equations.

5.2 Mathematical Modeling of the Serial Structure (2)

Further, will be made an analysis of the kinematic control for the serial robot (2), fixed on the mobile structure (1). According to [13], in the first phase the direct geometric model is established for the robot structure (2), the results being used as input data in kinematic modeling, ie the determination of motion control functions for speeds.

For analysis, the input data of robot (2) are presented in Fig. 12.

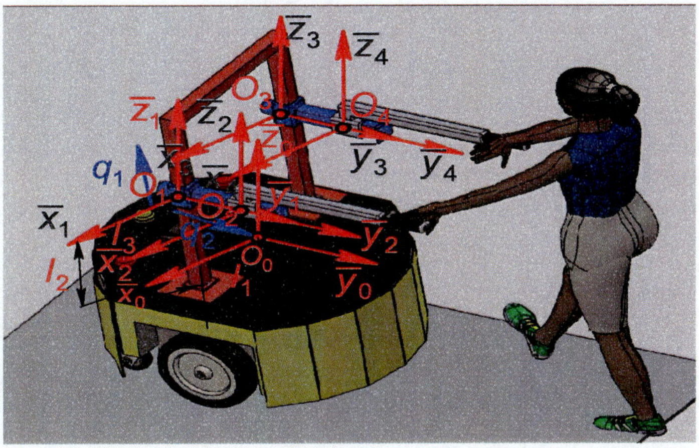

Fig. 12. The geometrical parameters of robot (2).

According to Fig. 12, and [13], the matrix of the nominal geometry, which describes the position and orientation of each kinetic link, is presented in Table 2, as:

Table 2. The situation parameters for the serial structure

Element $i = 1 \rightarrow 2$	Joint type $\{R; T\}$	$\{\overline{p}_{ii-1}^{(0)}\}^T = \overline{p}_i^{(0)} - \overline{p}_{i-1}^{(0)}$			$\overline{k}_i^{(0)T}$		
		$p_{xii-1}^{(0)}$	$p_{yii-1}^{(0)}$	$p_{zii-1}^{(0)}$	$k_{ix}^{(0)}$	$k_{iy}^{(0)}$	$k_{iz}^{(0)}$
1	T	l_1	0	l_2	0	0	1
2	T	0	l_3	0	0	1	0

Corresponding to initial state of the robot, in the geometric center of each kinematic link is attached a mobile system for each link $i = 1, 2$. According to Table 2 and Fig. 7, there are highlighted the following geometrical particularities:

$$\left(i = 1;\ \bar{k}_1 \equiv \bar{z}_1; \Delta_1 = 0\right); \left(i = 2; \bar{k}_2 \equiv \bar{y}_2; \Delta_2 = 0\right) \tag{7}$$

which serves to direct geometrical modeling of the robot, where $i = 1, 2$ is the joint, $\bar{k}_i$ is the versor of the axis along occurs the movement and Δ_i is an operator which takes into account the type of the axis ($\Delta_i = 1$ for rotation; respectively $\Delta_i = 0$ for translation), [13].

Following the algorithm of locating matrices [13], there is obtained the expression which describes the positioning of the arms of the structure (2):

$$ {}_2^0[T] = {}_1^0[T] \cdot {}_2^1[T] = \begin{bmatrix} {}_2^0[R] & \bar{p}_2 \\ 0\,0\,0 & 1 \end{bmatrix} = \begin{bmatrix} 1 & 0 & 0 & l_1 \\ 0 & 1 & 0 & l_3 + q_2 \\ 0 & 0 & 1 & l_2 + q_1 \\ 0 & 0 & 0 & 1 \end{bmatrix} \tag{8}$$

which includes both the rotation matrix ${}_2^0[R]$ for orientation, and the position vector $\bar{p}$, for positioning of the robot arms.

Considering that the orientation of the person is done with the help of the mobile robot (1), the serial structure (2) has no orientation. Thus, the column vector of the operational coordinates for structure (2) has the form:

$$ {}^0\bar{X} = \begin{bmatrix} \bar{p} \\ \cdots \\ \bar{\psi} \end{bmatrix} = \begin{bmatrix} \left(p_x \quad p_y \quad p_z\right)^T \\ \cdots\cdots\cdots \\ \left(\alpha_z \quad \beta_y \quad \gamma_z\right)^T \end{bmatrix} = \left\{ \begin{matrix} \begin{bmatrix} l_1 \\ l_3 + q_2 \\ l_2 + q_1 \end{bmatrix} \\ \cdots\cdots\cdots \\ \begin{bmatrix} 0 & 0 & 0 \end{bmatrix}^T \end{matrix} \right\} \tag{9}$$

and is characterizing the direct geometrical modeling for the robot (2).

In the literature (see [12, 13]), the expressions (9) are known as direct geometry equations, and express the position and orientation of the characteristic point of the final effector relative to the fixed reference system $\{O\}$ attached to the robot (1).

The kinematic analysis considers both the position and orientation of each joint required to describe the location of the final effector in the robot's workspace and the variation of joint speeds throughout the operating process. In the first stage of application of the iterative algorithm (described in [13]) it is considered that the linear and angular velocities and accelerations corresponding to the fixed base of the robot are defined as follows:

$$\left\{ {}^0\bar{\omega}_0 = \underset{(3x1)}{[0]},\ {}^0\dot{\bar{\omega}}_0 = \underset{(3x1)}{[0]},\ {}^0\bar{v}_0 = \underset{(3x1)}{[0]},\ {}^0\dot{\bar{v}}_0 = \underset{(3x1)}{[0]} \right\} \tag{10}$$

For $i = 1, 2$, there are established the angular and linear velocities that defining the absolute motion of each joint, of the structure (2). Thus, depending on the positioning Eqs. (8), the absolute movement of the arms are expressed by:

$$^{0}\dot{\bar{X}} = \begin{bmatrix} ^{0}\bar{v}_2 \\ \hdashline ^{0}\bar{\omega}_2 \end{bmatrix} = \begin{bmatrix} \left(0\ \dot{q}_2\ \dot{q}_1\right)^T \\ \hdashline \left(0\ 0\ 0\right)^T \end{bmatrix} \tag{11}$$

and are representing the kinematics equations, being useful in controlling the speeds in each $i = 1, 2$ joint of the robot (2).

6 Conclusions

The approach of this topic was determined by the need to design a support robot to perform activities or operations in domestic environments, so that communication and interaction between it and the human operator is done in a manner like human behavior. It is not desired to remove the necessary human contact involved in the care of the persons with elderly or mobility problems, however the aim is to facilitate the greater independence of the person within a caring environment.

The paper presents a concept of moving/recovery assistant robot for people with locomotors problems, to lift and move towards a chair or bed, or simply supporting a person with weak muscle structure. Given that, these people often need support and their weight can be an impediment for those who support them, this robot complements that. The concept is abbreviated ARPMD, and consists of a mobile structure, on which a serial robot is fixed. The prototype has been 3D modeled and analyzed by using specialized software, which provided data regarding the geometrical behavior. For each joint of the robot, the velocities are established, then the kinematic control functions, which are the base of the mechanical system functionality.

As it can be seen from paper, the mobility of the robot's arms allows the disabled person to use the robot to perform basic movements, such as sitting or lifting from a sitting position. We believe that this type of robot can have functionality in hospitals or homes, in helping elderly or people during a post-operative recovery.

Considering that this is just a prototype concept, no comments can be made regarding the acceptance of such a system by real users. Hopefully, in a near future, this concept can become a prototype and be tested by people with motor disabilities.

As an essential observation, the paper shows that man, through his creation, can assist other people by creating aids, in this case the creation of assistive robots, which can help, rehabilitate or even prolong a person's life.

Conflict of Interest. Hereby mention that, I have no conflict of interest regarding this paper.

References

1. https://www.semanticscholar.org/paper/Personal-Adaptive-Mobility-Aid-for-the-Infirm-and-Lacey-MacNamara/f418a73ee26505a95ff56dfce00b3dc665ad31a9. Accessed 10 Oct 2020
2. Shi, F., Cao, Q., Leng, C., Tan, H.: Based on force sensing- controlled human-machine interaction system for walking assistant robot. In: Proceedings of the 8th World Congress on Intelligent Control and Automation 2010, Jinan, China, pp. 6528–6533 (2010)
3. https://stepsbudapest.com/lokomat-pro-3d/. Accessed 01 Sept 2020
4. https://stepsbudapest.com/wp-content/uploads/Lokomat-Datasheet.pdf. Accessed 01 Sept 2020
5. Baciu, Cl.: Statica si mersul normal la omul bolnav. Munca Sanitara, Bucuresti (1985)
6. Buzescu, A.I., Scurtu, L.: Anatomie si biomecanica. Editura A.N. E.F.S. (1999)
7. Robanescu, N.: Reeducarea Neuro-motorie. Editura medicala Bucuresti (2001)
8. http://www.hotecmotor.com/worm-gear-motor/ht-wog58b.html. Accessed 09 July 2020
9. https://www.tutorialspoint.com/arduino/arduino_dc_motor.html. Accessed 11 July 2020
10. https://www.researchgate.net/figure/Mean-and-standard-deviation-values-of-the-body-measurements_tbl1_216031981. Accessed 12 July 2020
11. Negrean, I., Schonstein, C., Kacso, K., Negrean C., Duca, A.: Formulations about dynamics of mobile robots. In: The 5-th International CONFERENCE ROBOTICS 2010, Published in the Robotics and Automation Systems, Volumes 166–167 of Solid State Phenomena, pp. 309–314. ISSN 1012-0394
12. Schonstein, C.: Contributions in developing of a hybrid robot structure. Ph.D. thesis, Universitatea Tehnică din Cluj-Napoca (2011)
13. Negrean, I.: Mecanică Avansată în Robotică. Editura UT Press, Cluj-Napoca (2008). ISBN 978-973-662-420-9

Sensitivity to Mechanical Vibrations, Methods of Attenuation

Aurora Felicia Cristea(✉)

Buildings of Machines Faculty, Technical University of Cluj-Napoca, Cluj-Napoca, Romania
felicia.cristea@mep.utcluj.ro

Abstract. The paper seeks to highlight how it could influence the transmission of vibration in terms of sensitivity to the hand, respectively reducing it by mounting an additional vibration attenuator. Thus, the paper deals only with the experimental way of reducing them, when a vibration attenuator has mounted and the possibility of avoid the occurrence of occupational diseases due to these causes.

Keywords: Vibration attenuator · Transmissibility · Hand-arm system

1 Introduction

As explained in the summary of this paper, it is intended to highlight how it be might influence the transmission of vibrations in terms of sensitivity to the hand, respectively reducing it by mounting an innovation additional vibration attenuator. In this scope the special literature [1] imposed by the standard in force SR EN ISO 5349-1: 2003 [2–4] shows how the vibrations that are transmitted from the human hand to the human body, and it establish the displacements value so that we don't be affectated its. Mechanical vibrations are transmitted to humans from excitatory sources such as machine tools and mechanical devices and may change over time, if measures are taken to reduce and in this case will not be allowed to increase an occupational diseases of professional nature [1, 5, 6]. The term "Vibration disease is characterized by a polymorphism clinic with original evolution. The main etiological factor are often vibrations. Secondary and favored etiological factors could include:

Factors related to the human body such as the ability to adaptive compensatory capacity, vascular-peripheral nervous disorders, smoking, age, individual susceptibility;" [8].

"The unfavorable actions of vibration on body are the tremors at the frequencies between 30–1000 Hz; the one of 150 Hz less often causes pathology. Pneumatic hammer type instruments are very common in used of 30–50 Hz, or like as vibrating and rotating instruments with frequencies over 1000 Hz" [6].

Therefore, the cumulative effect of vibrations in hours and years, might be lead to occupational diseases [6], which is not desirable and should be avoided by taking precautions such as: avoiding used devices, respect break of times, protective equipment (specialty literature recommends only gloves in the case of the hands) etc.

S. Vlad and N. M. Roman (Eds.): MEDITECH 2020, IFMBE Proceedings 88, pp. 321–330, 2022.
https://doi.org/10.1007/978-3-030-93564-1_36

2 Mathematical Modeling

The present paper is based on a mathematical model of the hand-arm system [5, 10–12] with additional shock absorber of mechanical vibrations, the system takes into account the specifications of the literature. The paper will not insist on the mechanical and mathematical part of the research, this being briefly described. All these studies have been treated in the other papers, so the paper is part of the theme of engineering research with medicine implication and will focus only on the experimental demonstration that, the mechanical vibrations are transmitted from an excitation source and may be minimized.

The work is based on a basic model (Fig. 2), with concentrated masses, considering that soft tissues can be described phenomenologically, as a viscous-elastic medium.

Matrix form of the linear differential equations of motion can take the form of: present paper is based on a mathematical model of the hand-arm system [5, 10–12] with additional shock absorber of mechanical vibrations, a system that takes into account the specifications of the literature. Reminded that these studies have been treated in the other papers, so the paper is part of the theme of engineering research with medicine implication and will focus only on the experimental demonstration that mechanical vibrations are transmitted from an exciting source and can be minimized.

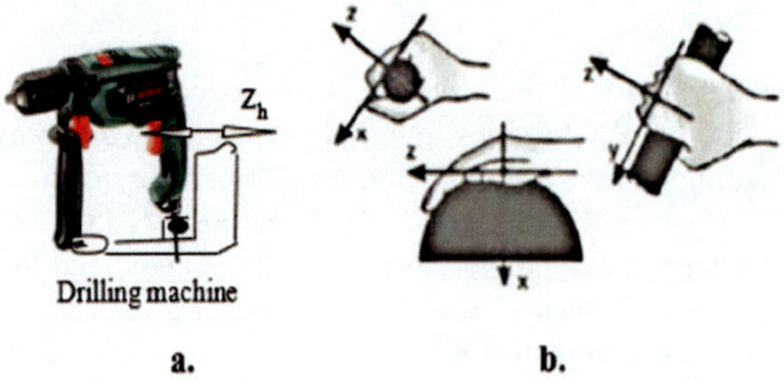

Fig. 1. Source of excitation (the device) that acts upon the hand - arm system [10]. a. The position through which the operator catches the tool fixed to the device, this has the main direction of vibration transmission, respectively the Oz axis; b. Anatomical reference system given by SR-EN ISO 5349/2003 [13].

Considering that soft tissues can be described phenomenologically, as a viscous-elastic medium.

Linear differential equations of motion in matrix form can take the form of:

$$[M] \cdot \{\ddot{z}(t)\} + [c] \cdot \{\dot{z}(t)\} + [k] \cdot \{z(t)\} = \{F(t)\}\ [14] \tag{1}$$

where, z (t) is the displacement after Oz axis of hand-arm system with three/four masses; $F(t) = M\ddot{z} \cdot \cos(\omega t)$ – excitation force;

ω = excitatory source pulsation [rad/s].

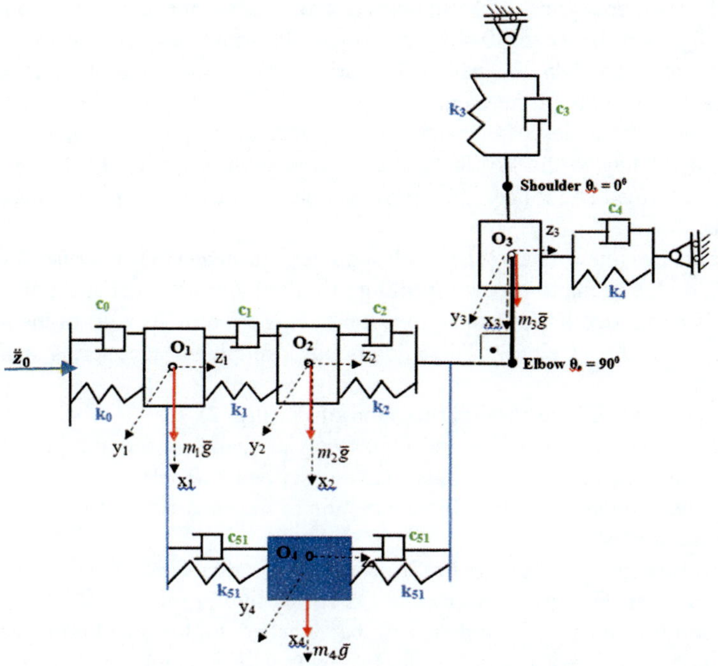

Fig. 2. Mechanical hand - arm system [10].

The matriceal Eq. (1), which characterizes the hand-arm assembly, is a second-order differential equation, linear, inhomogeneous and transcendent, in which:

z(t) respectively, z_i i = 1–4 represents the displacement according to the Oz axis, for the corresponding masses;

The notations in the Fig. 2 represent:

The amplitude of the excitation force, at the entrance to the system, and $\dot{z}_0 = 0.025$ m/s and $z_0 = 0.003$ m are measured values, along the Oz axis of the anatomical system [1], of the velocity and displacement, at the wrist, for the rotation of the machine at which maximums were recorded (1000 RPM). In the Fig. 2 the notations are following:

m_i – i = 1–4 represent the masses of the hand (palm), forearm and arm [kg];
k_i – i = 1–5 the longitudinal elasticity constants of the fingers, the hand, the forearm, the arm, attenuator along the z_h axis, respectively Ox at the arm [N/m];
c_i – i = 1–5 the damping constant of the fingers, hand, forearm, attenuator along the Oz axis, respectively Ox for the arm [Ns/m].

There are two possible situations to analyze. First, when there is no additional shock attenuator mounted on the forearm and the second when an additional shock absorber has mounted along the forearm; the second case is analyze in this paper.

The mechanical coefficients (ki, ci, i = 1–5) of the hand and the anthropometric parameters were determined experimentally by anthropometric measurements [1, 3].

These performed on a group of 15 subjects (men), of about the same height and weight, and their age was between 40–45 years, then the arithmetic average of these measurements has studied in the paper. The arithmetic average has used to determine mechanical coefficients of the hand-arm system.

The ethical principles of research (measurements) on human subjects have been observed, according to the World Health Organization, respectively the subjects had been prior informed and it been their provided physical well-being at the measurement conditions.

The joints of the wrist, elbow and shoulder are neglected, because the hand is positioned on the excitation source (drilling machine), during which a certain operation (drilling) is performed it The human operator having the arm vis a vis of the forearm at an angle of the 90° at elbow. These joints have simplified at a spring and a shock absorber.

It been considered the shoulder rotation of 0° (Fig. 2).

The conditions under which the work tool has handled and the position of the human operator during this time, there have described in Table 1.

It have been imposed (of build) the direction of transmission of the excitation, from the shock absorbers:

Masses: m1 (gives by c1) and m2 (gives by c2), after Oz axis;

Mass m3 after Oz (gives by c4) and Ox (gives by c3) (unstudied in this paper);

An additional shock absorber will be attached to the hand-arm system [13] (Fig. 2), along the forearm and it's the behavior will be studied following vibration damping along the system up to the arm, using the initial conditions like in the Table 1 (Fig. 1a).

Table 1. Initial conditions taken into account when simplifying the actual hand - arm model.

Vibration direction	Oz (in conformity with anatomical system given by SR EN ISO 5349/2003)
Excitation: harmonically displacement	F(t) z(t)
Subject position	Vertical (stand up)
Device diameter	42 mm
Frequency scale	0–200 Hz
Grip force	25 N, after ISO 15230/2007
Grip place	Hand (palm and fingers)
Shoulder angle	0°
Subjects	Men (over 40 years)

3 Experimental Measurements

In the second part of the research, it was effectuate acceleration RMS (Root Mean Square) measurements using a percussion-drilling machine [3]. The measurements were performed on the system without shock absorber and with the additional shock absorber mounted on the forearm (Fig. 3). The measurements were performed on a group of 15 people, approximately 40–45 years old men, weighing average 85 kg and 1.70–1.85 m tall. The measurements equipment included an triaxial vibration accelerometer type K_3, represented as element in the Fig. 3. This was connect to a vibration measuring device type SVAN 958. The paper presents only the vibration transmission after the Oz (Oz is along of the forearm). The accelerometer measurements vibration after wrist and elbow, it was fixe directly under the attenuator/damper bracelet by tightening it (Fig. 3) and taped directly on the skin. Sets of three measurements were performed and their arithmetic mean was used in the case analysis.

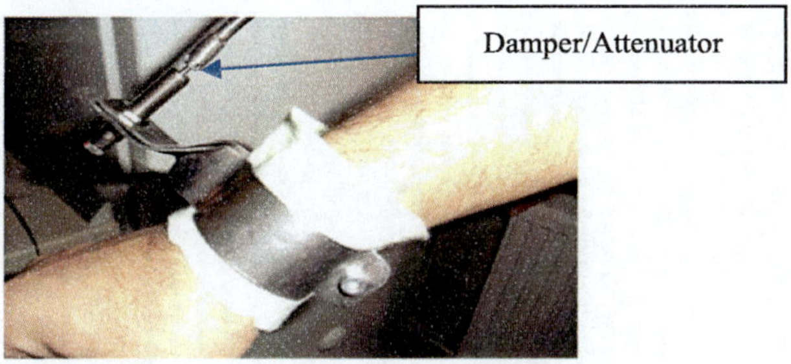

Fig. 3. Vibration damper [10].

The researches results are presented in the Fig. 4 for wrist anatomical area, where made measurements. In addition, they took in account the mode in which the bracelets have fixed on the skin (felt – Fig. 3). It mentions that, the interior padding of the metallic bracelets prevents skin lesions.

3.1 Mechanical Impedance

The measurements were performed for a frequency of 40 Hz (medium localized in the scale between 20–300 Hz) [5] and were measured the accelerations of the root mean square (a_{RMS}), next, after a double integration with special software and Runge-Kutta of order 4–5 method) they have been found the velocities and displacements corresponding of its.

Then, it has been calculated the mechanical impedance of the system, this been established to different frequencies, and for all anatomical masses of the system (hand-forearm-arm).

By definition, mechanical impedance is define as the ratio between the forces applied to a system and velocity. The mathematical relation that reproduces the mechanical impedance has the form:

$$Z = F_{excitation}/v \text{ [Ns/m] [9]} \tag{2}$$

The Fig. 4 shows the evolution of the impedance according to the excitation force and the velocities resulting, without additional attenuator, respectively at contact with the excitation source (drilling machine) and hand. The sensitivity has the highest value close 100% (gray curve), and is harmonic according to the excitation force. In the forearm it reduces its value (orange curve), but in jumps from 60%, 50% to 70% proving instability, but still, at the beginning of the movement has a peak of almost 100% (this being the elbow, where the muscular or adipose system does not there is), and from arm to shoulder the sensitivity is reduced until 10% (blue curve). This is due to the viscous-elastic properties of the skin, muscles and blood system. However, resuming these activities daily, for 8 h, the properties of these tissues become insensitive, and they no longer react in the same way, in the sense of damping the arousal.

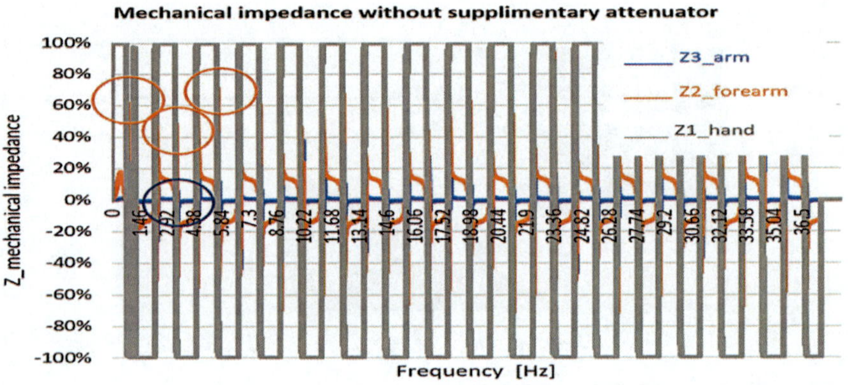

Fig. 4. Mechanical impedance for the hand-arm system without supplementary attenuator.

Unlike the sensitivity and behavior of the hand-arm system without shock absorber treated in Fig. 4, in Fig. 5 its behavior is analyze when an additional attenuator is mounted on the forearm, between the wrist and elbow (Fig. 3). Thus, it is possible to observe the evolution of the mechanical impedance depending on the excitation force and velocities. The sensitivity has the highest value close to 100% (gray curve) and it is harmonic according to the excitation force. In the forearm it reduces its value (orange curve), at first having a value of almost 100%, in elbow joint, then stabilizes at 58% lower than in the case of Fig. 4 (without attenuator) which had a value of up to 70%, in jumps. From arm to shoulder, it was reduced the sensitivity by around 5% (blue curve), much less than without the supplementary attenuator, which was 10%.

This minimization is due, first, to the additional attenuator mounted on the forearm. However, also in this case, too, special attention must paid to the resumption of these activities daily for 8 h.

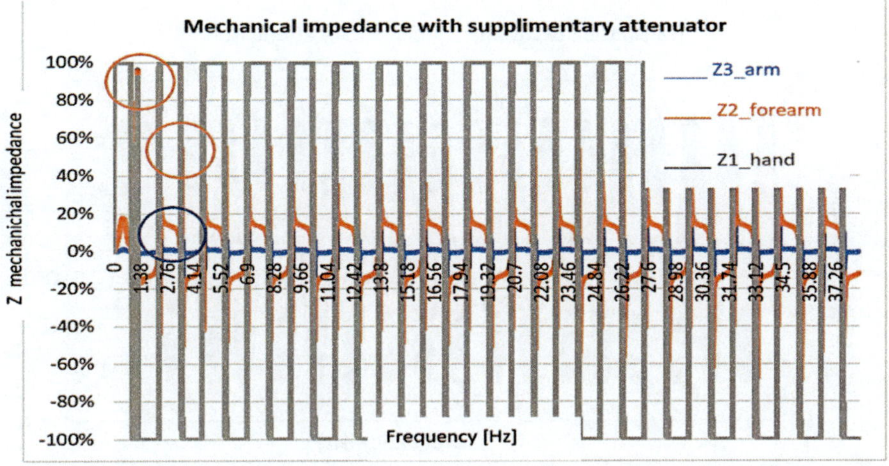

Fig. 5. Mechanical impedance with supplementary attenuator.

3.2 Mechanical Transmissibility

The present subchapter tries to validate the results obtained at sensitivity (Fig. 4, Fig. 5) by calculating the mechanical transmissibility of the hand-arm system. The transmissibility represents the ratio between the maximum transmitted force and the relation (3) gives the excitation force and it.

$$T = F_{max}/F_{excitation} \quad (3)$$

Figures 6 and 7 represent the evolution of mechanical transmissibility from the excitation source to the hand-arm system. The literature [5] says that in order for the system to be in equilibrium not to resonate or not to have disturbances, the transmissibility should be T <= 1.

As seen, in Fig. 6, we talk about the hand-arm system without additional damping and we discuss on the hand, it has a transmissibility greater than 1 at the beginning of the movement (blue curve) and stabilizes harmoniously around one smaller value. The same initial instability it has their forearm (transient movement) (orange curve) and the arm (gray curve).

The Fig. 7 brings stability by introducing the additional attenuator, so that this instability of the beginning of movement in Fig. 6 (in hand of 2.5 is reduced to 1.4, in Fig. 7, using an additional attenuator). The transmissibility for the arm is small very close of the axis zero, it does not be visible on the graph (Fig. 7). However, in comparison the Table 2 shows the reduction of the vibration from hand until arm, using or not a supplementary attenuator.

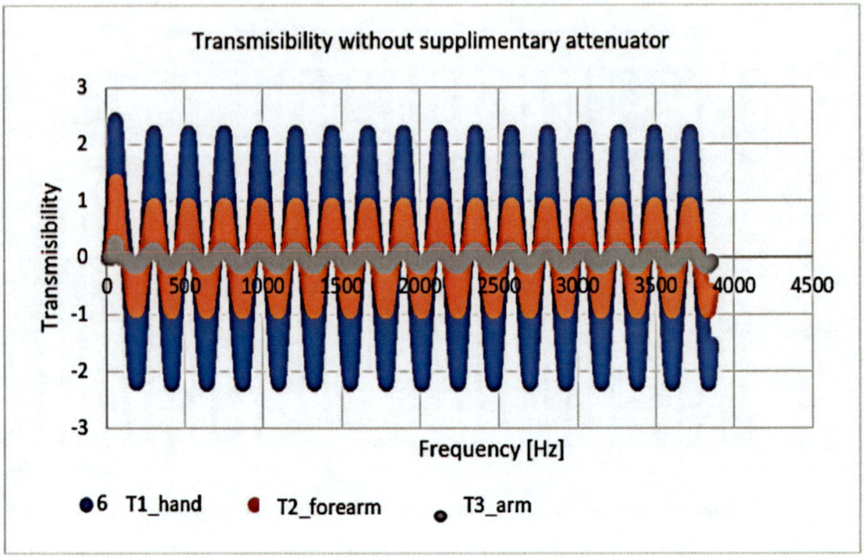

Fig. 6. Transmissibility without supplementary attenuator.

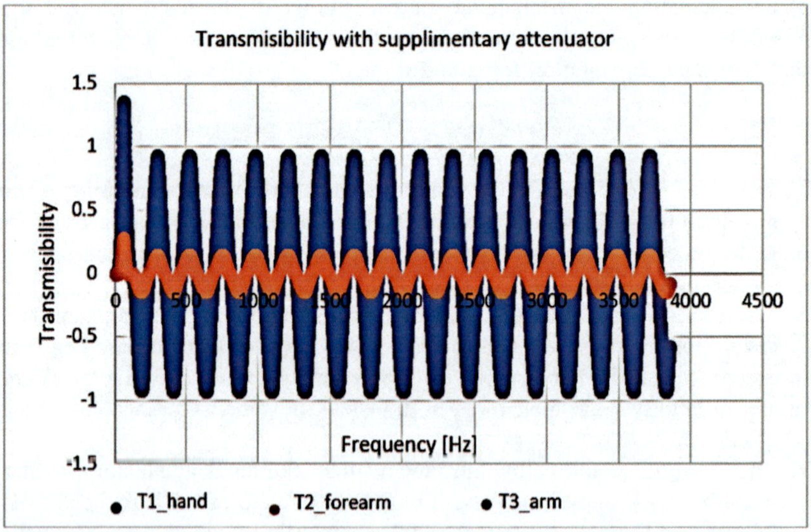

Fig. 7. Transmissibility with supplementary attenuator.

Table 2. Experimental vibration transmissibility.

Transmissibility	Without		With		Difference of transmisibility reducing [%]
Experimental	Hand	2.2	Hand	1	45%
	Fore arm	1	Fore arm	0.2	20%
	Arm	0.2	Arm	0.01	5%

4 Conclusions

The research [3, 10] carried out on the hand-arm system regarding the action of vibrations, it shows that:

Vibrations are transmit along of the system starting from the excitatory source through the hand and up to the arm. Under certain conditions, they grow, but as is normal, they attenuate through the skin muscles, to a certain extent, up to the shoulder. The paper highlights the fact that they decrease in a short time, with about 50% in the hand (Table 2), but they are reduced minimize with 20% for the forearm and with 5% to the arm, when is mounted on the forearm one supplementary attenuator, which makes the object of the study more interesting and validates it. Clear values of the human body's sensitivity to vibration are didn't yet known, but the standards give clear limits on the maximum acceleration (a_{RMS}) of the excitation source, respectively transmission of vibration at daily exposures of up to 8 h and measures to prevent adverse effects [2–4, 8]. Of course, the reduction of vibration transmitted to the system is beneficial to the health of the subject and comfort of his, but the mounted of a vibration attenuator along the forearm, makes them more "comfortable" because them being dissipated through the attenuator;

- It is imposed by the standards [2, 3] to be respected breaks during work, overtime being avoided, because this should could be to lead to signs of numbness in hand, this specifies the possibility of occurrence of occupational diseases. In this case, must be take supplementary protection measures.
- The professional affections of the human operator, due to the transmission of vibrations given by tools and vibrating devices can be:

"Washing with cold water or freeze water, produce the pain in the hand.

Nocturnal or rest pain in the hands accompanied by paresthesia (in the form of tingling).

After 10–15 min from the start of work, the pain usually disappears.

The pain differs according to the intensity and de pends on the degree of manifestation of the disease. They often cause sleep disorders" [5, 7].

In the next researches, the studies will done on equipment with high vibration (percussion), such as percussion hammers, and consecutive measurements will take several weeks. Regarding the additional attenuator, at it will done ergonomic improvements. In addition, a possible relationship with medicine institute regarding these results will done benefic.

Conflict of Interest. I mention that, I have no conflict of interest regarding this paper.

References

1. Harris, C.M., Crede, C.E.: Socks and Vibration, vol. I–II. Tehnica Bucuresti, Bucharest (1969)
2. https://lege5.ro/Gratuit/haztamzx/vibratii-mecanice-hotarare-1876-2005?dp=gi4dinzvguzta. Accessed 03 July 2020
3. https://e-standard.eu/ro/standard/71446. SR EN ISO 5349/2003. Accessed 03 July 2020
4. http://www.inpm.ro/files/publicatii/2013-05.07-ghid.pdf. Accessed 03 July 2020
5. 14th International Conference on Hand-Arm Vibration 21–24 May, Effect of shelf aging on *vibration transmissibility* of anti – vibration gloves finger-hand-arm system to vibration is approximately, Bonn, Germany (2019). 164 pp.
6. https://boliocupationale.usmf.md/sites/default/files/inline-files/05%20-%20MALADIA%20DE%20VIBRA%C8%9AIE_0.pdf. Accessed 05 July 2020
7. https://boliocupationale.usmf.md/sites/default/files/inline-files/boala%20de%20vibratie.pdf. Accessed 03 July 2020
8. https://eur-lex.europa.eu/LexUriServ/LexUriServ.do?uri=CONSLEG:2002L0044:20081211:RO:pdf. Accessed 03 July 2020
9. http://stiintasiinginerie.ro/wp-content/uploads/2014/07/26-34.pdf. Accessed 15 July 2020
10. Felicia, P.(Cristea)A., Mariana, A.: Vibration attenuator device attached to the hand-arm system of the human operator (2018). Patent no. 128900/2018
11. Pan, D., et al.: The relationships between hand coupling force and vibration biodynamic responses of the hand-arm system (2017). https://doi.org/10.1080/00140139.2017.1398843
12. Wen, J., Ma, W., Li, W.: A method for analyzing hand-transmitted vibration characteristics in the hand-arm parts when operating the portable pneumatic extinguisher. Special Issue, 10 pages (2019). https://doi.org/10.1155/2019/9402539. Article ID 9402539
13. http://ansp.md/wp-content/uploads/2014/05/Ghide-practic-1.pdf. Accessed 05 Oct 2020
14. http://www.mec.upt.ro/meca/poz10staff/LB/vibratiile_sistemelor_mecanice.pdf. Accessed 02 Oct 2020

Studies Regarding Vibration Transmitted, Using an Additional Damper, Mounted an the Hand-Arm System

Aurora Felicia Cristea(✉) and Claudiu Schonstein

Buildings of Machines Faculty, Technical University of Cluj-Napoca, Cluj-Napoca, Romania
{felicia.cristea, claudiu.schonstein}@mep.utcluj.ro

Abstract. The paper is a continuation of other studies and wants to demonstrate what negative effects (diseases) produce the vibrations transmitted to the hand until arm, without or with a supplementary vibration attenuator. Thus, the paper deals with both the theoretical and experimental mode of their diminution when mounted or not, a vibration attenuator. The attenuator will testing in terms of blood pressure in the hand-arm system, through the IGB index (ankle-wrist coefficient). Therefore, the paper is approaching from an engineering point of view through modeling, design, and the measurement and from the medical point of view through IGB index measurements as a validation measure.

Keywords: Supplementary attenuator · Transmitted vibration of the operator

1 Introduction

1.1 Generalities of Medical Nature

The paper comes with the supplementary additions and validates the results presented in other papers, on this studied. Such as, the vibrations on the hand-arm system are transmitting during the work process. This paper wish of new validate of the recently studies regarding the effect of the vibration about hand, in the work process.

Regardless of the industrial activity from which it comes, vibration acts on the human body and it can cause discomfort to the operator, might been changed its activity or even have negative influences on the health of the human operator, subjected to its. The action of vibrations with a frequency between 20–200 Hz [3, 9] determines osteo-musculoskeletal syndrome that evolves with pain, joint swelling, and limitation of movements, in this, the examination radiological highlights the narrowing of joint spaces and tendons and inflammatory phenomena occur in the muscles. Vibration exposure has found to be associated with a reduction in grip strength of the hand at the handling, and their neglect can lead to incapacity for work.

This paper might been compared with the similar theoretical studies, in terms of vibration transmissibility but without/with additional attenuator. The novelty of this paper consists in the fact that the obtained results was been wanted to validate other studies and to strengthen the fact that the vibrations of the hand-arm system are

S. Vlad and N. M. Roman (Eds.): MEDITECH 2020, IFMBE Proceedings 88, pp. 331–343, 2022.
https://doi.org/10.1007/978-3-030-93564-1_37

transmitted along for the forearm. In these cases, it was been measured and it has been calculated the IGB indices and displacements caused of the vibration in the work process. All this was been demonstrated using an additional vibration attenuator, which was been realized practical and patented.

In the paper, we will refer to human operators who are daily expose to vibrations through the regular use of portable machine tools (all these respecting the OSM protective rules regarding measurements). They might been present different forms diseases with different severity, from reducing peripheral circulation to stopping circulation in the extremities.

In the next subparagraph will been presented some possible occupational dis-eases related to the transmission of vibrations, like vascular disorders.

In this way, the two most serious conditions are relates to Raynoud syndrome, and another related to carpal tunnel syndrome. Raynaud's syndrome is a disorder of blood circulation in the fingers. It affects the reaction of the fingers to the cold producing the crisis of the "white fingers" by strongly reducing the circulation. At first, whitening attacks refer to the extremities of one or more fingers, but gradually if the exposure to vibration continues or increases in intensity this fact is dangerous (Figs. 1, 2, 3).

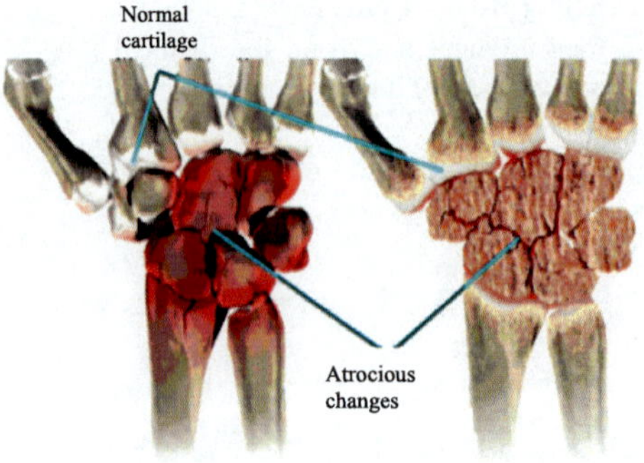

Fig. 1. Bone changes due to osteoarthritis [1].

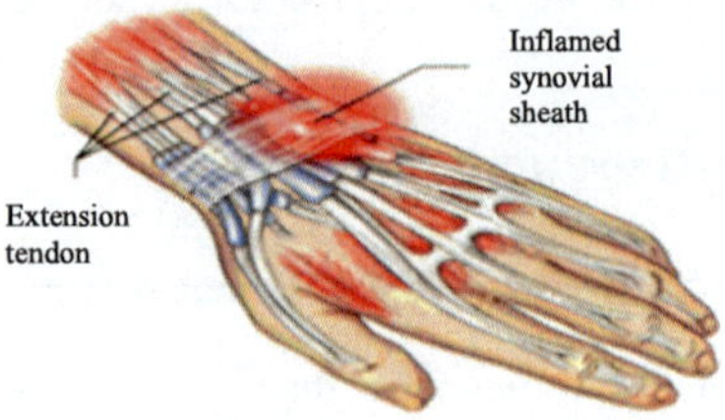

Fig. 2. Tenosynovitis (inflammation of the tendon sheath) [1].

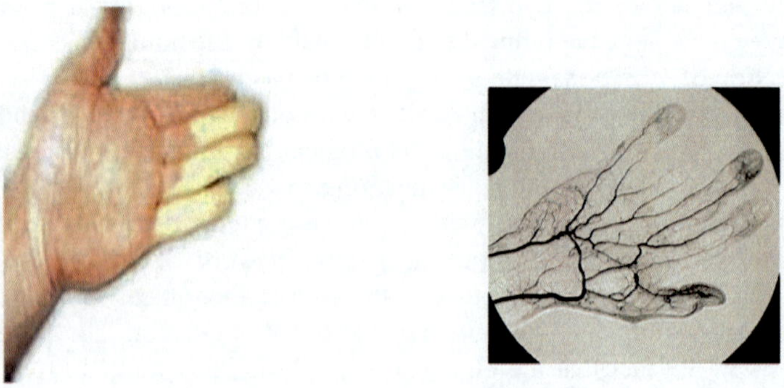

Fig. 3. Upper limb diagnosis [1].

One of effects of the illness caused by the action of vibrations on the hand could be the carpal tunnel syndrome [4], but this could be aggravate of the other work conditions like irritation of the median nerve (tingling, numbness in the thumb etc.), respectively dexterity (Fig. 4).

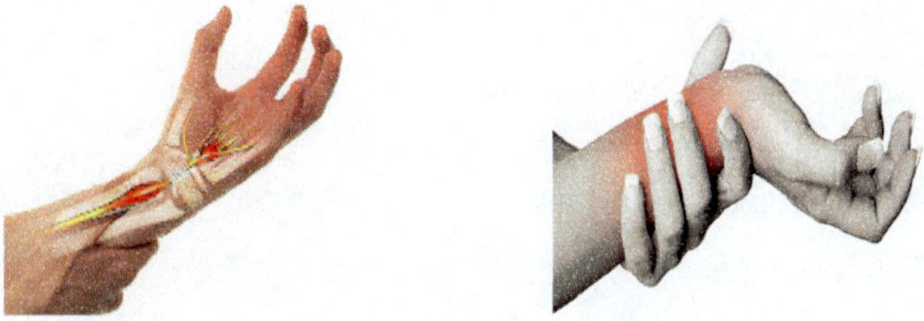

Fig. 4. Tendonitis and carpian tunnel syndrome [1].

2 Mathematical Model

For the hand-arm system, the soft tissues might been describe the phenomenological, as a viscous-elastic medium for which were substitute of the dampening and rigidity coefficients [3]. In this way, the differential equations of motion could take the form given by the relation (1) [2]:

$$[m] \cdot \{\ddot{u}(t)\} + [c] \cdot \{\dot{u}(t)\} + [k] \cdot \{u(t)\} = \{F\} \tag{1}$$

The mechanical hand-arm system is composed of 5 springs and 5 shock absorbers c0, c1, c2 located along the Oz axis, i.e. along the forearm and c3, c4 located along and perpendicular to the arm, and the springs that give the system rigidity and they are placed similar. The real model was been simplified to a model with concentrated masses in the center of mases, and the wrists were considered, in this case, rigid without elasticity or damping. If the additional shock absorber has been add, it mounted in parallel with the forearm and was fixe between the wrist and elbow. The supplementary attenuator has an attenuation and a stiffness c5, k5.

In the relation (1), u(t) is the displacement along the Oz / Ox axis of the three bodies that make up the hand-arm system.

The initial condition of the operator are there: position vertically, scale 0–20 Hz frequency, shoulder at 0° degrees, and gripping device is in palm.

Using as excitation source (drilling machine) and take in account SR-EN ISO 5349/2003, paper analyses only main vibration transmitted after Oz axis (along the hand-arm system).

Figures 5 represent the system solutions (1) obtained of the integration with Runge Kutta of 4–5 order.

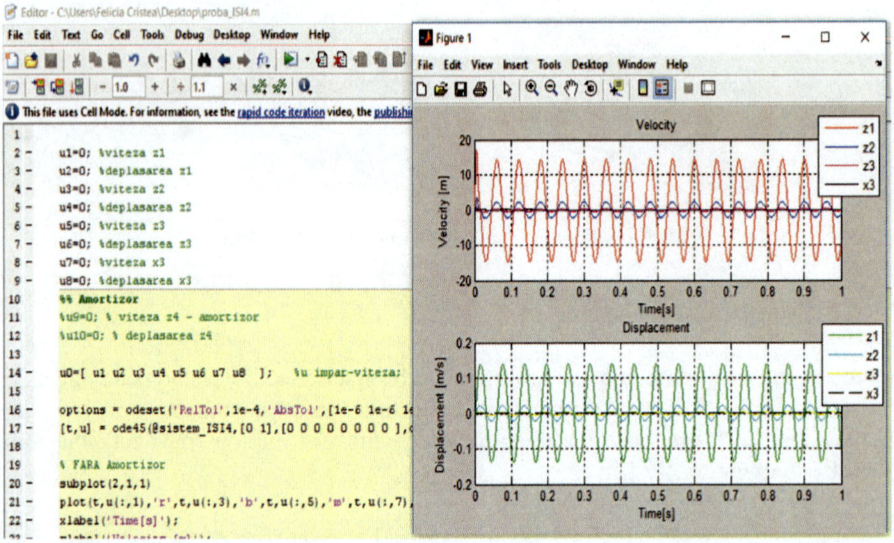

Fig. 5. Theoretical displacements and velocities of the system without supplimentary attenuator.

2.1 System with Supplimentary Damper Montated Along of the Forearm

The paper, in the next figures, will be analise the transmissibility of the hand-arm system (theoretical), in presens of not of supplimentary damper/atenuator.

Solving the Eqs. (1) by double integration to find out the velocities and displacements of the bodies, was performed with the Runge Kuta method of order 4–5, by running in Matlab [3, 5, 6].

Figure 6 represents the system solutions, there obtained from the integration.

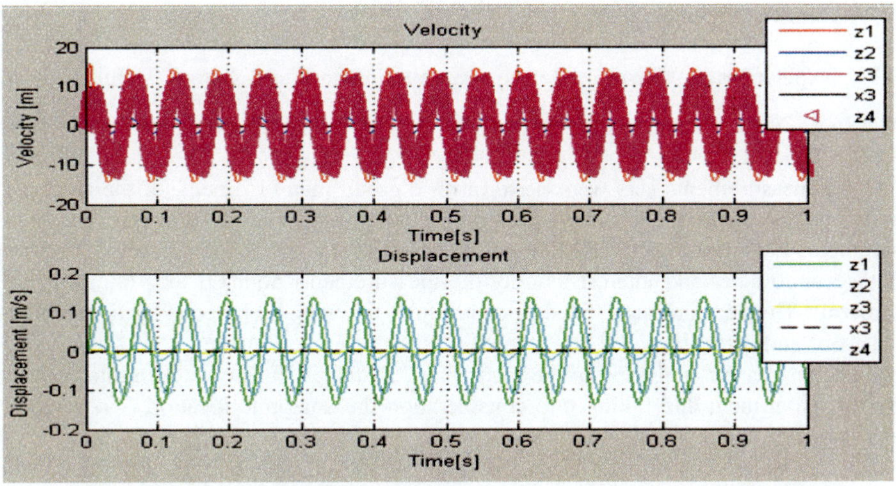

a.

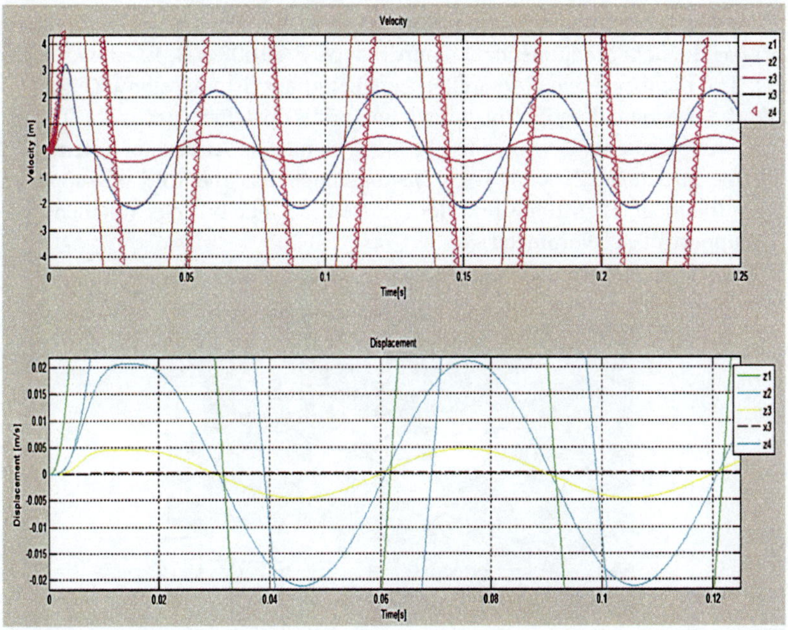

b.

Fig. 6. Theoretical velocities and displacements corresponding of the system with supplimentary attenuator. a. Results of displacements and velocities, b. ZOOM of velocity and displacement.

3 Measurements

3.1 Experimental Vibration Measurements on the Hand-Arm System

The study, it use the same drilling machine similar with it the other researches (1000 RPM).

The measurements sets were perform on a group men (15 men), of the appropriately age (between 40–50 years), weighing medium and over 1.70 m tall. The measurements have indicate the RMS (root mean square) accelerations displaying on the device vibrometer and interest us only the measurements after Oz axis (along of the forearm). The accelerometer fixed directly under the metallic bracelet by tightening it (Fig. 6b) and directly on the arm (taped).

The results are analyzing in the Figs. 9, 10, 11 for all the cases taken into account by the experiment and taking into consideration the anatomic location (wrist, elbow and shoulder.

In this paper is presented a new idea, this is the design of a vibration attenuator device, which will be fixed along of the forearm. The scope of this attenuator is minimizing the transmitted vibration from hand to the shoulder. The dissipater (attenuator) has the next components: 2 dampers type FA 1008 VB, fixed with a complex bracelets. The damper device in parallel, between the wrist and elbow (Fig. 7). It was respected the theoretical weight of the attenuator/dissipater, respectively 0.5 kg.

The vibration attenuator contains the next elements: Attenuator (**element 1**) are fixing in the structure (2 pcs.); It extension which has the role of supporting the attenuators (element 2); Mobile bracelet (element 3). Fix bracelet (element 4). Magnetically support of accelerometer.

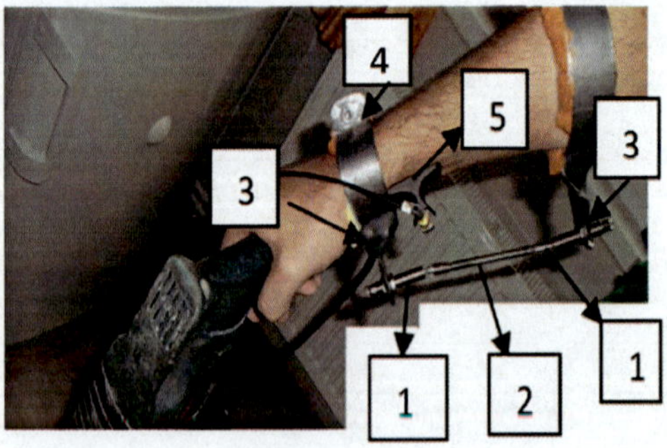

Fig. 7. Vibration damper device [10].

Firstly, theoretically, the hand-arm system is analyze in terms of transmitting vibrations from the excitation source using or not an additional shock absorber. Figure 8, referring to the hand, shows the blue curve (theoretical) 0.04 m without damping, as it is above the gray curve 0.015 m, with additional damping. Which indicates that the extra shock absorber reduces vibration, and to the theoretical form we are talking about in now [7, 8].

The Table 1 presents the experimental displacements for the hand-arm system without/with supplementary attenuator.

Table 1. Experimental displacement (peak).

	Without		With		Difference [%]
Theoretic	Hand	0.04	Hand	0.015	37%
	Forearm	0.006	Fore arm	0.002	33%
	Arm	0.058	Arm	0.0005	33%
Experimental	Hand	0.058	Hand	0.03	51%
	Forearm	0.008	Fore arm	0.004	50%
	Arm	0.0018	Arm	0.001	55%

The peaks of vibrations are reduced toward forearm using a damper and theoretically (according to the modeling), but also to the experimental mode (Table1) in percent. In this studies was observed that the reduction of vibrations with additional attenuator both in theory (33–37%) and in practice is minimized by more than 50–55%. This fact can only be gratifying and useful for the human operator who handles vibrating machines and devices. In this way, the chances of being affected by occupational diseases being small (White Fingers Syndrome).

The experimental displacements are stable, without the disturbances, from the beginning of the movement.

The measurements were performed in 8 h with the 2.5 s cycles and theoretical analize has been raportated at the them.

3.2 Theoretically and Experimentally Results – Without/With Attenuator

Figure 8 analyzes the behavior of the hand-arm system both theoretically and experimentally, when there is no additional attenuator and according with the Table 2. Its shows that the experimental vibrations on the hand - arm system have higher values than those resulting from the theory, but they respect the shape except for some jumps that mark imbalances, being superior in value to the theoretical ones.

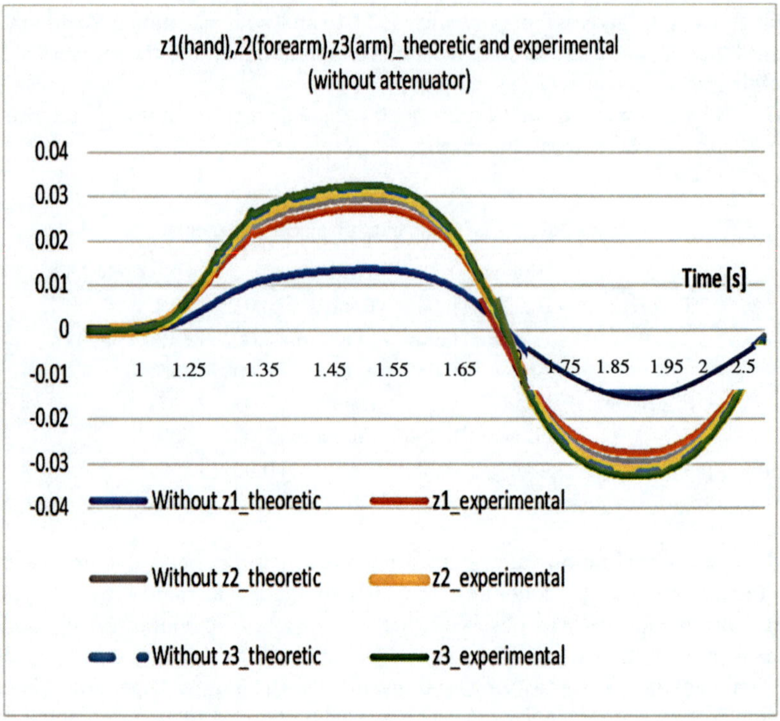

Fig. 8. Hand-arm system without supplementary attenuator.

Figure 9 analyzes the behavior of the hand-arm system when there is an additional attenuator and according with Table 1. Shows again that, the experimental vibrations on the hand-arm system have higher values than those resulting from the theory, but they respect the shape, and are superior in value, compared to the theoretical ones resulting from the presented simulation in the percentages of the Table 1. Those small jumps, in the shape of the graph, are the points where the percussion comes into operation, this contributing to the increase of the excitation values.

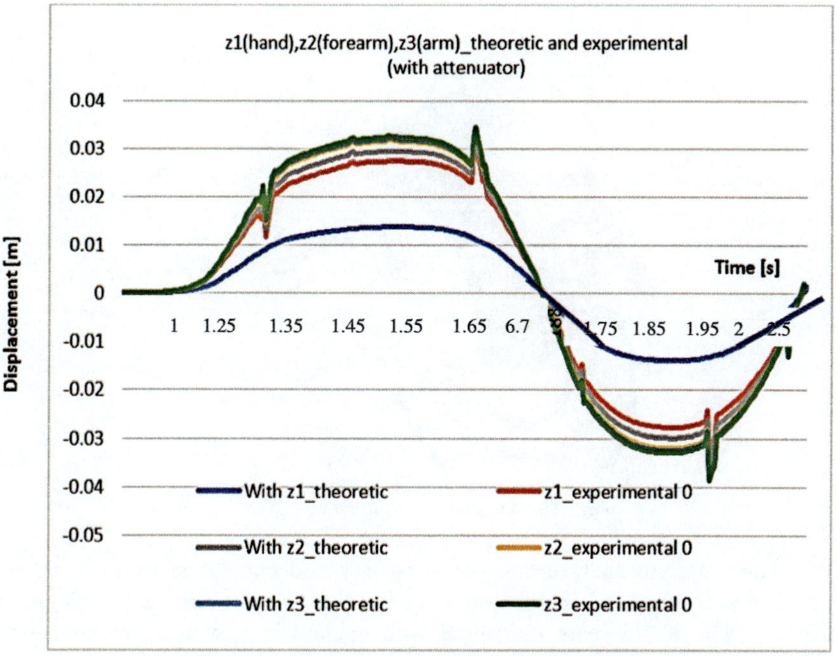

Fig. 9. Hand –arm system with supplementary attenuator.

3.3 Medical Investigation

To determine blood pressure in peripheral vessels, the Ankle-Arm Index (IGB) can be determined, which is a fast, noninvasive method used to diagnose arterial disease. This measurement helps the doctor to make a first assessment of the foot-hand sanguine circulation, but also vice versa and to monitor over time, its improvement or worsening.

The index is obtain by dividing the blood pressure from one ankle to that of the wrist.

$$IGB = \frac{P_{ankle}}{P_{forearm}} \tag{2}$$

Table 2. Diagnostic index [8].

Index	Diagnostic
<=1.41	uncompressible;
1.40–1.00	normal
0.99–0.91	in limit
0.90–0.41	unnormal
<=0.40	serious

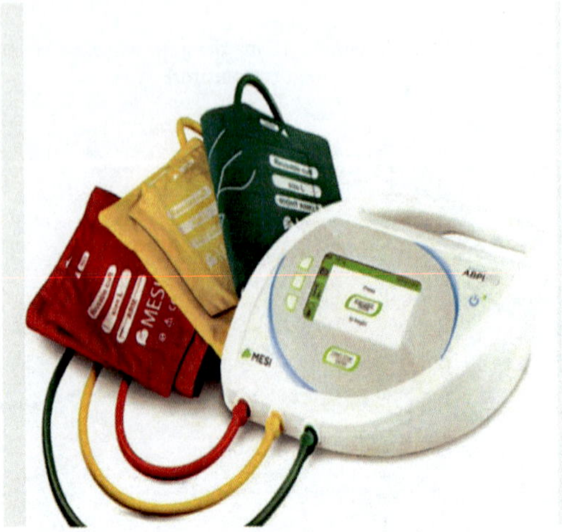

Fig. 10. Diagnostic equipment _ IGB.

The paper aims to analyze a set of tests on blood circulation to the hand-ankle (Fig. 10), at rest (horizontal position) and in operation (vertical position), for a group of 15 people, aged 40–45 years old, men, who operate in a work environment with vibration transmissibility (the subject handles with a percussion drilling machine). Analyzing the blood pressure in the wrist before using a vibration attenuator and after using it, during the drilling operation. It was, three sets of measurements performed, and was using their arithmetic mean in the paper.

Such as, while the human operator is lying horizontally, a technician measures blood pressure in both arms, using an inflatable cuff.

Then a doctor measures the blood pressure in two arteries at the ankles, using the inflatable cuff and a portable Doppler ultrasound device, which pressed on the operator's skin, he is already in a working position, vertical. The Doppler device uses sound waves to make sounds and allows the doctor to measure the pulse in the arteries of the ankle after the cuff is deflate.

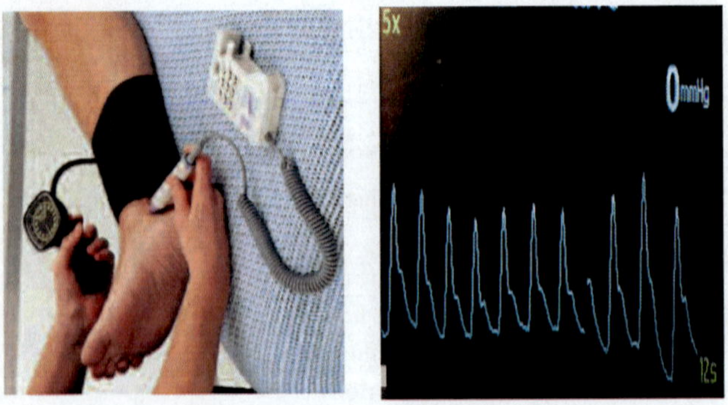

Fig. 11. Measurements of IGB.

Figures 11 show how to measure pressure. This presents an exception, respectively at rest the pressure is compare in the sitting position, and in the working position, it is compare in the vertical position.

The pressure in the blood vessels can vary depending on this position, this pressure is adding to fatigue, other disease, age and sex.

The IGB analysis for subjects is giving by the following graphs.

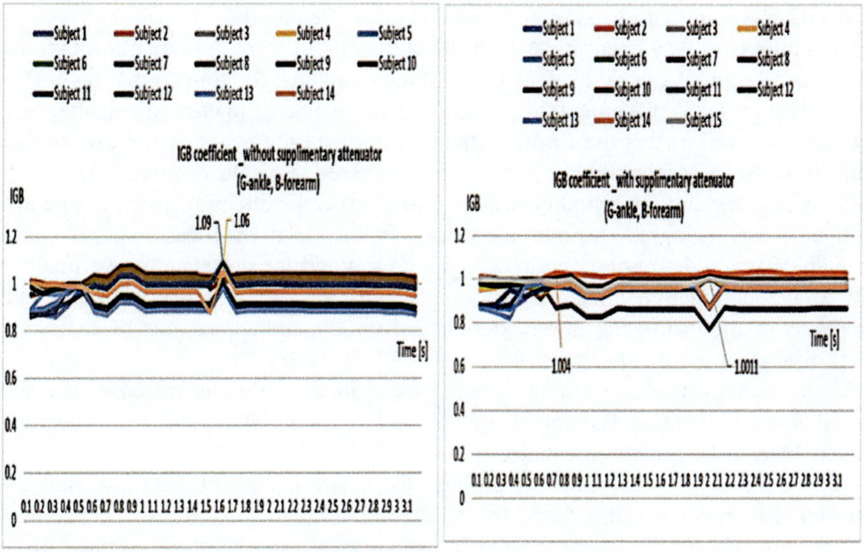

a. b.

Fig. 12. IGB index.

The Figs. 12a, b show us that IGB values without supplementary attenuator along of the forearm is bigger, and when this is mounted, they are these values are smaller, for all of the 15 subjects. In addition, we observe the peaks of the graphical shapes, that show instability in the pressure of blood in some moments, but these are ok, in case of attenuator supplementary.

4 Conclusions and Discussions

These conclusions are the result of some measurements performed with the installation of an additional vibration attenuation device (patented), which validated the fact that:

Vibrations were been transmitted from the source of excitation through the hand to the arm, and their effects over time could cause occupational diseases [3].

The paper monitors whether the measured values (Fig. 9) of the vibrations fall within certain limits, compared to those imposed by the standard SR EN ISO 5349/2003.

It is observed that the experimental results obtained are in the hand of 5.8 cm without additional attenuation (in the warning area according to the standards) and are reduced to 0.05 cm with additional attenuator, this being a significant variation and a confirmation of the fact that the additional attenuator has an important role in this transmission.

However, this reduction at the elbow is not brings significant with the additional attenuator (0.1 cm), it being quite small and without its use (0.18 cm). We can say that the additional attenuator has a vital role in attenuating the transmission of vibrations to the hand, where in fact occupational diseases occur frequently.

A comparison was been made with the standards in force (we mention that they refer to accelerations not to displacements. It mentions that the transformation of these accelerations into displacements by special software (using mathematical integration methods), showed us that the values obtained from measurements do not exceed these limits of the RMS acceleration of 5 m/s^2 (and expressed in displacements of 0.06 m) for 8 h daily at work, but are found near them (0.058 m) called the high-risk warning area.

In this case, taking into account the values obtained from the measurements, we can say with certainty that: non-compliance with breaks and the presence of overtime, can lead to the cumulative effect of vibrations, and this can degenerate at first with symptoms of numb fingers, low dexterity, and finally, the appearance of an occupational disease in the hand [1].

Under certain conditions at work, the transmitted vibrations increase, but as is normal, some of them are been attenuated by muscles, skin, blood system, so that up to the shoulder, to be greatly attenuated.

As this paper has shown, the big problem that arises is the fact that these vibrations transmitted largely to the hand and less to the forearm.

The paper also highlighted the fact that, they decrease in a short time, almost 100%, when using the additional vibration attenuator, which brings improvements in the comfort of the human operator and over time, prevents the occurrence of occupational diseases (the sanguine system-most often found in White Fingers Disease).

The maximum values must be avoided, and this is done by wearing protective equipment (specialized literature, provides rubber gloves) and respecting breaks in work. The other recommendation for prevention are:

From a medical point of view, we can say that carpal tunnel syndrome can be aggravated if in main, the wrist is repeatedly stretch and excessively.

Repeated use of equipment brings upload with a supplementary effort about of the wrist and contributes to the swelling and compression of the median nerve of this.

Minimizing repetitive handling when is possible.

It is recommend involving one or both hands for something difficult;

Avoid taking the object in the same position for a long time.

It is also, desired to ergonomic the vibration damping device so that it can be easily transported and handled.

Conflict of Interest. Hereby mention that, I have no conflict of interest regarding this paper.

References

1. https://e-standard.eu/ro/standard/714460 SR EN ISO 5349/2003. Accessed 05 July 2020
2. http://www.mec.upt.ro/meca/poz10staff/LB/vibratiile_sistemelor_mecanice.pdf. Accessed 05 Oct 2020
3. 14th International Conference on Hand-Arm Vibration, 21–24 May - Effect of shelf aging on vibration transmissibility of anti-vibration gloves finger-hand-arm system to vibration is approximately, p. 164, Bonn, Germany (2019)
4. Martinez, R., Assila, N., Goubault, E., Begon, M.: Sex differences in upper limb musculoskeletal biomechanics during a lifting task. Appl. Ergon. **86**, 103106 (2020). https://doi.org/10.1016/j.apergo.2020.103106
5. Wen, J., Ma, W., Li, W.: A method for analyzing hand-transmitted vibration characteristics in the hand-arm parts when operating the portable pneumatic extinguisher. Hindawi **10**. ArticleID 9402539 (2019). https://doi.org/10.1155/2019/9402539
6. https://lege5.ro/Gratuit/haztamzx/vibratii-mecanice-hotarare-1876-2005?dp=gi4dinzvguzta. Accessed 05 July 2020
7. Pan, D., et al.: The relationships between hand coupling force and vibration biodynamic responses of the hand-arm system (2017). https://doi.org/10.1080/00140139.2017.1398843
8. Disease of occupational diseases in the upper limbs. https://boliocupationale.usmf.md/sites/default/files/inline-files/05%20-%20MALADIA%20DE%20VIBRA%C8%9AIE_0.pdf. Accesses 05 July 2020
9. Canetti, E.F.D., Schram, B., Orr, R.M., Knapik, J., Pope, R.: Risk factors for development of lower limb osteoarthritis in physically demanding occupations: a systematic review and meta-analysis. Appl. Ergon. **86**, 103097 (2020). https://doi.org/10.1016/j.apergo.2020.103097
10. Pop (Cristea) Aurora Felicia, Arghir Mariana, Patent no. 128900/2018, Vibration attenuator device attached to the human operator's hand-arm system (2018)

Programing a Robotic Ambulance with Virtual Reality

Andrei Lavinia[1(✉)], Oniga Tudor[2], and Chindea Viorel[2]

[1] University of Medicine and Pharmacy Iuliu Hatieganu, CJ 3400 Cluj-Napoca, Romania

[2] Technical University of Cluj-Napoca, Bvd. Muncii, 400641 Cluj, Romania

viorel.chindea@auto.utcluj.ro

Abstract. By applying artificial intelligence and virtual reality, this study presents results and challenges for robotic ambulances. Programming vehicle dynamics and testing protocol is intended to support task of developing an autonomous ambulance and engineering efforts. Testing parameters are controlled with automated driver. Using software will diminish human input in driving. Actual values of robotic ambulance in testing are displacement, speed, and acceleration. Ambulance's velocity in testing on a virtual track with corners and straight lines is an important kinematic parameter that influences safety of transport and some program sections. Accelerations are also important to be programmed. Objective of this paper is to highlight sequences of programming robotic ambulance using virtual reality and artificial intelligence. Results are consisting in testing scenarios, ambulance automated driving program on virtual track, refined program code, solutions for challenges.

Keywords: Artificial intelligence · Automated · Programming · Robotic ambulance · Virtual reality

1 Introduction

Programing a robotic ambulance with virtual reality tools is not different than a conventional robot. For any type of propulsion in robotic vehicles, final product will be programmed as standard robot, if it relies on digital control.

There are two available solutions for applying artificial intelligence (AI) in robotic ambulance, either fuzzy logic or neural networks. Some programming tools [1] and automatic applications are used with fuzzy regulators or fuzzy techniques. Smart systems were used to measure and monitor levels of sound pollution [2, 3]. Thus, it was possible to make intelligent evaluations regarding some environmental parameters. Automatic and smart monitoring, along with corresponding control tools, in complex metropolitan environments [4, 5] can be used also with embedded artificial intelligence programs and fuzzy techniques. Complex systems were used in making automated tasks or in evaluating road events with more than one car [6]. Some updates and optimization steps of software and hardware components applied in testing [7] and AI, are required. In metropolitan areas, for automated road inspection, sound control and pollution evaluation, there are required autonomous, and self-driven robotic systems [8]. These are vital for performing complex tasks of control.

S. Vlad and N. M. Roman (Eds.): MEDITECH 2020, IFMBE Proceedings 88, pp. 344–354, 2022.
https://doi.org/10.1007/978-3-030-93564-1_38

Achievements were recorded in developing applied smart program for automated driving supported by fuzzy regulators with virtual reality software [9]. Also were made some advancements in programming, designing, and improving control of autonomous robots used as exploration tools [10]. Programing robotic systems and defining their inherent formalism, as well as implementation of artificial intelligence techniques, are close related. These are considered for improving automated robotic car's dynamics. They are also useful for improving energy efficient use in challenging competitions [11–13]. Specific considerations must be given toward three-dimensional development and virtual reality programing of complex environments. When making complex models, and thus creating graphical part of proposed scenario in VR environment, some tools must be used properly [14]. Along with researching energy use or its efficient distribution, it is possible to define and to implement some smart features in vehicle digital architecture. Presenting essential data on dashboard and on-board panels, concerning car's fuel economy in operational scenarios [15], as well as driving autonomy of automated robotic vehicle, is a step forward in creating smart applications. Programing these features and kinematic behavior will be actual development and incremental contribution in further definition of electronic control units [16, 17], both for automotive engineering and robotics. Thus, programing and software implementation are taken into consideration based on development required steps in practical action [18, 19]. It is obvious that high requirements for development and smart features implementation have been supported by detailed programming of electronic control modules of existing systems and applications [20].

Main objective of paper is to show how digital programing works in development and designing Virtual Reality environment when modelling an operational robotic car with Unity 5.6.4. Specific objectives consist in defining software programs and specific research materials. Other aims are programing in Unity 5 software, using a fuzzy regulator for car control, designing in VR application, architectural setting of digital environment, development of testing track, creation of other virtual objects, and operational test protocol for automated robotic car in VR developed environment, according to Fig. 1. Research of literature and lab work in VR designs are leading to development of a smart ambulance.

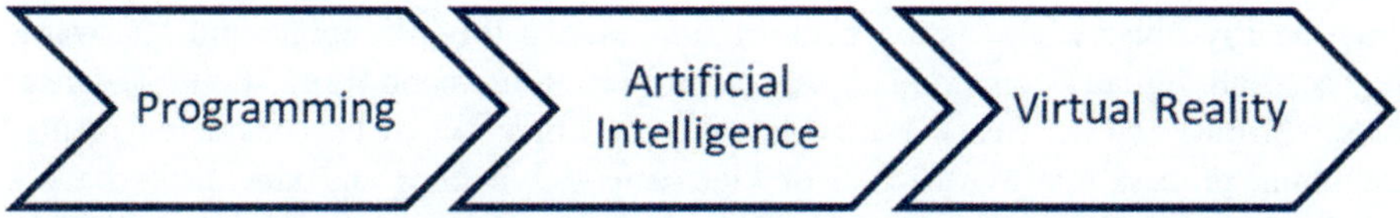

Fig. 1. Main steps in development of applied program for an automated ambulance.

2 Materials and Method

Researching program and virtual development presented in this paper are supported by existing materials and engineering methods used in conjunction with artificial intelligence, and corresponding techniques. Figure 1 shows structure of proposed robotic ambulance. There are electromechanical components assembled in present model. Wheel-system is made of wheels (front-right 1a, rear-right 1d, front-left 1b, rear-left 1c), from an assembly of tire and rim. Wheel train is assembled on robotic car's two axles. Between front and rear axles is determined wheelbase length (L).

Schematic of equipment and materials used in present research are intuitively positioned to facilitate operational aspects for research (see Fig. 2).

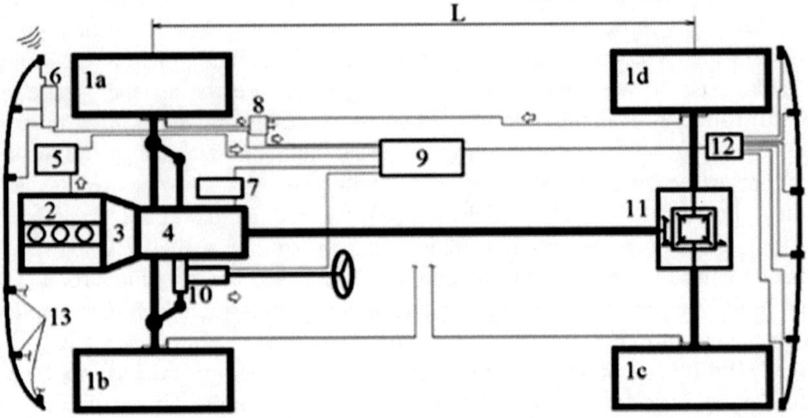

Fig. 2. Robotic ambulance. (1-wheels; 2-engine; 3-clutch; 4-transmission; 5-engine control unit; 6-parking control unit; 7-transmission control unit; 8-antilock braking system control unit; 9-central electronics; 10-steering unit; 11-differential; 12-rear parking control unit; 13-proximity sensors)

Engine 2 rotates clutch 3, and transmission 4. They transfer kinematic and dynamic parameters (power and torque) toward rear axle differential 11 to distribute them to motor wheels and to propel ambulance. Engine control module 5, frontal proximity sensor hub 6, transmission control module 7, antilock braking system control module 8, steering box control module 10, rear proximity sensors hub 12 receives and transfers signals to electronic system 9. Front proximity sensors 13 are converting recorded data about nearby objects and testing track in autonomous driving regime and afterwards they send all digitally converted signals toward electronic control module 9 via sensor hubs 6 (front) and 12 (rear). Propulsion is very important to be considered in programming process due to variation of kinematics. Dynamics and kinematics have a significant influence upon car propulsion. In conventional cars driven by human individuals fine tuning of parameters is made by driver during vehicle's exploitation. In robotic cars parameters must be completely programed.

Figure 3a presents schematic of step by step methodology that has been implemented in vehicle command unit supported by artificial intelligence. First there must be a basic comprehension of virtual reality program. Secondly, virtual reality animator adjustment must be realized. After module calibration, and further design of test track, robotic car, and surrounding environment, it is important to create scenarios for self-driving car in exploratory and critical conditions to validate or reject it from safety perspective. Figure 3b shows tool-panel for creating virtual scenario in Unity. Simulation is supported by a complex environment and render engine. If robotic car is safe in testing procedure further it may be developed at industrial level, but if it lacks security coverage, then, design and simulation must be reconsidered, as well as reprograming electronic control unit.

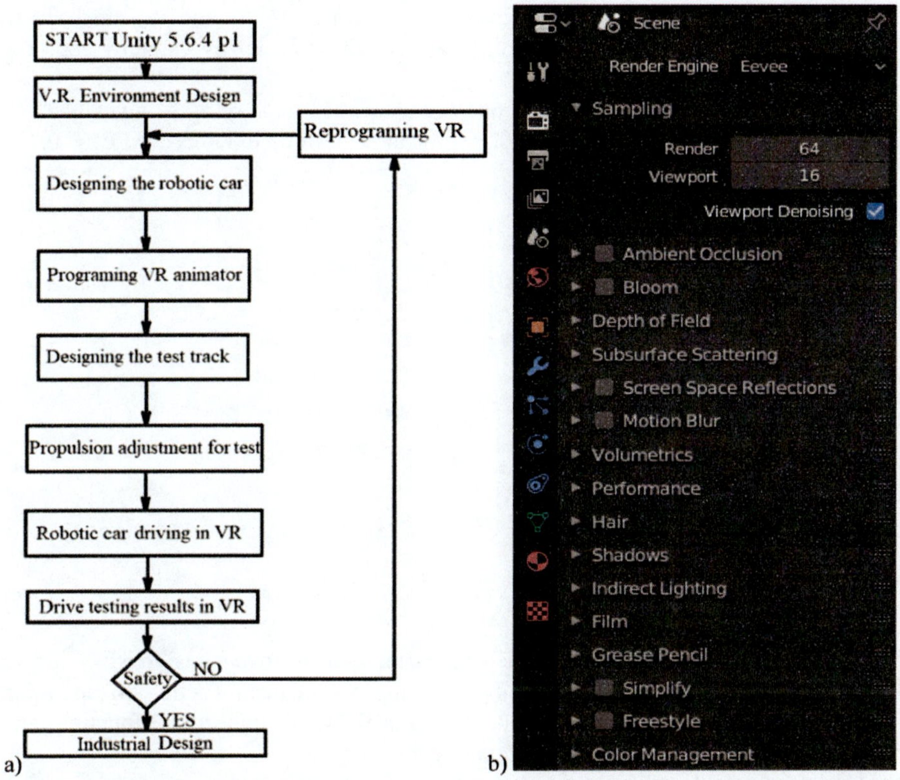

Fig. 3. Methodology for programing robotic ambulance (a) and settings package for designing with blender application (b).

3 Equations

Artificial intelligence (AI) offers support for controlling vehicle's behavior during testing and operation. Both AI solutions were considered. There is an increasing use of neural networks in AI applications. Also, fuzzy regulators and their inherent logic protocols are reliable.

Applied fuzzy logic protocol in current case is sustained by a specific applicable algorithm used for managing acquired data gained from proximity sensors. It consists in fuzzification step, for proper quantification or definition of terms used in protocol: "very" and "rather" [19].

$$\omega(very \qquad Y) = (\omega Y)^2. \tag{1}$$

$$\omega(rather \qquad Y) = \sqrt{(\omega Y)}. \tag{2}$$

where: ω is proximity signal transferred from robotic car's sensors to central electronic control unit 9; Y is specific fuzzy group. Logical models (1) and (2) are used in fuzzification. It supports transformation of a real value to a Fuzzy value by applying data from stored set.

Having a signal, picked up from road traffic, with a value of 80%, gives back an expression such as ω(very SAFE) = 0.36, but in same time ω(rather SAFE) = 0.775 based on (1), (2) becomes:

$$\omega_A(y) = \begin{cases} 0, & y \leq a \\ \frac{y-a}{n-a}, & a < y \leq n \\ \frac{b-y}{b-n}, & n < y < b \\ 0, & y \geq b \end{cases} \tag{3}$$

$$\omega_A(y) = \begin{cases} 0, & y \langle a \text{ or } y \rangle b \\ \frac{y-a}{b-a}, & a \leq y \leq b \\ 1, & b \leq y \leq c \\ \frac{d-y}{d-c}, & c \leq y \leq d \end{cases} \tag{4}$$

$$\omega_A(y) = e^{-\frac{(y-n)^2}{2k^2}} \tag{5}$$

in which: $\omega_A(y)$ is linguistic variable expression which covers real crisp value or logical signal at input part of system by applying one of membership corresponding functions (model (3) for triangular function, model (4) for trapezoidal function, and model (5) for gaussian function), recorded in memory.

Model which describes mathematically a Gaussian function used in decision making process of AI logic is applied in road like conditions, and it shows effects of three different constants on probability of gaining a signal:

$$f(y) = i \cdot e^{-\frac{(y-j)^2}{2k^2}} \tag{6}$$

where: i, j and k are random values; i represents minimum height of graph's pinnacle, j represents coordinates of average peak, and k represents graph's width.

4 Results

Mobile virtual fence (MVF), driving assistance applications, and interconnected vehicles are important transportation achievements. Programing robotic vehicles impacts especially urban road safety. Reduction of accidents in traffic is supported by efficiency of decision-making algorithm. This is based on a given fuzzy set. Defining road proximity objects and existing traffic infrastructure plays an important role for car programing, transport planning and logistical management. By programing and designing a robotic car in Unity 5 there were gained important data concerning kinematics and dynamics of robotic vehicle in road like conditions with multiple levels of challenges and obstacles.

Figure 4 presents access tool (a) to Unity 5.6.4 for entering RigidBody module (b). Figure 4c shows assets which allow user to access car.

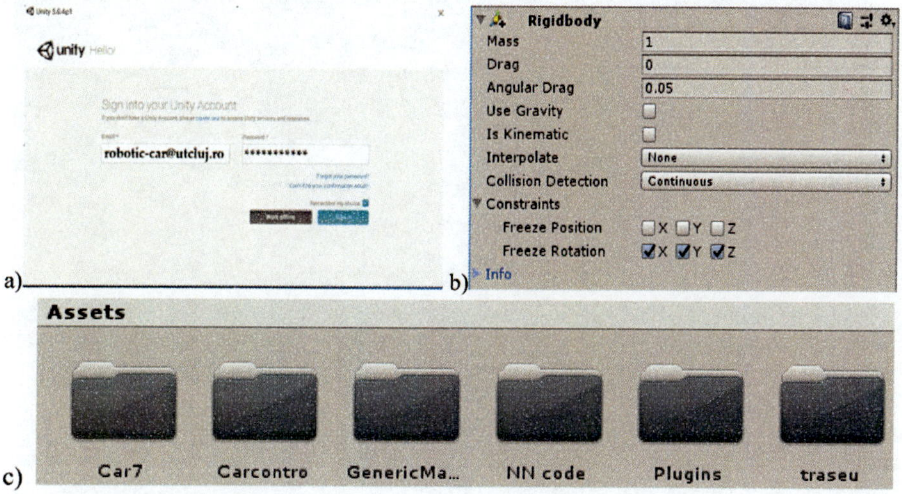

Fig. 4. Access window in Unity 5.6.4. application for entering program (a), parameter setting tool for rigid body - mass, kinematics, and collision detection (b), and main panel with assets from Unity Engine software showing available tools/devices (c)

Figure 5a shows testing track. A hilly environment is used for 3D map definition. Safety level of proposed test track is rather GOOD or SAFE. A little shift toward a colorful sight and virtual environment is completely defined. There is a starting point of test on virtual track and an end point. Tools for configuring testing procedure and SAFETY level are given in right part of window. complete reconfiguration of test track is possible but has a little significance at present moment. Figure 5b shows a sky-view of test-track in a simplified scheme, which may be apprehended one way or other. There are six corners and 12 curves which define the track on which robotic car must operate self-driving task. Virtual modelling of environment facilitates specific creation of objects and infrastructure to define the track. There is generated a multi-cornered test-track to evaluate how robotic car is performing on different road sectors. During tests on virtual track, vehicle has dynamic operation and automatically interacts with environment.

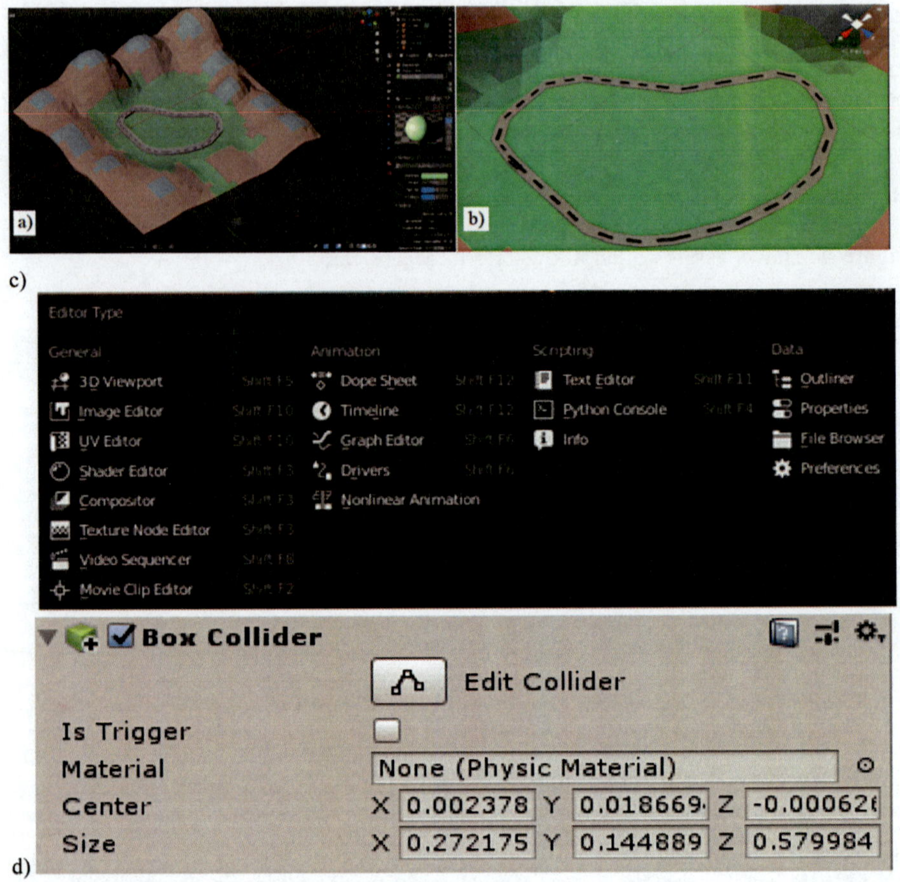

Fig. 5. Unity application window with specific tools for test track creation (a), sky-view of test track Unity window (b), editor type (c), and Box Collider (d).

Figure 6c shows Unity Editor module which allows access to 3D Viewport, Drivers package, Python console gate, Texture node editor and sixteen more shortcuts. There are four main groups of commands: General applications, Animation tools, Scripting editors, and Data managing instruments. In Fig. 6d is presented Box Collider setting tool. This is based on supposition that generated model has outer characteristics identical with its dimensions. test-track was created through Blender application, due to multiple technologies offered and advanced tools available compared to Unity Engine. Some parameters are collected in system with Rigidbody instrument tool. Boxcollider is important. In its absence Mesh collider would not record any impacts of automated vehicle. Mass, drag, angular drag, and gravity effect upon robotic vehicle are virtual values. They are optional features due to requirement on processing power. Insertion of any characteristic data in the model demands more processing power.

Figure 6a presents virtual ambulance in Unity program, for testing protocol. Figure 6b shows Scene view for project design in VR environment. It is unique method through which developer and designer may see and access virtual components in working stage of project.

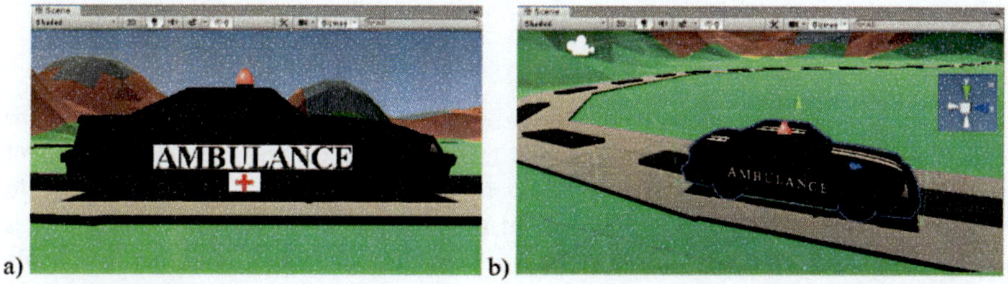

Fig. 6. Lateral view of virtual model used for programing robot-car. (a) and sky-view of test track Unity design (b).

Figure 7 shows code script lines that are programing automated ambulance.

a)

Car Model	Characteristic data			
	Parameter	*Value*	*UM*	*Observation*
Tourism [a]	Model	2020	-	Virtual
	Powertrain	3000	cm^3	GM
	Power output	200	kW	at flywheel
	Gear-box	6-speed	ratios	manual
	Body	Coupe	-	refined
	Length	4774	mm	-
	Width	1878	mm	-
	Height	1358	mm	-
	Curb weight	1524	daN	-

[a] A 5 seats passenger car.

b) Unity Script | 1 reference

```
public class CarController : MonoBehaviour
{
    private Vector3 startPosition, startRotation;
    private NNet network;
    [Range(-1f,1f)]
    public float a, t;
    public float timeSinceStart = 0f;
```

c) 1 reference

```
private void InputSensors() {

    Vector3 a = (transform.forward+transform.right);
    Vector3 b = (transform.forward);
    Vector3 c = (transform.forward-transform.right);

    Ray r = new Ray(transform.position,a);
    RaycastHit hit;
    if(Physics.Raycast(r, out hit )) {
        bSensor = hit.distance/30;
        Debug.DrawLine (r.origin, hit.point, Color.red);
    }
    r.direction = b;
    if(Physics.Raycast(r, out hit )) {
        aSensor = hit.distance/30;
        Debug.DrawLine (r.origin, hit.point, Color.red);
    }
    r.direction = c;
    if(Physics.Raycast(r, out hit )) {
        cSensor = hit.distance/30;
        Debug.DrawLine (r.origin, hit.point, Color.red);
    }
}
```

Fig. 7. Specific data (a) and program script for ambulance controller (b) and sensors (c).

Experimental research has been supported by a General Motors model of 2020 with powertrain organized as a rear wheel drive (RWD) designed on basis of technical data offered in Fig. 7a. Analyzing applied part of program, and its significant effects upon robotic vehicle's behavior and its interactions with virtual reality environment, supports high definition of entire design. It is offering reasonable arguments to further out some observations and conclusions on researched problem. fuzzy regulators and corresponding techniques were applied and proved quite useful in defining auxiliary strategies, plans and alternative solutions to digital programming problems. Even in situation of having a package with available information from test-track and from environment it has little impact on over-all program definition due to importance of kinematic and dynamic factors. Although foreseen outcome is vague at first glance it keeps challenge present: to create a decision-making algorithm and to facilitate machine learning process. Figure 7b is presenting programming code in C#, for digital definition of automated vehicle functions and environment features.

5 Conclusions

Programming a robotic car with virtual reality and using of artificial intelligence for control improves travelling experience and transportation process. Time management may also be optimized. Regarding operational conditions of robotic car programed to travel between starting point and end point of testing track in an automated process may be said that neural networks and Fuzzy Regulator are both admissible for simple applications. Each of them is a useful AI instrument for programing and controlling robot car on testing track in VR environment. Such virtual model of robotic application is suitable for prospective studies. Robotics and automation processes are useful for transportation and road traffic in high density urban areas.

Also, they are suitable for long trips with a clear plan made prior to travel and known input data. most important findings are specified as follow: 1. Using A.I. in programing of vehicle automated driving and control has a significant influence upon road traffic safety, by event reduction; 2. Programing robotic vehicle controls generated at some point required data for identification of optimal trajectories with high degree of safety coefficient; 3. Fuzzy techniques and regulators, as well as neural networks, applied in virtual reality programing supports soft direction changes in automation of robot car; 4. Variation of real values at input side, imposes application of permanent monitoring with proximity sensors; 5. Analyzed robotic vehicle has managed to follow test-track with minor revisions in programing and with few adjustments of proximity sensors on car; 6. Experimental method has provided all important information and procedures for robotic vehicle to follow test-track and to overpass challenges in a safe manner, using properly defined environment inside VR application; 7. In created group of developed scenarios were included some follow line procedures. There were applied strategies of avoiding road traffic events. Using proximity sensors and Mash collider were applied Kinematic changes to control robot's behavior. It was a process of driving along with autonomous robotic ambulance on a testing-track to study its capability to keep up on road line and to drive by near objects. process was done successfully and required minor revisions to operational parameters near curves; 8. More research, scientific considerations and development of robotic ambulance are required to be done to properly improve controlling process. Data quality and resolution are important and vital for AI performance.

Conflict of Interest. Authors declare that they have no conflict of interest.

References

1. Antal, T.A.: Programming Canvas X Pro 16 using scripting technologies. Acta Technica Napocensis Ser. Appl. Math. Mech. Eng. **58**(2), 151–156 (2015)
2. Arghir, M.: Sound pollution evaluation based on Fuzzy techniques. Acta Technica Napocensis Ser. Appl. Math. Mech. Eng. **56**(1), 7–12 (2013)
3. Arghir, M., et al.: Fuzzy techniques used for sound pollution evaluation. Acta Technica Napocensis Ser. Appl. Math. Mech. Eng. **58**(1) (2015)
4. Andrei, L., Băldean, D., Borzan, A.I.: Applied measurements and instrumentation for improving diagnostic devices and systems in metropolitan polluted environments with nitric and carbon oxides. In: Vlad, S., Roman, N.M. (eds.) 6th International Conference on Advancements of Medicine and Health Care through Technology; 17–20 October 2018, Cluj-Napoca, Romania. IP, vol. 71, pp. 45–49. Springer, Singapore (2019). https://doi.org/10.1007/978-981-13-6207-1_8
5. Băldean, D.L.: Studies and research of the N.V.H. aspects concerning exploitation mode of road vehicles. Ştiinţă şi Inginerie/Sci. Eng. **31**(60) (2017). http://stiintasiinginerie.ro/31-60
6. Borzan, A.I., et al.: Application of advanced engineering methods in studying a road traffic event between a 12-wheeler truck and a small tourism in a local junction from Cluj-Napoca. Romanian J. Autom. Eng. (RoJAE) **24**(3), 97–106 (2018). http://siar.ro/wp-content/uploads/2018/12/RoJAE-24_3.pdf#page=5. ISSN 2457–5275
7. Borzan, A.I., et al.: Development of a new interface for intelligent control of energy supply in dynamic environment. Procedia Manuf. **46**(1), 914–921 (2020)
8. Covaciu, F., Băldean, D.-L.: Contribution to research the applied engineering protocol to implement a fuzzy regulator for autonomous driving of an automotive model implemented in virtual reality. In: Dumitru, I., Covaciu, D., Racila, L., Rosca, A. (eds.) SMAT 2019, pp. 468–476. Springer, Cham (2020). https://doi.org/10.1007/978-3-030-32564-0_54
9. Covaciu, F., et al.: Developing communication of autonomous vehicles controlled with aid of artificial intelligence for person and capital safety. In: Personal Safety and Social Capital Construction, pp. 478–484. Universul Juridic, Arad (2020)
10. Covaciu, F., et al.: Design and control of a mobile robot of exploration, using virtual reality as a simulation environment. Ştiinţă şi Inginerie/Sci. Eng. **34**(43), 357–364 (2018). https://stiintasiinginerie.ro/34-43
11. Covaciu, F., et al.: Robotic art in creation and development of innovative shapes and programs for automated driven cars. J. Soc. Media Inquiry **2**(1), 22–39 (2020)
12. Chebly, A., et al.: Coupled longitudinal and lateral control for an autonomous vehicle dynamic modeled using a robotics formalism. IFAC Pap. **50**(1), 12526–12532 (2017)
13. Ferenti, I., et al.: Artificial intelligence implemented in rally vehicle for increasing energetic efficiency in competitions. Ştiinţă şi Inginerie/Sci. Eng. **34**(18), 137–146 (2018). http://stiintasiinginerie.ro/34-18
14. Hodor, A., et al.: Some considerations about 3D replication of complex surfaces. Acta Technica Napocensis Ser. Appl. Math. Mech. Eng. **56**(1) (2013)
15. Jovrea, S., et al.: Researching on-board display of esential informations concerning technical conditions in operation and fuel-economy of a motor-vehicle in operation. Ştiinţă şi Inginerie/Sci. Eng. **31**(67) (2017). http://stiintasiinginerie.ro/31-67
16. Marincaş, C., et al.: Contributions to the experimental research of electronic diesel control (EDC) module operation in relation with supply of the N47 engine from BMW 320d (e90) automobile. Ştiinţă şi Inginerie/Sci. Eng. **31**(82) (2017). http://stiintasiinginerie.ro/31-82

17. Moldovan, A., et al.: Experimental research of the management system from the SUV PEUGEOT 4007 vehicle. Ştiinţă şi Inginerie/Sci. Eng. **31**(71) (2017). http://stiintasiinginerie.ro/31-71
18. Wael, A.A., et al.: Investigation of Non-Binary Trellis Codes Designed for Impulsive Noise Environments. https://doi.org/10.13140/RG.2.2.15332.32647
19. Wang, X., et al.: Lateral control of autonomous vehicles. Control Eng. Pract. **34**(1), 1–17 (2015). https://doi.org/10.1016/j.conengprac.2014.09.015
20. LiDAR and Optical Filters – Helping Autonomous Vehicles See More Clearly, IRIDIAN, Special Technologies, Enabling autonomy. Accessed 7 Mar 2020

Current Trends in Assistive Upper-Limb Rehabilitation Devices

Anca Iulia NICU(✉) and Claudia Steluta Martis

Technical University of Cluj-Napoca, 400114 Cluj-Napoca, Romania
anca.nicu@ethm.utcluj.ro

Abstract. Due to the high incidence of health problems generated by strokes or other accidents that lead to loss of functionality of parts of the body, it is necessary to integrate patients into rehabilitation programs. These programs should include, in addition to classic treatment techniques, rehabilitation techniques with the use of specialized equipment, preferably portable and which can be used by the patient even at home. The present paper presents current trends in assistive upper-limb rehabilitation devices in order to identify new research topics in the development of such devices.

Keywords: Rehabilitation · Drive system · Actuator

1 Introduction

According to the latest statistics available from Institute for Health Metrics and Evaluation (2017) cerebrovascular accident (stroke) is the second common death illness in Romania after Ischemic heart disease [2], and Romania is on the third place in the top of the most new stroke cases and second on death and disability cases due to this affection in Europe.

"Current guidelines for the management of acute stroke recommend a course of treatment based on the diagnosis of ischemic stroke (versus haemorrhagic stroke) made using computed tomography (CT) scanners" [3]. In neurology, strokes are on the first-place due frequency, thus the urgent need for prevention. Both patients and doctors are focusing on rehabilitation technics to regain full mobility of the affected limbs.

1.1 Current Trends in Rehabilitation

The bulletin of World Health Organization (WHO) revealed that globally 87% of both stroke-related deaths and disability-adjusted life years occur in low- and middle-income countries. The stroke incidence in middle-income countries has more than doubled in the past four decades. The major problem is that it affects individuals that are at the peak of their productive life. Although there is a major impact on the socio-economic development of the middle-income countries, there is still very little attention to this subject [3].

In Romania, the free rehabilitation programs offered to patients in need covers only 30% of the total number of patients that must regain limb mobility after stroke. Alternative medical devices that can be partially or totally wearable is a good

S. Vlad and N. M. Roman (Eds.): MEDITECH 2020, IFMBE Proceedings 88, pp. 355–361, 2022.
https://doi.org/10.1007/978-3-030-93564-1_39

opportunity for partial or total recovery of the limb functionality with assisted help of the qualified medical personnel [1].

Since 1960 there has been an interest in improvement of actuators technology used for rehabilitation purposes, but the scientific community encountered a problem in meeting the appropriate energy requirements. Due to the poor research founding in this area there were no major discoveries until the beginning of 2000.

In the past two decades there has been an increasing interest toward rehabilitation and regaining movement of body parts after strokes or other types of injuries to accelerate the healing process of the human body. Unfortunately, most of the scientific papers presents lack of equipment details, standardization, and result evaluation after including the rehabilitation device into the patient daily recovery routine [4–7]. In [8] Maciejasz et al. have conducted a major review of 120 medical assistive upper limb devices to determine the trend from no wearable devices to wearable devices, without considering the computer systems used to drive those devices.

1.2 Benefits of Upper Limb Rehabilitation Devices

The first question to ask is "what is it good for? What is the purpose of using rehabilitation devices?" The answer is simple: the need of regaining full/partial mobility of the individual upper limb after stroke or injuries increased innovation in assistive technology. Those systems could provide benefits both to the patients and the doctors when rehabilitation is needed for long periods.

The study performed by Albert [20] revealed that only 40% of all patients recovering after stroke could return to a full and complete life and one of three remains dependent on support and care. In both cases, the goal is to recover partial or total function of the affected limb and efficient and highly qualified medical staff is required. Due to the reduced number of specialised staff the rehabilitation devices can be a viable alternative for long time recovery periods.

The general aim in using robotic systems for rehabilitation purposes is related to "increase efficiency reproducibility of the treatment methods while ideally improving the economic situation in rehabilitation" [15].

2 State of the Art, Material and Methods

Innovation in rehabilitation technology is the key word in our days. Due to the impropriate use of exoskeleton, mechatronic device, mechanical assistive device, wearable assistive mechatronic device etc., there are a lot of devices that are difficult to include in certain standards. However, there were several attempts to standardize those devices using different criteria [11].

There are two main direction in this area of research: assistance (A) and rehabilitation (R) of the individual. Thus, the developed devices can be used for rehabilitation (R), assistance (A) or both (A&R).

In [11] Varghese et al. provided a state of the art in design and actuation, reviewing the robotic devices used in upper limb rehabilitation and assistance (see Fig. 1). The study reviewed 198 scientific papers from 1998 until 2016.

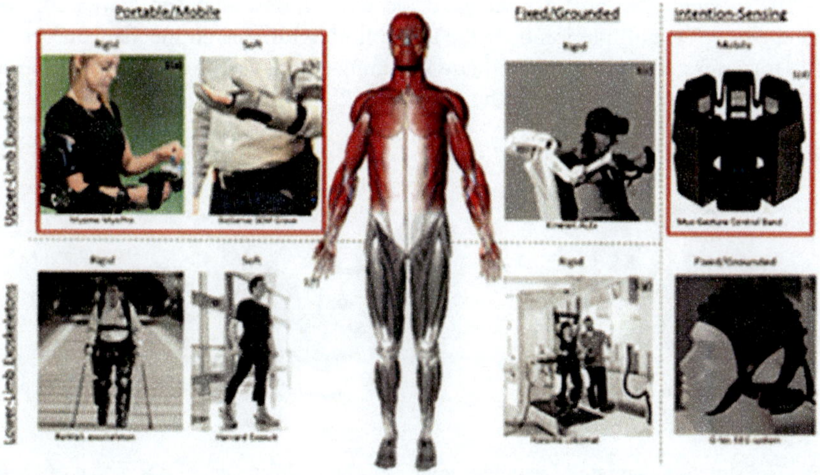

Fig. 1. Types of exoskeletons used for rehabilitation and new intention-sensing technologies [11]

They concluded that only a few systems for upper-limb rehabilitation are fully wearable and portable. Thus, the trend in rehabilitation domain is to develop equipment to be more efficient, portable, easy to manipulate and able to assist individuals in completing activities of daily living (ADL) at home.

The reviewed characteristics of the systems were: (i) type (rigid/flexible); (ii) if and how it supports movements; (iii) the actuation scheme used; (iv) type of application designed for. In this purpose, a selection of devices is presented in Table 1.

Table 1. Selection of rehabilitation shoulder devices

Bibliography	Year	Country	Type	Support movements	Actuation scheme	Application
[11]	2013	Italy	Rigid	Shoulder – Flexion/Extension (FE), Abduction/Adduction (AA) Elbow – (AA)	Elastic elements or DC brakes, FES (optional), DC motor (optional hand orthosis)	Assistance
[12]	2012	USA	Soft exoskeleton	Shoulder – A/A	Electric; 2 × DC motor + series elastic element (compliant brace and system)	Assistance & rehabilitation
[13]	2006	China	Rigid, Orthosis Prototype	Shoulder – FE, AA; Elbow – FE; Forearm – PS	Electric; 3 × AC motors + 2 × DC motors	Rehabilitation
[18]	2012	Portugal	Soft-Orthotic System Prototype	Shoulder – FE, AA	Actuation and controller located on table; bowden cables to minimize weight	Assistance
[19]	2020	China	Exoskeleton	Shoulder – FE, AA Elbow – FE,	cable-driven motors, servomotors (57 for shoulder, 42 for elbow)	Passive rehabilitation training

In 2020, Pang, Z et al. [19] designed, developed, and analysed a wearable rehabilitation robot for multiple joints of the upper limb with tension mechanism based on flexible transmission during rehabilitation training process. The diagram for the joint used in the designed prototype is shown in Fig. 2.

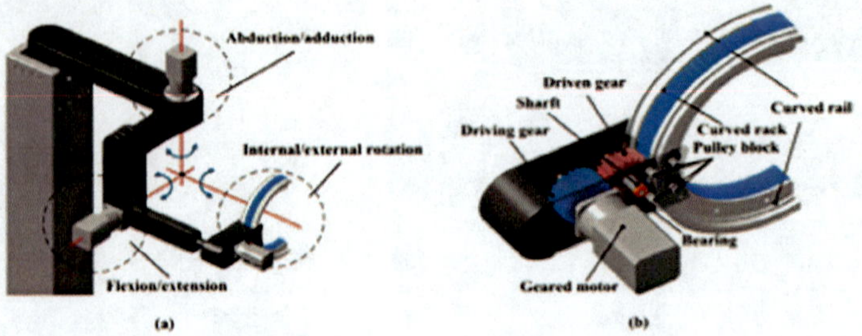

Fig. 2. The structure diagram of shoulder joint. (**a**) Three-dimensional (3D) model of shoulder joint; (**b**) 3D model of shoulder joint rotation degree of freedom [19]

2.1 Control Systems

The control system used in manipulating the rehabilitation device is mandatory for enhancing the appropriate results. It manages the actuator behaviour, using the joint position as biomechanical signal. To control velocity, torque, and force outputs the system includes both open and close loops (Fig. 3).

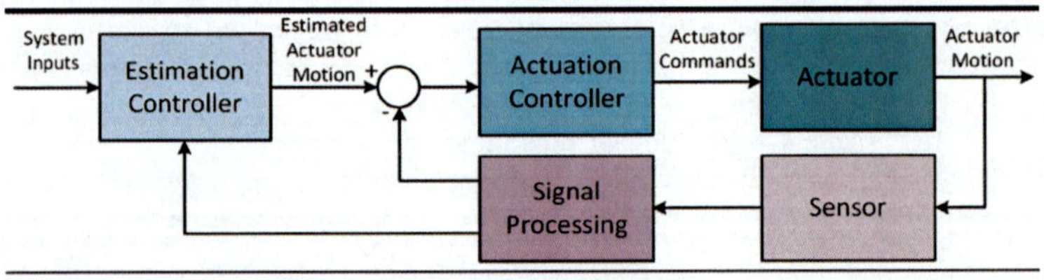

Fig. 3. A generalized control system block diagram for wearable assistive devices [5]

In [10] the PID control was performed in closed loop. There are other ways to control the system, for example complex algorithms that consider EMG signals to command a fully exoskeletal arm for daily base usage [15].

In [18] EPOS-2 24/2 digital positioning controller from Maxon Motors was used to control the position of the device. The misalignment identification process makes the brace easy to put on, and it does not require an exact position of placement on the individual upper limb.

2.2 Computer System Integration and Software

There are few articles that briefly describe the computer system used to command the rehabilitation device, no mater of the application [16]. The communication protocols are described as standard, and no explanations are given. The laptops and personal computers are generally integrated into the system, and more often to visualize the whole process programs like MATLAB or LabVIEW are used by engineers.

3 Discussions

Most of the applications related to upper limb rehabilitations are used for finger rehabilitation (50%) or elbow rehabilitation. Few are developed for shoulder and so far, we did not encounter in the scientific literature a full rehabilitation system for the upper- limb. The observed results revealed that there is a trend in developing finger rehabilitation devices, since it can be designed to be wearable, used in home environment with minimum assistance form the specialized personnel while the sophisticated rehabilitation device used for shoulder injuries is rare.

To increase portability for wearable devices two types of batteries were used LI-Io and Li-Po, with good results in powering the electronics and the motors of the described devices. There is still much to work in reducing weight and size of the upper limb rehabilitation device if portability is one of the major desires of the future trend.

Patient satisfaction with respect to medical care is also a very important part of rehabilitation process [21, 22].

The number of applications increased during the last decade, but most of the devices are at the prototype level and there are no reliable results obtained after usage. The market in this field is growing and the research funding can offer the opportunity to obtain good results, but the medical staff are still sceptical in using devices that are still in testing phase. So far, there is only one commercially available upper-limb wearable assistive device MioPro [17].

Since most of the scientific results do not present sufficient evidence of how the subsystems are integrated into the rehabilitation device, it is unlikely to be able to duplicate the research using only the giving data. Lack of the statistical data after usage, and poorly integration of the rehabilitation devices into clinic daily rehabilitation processes makes this field of research less attractive. A significant attention should be focused on acquiring the repeatability of the movement.

Discussions about safety in using such devices are most of the times excluded from the scientific papers, being considered as later-stage requirements although it is an important part of the already developed device. Creating interdisciplinary teams of doctors and engineers could conduct to a different result.

4 Conclusion

The paper presents current trends in assistive rehabilitation device used in shoulder rehabilitation. Although many scientific articles were at the base of this research only sixteen researched upper-limb rehabilitation devices, most of the scientific papers being focused on wrist and finger rehabilitation techniques.

The increasing focus on the assistive technology is due to the need of a healthy, well-functioning healthcare system where even elderly or individuals with motor disabilities could find their independence.

Currently, there are only few approved systems on the market but there are many devices developed and realized for research purposes.

Lack of standardization and poor statistic data makes this subject delicate to conclude. In the last two decades, engineers have tried to develop furthermore assisted devices that can help in shortening the rehabilitation periods mainly for athletes but also for patients recovering from strokes.

To have results, the already developed devices should be placed into clinical trials and if the results are as expected, they should be put on market. There is still a lot of insecurity regarding the integration of these assisted devices into the day-to-day paces for rehabilitation but with proper promotion and further development, it could be the future of rehabilitation field.

After reviewing 37 devices presented in scientific papers, only 5 devices were designed for shoulder rehabilitation.

Conflict of Interest. The authors declare they have no conflict of interest.

References

1. State of Health in the EU Romania Country Health Profile 2019. https://www.euro.who.int/__data/assets/pdf_file/0009/419472/Country-Health-Profile-2019-Romania.pdf. Accessed 15 Sept 2020
2. IHME. http://www.healthdata.org/romania. Accessed 1 Oct 2020
3. Johnson, W., Onuma, O., Owolabi, M., Sachdev, S.: Stroke: a global response is needed. Bull. World Health Organ. **94**, 634A (2016). https://doi.org/10.2471/BLT.16.181636
4. Manna, S.K., Dubey, V.N.: Comparative study of actuation systems for portable upper limb exoskeletons. Med. Eng. Phys. **60**, 1–13 (2018). https://doi.org/10.1016/j.medengphy.2018.07.017
5. Despenser, T., Zhou, Y., Edmonds, B.P.R., Lidka, M., Goldman, A., Trejos, A.L.: Rehabilitative and assistive wearable mechatronic upper-limb devices: a review. J. Rehabil. Assist. Technol. Eng. (2020). https://doi.org/10.1177/2055668320917870
6. Lum, P.S., Burgar, C.G., Shor, P.C., et al.: Robot-assisted movement training compared with conventional therapy techniques for the rehabilitation of upper-limb motor function after stroke. Arch. Phys. Med. Rehabil. **83**, 952–959 (2002)
7. Volpe, B.T., Lynch, D., Rykman-Berland, A.: Intensive sensorimotor arm training mediated by therapist or robot improves hemiparesis in patients with chronic stroke. Neurorehabil. Neural Repair **22**, 305–310 (2008)

8. Maciejasz, P., Eschweiler, J., Gerlach-hahn, K., et al.: A survey on robotic devices for upper limb rehabilitation. J. NeuroEng. Rehabil. **11**, 1–29 (2014)
9. Brackbill, E.A., Mao, Y., Agrawal, S.K., et al.: Dynamics and control of a 4-DOF wearable cable-driven upper arm exoskeleton. In: IEEE International Conference on Robotics and Automation, Kobe, Japan, 12–17 May 2009, pp. 2300–2305 (2009)
10. Varghese, R., Freer, D., Deligianni, F., Liu, J., Yang, G.-Z.: Wearable robotics for upper-limb rehabilitation and assistance (2018). https://doi.org/10.1016/B978-0-12-811810-8.00003-8
11. Pedrocchi, A., et al.: MUNDUS project: MUltimodal Neuroprosthesis for daily upper limb support. J. Neuroeng. Rehabil. **10**(1) (2013). Article number: 66. https://doi.org/10.1186/1743-0003-10-66
12. Li, Q., Wang, D., Du, Z., Song, Y., Sun, L.: sEMG based control for 5 DOF upper limb rehabilitation robot system. In: 2006 IEEE International Conference on Robotics and Biomimetics, ROBIO 2006, pp. 1305–1310 (2006)
13. Vaca Benitez, L.M., Tabie, M., Will, N., Schmidt, S., Jordan, M., Kirchner, E.A.: Exoskeleton technology in rehabilitation: towards an EMG-based orthosis system for upper limb neuromotor rehabilitation. J. Robot. **2013**, 1–13 (2013). https://doi.org/10.1155/2013/610589
14. Sivan, M., et al.: Home-based Computer Assisted Arm Rehabilitation (hCAAR) robotic device for upper limb exercise after stroke: results of a feasibility study in home setting. J. Neuroeng. Rehabil. **11** (2014). Article number: 163. https://doi.org/10.1186/1743-0003-11-163
15. MYO PRO. https://myomo.com/what-is-a-myopro-orthosis/. Accessed 1 Oct 2020
16. Galiana, I., Hammond III, F.L., Howe, R.D., Popovic, M.B.: Wearable soft robotic device for post-stroke shoulder rehabilitation: identifying misalignments. In: 2012 IEEE/RSJ International Conference on Intelligent Robots and Systems, Vilamoura, Algarve, Portugal, 7–12 October 2012, pp. 318–322 (2012). ISBN: 978-1-4673-1735-1/12/S31.00
17. Pang, Z., Wang, T., Wang, Z., Yu, J., Sun, Z., Liu, S.: Design and analysis of a wearable upper limb rehabilitation robot with characteristics of tension mechanism. Appl. Sci. **10**, 2101 (2020). https://doi.org/10.3390/app10062101
18. Albert, S.J., Kesselring, J.: Neurorehabilitation of stroke. J. Neurol. **259**, 817–832 (2012). https://doi.org/10.1007/s00415-011-6247-y
19. Constantinescu-Dobra, A., Sabou, A., Coțiu, M.C.: The influence of socio-demographic, psychological and medical variables on patient satisfaction with diabetes care in the hospital setting in Romania. In: Vlad, S., Roman, N.M. (eds.) 6th International Conference on Advancements of Medicine and Health Care through Technology, 17–20 October 2018, Cluj-Napoca, Romania. IP, vol. 71, pp. 277–281. Springer, Singapore (2019). https://doi.org/10.1007/978-981-13-6207-1_42
20. Cepisca, C., Adochiei, F.C., Potlog, S., Banica, C., Seritan, G.: Platform for bio-monitoring of vital parameters in critical infrastructures operation. In: 7th International Conference on Electronics, Computers and Artificial Intelligence (ECAI). International Conference on Electronics Computers and Artificial Intelligence, pp. E7–E10 (2015). ISBN: 978-1-4673-6647-2, WOS: 000370971100009. https://doi.org/10.1109/ECAI.2015.7301144

Health Technology Assessment; Clinical Engineering Assessment; Miscellaneous Topics

Numerical Study on Electropermeabilization of Cell Membranes in Sine-Wave Electric Field of Variable Frequency

A. M. Sandu[1(✉)], M. A. Ungureanu[2], M. Morega[3], V. L. Călin[4], M. G. Moisescu[4], I. A. Paun[2,5], and M. Mihailescu[2]

[1] CAMPUS Research Center, Doctoral School of Electrical Engineering, University Politehnica of Bucharest, 313 Splaiul Independentei, Bucharest, Romania
ana_maria.sandu@upb.ro

[2] Physics Department, University Politehnica of Bucharest, Bucharest, Romania

[3] Faculty of Electrical Engineering, University Politehnica of Bucharest, Bucharest, Romania

[4] Research Excellence Center in Biophysics and Cellular Biotechnology Department, University of Medicine and Pharmacy Carol Davila, Bucharest, Romania

[5] CETAL, National Institute for Laser, Plasma and Radiation Physics, Magurele, Romania

Abstract. Electropermeabilization represents the general process of increasing the permeability of lipid membranes to various substances, facilitating their transition between the indoor and outdoor cellular environment, by exposure to variable electric field; the process is highly dependent, on both the electric field features (amplitude, waveform, frequency and space-distribution) and the characteristics of the target (cell morphology and dielectric properties of the cell membrane). Some of these aspects are evaluated in this study, using numerical simulation with the finite element analysis. The behavior of realistically shaped cells (computation domains built based on holographic imaging techniques) and idealized circular cells is comparatively assessed under uniform distributed, time-harmonic electric field, within a large frequency spectrum. Our results confirm that simulation is a valuable tool for the predetermination and adjustment of the optimal settings for an experimental approach, taking into account the particularities of the used field or cell sample.

Keywords: Electric field · Electropermeabilization · Induced transmembrane voltage · FEM analysis · Realistic cell models

1 Introduction

Natural resting state of a cell is characterized by the transmembrane electric field (TEF) of approx. 10 MV/m [1, 2], a value comparable to the *dielectric strength* for oil or other lipids (10…15) MV/m in static electric field. Physiologic variations of the corresponding transmembrane voltage (TV), due to local ionic transport through the

S. Vlad and N. M. Roman (Eds.): MEDITECH 2020, IFMBE Proceedings 88, pp. 365–373, 2022.
https://doi.org/10.1007/978-3-030-93564-1_40

membrane channels, produce small fluctuations of the natural TEF, but do not cause major magnitude changes of this electrical permanent stress. An externally applied electric field is, however, capable to enhance the stress up to thresholds that favor the formation of new transmembrane aqueous pores, used by permeant and nonpermeant substances (ions, water, drugs, genetic material etc.) to cross the lipid bilayer. This biophysical process is known by the generic name of electropermeabilization (EP) and, depending on the strength, waveform, frequency, duration of the applied electric field and other local settings, it can be reversible (useful for various bio-medical interventions) or irreversible (up to the destruction of the cell) [3]. During successful nondestructive EP, the externally *induced transmembrane voltage* (ITV) might rise up to 1 V and it overlaps the natural TV, commonly oscillating between resting and active states, within the range (−0.07…0.03) V. Although the externally applied electric field sums up (or decreases from, depending on polarity) the natural electric field of the membrane, EP technologies assume that the two phenomena are based on different biophysical grounds and they act on different magnitude scales, so they could be independently treated [1].

Experimental research shows that the amplitude of the applied electric field strength (E), able to initiate EP of the plasma membrane is frequency-dependent; magnitudes in the order of 0.1 MV/m are needed at low frequencies (LF), *i.e.* up to approx. 100 kHz and magnitudes in the order of 10 MV/m are effective for high frequencies (HF), *i.e.* over approx. 10 MHz, showing that membrane bio-electric behavior is highly influenced by the frequency of the applied field [4, 5]. For medium frequencies (MF), *i.e.* within 100 kHz…10 MHz range, the relationship between the effective E amplitude and frequency shows a nonlinear trend.

Most EP experiments are conducted under pulsed electric applied stress [2, 6]; if the magnitude of E is high enough (above the EP threshold), EP process is triggered rapidly and pores generation occurs on the rise time and plateau of the first pulse. For this reason, the relatively wide harmonic content of the respective interval should be associated with the spectral behavior. Micro-pulses trigger EP by their LF and MF content and they target the lipid membrane, while nano-pulses induce EP under their HF spectral content and penetrate inside the cell; the latter triggers the EP of inner membranes, such as those of endoplasmic reticulum or lipid vesicles introduced in the cell for drug delivery [1, 5, 7]. Excitation in continuous wave (CW) is only recently considered as an efficient alternative to pulses, although the advantage of its narrow bandwidth makes the spectral analysis more precise [5, 7, 8]. Since the cellular components (membrane, cytoplasm, extracellular fluid) are characterized by frequency-dependent dielectric properties, it explains the observed variations in membrane sensitivity to electric stress and the variety of associated bio-medical applications [9, 10].

The paper aims to explore, by an electric field analysis and some numerical experiments, the influence on EP conditions of the cell morphology (shape and size) and of the CW electric field frequency. The physical quantity, which has been correlated to the EP triggering is the transmembrane voltage, induced before starting the dynamic EP process (ITV, as mentioned before).

2 Numerical Models

Numerical simulation could show its utility for the optimal predetermination of the experimental EP procedure settings; it could also offer hints for an efficient adjustment of the selected features. Mathematical modeling provides deterministic, predictable and repetitive solutions, and, according to its complexity, it could simulate personalized realistic structures and exposure conditions, giving a rapid access to post-processing outputs; these advantages are precious especially in cellular micro-dosimetry.

2.1 Holographic Cell Images

Digital holographic microscopy (DHM) represents a noncontact, marker free technique suitable for monitoring of rapid cellular processes and for visualization of transparent samples without scanning (such as cells in their natural environment) [11]. B16F10 murine cells, an in vitro model for cutaneous melanoma were cultured in high glucose Dulbecco's modified Eagle's medium (Sigma, EU), supplemented with 10% fetal bovine serum (Sigma, EU). 24 h prior to experiments, the cells were detached with trypsine (0.5% Porcine Trypsin 4NaEDTA, Gibco, EU) and seeded at low density on microscopic slides. Cultures with high ratio of cells in M and G1-S cell cycle phases were obtained by a synchronization treatment (100 ng/l Nocodazole or 1 mM Mimosine, respectively, were added to the culture medium for various periods of time). While progressing through different cell cycle phases the shape of the cells was changing.

Holograms of these cells were acquired in an experimental setup for DHM, based on a Mach-Zehnder interferometer, in *off-axis* configuration, with the sample inserted in one arm. The CCD camera records holograms formed at reference and object beams superposition. These holograms become phase images (PI) by digital reconstruction using dedicated commercial software Koala^{R} based on the diffraction integral. PIs contain in every pixel values of the optical phase, parameter that is proportional to the refractive index and the height of the cell. PIs differ from images obtained by other microscopic techniques where the values in each pixel are proportional with its intensity and color [12]. PIs were used as starting point to build the numerical models for EP simulations (Fig. 1) [13].

Cell_A Cell_B

Fig. 1. PIs resulting from holograms reconstruction and after segmentation for two cells; the morphologies presented here are ready for transfer to the FEM software. The cells have different shapes and dimensions: Cell_A is almost circular, with a characteristic diameter of approx. 5 μ, while Cell_B has an irregular shape and covers approx. half of the area of Cell_A (not evidenced here, due to scale differences).

2.2 Computational Domain and Electric Field Problem

The PIs were processed and imported for FEM numerical analysis in COMSOL Multiphysics, following the steps: 1) segmentation using MATLAB software (Fig. 1); 2) cell membrane selection; 3) graphic image file conversion in Autocad software; 4) simulation by the FEM analysis software.

Our numerical model of the cell is a 2D representation in Cartesian coordinates, previously assessed [13] with reference to analytical models [14]. Validation and analysis processes were conducted for LF time-harmonic operating conditions, described by Laplace equation for the electric potential (phasor form),

$$-\nabla \cdot (j\omega\varepsilon_0\underline{\varepsilon}_r\nabla\underline{V}) = 0 \tag{1}$$

with the angular frequency $\omega = 2\pi f$, the complex dielectric constant $\underline{\varepsilon}_r$, the vacuum permittivity $\varepsilon_0 = 8.854 \cdot 10^{-12}$ F/m and the complex number $j = \sqrt{-1}$. This model is adequate for the analysis of EP phenomena under CW excitation, over a wide spectral range (LF, MF and HF) as defined in the first section.

The computational domain is built in COMSOL Multiphysics; the cell sub-domain (cytoplasm and membrane) is imported inside a square enclosure (square box of 50 × 50 μ) filled with a solution similar to an extracellular fluid, where the uniform electric field is applied. The 5 nm wide membrane is modeled as a distinct subdomain, which represents a more realistic alternative to the simulation of the membrane by an inner boundary with distributed impedance condition [5, 7].

Electric field penetration inside the cells is studied here for two realistic cell models with particular morphology and size, called Cell_A and Cell_B (Fig. 1). Corresponding circular pairs, built with the same area, are generated for comparison of their behavior. Cell_A is paired with Circular-Cell_A (radius 5.28 μ) and Cell_B is paired with Circular-Cell_B (radius 2.9 μ).

The boundary conditions are set on the walls of the enclosure: constant value of the electric potential V_0 for the top, and ground for the bottom, while the lateral walls limit the space by electric insulation. V_0 is correlated with the dimensions of the box, such that the applied uniform E-field has 1 kV/cm in amplitude.

2.3 Dielectric Properties

Dosimetry approach for assessing the effects of the electric field on human tissues requires considering the influence of frequency on dielectric properties. Mathematical models have been determined from the experimental results by quasi-empirical means, and a good synthesis of these relaxation models is presented by [15]. Debye model, detailed in the same cited work is the most popular relaxation model and defines the complex dielectric constant $\underline{\varepsilon}_r$ as a function of frequency f, with the expression

$$\underline{\varepsilon}_r(f) = \frac{\sigma_s}{j\omega\varepsilon_0} + \frac{\varepsilon_s - \varepsilon_\infty}{1 + j(f/f_r)} + \varepsilon_\infty \tag{2}$$

Based on the Debye general relaxation model, specific properties for the cell components were determined by [16] and validated by [4] within MF and HF ranges. The other quantities in Eq. (2) are defined in Table 1 for the cellular compartments, as proposed by [16] and are adopted in our study too.

Table 1. Dielectric properties of cell components (Debye model) [16].

		Membrane	Cytoplasm & extracellular fluid
Relaxation frequency	$f_r[Hz]$	$179.85 \cdot 10^6$	$17.9 \cdot 10^9$
Static electrical conductivity	$\sigma_s[S/m]$	$1.1 \cdot 10^{-7}$	0.55
Static relative permittivity	ε_s	11.7	67
Residual relative permittivity	ε_∞	4	5

Since the numerical model requires the definition of effective dielectric properties for each sub-domain at respective operating frequency, the expressions

$$\sigma(f) = \sigma_s + \frac{\omega\varepsilon_0(\varepsilon_s - \varepsilon_\infty)}{1 + (f/f_r)^2}(f/f_r) \tag{3}$$

and

$$\varepsilon_r(f) = \varepsilon_\infty + \frac{\varepsilon_s - \varepsilon_\infty}{1 + (f/f_r)^2} \tag{4}$$

are derived for the effective electrical conductivity and dielectric constant, respectively. Figure 2 shows the two nominated functions, for each cell compartment, over the whole nonionizing radiation frequency spectrum.

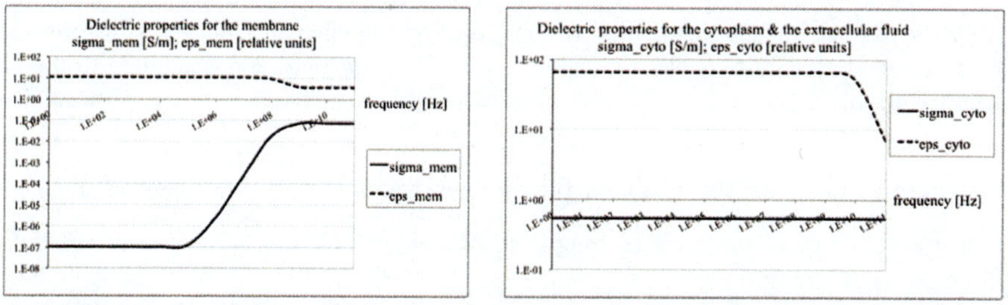

Fig. 2. Frequency-dependent dielectric properties of cell components (Debye models) by [16] (the vertical axis is valid for both quantities; values are in [S/m] for conductivity and dimensionless for dielectric constant and the scales are logarithmic).

3 Results and Discussion

Following the objectives addressed by the present EP study, the frequency influence on the CW electric field penetration inside one single exposed cell is analyzed. A set of characteristics, like: the morphology of the exposed cell and the orientation of the cell (with irregular shape) with respect to the orientation of the field, are considered.

Four structures (Cell_A, Cell_B and their circular pairs described earlier) are analyzed while exposed to the same uniform electric field (CW, 1 kV/cm), with frequency taken as parameter within the range 10 kHz…10 GHz.

Figure 3 illustrates a comparison among the four cell samples with different morphologies; the frequency dependence of the maximum ITV value (ITVmax), identified and calculated in the upper region of each cell is displayed over the entire spectrum (left graph) with a zoom for the HF domain (right graph).

As already known [13], ITVmax values occur in the polar regions, where the E-field vector is perpendicular to the cell's surface. The frequency dependence of ITVmax has a similar trend for all cell morphologies, because it is mainly determined by the membrane dielectric properties. Less revealing, and only for the HF range, are the properties of the other cell components. Dielectric properties confer strong insulating characteristics to the membrane at LF, and lossy dielectric characteristics (with a dramatic rise of the electrical conductivity) at HF, connected by a non-linear trend within the MF spectrum.

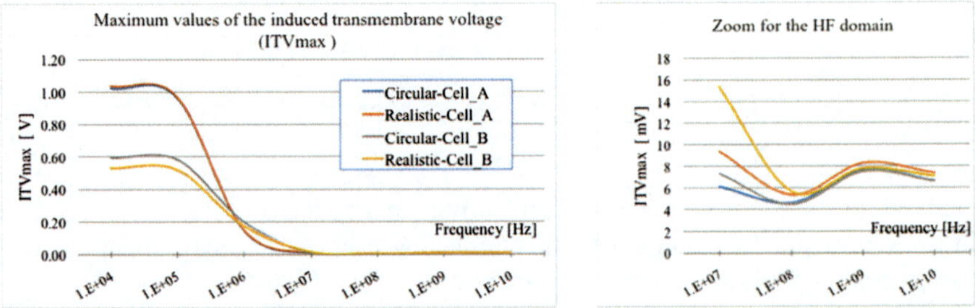

Fig. 3. ITVmax variation with the frequency of the applied electric field (CW, 1 kV/m uniform field) for the four analysed morphologies (two realistic and two circular samples, and respectively two size categories). ITVmax over the entire frequency range (left) and magnified in the HF domain (right)

As Fig. 3 illustrates, the ITVmax function on frequency is characterized by:

(i) a difference of two orders of magnitude between ITVmax values over LF compared to HF range;
(ii) although ITVmax values seem to vanish in the HF domain (Fig. 3 left), they actually are in the order of mV, as shown by the magnified curves (Fig. 3 right); over the HF range, ITVmax lowers slightly nonuniform with the frequency rise, most probably due to the frequency influence on the dielectric constant of cytoplasm and extracellular fluid (see Fig. 2 right).

An improved illustration of this type of membrane response to variable electric field at different frequencies is shown in Fig. 4, presenting the ITV graphs along the entire contour of each cell, at seven frequency values; enlarged scale diagrams of the same ITV highlight the gradual but significant decrease of the ITV amplitude in the HF range.

A complementary observation could be made regarding the electric field distribution in the presence of the cell, for two extreme frequencies: 10 kHz and 1 GHz, considered representative for LF and HF respectively. Figure 5 shows the spectra of the electric quantities for Cell_A exposed to electric field as specified before.

When the EP protocol is designed for a certain cell sample, one could perform such numerical simulations based on the realistic image of that cell, and get a qualified prediction to the outcomes and the necessary settings for the live-cell experiment. Characteristics like those of Fig. 3 offer information on the adjustment of the applied E-field amplitude and frequency, in order to result a certain threshold level in the membrane region targeted by the EP procedure.

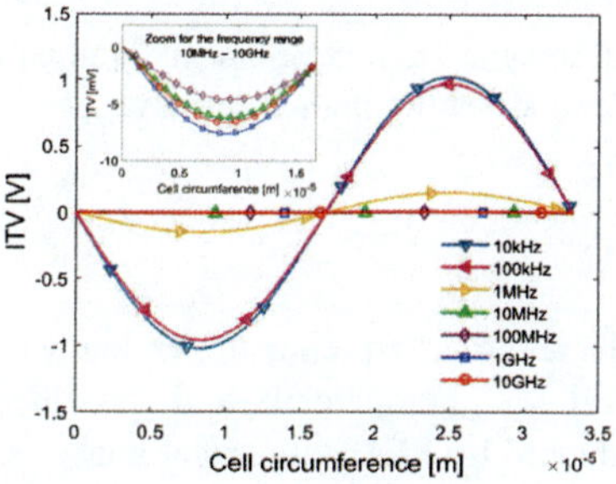

a) Circular-Cell_A

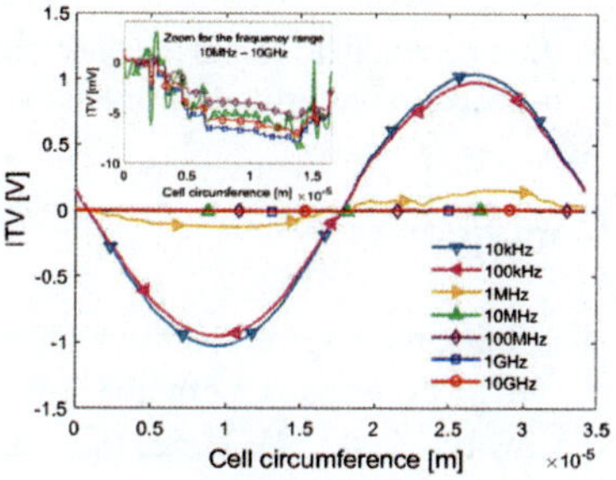

b) Realistic-Cell_A

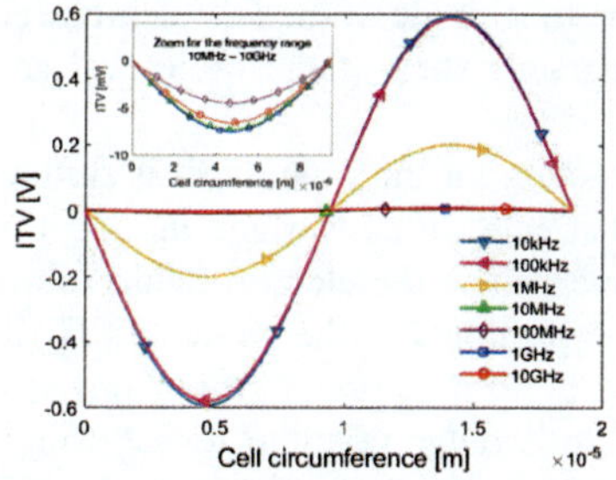

c) Circular-Cell_B

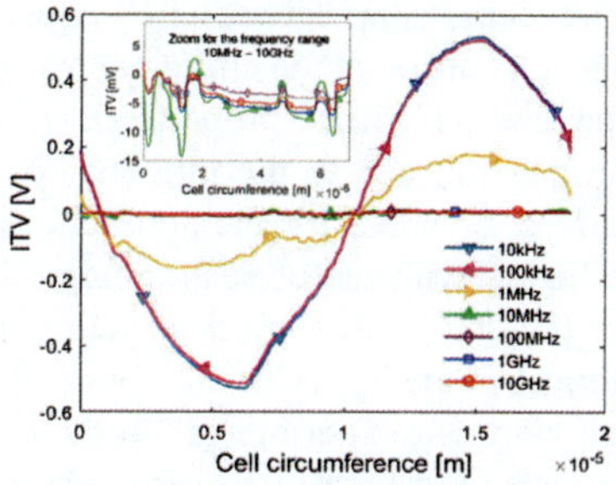

d) Realistic-Cell_B

Fig. 4. ITV along the entire contour of each analyzed cell, under uniform E-field (CW, 1 kV/cm); field frequency is a parameter of the study.

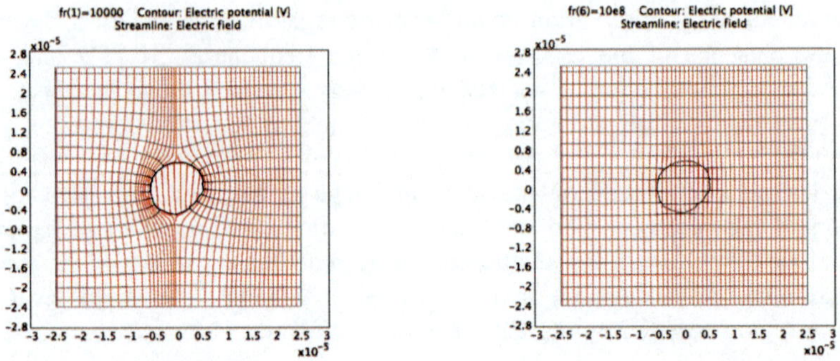

a) 10 kHz

b) 1 GHz

Fig. 5. Equipotential lines (black) and electric field strength (red streamlines) for Realistic-Cell_A exposed to uniform electric field (CW, 1 kV/cm) at low *vs.* high frequency.

4 Conclusions

ITV distribution along the contour of realistic cells and the frequency dependency of its values are in consensus with the results obtained for circular cells and for different frequencies by [1, 4, 14]. Here, the study is conducted by FEM numerical analysis, on realistic cell models of murine cells, obtained from holographic microscopy images after CAD processing. ITV is considered a significant quantity that helps identify the relative electric stress threshold able to trigger EP. The electric field CW is assumed and the relationship between its frequency and the magnitude of the ITV is highlighted, taking into account the influence of specific characteristics of the biological sample (shape, size, dielectric properties).

In general, due to the influence of various factors on the process, it is difficult to identify a threshold for the applied electric field strength for EP triggering, or a direct relationship with the corresponding ITV. However, since the electric field problem is linear (*i.e.* the applied electric field strength is proportional to the generated ITV), the numerical modeling (like the approach presented here) may predict the major settings for a living-cell experimental study, *e.g.* E-amplitude range required for EP triggering at a given frequency in case of specific cellular characteristics; minimal data for calibration are required. These data can be obtained from similar experimental tests, or from an associated statistical simulation of the pore generation process, such as the asymptotic model introduced by [17].

Conflict of Interest Statement. The authors declare that they have no conflict of interest.

References

1. Kotnik, T., et al.: Cell membrane electroporation. Part 1: the phenomenon. IEEE Electr. Insul. Mag. **28**(5), 14–23 (2012)
2. Yarmush, M.L., et al.: Electroporation-based technologies for medicine: principles, applications, and challenges. Annu. Rev. Biomed. Eng. **16**, 295–320 (2014)
3. Kotnik, T., et al.: Membrane electroporation and electropermeabilization: mechanisms and models. Annu. Rev. Biophys. **48**, 63–91 (2019)
4. Liberti, M., et al.: Microdosimetry in the microwave range: a quantitative assessment at single cell level. IEEE Antennas Wirel. Propag. Lett. **8**, 865–868 (2009)
5. De Angelis, A., et al.: A microdosimetric realistic model to study frequency-dependent electroporation in a cell with endoplasmic reticulum. In: 2019 Proceedings of the 49th European Microwave Conference, Paris, France, 1–3 October 2019, pp. 212–215 (2019)
6. Goldberg, E., et al.: Cell membrane electroporation modeling: a multiphysics approach. Bioelectrochemistry **124**, 28–39 (2018)
7. Caramazza, L., et al.: Numerical investigations of CW electric fields on lipid vesicles for controlled drug delivery. In: Proceedings of the 49th European Microwave Conference (EuMC 2019), Paris, France, October 2019
8. Garcia-Sanchez, T., et al.: Sine wave electropermeabilization reveals the frequency-dependent response of the biological membranes. Biochimica&Biophysica Acta Biomembr. **1860**(5), 1022–1034 (2018)
9. Markx, G.H.: The use of electric fields in tissue engineering: a review. J. Organogenesis **4** (1), 11–17 (2008)
10. Amini, M., Hisdal, J., Kalvoy, H.: Applications of bioimpedance measurement techniques in tissue engineering. J. Electr. Bioimpedance **9**, 142–158 (2018)
11. Scarlat, E., et al.: Discriminating the main representatives of the white blood cell species on the basis of the fractal properties of the DHM phase profile. UPB Sci. Bull. **75**(2), 147–154 (2013)
12. Mihailescu, M., et al.: Digital off-axis holographic microscopy: from cells visualization to phase shift values ending with physiological parameters evolution. Rom. J. Phys. **61**(5–6), 1009–1027 (2016)
13. Sandu, A.M., et al.: Realistic models of cultured cells for electroporation simulations starting from phase images. In: Proceedings SPIE 11718, Advanced Topics in Optoelectronics, Microelectronics, and Nanotechnologies X, p. 117180W, 31 December 2020. https://doi.org/10.1117/12.2571092
14. Grosse, C., Schwan, H.P.: Cellular membrane potentials induced by alternating fields. Biophys. J. **63**(12), 1632–1642 (1992)
15. Foster, K.R., Schwan, H.: Dielectric permittivity and electrical conductivity of biological materials. In: Polk, C., Postow, E., (eds.) CRC Handbook of Biological Effects of Electromagnetic Fields, pp. 27–98, CRC Press Inc. (1986)
16. Merla, C., Liberti, M., Apollonio, F., D'Inzeo, G.: Quantitative assessment of dielectric parameters for membrane lipid bi-layers from RF permittivity measurements. Bioelectromagnetics **30**(4), 286–298 (2009)
17. Krassowska, W., Filev, P.D.: Modeling electroporation in a single cell. Biophys. J. **92**, 404–417 (2007)

On the Design of an Isolated Booth for Tuberculosis Sample Collection

Ionut Ulinici[1], Ferenc Puskas[2], Bogdan Gherman[1], Cecilia Roman[3], Iosif Birlescu[1], and Doina Pisla[1](✉)

[1] Research Center for Industrial Robots Simulation and Testing, Technical University of Cluj-Napoca, 400114 Cluj-Napoca, Romania
doina.pisla@mep.utcluj.ro
[2] S.C. Electronic April, 400349 Cluj-Napoca, Romania
[3] Research Institute for Analytical Instrumentation, ICIA, 400293 Cluj-Napoca, Romania

Abstract. The paper aims to present a solution for the safe, sterile sampling and testing procedures of suspected tuberculosis patients, under the form of a robotic assisted automated process. Starting from the current state of tuberculosis management around the world, as well as the current problems which the medical personnel face when dealing with tuberculosis, a novel booth was developed as a response to many of the requirements necessary during the sample collection of airborne pathogens from infected individuals. The booth is designed to provide a much needed addition to the medical environment.

Keywords: Isolation booth · Tuberculosis · Robotic manipulation

1 Introduction

Pulmonary tuberculosis, from hereinafter referred to as TB or TBC, is a pulmonary disease caused by the Mycobacterium tuberculosis bacteria. This bacteria usually proliferates in the lungs (due to the favorable environment), but is not limited to them. If the bacteria spreads through the bloodstream to other organs then TB becomes Miliary TB which is a more advanced stage of the disease [1]. According to the world health organization (WHO) TB is one of the top ten causes of death worldwide, with a total of 1.5 million deaths registered in 2018, including a number of 251000 cases coming from people with HIV. These numbers also make TB the leading cause of death from a single infectious agent, topping more widely known diseases such as HIV/AIDS. The 2018 numbers registered by WHO, also show that in that year 5.7 million men, 1.1 million children and 3.2 million women were positive for TB. Fortunately though TB is curable and preventable [1].

One major issue in TB management is the prospect of TB tests. As TB is a transmittable disease, it is highly important that saliva samples are corrected safely and in a contaminant free environment to ensure the validity of the test and the safety of the sample collecting party (the medical personnel). TB testing, via buccal cavity swabs, is the go to detection method for tuberculosis infections. This method though places

S. Vlad and N. M. Roman (Eds.): MEDITECH 2020, IFMBE Proceedings 88, pp. 374–381, 2022.
https://doi.org/10.1007/978-3-030-93564-1_41

medical personnel at risk, during the sampling procedure and if not done in a sterile environment, can also contaminate the sample which in turn could affect the test validity.

A possible solution to this is the usage of robotic/remote operated devices, for sample handling. While the sample gathering procedure would be left up to the patient, the handling of the testing swab can be done by a sterile robotic device in a sterile environment, thus reducing the chance of sample contamination greatly. One example of such a device is the Mycrobacterium tuberculosis automatic testing system [2], a three degree of freedom device developed for test sample transfer operations. The device is generally composed of a fixed base a 360° rotating table, a lifting joint a forearm rotary joint and the end effector, the device aims to have a 0.25 mm end effector accuracy level. Other non TB related robotic sampling mechanisms such as [3] have been developed, and most of these devices encompass simple robotic structures with high precision gripper designs, used for the manipulation of glass vials or slides. Most of these devices present the robotic manipulation mechanism, with some having incorporated sample depositories.

The paper is structure as follows: Sect. 2 presents the general requirements for the TB sampling collection booth using Analytical Hierarchy Processes (AHP) and decision matrices (PUGH matrix); Sect. 3 describes the collection booth and Sect. 4 presents the conclusions.

2 General Requirements and Justification

The most important characteristics of sterilization methods that must be used to ensure sterile sample collection (used as selection criteria) are presented (Table 1).

Table 1. Selection criteria

Name	Description
Human toxicity	Indicates the level of toxicity (harm) direct/indirect contact with substance/material is to humans
Disinfection cycle	Indicates how long the disinfection of an area/surface takes (ex. A rapid cycle would be between 30–45 min)
Material compatibility	Indicates what materials the substances/materials can interact with, safely
Spatial/surface application efficacy	Indicates the efficiency of the solution to sterilize as large an area as possible
Clinical experience	Indicates the existence of clinical trials for each respective solution
Cost effectiveness	Encompasses the upfront installation, operation and maintenance costs
Environmental friendliness	Indicates the estimated impact level upon the environment
Activation	Indicates the necessity for additional preparation of the solution post installation
Automated application	Indicates the ease of automation of the sterilization process
Odor	Indicates the level of olphactory discomfort/irritation caused to humans
Storage	Indicates the need for special requirements for storage (with regard to expiration date, environmental conditions etc.)

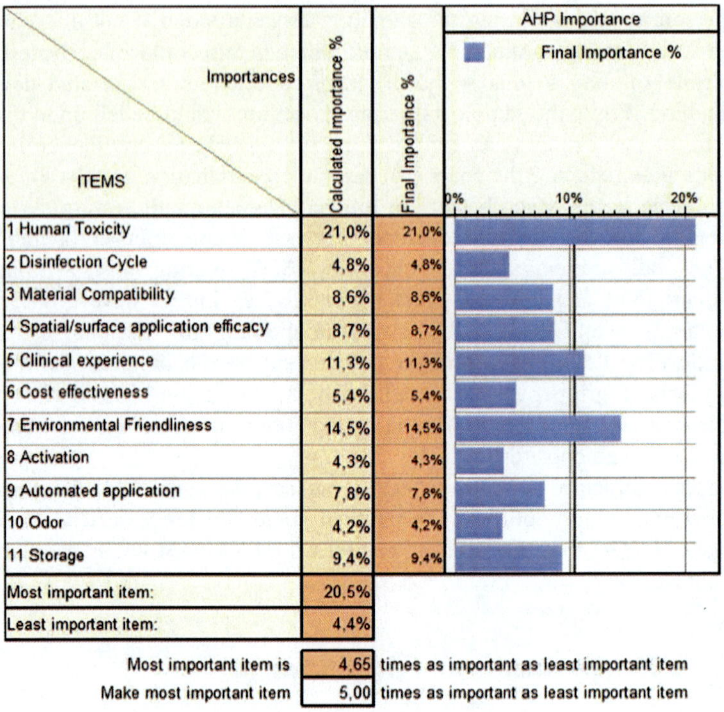

Fig. 1. Selection criteria AHP analysis

Using the Qualica [4] QFD analysis [5] software, an analytic hierarchy process to determine the importance of the aforementioned characteristics was performed (Fig. 1), and the PUGH hierarchization [5] of several sterilization methods was achieved (Fig. 2). From these the best options to use in concordance with the proposed structure have been chosen as system characteristics. Calculated importance of most important item is 4.65× importance of least important item, compared to our desired.

The sterilization methods chosen as possible contenders, were selected from a series of reviews on sterilization technologies [6–10] used in medical environments and are: heavy metal coating (Cu, Ag, etc.), peracetic acid/hydrogen peroxide mixture, glutaraldehyde, hydrogen peroxide, ortho-phtalaldehyde, peracetic acid, improved hydrogen peroxide (2%), steam, hydrogen peroxide gas plasma, 100% ethylene oxide (ETO), ETO mixtures (8.6%, 10%, 8.5% ETO with 91.4%, 90%, HCFC and 91.5% CO_2 respectively), vaporized hydrogen peroxide, ozone, UV light.

Following the results obtained from the PUGH matrix, it has been determined that the technologies most fitting to the predefined selection criteria are, in no particular order: Heavy metal coating, Improved hydrogen peroxide (2%), Steam and UV light. Given the particularities of the application described in this article, the heavy metal coating of all surfaces would be unrealistic as for proper sterilization; all of the inner surfaces of the booth would have to be coated. In a similar way, steam powered sterilization would pose a threat to the robotic system, due to the tendency to leave wet surfaces post-application. As a result UV light and ultrasonic improved hydrogen peroxide (2.0%) pulverization seem to be the most fitting, as they are efficient, and compatible with an automated deployment methods.

ITEMS / New Concepts	1 Heavy metal coating	2 Peracetic acid/Hydrogen peroxide	3 Glutaldehyde	4 Hydrogen Peroxide	5 Ortho-phthalaldehyde	6 Peracetic acid	7 Improved hydrogen peroxide (2%)	8 Steam	9 Hydrogen peroxide gas plasma	10 100% ethylene oxide (ETO)	11 ETO mixtures	12 Vaporized Hydrogen peroxide	13 Ozone	14 UV light	Importance
1 Human Toxicity	++	+	O	+	+	--	+	++	-	--	--	++	++	++	21,0%
2 Disinfection Cycle	++	O	O	O	++	--	++	+	-	--	--	O	O	++	4,8%
3 Material Compatibility	-	-	++	-	++	+	-	-	-	+	+	-	--	++	8,6%
4 Spatial/surface application efficacy	--	+	+	+	+	-	+	O	--	+	+	O	O	-	8,7%
5 Clinical experience	--	-	++	O	-	++	-	O	-	-	-	-	--	O	11,3%
6 Cost effectiveness	O	O	+	O	+	++	O	O	O	O	O	O	O	+	5,4%
7 Environmental Friendliness	+	O	O	+	-	++	++	++	++	--	--	++	++	++	14,5%
8 Activation	++	++	O	++	++	-	++	++	++	O	O	-	O	++	4,3%
9 Automated application	-	O	O	O	O	O	++	++	O	O	O	+	++	++	7,8%
10 Odor	++	+	--	++	++	O	++	++	++	-	-	++	++	++	4,2%
11 Storage	++	O	O	+	-	--	O	++	++	--	--	-	++	++	9,4%
Positive Effects	6	4	4	6	7	4	7	7	4	2	2	4	5	9	
Negative Effects	4	2	1	1	3	5	2	1	5	6	6	4	2	1	
Neutral Effects	1	5	6	4	1	2	2	3	2	3	3	3	4	1	
Net Effect	2	2	3	5	4	-1	5	6	-1	-4	-4		3	8	
Positive Priorization	48,5%	15,6%	24,6%	26,3%	33,6%	34,1%	45,5%	62,8%	32,3%	5,8%	5,8%	42,3%	56,9%	76,4%	
Negative Priorization	-25,5%	-6,6%	-4,2%	-2,9%	-11,7%	-39,5%	-6,6%	-2,9%	-23,9%	-54,8%	-54,8%	-11,2%	-19,9%	-2,9%	
Net Effect	23,0%	8,9%	20,4%	23,5%	21,9%	-5,5%	38,8%	59,9%	8,4%	-49,1%	-49,1%	31,1%	37,0%	73,5%	

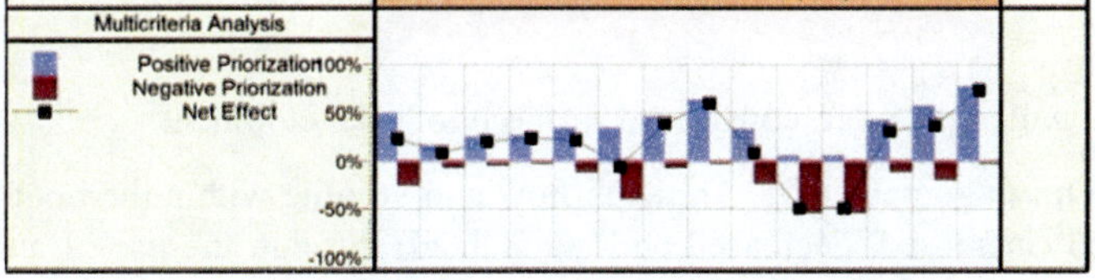

Fig. 2. Sterilization technologies (Pugh matrix)

3 Isolation Booth Description

Based on the AHP analysis (Fig. 1) and the Pugh matrix (Fig. 2) the construction of the sterile booth is in fact the most important aspect in ensuring no sample contamination and providing a safe environment for sample collection. Consequently, the booth was designed in a modular manner in (with modules such as booth unit, filtering and extraction module, automated manipulation and control module) in order to be capable to carry out a fully automated sterilization procedure after every sampling run.

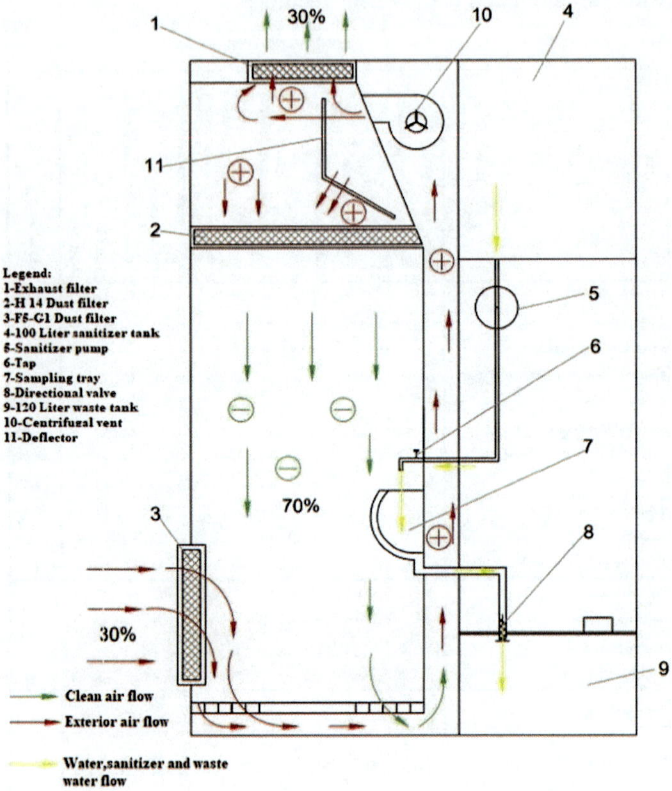

Fig. 3. SAFE booth, functioning principle (**schematic**)

Several challenges were considered in the booth development:

1. **Air flow management** (Fig. 3); the airflow and volume within the booth should be controlled (increased, decreased and maintained) through the use of a digital PLC controller. The air flow system will also manage the monitoring of any defects that may occur to the ventilation and air recycling mechanisms. The values for the air flow speeds must also be monitored; this can be achieved via a 7 in. LCD incorporated within the booth's construction that allows the reading and programming of air flow speeds.
2. **Air quality**; ensuring air quality, is also one of the aspects that have been considered, i.e. 70% of the air within the booth will be recycled through a HEPA filter whereas the remaining 30% will be evacuated through a HEPA 14 filter that provides 99,9995% efficiency for particles smaller than 0.3 nm.

3. **Booth sterilization;** for proper booth sterilization UV light and ultrasound pulverizer will be used, all of which are automated and controlled via a touch screen display.
4. **Energy efficiency**; efficient lighting system that does not alter the interior atmosphere through thermal radiation is used. This is achieved via neon lights mounted on the inner front side of the booth. To ensure the booth is not being used past its functioning quota, a timer system must be implemented that warns the users when the HEPA filters need to be changed, following a set number of working hours.

As such the operational principle should be as follows (Fig. 3): Exterior air enters in the booth while passing through the F5-G1 Dust filter (3), it is later sucked in through the air flow ducts along with interior booth air via the centrifugal vent (10), the air is then partially vented out through the exhaust filter (1) and recirculated and cleansed through the H14 Dust filter, and reintroduced inside the booth as clean filtered air. The booth will also present a liquid sanitizer circuit for body hygiene where sanitizer is pumped from the tank (4) via the pump (5) through tap (6) into sampling tray (7), waste sanitizer is then ejected into waste tank (9) through the directional valve (8).

The SAFE system aims to use a robotic arm (Fig. 4) for sample manipulation. This arm handles the sample vial, by placing the vial inside a sterile container through a door inside the booth. The device also handles the sterilization of the booth preparing it for the next patient. The robotic device presents 6 degrees of freedom, and is actuated by electric servomotors. Robot programming achieves personalized implementation (considering the requirements of each medical environment using the SAFE system). The robot arm is developed in such a way that it is easily sterilized, silent and non-encumbering, i.e. a perfect fit for a medical environment.

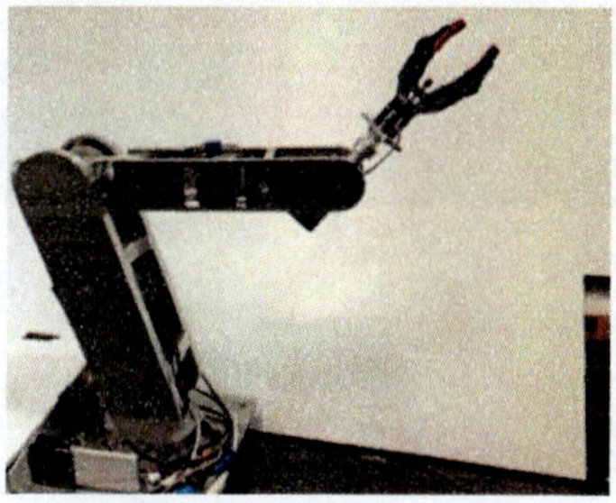

Fig. 4. The robotic arm used in the SAFE system

Among the other necessary components for an automated process (air flow, robotic arm, sterilization, lighting etc.), the SAFE booth (Fig. 5) is comprised of transparent panels which can be mounted to fit the space capabilities of the client, being large enough to fit only one person within.

The atmosphere within the booth is controlled and monitored in such a way as to ensure that the ejected air is filtered through a HEPA filter, before it reaches the exterior atmosphere, this means that 99.999% of bacteria are retained by the filter. The air flow speed will be stabilized via the use of a servo-controller with an incorporated micro-processor, which can also maintain a negative pressure inside, thus reducing the risk of contamination.

The booth ensures the user's comfort through the implementation of minimal sterilizable furniture, flowing clean water, a sink with a running drain connection to the sewage system, and transparent side panels that reduce the risk of a claustrophobia sensation and also allow monitoring during operation.

High importance is also attributed to the materials from which the booth is built, as well as the means by which fittings are made as to not provide any hard to sterilize surfaces where bacterial pathogens could deposit.

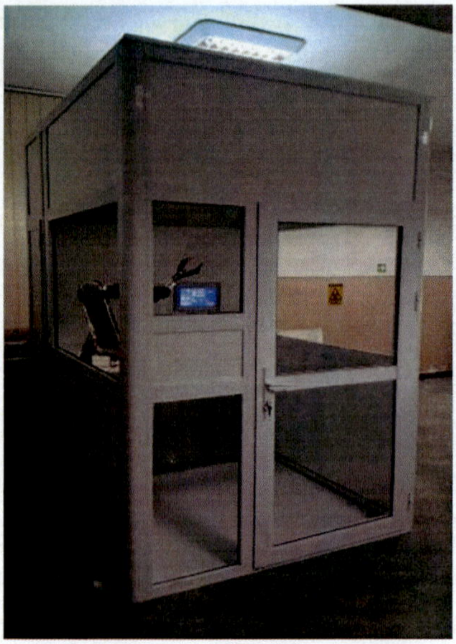

Fig. 5. The SAFE booth with the robotic arm introduced within it.

4 Conclusion

The SAFE booth, concept for a novel system for sterile, contaminant-free tuberculosis phlegm sample collection has been presented. The system showcased an automated, self-sterilizing booth with an included robotic arm for safe and contaminant free sample manipulation. The booth itself will be outfitted with programmable air flow systems as well as HEPA filters for preventing contaminants and bacterial pathogens within the booth from escaping into the exterior atmosphere. The booth also uses UV light and ultrasound pulverizers for sterilization after each use. The robotic device to be implemented within the booth has also been developed in such a way that it permits the full sterilization of all its components, additionally the arm itself could be programmed to aid with the sterilization process. The design of the booth was done in such a way as to not cause any discomfort to the patient using it (transparent panels to eliminate claustrophobic tendencies), while also allowing the sampling procedure to be observed by medical personnel. In this context, the full development of the SAFE booth and eventually its introduction on the market has the potential to reduce the number of infections that occur due to air-borne pathogens released by infected people during sample gathering/testing procedures in the case of tuberculosis mainly, but also HIV, anthrax, etc. Consequently, the SAFE system may have an indirect influence in the reduction of costs that occur in the medical system due to nosocomial infections (infections occurring in a hospital environment).

Acknowledgement. This work was supported by a grant of the Romanian Ministry of Education and Research, CCCDI - UEFISCDI, project number PN-III-P2-2.1-PTE-2019-0160, within PNCDI III, project 65PTE/01.09.2020.

Conflict of Interest. The authors declare that they have no conflict of interest.

References

1. https://www.who.int/news-room/fact-sheets/detail/tuberculosis. Accessed 7 Sep 2020
2. Wang, Y., Guanyang, L., Chen, C.: Development and study of a robot for transferring mycobacterium tuberculosis samples. In: Proceedings of the 2012 IEEE International conference on Robotics and Biometrics, Guangzhou China, December 11–14 (2012)
3. Prabhakar, P., et al.: Autosampler with robot arm. Patent US005948360A, 9 Sep 1999
4. https://www.qualica.net/software.html. Accessed 7 Oct 2020
5. Vaida, C., et al.: Smart furniture-Qua Vadis. In: 3rd International Conference on Quality and Innovation in Engineering and Management, International Conference on Production Research-Africa, Europe and Middle East (2014)
6. Rutala, W., et al.: Healthcare Infection Control Practices Advisory Committee. Guideline for Disinfection and Sterilization in Healthcare Facilities, Atlanta (2008)
7. Rutala, W., et al.: Cleaning, disinfection and sterilization. In: APIC Text of Infection Control and Epidemiology, 3rd edn., pp. 1–7. Association for Professionals in Infection Control and Epidemiology, Inc., Washington DC (2009)
8. Rutala, W., et al.: Disinfection and sterilization in healthcare facilities. In: Practical Healthcare Epidemiology, 3rd edn., pp. 61–80. University of Chicago Press, Chicago (2010)
9. Rutala, W., et al.: Sterilization, high-level disinfection, and environmental cleaning. Infect. Dis. Clin. N Am. **25**, 45–76 (2011)
10. Weber, D., Rutala, W.: Self-disinfecting surfaces. Infect. Control Hosp. Epidemiol. **33**(1), 10–13 (2012). https://doi.org/10.1086/663648

Study on the Prevention of the Installation of Physical Deficiency Scoliosis in Children of Prepubertan Age

Vizitiu Elena, Constantinescu Mihai, Dan Laurențiu Milici(✉), and Vicoveanu Dragoș

Ștefan Cel Mare University of Suceava, Suceava, Romania
dam@usm.ro

Abstract. In this paper we will present some theoretical-methodical aspects of identifying at an early stage of postural spinal deficits in children in the pre-pubertal period and the development of prophylactic programs in order to maintain and develop a correct body posture. The methods and means applied to the persons involved in the program are part of the kinetic recovery in water and on land. The aim of the paper is to identify in real time the physical deficiency of scoliosis, at the level of vicious attitude and to prevent its development. Water is a suitable environment to facilitate movement and provides extremely effective conditions for obtaining optimal health during growth and development. In this sense, we propose kinetic recovery programs on land, hydrokinetic therapy and therapeutic swimming.

Keywords: Scoliosis · Patients · Therapeutic swimming · Evaluation

1 Introduction

One of the problems that modern children face more and more often are posture deficiencies, given the fact that "movement" is increasingly replaced in our daily lives (means of transport, bench position, backpack weight, lack of activities physical activity, inadequate nutrition, incorrect allocation of sleep periods, sedentary lifestyle, etc.). At present, the physical deficiency of scoliosis, like other physical deficiencies, has come into the attention of specialists in the field of recovery, because these cases, unfortunately, have grown.

In this sense, research is still being done, the etiology of the installation of postural deficits is not established, and most are called idiopathic [2, 4, 12].

The complexity of this deficiency requires a thorough and long research from all points of view, bio-psycho-emotional to discover the causes, recovery and above all the prophylaxis of its installation.

All techniques used in the recovery scheme must start from a positive diagnosis, accompanied by a clinical picture and objective investigations, the functional residual, the current condition and the presumptive prognosis of the deficit, and the setting of immediate and long-term goals. The elaboration of recovery programs, both on land and in water, addressed to the deviation of the spine (scoliosis), must be approached

S. Vlad and N. M. Roman (Eds.): MEDITECH 2020, IFMBE Proceedings 88, pp. 382–387, 2022.
https://doi.org/10.1007/978-3-030-93564-1_42

from a prophylactic point of view, aiming to limit as much as possible the installation of a vicious postural attitude [1, 3, 5, 9].

Hydrokinetotherapy has close links with aquatic therapy, balneoclimatotherapy, kinetotherapy, hydrotherapy, adapted sports, occupational therapy, sports therapy and of course with the therapeutic swimming from which the learning exercises of sports swimming are taken. The great advantage of hydrokinetic therapy and therapeutic swimming is that it is practiced in water. At the same time, the influence of water on the body's attitude is determined by the practice of swimming technique regularly by the demands imprinted on the musculoskeletal system, especially the musculoskeletal system, the joints being released by body weight. The muscular effort in this case can be reduced or increased depending on the means approached in the work programs according to the proposed objectives [6–8].

Specific to the movements performed in the aquatic environment, the horizontal position of the body requires the body to spend 5–9 times more energy than on land, ensures the relaxation of skeletal muscles and frees the spine from body weight. The pressure of the water on the chest makes it difficult to breathe, causing the development of the muscles of the cardio-respiratory function. Working in water stimulates adaptive psychomotor processes, improving motor and psychomotor patterns stored in memory, especially during the growth and development of children [10, 11].

2 Material and Method

The aim of the paper is a detailed analysis of the pathophysiology of physical deficiency of thoraco-lumbar scoliosis (left), in children of prepubertal age and the identification of the means of optimal recovery through aquatic activities [12].

General objectives of the recovery program (Table 1):

1. Maintaining a correct posture,
2. Maintaining and developing joint mobility of the spine,
3. Maintaining and developing muscle tone in the spine,
4. Maintaining a correct posture in static and dynamic,
5. Development of motor skills for endurance and effort.

Table 1. xxxxxxx

<table>
<tr><td colspan="4">Exercise program on adaptation to the aquatic environment – stage 1</td></tr>
<tr><td>Stage I. Accommodation program with the water</td><td>Duration of the program: 2 weeks</td><td>2 lession /week</td><td>Volume 150-200m</td></tr>
<tr><td colspan="4">Objectives</td></tr>
<tr><td colspan="4">- Maintaining the well-being of local and general relaxation of the body;
- Awareness of the effects of swimming on the body from a sanogenetic point of view;
- Optimizing the processes of growth and harmonious development of the body;
- Water accommodation.</td></tr>
<tr><td colspan="4">Floating and sliding</td></tr>
<tr><td>Means</td><td>25 min on land
25 min in water</td><td colspan="2">Verbal methods;
demonstration;
analytical;
iteration;</td></tr>
<tr><td colspan="4">Floating on the chest; Floating on the back; Vertical floating and ironing; Underwater breathing; Sliding on the chest and back; Sliding on the chest and back with leg movements craul; Sliding on the chest and back with arm movements under water; Sliding on the chest and back in coordination with breathing; Water games.</td></tr>
<tr><td colspan="4">Hydrokinetotherapy program - stage II</td></tr>
<tr><td>Stage II. program of static and dynamic exercises in small water</td><td>Duration of the program: 5 weeks</td><td>2 lessons / week</td><td>Duration 35-50 minutes</td></tr>
<tr><td colspan="4">objectives</td></tr>
<tr><td colspan="4">- Maintaining joint mobility of the spine;
- Toning in conditions of shortening the muscular groups of the back on the side of the convexity;
- Formation of the reflex of correct attitude of the body;
- Optimizing the processes of growth and harmonious development of the body in order to maintain a correct body posture.</td></tr>
<tr><td colspan="4">static and dynamic exercises of the scapulo-humeral belt / upper limbs; Free and heavy exercises of the pelvic girdle / lower limbs; Exercises for developing the strength of the paravertebral and abdominal muscles.</td></tr>
<tr><td colspan="4">Therapeutic swimming program - stage III</td></tr>
<tr><td>Stage III. exercise program on different distances and with different auxiliary material</td><td>Duration of the program: 5 weeks</td><td>2 lessons / week</td><td>Volume 200-400m</td></tr>
<tr><td colspan="4">objectives</td></tr>
<tr><td colspan="4">- Increasing the capacity of effort;
- Development of proprioceptive abilities in statics and dynamics;
- Formation of the reflex of correct attitude of the body in static and dynamic;
- Introduction of swimming elements and procedures.</td></tr>
<tr><td colspan="4">Exercises to maintain body posture in weight conditions; Exercises to maintain balance in static and dynamic; Coordination exercises over different working distances; Exercises for awareness and self-control of body posture.</td></tr>
</table>

3 Description of Tests

In order to analyze the data obtained on the study, the following statistical processing methods were used: arithmetic mean, standard deviation, mean error; coefficient of variability, comparison criterion "t" by Student (Table 2).

Table 2. xxxxxxxx

	Initial anthropometric evaluation				Final anthropometric evaluation			
	Heigh (cm)	Weigh (kg)	IMC (uc)	Basic metabol-ic rate /kcal /day	Heigh (cm)	Weigh (kg)	IMC (uc)	Basic metabolic rate /kcal /day
Mean	152.50	46.20	19.78	1348.90	152.50	45.90	19.55	1344.80
Standard deviation	6.15	7.05	2.08	114.08	6.15	7.13	1.75	117.43
Coeffcient of variability	34.05	43.52	3.07	11841.68				

The basal metabolic rate is an estimate of the minimum number of calories a person needs each day to maintain basic functions (breathing, circulation and digestion) at rest.

1. Ground finger length in flexion-mobility of the spine in flexion from an orthostatic position,
2. Soil acromion length/soil sias length - are anthropometric measurements related to the total height of the body and may lead to the identification of deviations of the segments from the proportionality ratio that exist between the segments,
3. Lateral inclination left/right - frontal mobility of the spine from the orthostatic position (Table 3).

Table 3. XXXXXX

	Initial assessment of joint balance spine				Final assessment of joint balance spine			
	Length of the ground fingers in flexion (cm)	Soil acro-mion length (cm)	Length sias - floor (cm)	Left / right (ground fingers) (cm)	Length of the ground fingers in flexion (cm)	Soil acro-mion length (cm)	Length sias - floor (cm)	Lateral tilt Left / right (ground fingers) (cm)
Mean	5.40	124.00	89.10	38.1/35.3	2.30	124.00	89.10	36.30/34.90
Standard deviation	5.68	4.76	5.13	5.92/4.30	3.16	4.76	5.13	4.88/4.53
Coefficient of variability	15.68	20.40	23.69	25.5/16.7				

1. Cobb method - this method can accurately calculate the angle of the spine after performing X-rays.
2. Lift of the trunk from the ventral decubitus - to evaluate the paravetrebralatoraco-lumbar muscles,
3. Lifting of the trunk from supine position - for the evaluation of the abdominal muscles,
4. Muscle testing (F1–F5) -measuring the strength of the paravertebral and abdominal muscles.

We used these tests because they have a direct relevance from the somato-functional point of view of the spine in order to maintain and recover a correct postural status (Table 4).

Table 4. xxxxxxx

	Initial evaluation effort test				Final evaluation effort test			
	Cobb method values, calculated on rgf. (G)	Trunk lifts from supine position (no. of repetitions / 20 sec)	Paravertebral muscle testing (F1-F5) abdominal	Trunk lifts from supine position (no. of repetitions / 20)	Cobb method values, calculated on rgf. (G)	Trunk lifts from supine position (no. of repetitions / 20 sec)	Paravertebral muscle testing (F1-F5) abdominal	Trunk lifts from supine position (no. of repetitions / 20)
Mean	17.00	11.10	11.00	F4/F4	15.90	15.30	16.90	F5/F5
Standard deviation	7.53	2.28	1.76	0.67/0.48	6.84	1.89	1.66	0.32/0.00
Coefficient of variability	44.30	2.57	-0.80	0.13/0.00				

4 Conclusions

An important role in the process of raising and developing children is attributed to specialists in the field of physical therapy, who have been concerned about the implementation of exercise at different ages of the spine, at the same time we found an information deficit in order to prevent these deficiencies mainly in prepubertal age.

In the process of conducting the study was established the influence of specific kinetic means on land, hydrokinetotherapy and therapeutic swimming on somato-functional, motor and psychomotor parameters, statistical results showed a statistically significant evolution:

- The averages mean of the grous have changed in order to improve or maintain the assessed characteristics after the recovery program,
- The standard deviation indicates a homogenization of the group in the case of all final tests performed following the proposed recovery program,
- The coefficient of correlation calculated between the anthropometric values and the articular balance of the spine, between the articular balance of the spine and the

evaluation of the effort test, between the evaluation of the effort test and the anthropometric values, leads to the idea of a much better correlation between the evaluations performed on the end of the recovery program against the same values at the beginning of the preparation [13].

The general conclusion is that the recovery program of the pathophysiology of physical deficiency of thoraco-lumbar scoliosis proposed by the team and carried out over a period of 6 months gives the expected results for young people aged between 10 and 13 years and the tests proposed in the three evaluation batteries are capable to highlight the positive results obtained in the training program.

Conflict of Interest. The authors declare that they have no conflict of interest.

References

1. Cîrlă, L.: Swimming - Psychomotor Skills and Technical Training. Ed. Printech, Bucharest (1999)
2. Constantinescu, M.: Instrumental somatoscopy of functional physical deficiencies of the spine. The Annals of the "Ştefan cel Mare" University, Suceava (2014)
3. Constantinescu, M.: swimming applications in the scoliosis prophylaxis, p 6, the annals of the "Ştefan cel Mare" University Physical Education and Sport Section the Science and Art of Movement (2018). ISSN:1844-9131, eISSN:2601-341X Volum XI issue 2
4. Constantinescu, M.: Book of Practical Works- Physical Therapy in Pediatric Diseases - Suceava: "Ştefan cel Mare" University Publishing House (2019). ISBN:978-973-666-517-2
5. Fozza, C.A.: Guidance for the Correction of Physical Deficiencies. Romania of Tomorrow Foundation Publishing House, Bucharest (2002)
6. Penţia, D.: Swimming, Sanogenic Activity. Mirton Printing House, Timisoara (1995)
7. Raţă, E.: Theory and Practice in Water Sports: Swimming. Practical Workbook. "Ştefan cel Mare" University, Suceava (2014)
8. Romano, M., et al.: The SEAS Concept of Exercises for Scoliosis, Physiotherapy Theory and Practice. Copyright & Informa Healthcare USA Inc., Miami Beach (2011)
9. Sima, D.: Medical recovery through aquatic therapy. In: Priorities and Perspectives in Physical Education and Sport: International Session of Scientific Communications. University Publishing House, Bucharest (2014)
10. Vizitiu, E., Constantinescu, M.: Prevention of physical deficiencies of the spine in the prepubertal period by swimming. "Ştefan cel Mare" University Publishing House, Suceava (2019). ISBN: 978-973-666-562-2
11. Milici, D.: Computerized system for testing and formation the speed of backward pusf of sportmen. In: The 13th International Symposium on Measurements for Research and Industry Applications org. by International Measurement Confederation IMEKO TC4, pp. 673–677 (2003)
12. Milici, L.D., Plăcintă, V.M., Bujor, L., Milici, M.R.: System for highlighting the emotional states, used in assessing the teaching methods. In: The 9th International Symposium on Advanced Topics in Electrical Engineering (ATEE), pp. 965–968 (2015)
13. Milici, D., Milici, M., Cernomazu, D., Popa, C.: Modeling of physical and psychological human performance evolution. In: Proceedings of the 6th International Conference on Electrical & Power Engineering, pp. 281–286 (2021)

Assessment of Magnetic Fields Produced in Clinical Environment "Operating Rooms"

Hussain Alnamir(✉)

University "Politehnica" of Bucharest, Doctoral School of Electrical Engineering, Bucharest, Romania

Abstract. The occupational exposure nature of the electromagnetic fields was totally distinctive for different occupations and work spaces. Moreover, the technologically-innovative equipment in hospitals is distributed on high-density, while the medical equipment is becoming more electrically complicated due to the advancements in technology. Often, the operating rooms were full with numerous electronic devices that are packed in little space. A lot of operating room staff, such as anesthesiologists, surgeons and specialized nurses, work daily in such an environment and for lengthy time periods.

The initial phase of this work provides evaluation of the magnetic field strengths in operating rooms at the standing position of operating room personnel. Then, MF intensity was recorded at 0.1, 0.5, and 1 m far from the device while just one of the systems was on and the others were off.

The results of this work are indicating that the maximal rates that are related to MF intensity levels at standing positions of the staff have been find to shoulder arthroscopy operating environment and Endoscopy surgery, which were 5.4 mG and 4.92 mG, respectively, and is lower than advised level by ICNIRP Based on the results, it doesn't appear that occupational exposure which is related to staff in operating rooms would exceed standard provisions; thus, there aren't significant risks and no worries regarding personnel overexposed to non-ionizing radiation.

Keywords: EMFs of low frequency · Operating room · ICNIRP

1 Introduction

The magnetic fields (MFs) with a maximum frequency of 3000 Hz are defined extremely low-frequency magnetic fields (ELF-MFs), these are carefully examined in connection to public-health, yet up to now, there isn't a confirmation regarding the way that being exposed to such fields was unsafe to humans [1], a few epidemiological types of research indicated a possible relation between the occupational exposure to the ELF-EMFs and the growing number of different pathologies, involving testicular cancers, brain cancers, breast cancers and leukemia [2]. Because of the new and changing lifestyles, the majority of individuals are exposed to unwanted MFs radiations, that are mainly ELF-MFs [3]. A few of the major sources to generate low frequency MFs (especially power frequency) are devices including electric chargers for wireless devices, domestic LCDs, TVs, hairdryers, electric engines, electric heated

S. Vlad and N. M. Roman (Eds.): MEDITECH 2020, IFMBE Proceedings 88, pp. 388–398, 2022.
https://doi.org/10.1007/978-3-030-93564-1_43

mattress, and so on [4]. Various types of electronic and electrical systems were utilized in the operating room environment. Typically, such equipment creates electromagnetic fields (EMFs) [5]. Thus, patients and other operating room personnel were always exposed to the EMFs with different strength values. The impacts of such EMFs on human health haven't been identified in a clear way.

Also, there were indications of bad impacts related to ELF-MFs on neurotransmitter effects, calcium metabolism changes, blood-brain barrier, and elevated dangers of brain tumors [6]; the surgical personnel work daily inside the same magnetic field environment and the exposure is of professional significance; moreover, the surgical equipment and the magnetic field emissions are very close to the surgeons' body (in their hand, close to the trunk) and the exposure level is highest considering all the possible scenarios.

The magnetic field (and not the electric field) is considered because the potential biological effects are related to the electric fields induced inside the body. These electric fields induced by exposure to time varying magnetic field are significantly higher than any electric fields induced in the body by exposure to environmental electric fields; this behavior is explained by the dielectric properties of the tissues (only a small fraction of the external electric field does penetrate in the anatomical regions).

The cumulative aspect of the exposure to magnetic field is not clarified by the literature in studies addressing biological and health impact, but it is of common sense that more exposure could only worsen the stress.

The absence of data regarding the MF levels in operating rooms of hospitals was the major motive of the presented study.

2 Exposure to Magnetic Fields from Medical Sources

Electromagnetic, magnetic, and electric fields are mainly used in therapeutic applications and medical diagnostics. EMF with adequate strength might have biological impacts as well as a few medical devices using these for therapeutic or diagnostic aims. Current technological developments elevated the diversity and the possible strength of the EMF from the medical sources and thus raising doubts regarding occupational safety. In addition, the EMF created via the medical sources in hospital environments might be fairly categorized into 2 classes: high-frequency field sources and sources of static and low-frequency fields. Based on fields' strength, that type of exposure might result in the retinal stimulation (otherwise known as the magnetophosphenes), muscle contractions, pains, tingling sensation, and vestibular disturbances. The high-frequency EMF which was adequately strong might result in extreme heating as well as tissue damage (Table 1) [7]. For safely using EMF, ICNIRP specified the major limitation with regard to induced electric field strengths and the body's SAR (specific absorption rate), below which such health and sensory impacts won't happen. The reference levels with regard to the strength related to external EMF were obtained from such major restrictions. In the case when workers were exposed to EMF that is weak compared to reference levels, the major limitations won't be exceeded [8]. Furthermore, the European Union is using the reference levels and major restrictions of ICNIRP in its occupational health as well as safety legislations, where the degrees of the reference were indicated as 'action levels' and major restrictions as 'values of the exposure limit' (Tables 2 and 3). With regard to EMFs of low-frequency, it is distinguishing the low-action levels, which are associated with sensory impacts exposure limits, also the limb action and the high-action levels, which are associated with the health impacts exposure limit values (i.e. the nerve stimulations).

In addition, the strong high-frequency EMF was intentionally utilized for the purpose of heating the patient tissues in the hyperthermia and therapeutic diathermy [9].

Table 1. Classifications regarding the medical devices based on frequency range [7]

Frequency range (Hz)	Potential biological effect	Medical device
$f = 0$	Disturbed equilibrium, reduced blood flow, cardiac arrhythmia	– MRI (static field)
$0 < f < 10^5$	Vertigo, nausea, metallic taste, magnetophosphenes, nerve stimulation	– MRI (movement) – MRI (gradients) – Magnetotherapy – Power frequency equipment
$10^5 < f < 10^{11}$	Heating	– MRI (radiofrequency field) – Diathermy – Electrosurgery – Hyperthermia – Microwave imaging – Radar monitoring – Therapeutic ultrasound

Table 2. Action levels for magnetic field in Directive 2013/35/EU [9]

Frequency	Low AL B (μT)	High AL B (μT)	Limbs AL B (μT)	Thermal AL B (μT)
$1 \leq f < 8$ Hz	$2.0 \times 10^5/f$	$3.0 \times 10^5/f$	$9.0 \times 10^5/f$	-
$8 \leq f < 25$ Hz	$2.5 \times 10^4/f$	$3.0 \times 10^5/f$	$9.0 \times 10^5/f$	-
$25 \leq f < 300$ Hz	1.0×10^3	$3.0 \times 10^5/f$	$9.0 \times 10^5/f$	-
300 Hz $\leq f < 3$ kHz	$3.0 \times 10^5/f$	$3.0 \times 10^5/f$	$9.0 \times 10^5/f$	-
$3 \leq f \leq 100$ kHz	1.0×10^2	1.0×10^2	3.0×10^2	-
100 kHz $\leq f < 10$ MHz	1.0×10^2	1.0×10^2	3.0×10^2	$2.0 \times 106/f$
$10 \leq f < 400$ MHz	-	-	-	0.2
400 MHz $\leq f < 2$ GHz	-	-	-	$1.0\ 10^{-5}\sqrt{f}$
$2 \leq f < 300$ GHz	-	-	-	$4.5 \times 10–1$

Table 3. Values of exposure limits in the Directive 2013/35EU, low frequency field

Frequency	Sensory effects ELV		Health effects ELV	
	B(T)	E (V/m)	B(T)	E (V/m)
$0 \leq f < 1$ Hz	2	-	8	-
$1 \leq f < 10$ Hz	-	0.7/f	-	1.1
$10 \leq f < 25$ Hz	-	0.07	-	1.1
$25 \leq f \leq 400$ Hz	-	0.0028 f	-	1.1
$400 \leq f \leq 3$ kHz	-	-	-	1.1
3 kHz $\leq f \leq$ 10 MHz	-	-	-	3.8×10^{-4}f

2.1 Equipment of the Operating Room

All operating rooms have special devices, like patient monitors, emergency resuscitative devices, as well as tools for diagnostics (see Fig. 1).

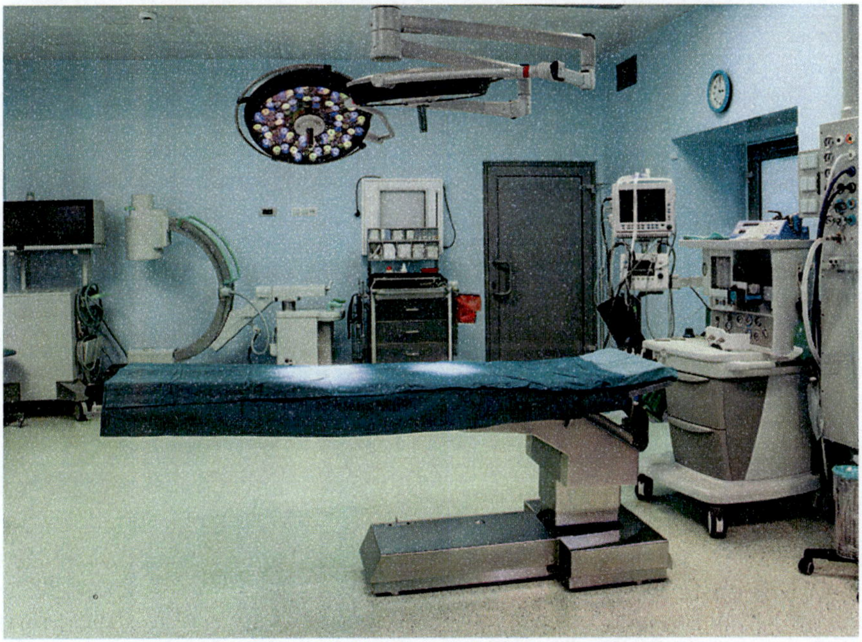

Fig. 1. Equipment of the operating room

3 Materials and Methods

3.1 Materials

For the present work, we used the EXTECH 480823 single axis magnetic field meter (see Fig. 1). This meter is designed to directionally measure the levels of low frequency magnetic fields, measuring range 0.1 to 199.99 mG or 0.01 to 19.99 μT, and frequency range 30 to 300 Hz [10]. The main purpose of the measurements is the general assessment of the medical environment, not the accuracy of the measurements technique (Fig. 2).

Fig. 2. EXTECH 480823 magnetic field meter

3.2 Methods

The MF density in various parts of the operating rooms is estimated utilizing EXTECH 480823 single-axis MF meter, whereas all equipment is working. On a certain day, the amplitude levels related to MF in twelve operating rooms have been evaluated from start of the first surgery of the day till the last surgery's end, involving periods between the surgical cases. In addition, the evaluation was achieved at a standing position regarding the operating room personnel as well as in the closest position to the patient, in the anesthesiologist's position as well. In each one of the cases, data acquisition has been achieved at 120 cm height above the floor. Furthermore, the data have been indicated as the resulting values. The MF density on spatial axes x, y, and z (RMS values), is typically recorded and the resulting value is then evaluated. After that, the MF density is recorded in various orientations, at distances of 0.1, 0.5, and 1 m from the device, whereas only one of the electronic and electrical systems in the room is ON and the rest of them are OFF (for example, in the case when cautery machine only has been on, whereas the rest of the electric systems are off, for instance, fluoroscopy, suction machine, LCD, pump, and so on).

4 Results

The MF density averaged values [in (mG)] in the nearest location to patients and at standing positions of the personnel of the operating room as well as the average MF in the position of anesthesiologist are shown in the Table 4, the frequency of the magnetic field measured is 30 to 300 Hz.

Table 4. The average values of the mf densities [in (mg)] in operating rooms

Nr	Operating rooms	Nr. of measurements	Magnetic field densities (average mG)		
			In the nearest at the closest distance to the patient	At the standing position of specialists and personnel	In the position of the anesthesiologist
1	Laparotomy surgery	87	1.25	1.10	1.94
2	General surgery	98	0.90	0.58	1.78
3	Knee arthroscopy surgery	79	0.79	0.45	2.01
4	Tracheostomy surgery	88	0.30	0.22	2.01
5	Prostate biopsy surgery	96	1.45	1.30	1.75
6	plastic surgery	104	1.60	0.98	2.32
7	Gastrostomy surgery	96	0.75	0.80	4.71
8	Parotid surgery	80	0.67	0.50	3.87
9	Shoulder arthroscopy surgery	96	1.23	5.40	2.05
10	Orthopedic surgery	72	0.84	0.67	1.95
11	Artery repair surgery	67	0.59	0.81	3.10
12	Endoscopy surgery	74	1.15	4.92	2.00

Table 4 is showing that the maximum exposure levels at personnel's standing position were in a shoulder arthroscopy operating room nr. 9 and endoscopy operating room nr. 12 which is 5.4 and 4.92 mG. In addition, it has been indicated that the anesthesiologists in the majority of operating rooms were under an exposure to the ELF-EMF which has exceeded the Swedish Board for Technical Accreditations for the recommended value of the computer screens [11].

The average of the MF density at various distances from every one of the active systems has been displayed in Table 5. Whereas only one of the electronic and electrical systems in the room is ON and set for operation in the same way as they were connected to a patient, and the rest of them are OFF.

Table 5. Average of the MFs density levels at various distances from every active system

Equipment	Average of magnetic field intensity (mG)		
	At 10 cm	At 50 cm	At 100 cm
Operating theatre light	3.47	0.72	0.33
Anesthesia trolley	6.40	1.00	0.48
Ventilator	1.99	0.72	0.29
Anesthesia LCD monitor	11.32	2.41	0.98
Dry-heat sterilizer	1.97	0.91	0.36
Fluoroscopy	1.89	0.75	0.22
Oximeter pulse	1.31	0.62	0.37
Laparoscope	18.37	1.59	0.49
High-voltage power supply	52.06	14.32	6.01
Defibrillator	0.89	0.26	0.20
Negatoscope	41.25	2.31	0.96

As can be seen in Table 5, from every active system in the operating room, the maximal rate which is related to the value of the MF density has been identified around High-voltage power supply in a 10 cm distance away from these systems; the estimated value was approximately 52.06 mG. In addition, the magnitude levels for the other systems, like anesthesia LCD monitor (11.32 mG) and laparoscopy (18.37 mG), were in the next ranking. Graphical representation of the magnetic field distribution from each active device represents in the following figures (Figs. 3, 4 and 5).

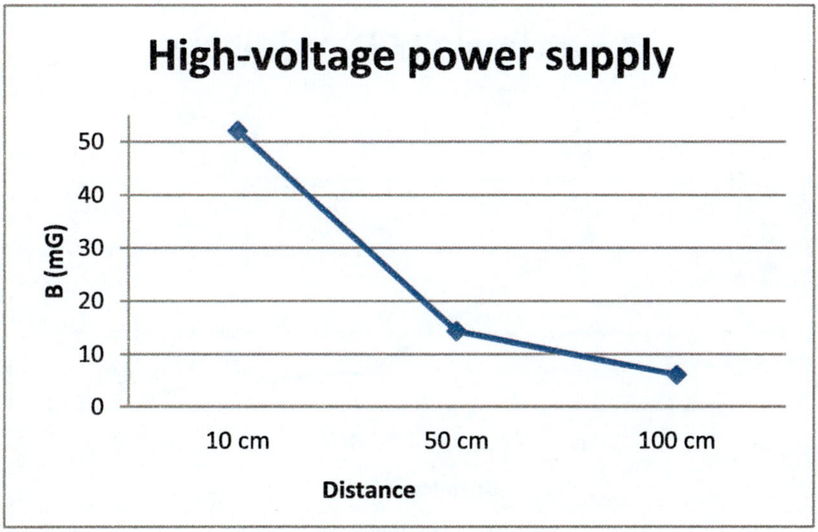

Fig. 3. Magnetic field distribution in different distances from high –voltage supply

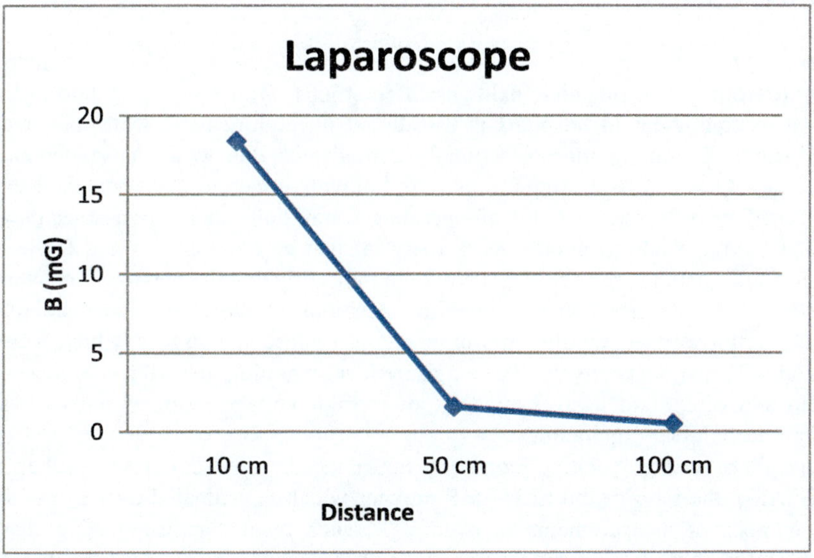

Fig. 4. Magnetic field distribution in different distances from laparoscope.

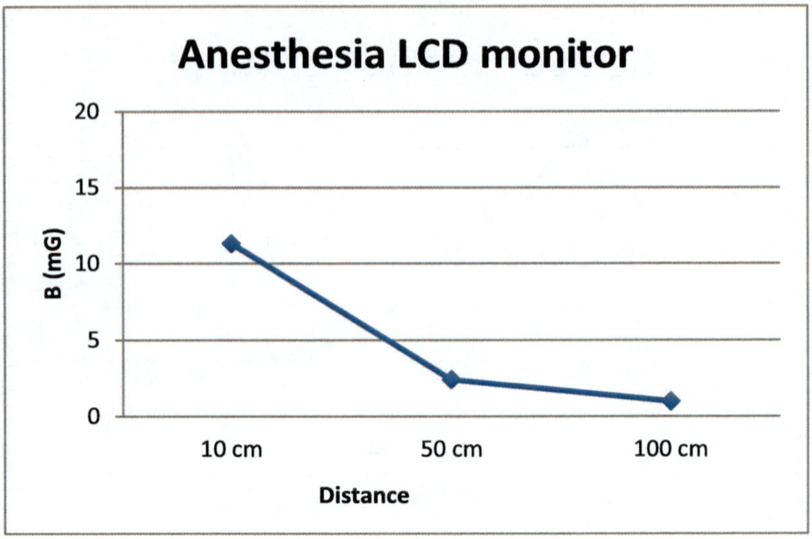

Fig. 5. Magnetic field distribution in different distances from Anesthesia LCD monitor

5 Conclusions

Presently, there are in progress a few projects for the study of the occupational exposure from electro-magnetic fields in different jobs. Moreover, the technologically-innovative equipment in hospitals is distributed on high-density, while the medical equipment is becoming more electrically complicated due to the advancements in technology. Often, the operating rooms were full with numerous electronic devices that are packed in little space. A lot of operating room staff, such as anesthesiologists, surgeons and specialized nurses, work daily in such an environment and for lengthy time periods. Also, the results of this work are indicating that the maximal rates that are related to MF intensity levels at standing positions of the staff have been find to shoulder arthroscopy operating environment and Endoscopy surgery, which were 5.4 mG and 4.92 mG, respectively. The exposure level regarding the MFs is highest up to 0.50 m and often 1 m away from the power high voltage systems and the values compare to ICNIRP's guidelines.

In eight of twelve operating rooms (i.e. 66%), the average MF density level exceeds 2 mG. Also, the ratio of the ELF-EMF number which is estimated over 2 mG to the total number of measurements is over 60%, such results are suggesting that the anesthesiologist might be exposed to MFs that exceed 2mG, Swedish Board for Technical Accreditations for computer monitors upper limit for the production of the EMF through computer monitors which was detectable at 30cm, for over 60% of the working time. In addition, the limitations regarding that work possible to exceed the exposure classification, since various electronic system types are producing high level electro-magnetic waves in the operating rooms. Standardization regarding technologies and processes used for exposure assessment was developed, that makes it complicated to do direct comparison related to results from various researches implemented through using many strategies of assessment. An additional limitation was that the source publications aren't providing enough information for assessing the effect of measurement uncertainty in the case of comparing the measured values with the action levels.

Based on the results, it doesn't appear that occupational exposure which is related to staff in operating rooms would exceed standard provisions; thus, there aren't significant risks and no worries regarding personnel overexposed to non-ionizing radiation. In conclusion, the study indicated that MF strength in operating rooms is dependent on the type and number of machines and monitors, various surgery types, the location regarding the major power input lines, distance of operating room staff, including anesthesiologists from the operation site, and the illumination related to operating rooms, among other factors. The magnetic safety should be one of the priorities of operative settings. It has been proposed to personnel for keeping a safe distance from the supplies of the high voltage and for the purpose of reducing the intensity regarding electric and MF of the operating rooms, it is importance to installing electrical system power supplies out of operating rooms. Furthermore, every practitioner, regardless their practice settings, must use the concept of dose reduction, distance, and shielding.

Acknowledgment. The author would like to express thanks to his advisor Professor Dr. Eng. Mihaela Morega, for the guidance, encouragement and advice.

The author would like to express his gratitude for the help provided by the Iraqi Ministry of Health for approving the measurements campaign in several hospitals and clinical environments in Baghdad, and to the Doctoral School of Electrical Engineering of the University "Politehnica" of Bucharest for assistance in elaborating this paper.

Conflict of Interest. In my ethical obligation as a researcher, I am reporting that the author declare that there is no conflict of interest.

References

1. Morega, M., Baran, I.M., Morega, A.M., Hussain Alnamir, K.L.: On the Assessment of Human Exposure to Low Frequency Magnetic Field at the Workplace, vol. 63, pp. 162–171. Bucharest (2018). ISSN: 0035-4066
2. Ahlbom, M.: Neurodegenerative diseases, suicide and depressive symptoms in relation to EMF. Bioelectromagnetics **22**(Suppl 5), S132–S143 (2001)
3. Hareuveny, R., Kavet, R., Shachar, A., Margaliot, M., Kheifets, L.: Occupational exposures to radiofrequency fields: results of an Israeli national survey. J. Radiol. Prot. **35**, 429 (2015)
4. Akdag, M.Z., Dasdag, S., Aksen, F., Isik, B., Yilmaz, F.: Effect of ELF magnetic fields on lipid peroxidation, sperm count, p53, and trace elements. Med. Sci. Monit. **12**, BR366–BR371 (2006)
5. Selmaoui, B., Lambrozo, J., Touitou, Y: Endocrine functions in young men exposed for one night to a 50-Hz magnetic field. A circadian study of pituitary, thyroid and adrenocortical hormones. Life Sci. **61**, 473–486 (1997)

6. Forgacs, Z., Thuroczy, G., Paksy, K., Szabo, L.D.: Effect of sinusoidal 50 Hz magnetic field on the testosterone production of mouse primary Leydig cell culture. Bioelectromagnetics **19,** 429–431 (1998)
7. Stam, R., Yamaguchi-Sekino, S.: Occupational exposure to electromagnetic fields from medical sources. Ind. Health **56**, 96–105 (2017)
8. International Commission on Non-Ionizing Radiation Protection: Guidelines for Limiting Exposure to Time-Varying Electric and Magnetic Fields (1 Hz to 100 kHz). Health Phys. **99**, 818–836 (2010)
9. European Parliament and Council: Directive 2013/35/EU of the European Parliament and of the Council of 26 June 2013 (2013)
10. EXTECH Instruments, Electromagnetic Field Meter, Model 480823, User manual 480823-en-GB_V2.3
11. Svenska, E.K.: Computers and office machines: Measuring methods for electric and magnetic near fields. 2nd edn., Stockholm. Report No. SS 436-14-90 (1995)

Indoor Air Quality Monitoring System for Healthcare Facilities

Adela Puscasiu[1(✉)], Alexandra Fanca[1], Honoriu Valean[1], Dan-Ioan Gota[1], Ovidiu Stan[1], Nicolae M. Roman[2], Iulia Clitan[1], and Vlad Muresan[1]

[1] Faculty of Automation and Computer Science, Department of Automation, Technical University of Cluj Napoca, Cluj Napoca, Romania
adela.puscasiu@aut.utcluj.ro

[2] Faculty of Electrical Engineering, Electrical Engineering and Measurements Department, Technical University of Cluj Napoca, Cluj Napoca, Romania

Abstract. The human body is not able to detect whether the air is polluted or not, which is why it is important to have a device that can measure the air quality, both in the closed environment and in nature. Thus, air pollution is defined as a harmful phenomenon to the ecological system and to the normal conditions of human existence and development when certain substances in the atmosphere exceed a certain concentration. The paper proposes an indoor air monitoring system for healthcare facilities, using an Arduino Mega2560 board, temperature sensor, humidity sensor, eCO_2 (equivalent CO_2 which designates global warming potential), TVOC (Total Volatile Organic Compounds) and atmospheric pressure sensor. The technologies used for the software implementation are Arduino IDE for the acquisition of data from sensors and Visual Studio for the graphical interface implementation.

Keywords: Indoor · Air quality · Monitoring system · IoT

1 Introduction

Indoor air pollution has become a serious problem that is damaging public health. An indoor air quality monitoring system helps to detect and improve indoor air quality. Such a system is useful to better understand the current state of air quality, as well as to study the impacts of poor long-term air quality on public health. Currently, the monitoring systems available are very expensive.

Every day, people respire 14,000 L of air. Indoor Air Quality (AQ) is an indicator of air quality by detecting volatile compounds such as paint exhaust, office chairs, trash, cooking compounds, feeding, exhaling and/or sweating. Poor AQ may lead to headaches, tiredness, face and skin irritation or long-term effects on health.

The AQ closely refers to occupants' health and comfort and thus AQ must always be maintained at the correct standard in a hospital with multiple health services, such as an emergency room, admission ward, hospital room, etc. Some harmful chemical compounds have been reported to be found in hospital air, e.g. anesthetic gasses (operating rooms), formaldehyde or glutaraldehyde and ethylene oxide (disinfection

S. Vlad and N. M. Roman (Eds.): MEDITECH 2020, IFMBE Proceedings 88, pp. 399–408, 2022.
https://doi.org/10.1007/978-3-030-93564-1_44

units), while these pollutants are less concentrated outside of the air. These gasses may be harmful not only to patients, but also to medical practitioners and staff.

AQ must be assessed and inspected periodically for identification of unintended gas sources. AQ-measuring equipment is expected to measure air continuously throughout the day in order to investigate efficiently AQ. Sadly the AQ measuring equipment is costly. It is then very restricted to track AQ simultaneously and continuously in several places. Happily, several low-cost AQ detection sensors have been produced commercially and may be running in many embedded systems platforms. AQ monitoring is then more possible than ever in all dangerous places in a hospital.

The aim of this document is to build a cheap AQ measuring system. The rest of the paper is organized as it follows: next section presents the state of the art of monitoring air quality systems, Sect. 3 presents the global system architecture, the use case diagram, the sequential diagram for a data acquisition scenario and humidity graph display, the hardware and software presentation, the results are presented in Sect. 4, and Sect. 5 presents the discussion.

2 State of the Art

Currently, there are already on the market many devices equipped with sensors for collecting and monitoring of indoor environment quality parameters. These devices include Foobot [2], Netatmo [3], Speck [4], and SAMBA [5]. The last one is a state-of-the-art monitoring station for continuous, real-time measurements, which combines hardware and a software solution designed to automatically analyze and visualize the data interpretation.

[6] presents other experiments on air quality that were carried out on a bus. These were made during an outbound and return journey with 24 passengers plus a driver. The target parameter for monitoring was the concentration of carbon dioxide. The allowed value was exceeded by 1000 ppm CO_2 during the trip, and, at the beginning of the return trip, the maximum measured concentration was 2400 ppm. Calculations of the minimum quota of fresh air were taken to provide what was required by the standard conditions.

An IoT-based air quality monitoring system and proposes a complete method of indicating the indoor air quality level, in real-time, effectively coping with dynamic changes, and is efficient in terms of processing and memory is presented in [7].

An IAQ (Indoor Air Quality) system based on an IoT (Internet of Things) that incorporates in its construction ESP8266 (used to transmit the sensing data to the terminal), Arduino and XBee technologies for processing, data transmission and microsensors for data acquisitions are presented in papers [8] and [9]. This system allows the user to know, in real-time, a variety of environmental parameters such as air temperature, relative humidity, carbon monoxide (CO), CO_2, and luminosity.

Another similar system is presented in [10]. Apart from the CO, this system measures, in addition to the system previously presented, the noise level. The sensors used in this experiment are DHT11 Humidity & Temperature sensor, MG-811 CO_2 sensor module, TEMT6000 Ambient Light sensor, microphone amplifier, 1-Channel

Flame sensor module, the transmitter, and the receiver. The information transmitted by the sensor is given by the Serial Monitor.

[11] developed the PM1, PM2.5, PM10, CO, CO_2, VOCs, temperature, and humidity monitoring system to create a safe work environment and living environment. Apart from the air quality sensors, the system is equipped with wireless communication modules, displays, and smartphone applications. Data from the developed system is forwarded to the smartphone application, then to the cloud server by using wireless communications as Bluetooth and Wi-Fi for short-distance communication and Lora for long-distance communication. The mobile application provides information on indoor and outdoor air qualities, adequate actions, pollution warnings, and weather conditions.

The indoor environment monitoring system consists of a remote terminal that uses an Android mobile phone, an APP to generate data display and control decisions, control node, ZigBee wireless network which includes a main node and sub-nodes and sensors and is presented in [12]. Through the Zigbee network, each sensor node plays the role of the coordinator. The collection of various parameters of the room environment, such as temperature, humidity, PM2.5, and light intensity is collected by the sensor nodes distributed in the room. The collected data is sent to the control node via the ZigBee network, the data is converged and forwarded to the remote terminal, and the alarm data is displayed, recorded, and set by the line data.

The thermal comfort of people in an indoor environment under the system architecture of a wireless sensing network is presented in [13]. To design the man-machine monitoring interface and the control model Visual C# software is used. Also, on the interface parameters such as temperature, humidity, and wind speed can be selected. The system proposed emphasizes wireless transmission and rapid analysis of environmental parameters.

3 Core Concept

The system allows the user to acquire and to monitor the data collected from sensors, and to view the evolution of the parameters in a graphic form. In Fig. 1 is presented the global system architecture.

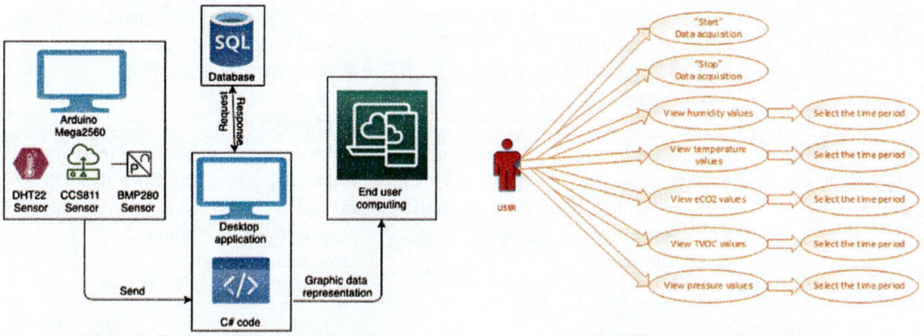

A. The global system architecture **B.** Use-case diagram

Fig. 1. Core concept of the proposed indoor air quality monitoring system

The global system architecture (Fig. 1(A)) presents the Arduino Module, which uses DHT22 (temperature and humidity sensor), CCS811 (gas sensor), BMP280 (pressure sensor) sensors and reads their data through a sketch. Windows Form is a class library within C#, which allows the data collection or stopping data collection for database storage and allows the graphical representation of data measured by sensors. Also, SQL represents the database, in which the values of the parameters measured by the sensors are stored.

3.1 Software Design

As can be seen in Fig. 2(B), the use case diagram is presented and in Table 1 the role of the actors are detailed It emphasizes all the actions that the user can do are start or stop the acquisition of real-time data, to view an evolution graph for each presented parameter and to set a certain period for which the user wants to view the evolution of the selected parameter.

Table 1. Use case scenario.

Use case	Main actor	Prerequites	Post conditions		Steps
			Success scenario	Error scenario	
Start data acquisition	User	The Arduino noard must be connected to the PORT	The data purchased from the sensors are saved in the database, along with a success message The data is displayed on the screen	If the connection to the Arduino board is invalid, an error message is displayed	The user must press the Start button next to the Data Acquisition text
Stop data acquisition	User	The user must have pressed the Start button next to the Data Acquisition text	Stops collecting data from sensors		The user must press the Stop button next to the Start button
Open humidity chart	User	If the Start button has been pressed for data acquisition, the user must press the STOP button	A new window opens and present the Humidity graph	If the connection to the DB is invalid or already open, an error message is displayed	The user must press the Open button next to the Moisture graph
Open temperature chart	User	If the Start button has been pressed for data acquisition, the user must press the STOP button	A new window opens and present the Temperature graph	If the connection to the DB is invalid or already open, an error message is displayed	The user must press the Open button next to the temperature graph

(continued)

Table 1. (*continued*)

Use case	Main actor	Prerequites	Post conditions		Steps
			Success scenario	Error scenario	
Open eCO_2 chart	User	If the Start button has been pressed for data acquisition, the user must press the STOP button	A new window opens and present the eCO_2 graph	If the connection to the DB is invalid or already open, an error message is displayed	The user must press the Open button next to the eCO_2 graph
Open TVOC chart	User	If the Start button has been pressed for data acquisition, the user must press the STOP button	A new window opens and present the TVOC graph	If the connection to the DB is invalid or already open, an error message is displayed	The user must press the Open button next to the TVOC graph
Open pressure chart	User	If the Start button has been pressed for data acquisition, the user must press the STOP button	A new window opens and present the pressure graph	If the connection to the DB is invalid or already open, an error message is displayed	The user must press the Open button next to the Pressure graph
Select data	User	A window with any graph (humidity, temperature, eCO_2, TVOC, Pressure) must be oppend and the user must press the Select Data button	The graph of the day selected by the user is displayed	If there is no measurement make on the chosen day, nothing is displayed	The user must select the data from the calendar to the right of the window and then must press the Select Data button

In the sequence diagram, user interaction with the system is presented in Fig. 2, by acquiring data and displaying a humidity graph. The user starts the data acquisition by pressing the Start button, next to "Data acquisition", Windows Form sends a request to the Arduino to calculate the data, and data is then sent back to Windows Form. Windows Form sends the measured data to the SQL database and will display it in the main window. This operation is being in a loop until the user wants to stop data acquisition. The user stops the data acquisition by pressing the Stop button, thus interrupting the loop.

The user can see the graph for humidity by pressing the Open button. The user chooses a date from the calendar, by pressing the SelectPeriodOfTime button. After these operations, Windows Form sends a request with the selected date to SQL and it sends the data for the selected date. Windows Form generates the graph for the user-selected data and displays it.

The database is a simple one, consisting of only one table, represented by 7 columns, described in Table 2.

Table 2. Database structure

Tabel columns name	Role
ID	Represents the number of measurements, in the whole number and having autoincremented
Humidity	Represents the humidity measured from the sensor, represented in float data type
Temperature	Represents the temperature measured on the sensor, represented in float data type
eCO_2	Represents the equivalent carbon dioxide measured from the sensor, represented in float data type
TVOC	Represents the total of volatile organic compounds measured from the sensor, represented in float data type
Pressure	Represents the atmospheric pressure measured from the sensor, represented in the float format
Date	Represents the date on which the measurement was made, represented in DataTime, which saves each time in the format "month-day-year hour-minute-second" (MM-DD-YYYY hh-mm-ss)

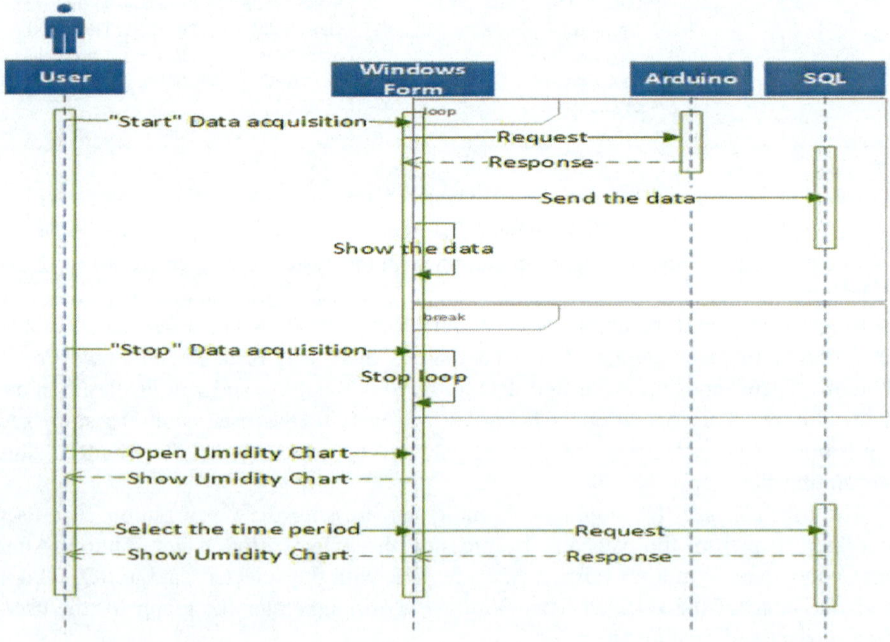

Fig. 2. Sequence diagram for a data acquisition scenario and humidity graph display

Each graph represents the data measured on a specific user-selected day in the calendar, implemented with the DateTimePicker class. The DateTimePicker format is a custom MM-dd-yy format, so you can make selecting the date from the database easier and straightforward in the standard format. These data were added one by one in a list, we used a list because we do not know its final size. The size of the list increases dynamically as needed. To generate the graph the user must select the date in the calendar and press the SelectPeriodOfTime button. After accessing the button, all the points that were on the graph are deleted and the new points for generating the new graph, selected on the calendar day, are generated. The data is retrieved by the GeteCO_2() and GetDate() methods, which access the database to create a new measurement list, on the date chosen by the user from the graphical interface.

3.2 Hardware Design

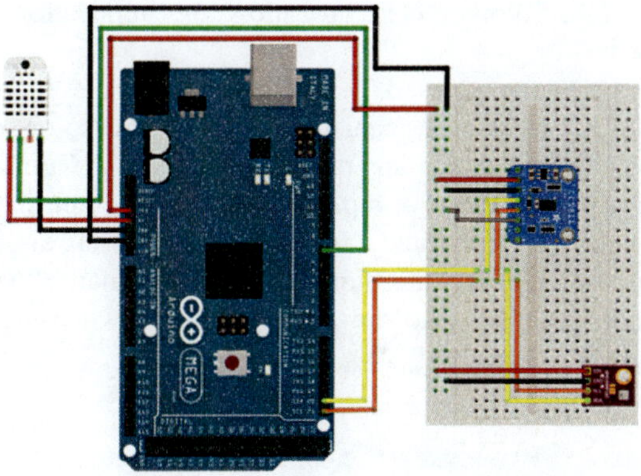

Fig. 3. Arduino hardware assembly

The hardware components are Arduino Mega (the microcontroller), DHT22 (temperature and humidity sensor), CCS811 (gas sensor), and the BMP280 (pressure sensor). Figure 3 presents the Arduino hardware assembly, on the left side of the picture is the Arduino board, and on the other side is the breadboard with the connection presented above.

4 Results

The graphical user interface (GUI) allow users to acquire, view and generate sensors data graph (as show in Fig. 4). To start acquiring data, the user must press the Start button. By pressing this button, a serial connection of the port is done, to be able to connect the Windows Form application with the Arduino Mega board and by default with the application loaded on this microcontroller. The reading is done line by line, the first measurements being skipped to ensure a better reading of the data. The data, being read through Serial via ReadLine() method, is in string form and is converted to double form to be ready for saving in the SQL database.

If the measured levels of eCO_2 or TVOC exceed the maximum permissible values, the application warns the user through a message, a MessageBox, with the text "Danger! LOW CO_2 LEVEL! Open the window!", or, if the TVOC value exceeds the maximum permissible values, the MessageBox with the text "Danger! A LOW TVOC LEVEL! Open the window!" is shown. The measured data is then represented on the main window next to the date and time at which the measurements were taken.

The connection to the database is made through SqlConnection, the data being saved in the SQL database, in the Measurements table. After saving the data, the user is greeted with a successful message "Added Data", and the connections to the port and the database respectively are closed for security. This is done in a loop once every 10 min until the user wants to stop the data acquisition.

By accessing the Stop button, the data acquisition is stopped. If the data acquisition is open (it is in a continuous loop) and at the same time the Stop button is pressed we must use the BackgroundWorker class, to allow the application to be used while another thread is active.

After the Stop button is pressed, a successful message for stopping the data acquisition appears, like "Stop data acquisition". To be able to open a graph, the user must access the Open button, next to any tag of the graph (e.g. Humidity Graph), which allows the opening of a new form of application for each graph individually.

In the next subsection are presented a few indoor monitoring air quality cases. Also, it can be seen in the next figures, the humidity generated daily, the temperature evolution after the window is open and the eCO_2 evolution when the freshener is used closer to the sensors.

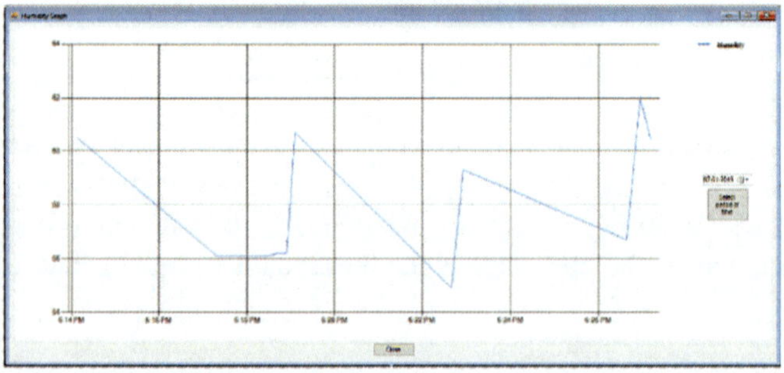

Fig. 4. Humidity graph generated on 1/7/2019

4.1 Opening the Window Case Scenario

The sensors were placed for this test near the window. The graph generated after opening a window is shown in Fig. 5. It can be easily observed that the humidity drops inside the room, from 73% to about 56%, only by opening a window. In the case of the temperature, it can be easily observed that an increase of up to 4.5 °C can be detected, the temperature being somewhere at 26° in the room, and the outside temperature is over 34°.

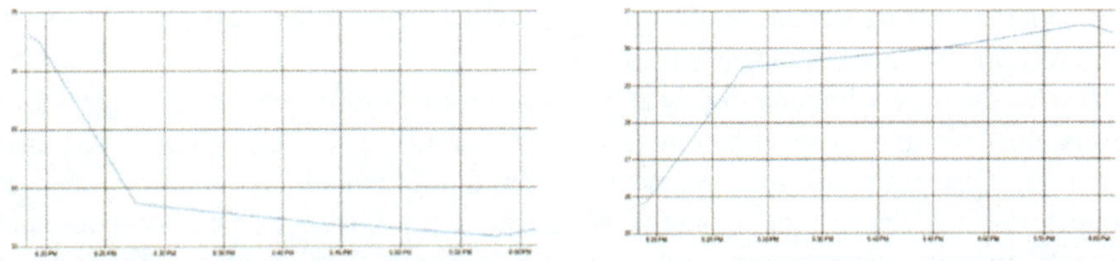

A. Humidity graph **B.** Temperature graph

Fig. 5. Recorded values after opening a window (A - himudity, B - temperature)

4.2 Using an Air Freshener Scenario

For this situation, a freshener was used to see the level increase of carbon dioxide equivalent (eCO_2). The graph generated in this case can be seen below in Fig. 6.

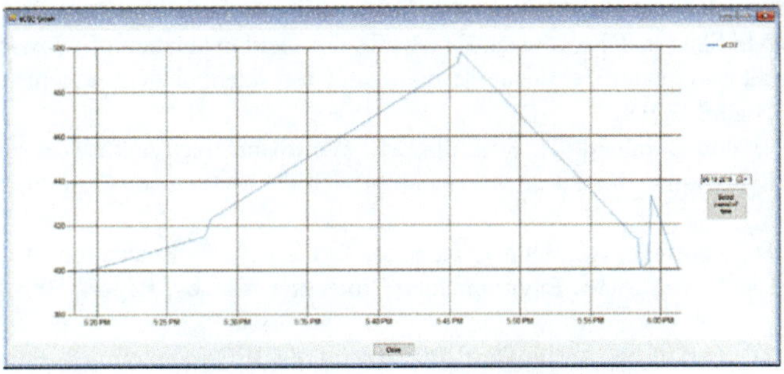

Fig. 6. eCO_2 graph using an air freshener

5 Discussions

Following the development of this project, a system of indoor air quality monitoring was developed for anyone who wants to track the graphical evolution of temperature, humidity, pressure, or gases emitted. This system is intended for use in indoor environments. Data from the Arduino board sensors are transmitted and stored in a database. The aim of this project is the development of an air quality monitoring system with a low cost and easy to use, the total cost being about 60 euros.

The software implementation has an easy to use interface for anyone, being very intuitive. As a result of this implementation, the application manages to acquire the data measured by the sensors that populate the Arduino board, display them and store them in a database, which can then be used to create graphical day-to-day monitoring.

For the future development of the application, a data prediction algorithm can be implemented. Higher performance hardware with better accuracy can be used, but this would mean a higher cost. Also, a control system, for example, ventilation control, an automated system (or remote control) of opening the window when the normal limits of the parameters are exceeded, could be implemented. Also, a wireless shield could be added to remotely communicate with the device.

The hardware implementation was made using the Arduino board and specialized sensors at a low price. This choice was made because, their configuration is very easy, having the possibility of improvement in the future by adding other sensors, for example, formaldehyde sensor, gas resulting from the burning of carbon-containing materials, such as cigarette smoke; dust sensor; carbon monoxide sensor; ozone sensor and sulfur dioxide sensor.

Conflict of Interest. The authors declare that they have no potential conflict of interest relevant to this article.

References

1. Zender-Swiercz, E., Telejko, M.: Indoor air quality in kindergartens in Poland. In: IOP Conference Series Materials Science and Engineering, vol. 471, no. 9, p. 092066 (2019)
2. Rangel, A.M., Sharpe, T.R., Musau, F., Mcgill, G.: Field evaluation of a low-cost indoor air quality monitor to quantify exposure to pollutants in residential environments. J. Sens. Sens. Syst. **7**, 373–388 (2018)
3. Worner, D., Von Bomhard, T., Roschlin, M., Wortmann, F.: Look twice: uncover hidden information in room climate sensor data. In: IEEE Conference, Cambridge, MA, USA (2015)
4. Williams, R., Kaufman, A., Hanley, T., Rice, J., Garvey, S.: Evaluation of elm and speck sensors. EPA, United States Environmental Protection Agency, Report, EPA/600/R-15/314 (2015)
5. Parkinson, T., Parkinson, A., de Dear, R.: Continuous IEQ monitoring system: context and development. Build. Environ. **149**, 15–25 (2019)
6. Gajewski, A.: Indoor air quality in a bus. WIT Trans. Built Environ. **134**, 749–757 (2013)
7. Kang, J., Hwang, K.: A comprehensive real-time indoor air-quality level indicator. Sustainability **8**(9), 881 (2016)
8. Marques, G., Pitarma, R.: An indoor monitoring systems for ambient assisted living based on Internet of Things architecture. Int. J. Environ. Res. Public Health **13**(11), 1152 (2016)
9. Abraham, S., Li, X.: A cost-effective wireless sensor network system for indoor air quality monitoring applications. In: The 9th International Conference on Future Networks and Communications (FNC-2014) (2014). Procedia Comput. Sci. **34**, 165–171 (2014)
10. Kalinin, Y.S., Velikov, E.K., Markova, V.I.: Design of indoor environment monitoring system using Arduino. Int. J. Innov. Sci. Mod. Eng. **3**(7), 46–49 (2015)
11. Sung, Y., Lee, S., Kim, Y., Park, H.: Development of a smart air quality monitoring system and its operation. Asian J. Atmos. Environ. **13**(1), 30–38 (2019)
12. Ji, H., Zhang, H., Zhou, K., Liu, P.: Design of indoor environment monitoring system based on Internet of Things. In: IOP Conference Series Earth and Environmental Science, vol. 252, p. 042012 (2019)
13. Sung, W.-T., Hsiao, S.-J., Shih, J.-A.: Construction of indoor thermal comfort environmental monitoring system based on the IoT architecture. Sensors **2019**, 1–16 (2019)

Automated Sorting of Pharmaceutical Waste Using Machine Vision Technology

Laszlo Rapolti[1(✉)], Rodica Holonec[1], Laura Grindei[1], and Oana Viman[2]

[1] Technical University of Cluj-Napoca, Cluj-Napoca, Romania
laszlo.rapolti@ethm.utcluj.ro
[2] University of Agricultural Sciences and Veterinary Medicine of Cluj-Napoca, Cluj-Napoca, Romania

Abstract. There is an increasing concern regarding the adequate disposal of pharmaceutical waste. Since the detection of active pharmaceutical waste in the landfill soils and waters, several policy changes have been implemented. Most of the pharmaceutical waste is gathered from households, thus probably a new recycling policy should be implemented. This paper aims to prevent the pharmaceutical waste from getting into the soil by sorting any incoming pills in different categories so these can be disposed of properly without the risk of contaminants getting in the environment. The pills sorting system uses a machine vision based system. Pills are put on a conveyor belt which transports them through the sorting area. Pills images are analyzed and consequently they will be placed into different bins with the help of multiple palettes driven by stepper motors. This method ensures an efficient solution that could be implemented in any landfill sorting station.

Keywords: Pharmaceutical waste · Pill sorting · Waste management

1 Introduction

Pharmaceutical waste is one of the main types of residues that mankind has not acknowledged for a long time. Although till this day the short- and long-term effects of these substances are not known. It seems that these are starting to be noticed in aquatic environments and are triggering the focus of researchers to study them further since these substances could potentially be a dangerous contaminant to the environment [1–3]. The majority of used pharmaceutical pills end up in a landfill directly through municipal or industrial waste and potentially also in the surrounding soil [4–6]. As for the wastewater facilities these are not designed to remove pharmaceutical compounds [7]. To remove these compounds, a nanofiltration of reverse osmosis treatment would be needed, which are high cost processes and are difficult to implement. Pharmaceutical specialist have the potential to be on the front line of this developing problem as medical professionals and drug specialist are in excellent position to instruct the understanding about safe medication disposal [8]. Appropriate patient advising on safe medicine disposal can have a noteworthy effect to the general wellbeing and climate. A handy methodology must to be there to join this significant issue for example in the

S. Vlad and N. M. Roman (Eds.): MEDITECH 2020, IFMBE Proceedings 88, pp. 409–416, 2022.
https://doi.org/10.1007/978-3-030-93564-1_45

educational program as this cause has a great importance. There ought to be a few standards and tougher rules implemented for the disposal of medicine. Cautious and legitimate removal of prescriptions can assist with diminishing ecological heap of medications [9].

This paper aims to deliver a method to avoid soil and water contaminates by sorting waste pharmaceutical pills before getting to a landfill with the help of Machine Vision technology [10]. The general idea behind a machine vision system is to automize the whole process with the help of cameras and by using different types of robotic arms. In this case a conveyor belt with rollers was used to transport the pharmaceutical pills underneath a camera for classification and identification [11, 12]. After the classification process the pills travel further ahead on the conveyor belt and with the help of stepper motor-controlled palettes, they are pushed into their own sorting bins. Thus, finalizing the sorting process without the need of human input.

2 System Description

The Fig. 1 illustrates the block diagram of the experimental system. This consists of:

- A Microsoft LifeCam web camera
- Four NEMA 17 stepper motors
- A A4988 drivers for each stepper motors
- An Arduino UNO board

The camera is used for a visual inspection of the pills. The conveyor belt carries the pharmaceutical pills from the loading point to the web camera, where it stops to allow the system to run the classification algorithm. After the classification algorithm has ended, the exact steps needed for the pills to travel on the conveyor is sent to the Arduino board, so for them to get to their exact location to be sorted.

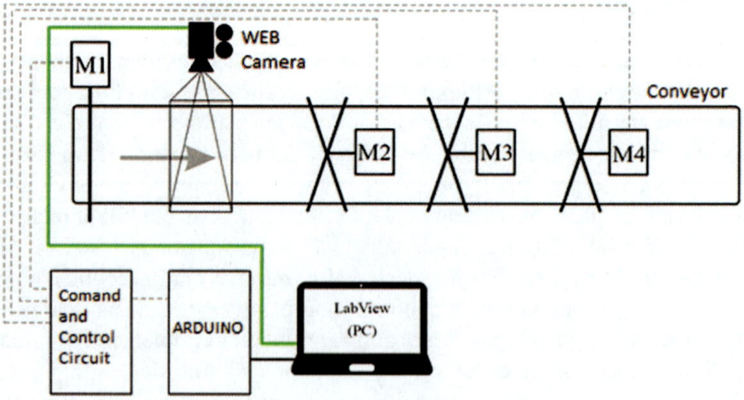

Fig. 1. Block diagram of the experimental system

The stepper motors were chosen because of the high precision control. They are equipped with pallets (Fig. 2a) which removes the classified pills from the conveyor belt. The palettes are arranged in a cross pattern seen in Fig. 2b. This way there is no need for the them to return to a standard or "home" position. The palettes rotate clockwise or counterclockwise as needed. The turn signal is provided by the main controller. By using step motors, we precisely controlled the number of degrees needed to rotate and to to push the pill off the conveyor belt. This way the pills end up in the proper sorting bin.

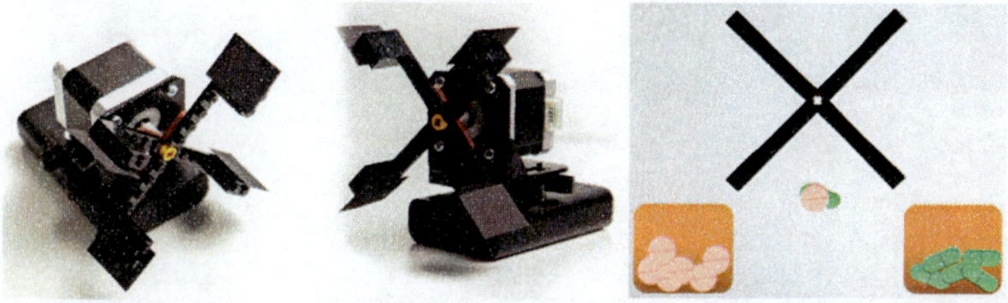

(a) (b)

Fig. 2. a. (left) Stepper motors equipped with palettes; b. (right) The palettes cross pattern.

Our experimental system is designed in order to classify up to five types of pills and includes a sixth bin for unknown or unclassified pills. The control of the motors is realized by using four dedicated drivers (Fig. 3) connected to the Arduino board. The drivers supply the voltage needed to power the stepper motors and control their speed and rotation direction.

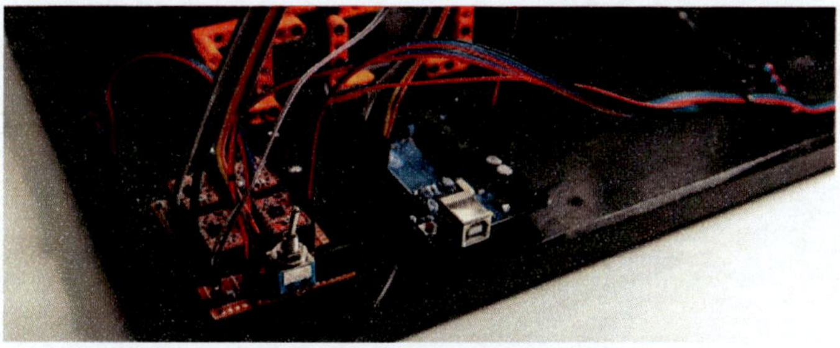

Fig. 3. The experimental systems control units

With the help of LabVIEW software, the communication between the PC running the classification algorithm and the Arduino was established using Serial communication with the help of a USB cable.

2.1 Sorting Algorithm Structure

The main objective of the project was to sort the incoming pills on a conveyor belt. (Fig. 4) The camera positioned above the conveyor belt is used to determine the class of the pill positioned underneath the camera. In order to raise the precision of the vision algorithm a light source was placed above the camera. This ensures that no foreign objects would be detected by the software.

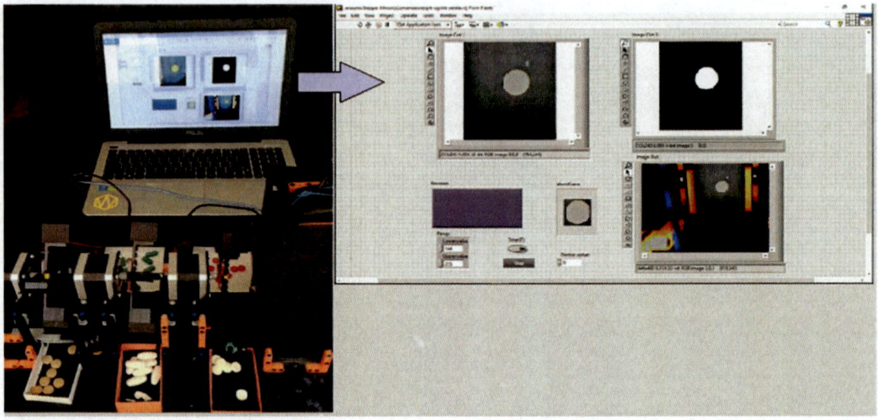

Fig. 4. Experimental stand for pharmaceutical pill sorting and the user interface

The presence of the pill under the camera is determined after the next image processing steps (Fig. 5):

- The application of a mask for selecting the area of interest
- The extraction of the HSV value color plane for obtaining a greyscale image
- The application of a threshold (for bright objects) for obtaining a binary image
- The application of an edge detection algorithm by applying a horizontal line for determine the number of edges. If these number is equal with two it results that the pill is present on the conveyor belt and a command is given for stopping the belt movement.

In the next step of the algorithm, a "Color pattern Matching" operator is applied and according to the pill colored learned templates the pill is recognized respecting the score imposed by the user (Fig. 6). In this moment the conveyor will transport the pills next to the pallets that will push them into the proper bin.

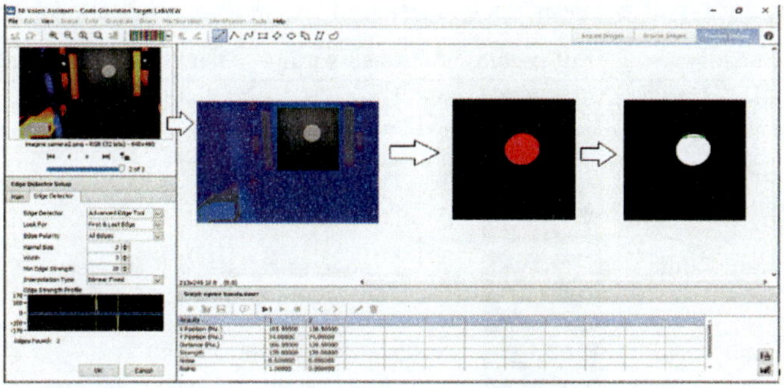

Fig. 5. Pill detection algorithm

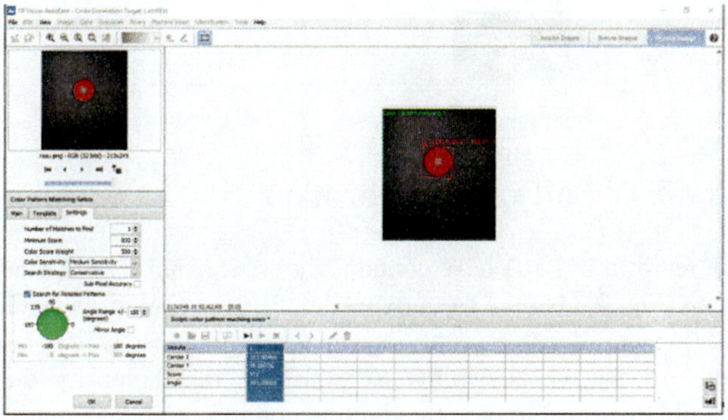

Fig. 6. Color pattern matching algorithm.

2.2 Speed Control and Distance Calculation

The NEMA 17 stepper motors where chosen to be used in this experiment because of their high precision and for the possibility of them to be controlled in an open loop. For the motor to take a single angular step, the command must be equal to an angle of 0,45 degrees. A full rotation of the rotor means 800 steps. A classical sorting system would need multiple position locating sensors to determine in what position and where the actual object is which increases the cost. To reduce this cost, we decided that the image processing algorithm can be used to provide the exact position of the pharmaceutical pills. Meaning that the exact distance starting from the pills position in the processed image till the mechanic pallets can be precisely calculated. (Table 1.)

Table 1. Travel distances (motors steps) of all the recognized pill categories

Pill category	Pill picture	Actioned sorting pallet	Distance to travel
Round and White		Nr. 1 move to right	550 steps
Round and Red	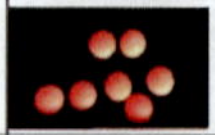	Nr. 1 move to left	550 steps
Oval and White		Nr. 2 move to right	1200 steps
Oval and Green		Nr. 2 move to left	1200 steps
Round and Yellow		Nr. 3 move to right	1850 steps
Unknown	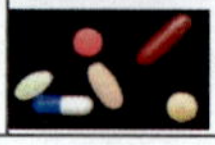	Nr. 3 move to left	1850 steps

3 The LabVIEW Software Application

The application realized in LabVIEW contains the front panel for the user interface and its block diagram (Fig. 7). On the front panel the user can find three displays that show the image recognition process and the one that tells the user which is the identified class. The user also has the possibility to adjust the thresholds for the image binarization. This could be useful in case the lighting conditions change. On the other hand, the block diagram contains the actual program sequences. The program contains a sequence structure with five different frames for each step of the sorting process. In the first frame the user's input is waited to start the sorting process. When the user activates the sorting, a 3 s delay is applied for the user to confirm every setting is set, as necessary. In the second frame of the program the conveyor belt starts to run until the image processing software detects an objects presence. The software detects the objects with the help of the "Edge detection" algorithm that there are pharmaceutical pills present on the conveyor, sending a stop signal to the conveyor belt. This triggers the next frame of the sequence which takes a picture of the pills and applies the classification algorithm on it. Once the pill is classified a number is memorized from -1 to 4. The following sequence provides a one second delay which ensures that all the numbers have been memorized and the vision software has ended its cycle. The memorized number signals the succeeding sequence which class it has found and how many steps are needed for the pill to be put in a sorting bin. In the last sequence the Arduino is signaled to send a specific number of steps to the conveyor belt and in which direction to rotate the stepper motors palettes. If the algorithm doesn't recognize the current pill the value of -1 is selected which means that the current pill must be discarded in the sixth recycle bin. To control the motors, a separate case structure was used with three timed sequences to ensure the ending of each process.

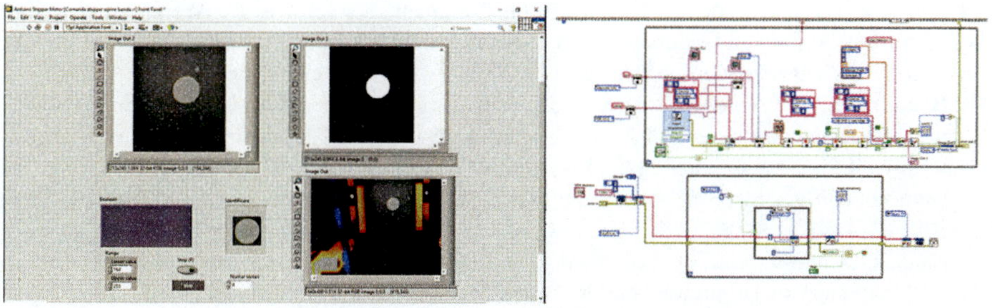

Fig. 7. Front panel and block diagram of the program

4 Conclusions

Pharmaceutical waste management is in general a special problem because it must be solved with minimum risk. Quantities of dangerous chemical substances contained, for example, in different medications stored at homes, pharmacies or hospitals can be thrown into the municipal waste stream. Proper management of such type of waste is often a big challenge.

We designed and implemented a sorting system able to perform solid pills detection and classification on a conveyor belt. The idea comes from our previous work and experience gained in a research project regarding the sorting of electrical and electronic waste.

The system was tested for 100 different expired pills. The drugs where placed randomly, one by one, on the conveyor belt. The images templates were learned and the pills were recognized based on a color pattern matching algorithm. The vision-based sorting system captures images, classifies and sorts according to the matched class. A 97% sorting accuracy was achieved.

For future work we plan to improve the sorting system by using a more complex and rapid machine vision system equipped with multiple sorting robotic arms.

Acknowledgments. This work was supported by a grant of the Romanian Ministry of Research and Innovation, CCCDI-UEFISCDI, project number PN-III-P1-1.2-PCCDI-2017-0652/84PC CDI/2018, within PNCDI III.

Conflict of Interest. The authors declare that they have no conflict of interest.

References

1. Heberer, T.: Occurrence, fate, and removal of pharmaceutical residues in the aquatic environment: a review of recent research data. Toxicol. Lett. **131**, 5–17 (2002)
2. Ternes, T.A.: Occurrence of drugs in German sewage treatment plants and rivers. Water Res. **32**, 3245–3260 (1998)

3. Carlsson, C., Johansson, A.-K., Alvan, G., Bergman, K., Kühler, T.: Are pharmaceuticals potent environmental pollutants? Part I: environmental risk assessments of selected active pharmaceutical ingredients. Sci. Total Environ. **364**, 67–87 (2006)
4. Bound, J.P., Voulvoulis, N.: Pharmaceuticals in the aquatic environment–a comparison of risk assessment strategies. Chemosphere **56**, 1143–1155 (2004)
5. Ruhoy, I.S., Daughton, C.G.: Types and quantities of leftover drugs entering the environment via disposal to sewage - revealed by coroner records. Sci. Total Environ. **388**, 137–148 (2007)
6. Zhao, Y., Chen, Z., Shi, Q.: Monitoring and long-term prediction for the refuse compositions and settlement in largescale landfill. Waste Manage. Res. **19**, 160–168 (2001)
7. Bungau, S., et al.: Aspects regarding the pharmaceutical waste management in Romania. Sustainability **10**, 2788 (2018)
8. Bokhoree, C., Beeharry, Y., Makoondlall-Chadee, T., Doobah, T., Soomary, N.: Assessment of environmental and health risks associated with. APCBEE Procedia 9, 36–41 (2014)
9. Kadam, A., Patil, S., Patil, S., Tumkur, A.: Pharmaceutical waste management an overview. Ind. J. Pharm. Pract. **9**, 2–8 (2016)
10. Zhi-Ying, Y., Yue, W., Cheng-lin, W.: Research on Intelligent Sorting System Based on Machine Vision, School of Logistics, Beijing Institute of Material Key Laboratory of Logistics System and Technology Beijing China (2009)
11. Sahu, R., Swain, M.K., Bihari Swain, K., Patnaik, R.K.: Productive and economical sorting of objects by low resolution camera. In: IEEE International Conference on Smart Technologies and Management for Computing, Communication, Controls, Energy and Materials (ICSTM), Veltech. Dr. RR & Dr. SR University, Chennai, 2–4 August 2017, pp. 290–294. Paginile (2017)
12. DevisionX: https://devisionx.com/industrial-applications/pharmaceutical. Accessed 7 June 2020

Pain Points in the Experiences of Diabetes Patients with Ambulatory Care. An Analysis of Patients Fears and Recommendations for Service Improvement in Romania

Madalina-Alexandra Cotiu[1(✉)], Anca Constantinescu-Dobra[1], and Adrian Sabou[2]

[1] Faculty of Electrical Engineering, Technical University of Cluj-Napoca, Cluj-Napoca, Romania
madalina.cotiu@enm.utcluj.ro
[2] Faculty of Automation and Computer Science, Technical University of Cluj-Napoca, Cluj-Napoca, Romania

Abstract. Diabetes is a chronic metabolic disease with complex management where treatment adherence is crucial for positive outcomes. Research indicates patient satisfaction is associated with patients' propensity to follow medical recommendations and better treatment adherence. The paper investigated the main fears patients have when accessing medical care for diabetes in an ambulatory setting, as well as their suggestions for service improvement. This was done to contribute to a better understanding of patients' perceptions regarding the medical care they receive, and the pain points they have in this regard. Healing pain points can lead to higher patient satisfaction. A questionnaire comprising two open questions regarding patients' fears and suggestions for improvement was distributed in Romania. Demographic, medical, and attitudinal patient data were also collected. The study sample size was 324 respondents. Text analysis and crosstabulation were performed. Results indicated patients most feared a worsening condition and waiting times. Suggested improvements focused on reducing waiting times, increased access to resources, and better clinic facilities.

Keywords: Diabetes patients · Fear · Service improvement

1 Introduction

Diabetes is a chronic, metabolic disease characterized by elevated levels of blood glucose. There are two main types of diabetes: type 1 – juvenile or insulin-dependent diabetes in which the pancreas produces little or no insulin by itself and type 2 diabetes which occurs when the body does not produce enough insulin or becomes resistant to it. Type 2 diabetes usually occurs in adults and has been linked with obesity and lack of physical exercise [1]. Over time diabetes can lead to complications including serious damage to the heart, kidneys, eyes, blood vessels and nerves [1]. Approximately 463 million adults were suffering from this disease in 2019, and the number is projected to

S. Vlad and N. M. Roman (Eds.): MEDITECH 2020, IFMBE Proceedings 88, pp. 417–423, 2022.
https://doi.org/10.1007/978-3-030-93564-1_46

rise to 700 million by 2045. Diabetes caused at least 760 million USD in health expenditure in 2019, which represents 10% of the total healthcare spending on adults [2].

Diabetes management involves constant care and medical supervision, as well as sustained efforts on behalf of each patient and their family and friends. To achieve positive results and prevent complications, both treatment adherence and increased interest in following medical recommendations are required. Research has indicated that patients who are satisfied with the care they receive, tend to be more inclined to follow medical recommendations and have better treatment adherence. At the same time, patient satisfaction has also been linked with increased staff morale and better medical outcomes [3, 4]. Patient satisfaction refers to patients' attitudes regarding different aspects of care [5]. Although research in the area is vast, there is no commonly agreed model on what determines patient satisfaction, or the satisfaction of patients with diabetes. Previous studies suggest determinants of patient satisfaction with diabetes care include interaction with the medical staff, clinic facilities, waiting times, procedures for accessing treatment, demographic factors, as well as prevention and educational services provided, complications associated with the disease, patient autonomy, and disease management [6, 7, 8].

The study we propose aims to contribute to existing research on patient satisfaction by identifying patients' fears regarding medical services received in the ambulatory setting, as well as their proposals for service improvement. By acknowledging patients' pain points regarding ambulatory medical services, organizations will be able to better direct their efforts for service improvement. The study is relevant to medical professionals and representatives of medical facilities who wish to improve the quality of the services they offer.

2 Methodology

We performed a quantitative research with convenience sampling. A questionnaire eliciting responses regarding different aspects of care as well as demographic data was distributed in printed and online formats between July 2014-February 2015 among patients with type I and type II diabetes in Romania. 339 valid responses were collected, representing a 39.5% response rate. Two open ended questions asked respondents to indicate their worst fear connected to accessing ambulatory medical care and suggest one improvement they would make, should they have the resources to. Demographic, medical, and attitudinal data were also collected.

Data collected were processed using IBM SPSS Statistics v. 23. Qualitative and quantitative analyses were performed. Descriptive statistics were used to analyze the sample structure. Responses for the two open-ended questions were coded into group categories based on the theme they referred to. Crosstabulation, and Fisher's Exact test (suitable for sample size <1000) [9] were performed to identify correlations between the different group categories and the following demographic, medical and attitudinal data: age, income level, diabetes type, associated complications, treatment duration, the length of their relation with the current physician, patients' attitude towards diabetes management. A relevance threshold of 5% as the percentage of patients indicating a certain category was used to determine whether a particular group category would be further analyzed via crosstabulation.

3 Results

3.1 Sample

Most respondents (51,5%) were aged between 36 and 64, 17,9% were under 36 and 30,6% were over 65 years old. Respondents tended to have low income, 62,1% earning less than the minimum wage.

In terms of their medical condition, 33,7% of patients suffered from type I diabetes and 62,3% declared they also suffered from complications associated with the disease. Approximately one third of respondents (35,1%) had been diagnosed less than five years before, while most patients (55,5%) had been diagnosed between 5 and 18 years prior to the study. Respondents tended to have long-term relationships with their physicians as 48,2% of patients had been treated by the same doctor for more than 5 years.

Most patients only considered themselves somewhat careful regarding diabetes management (56,3%), while only 38,6% declared they were very careful in managing their disease.

3.2 Patients Pain Points Regarding Ambulatory Care Services

Patients' pain points were investigated via an open question inviting respondents to share their worst fear related to appointments for ambulatory medical care. Following text analysis, patients' responses were grouped in the following five categories: waiting time, worsening disease, unpleasant interactions with the medical staff, lack of resources and personal-related fears. The latter was only indicated by 2.4% of patients so no further analyses were performed on this category as it did not meet the 5% relevance threshold.

Waiting time referred to patients' fear that after arrival in the ambulatory clinic they would have to wait considerable amounts of time before being seen by their physician. Patients indicated waiting times of up to 2–3 h with negative impact on their daily routines (e.g. "I wait a lot and this messes my schedule."). This fear was shared by 15,9% of patients. No relationships were identified between this category and demographic data indicating this is a pain point shared irrespective of patient demographics.

Worsening disease referred to patients' fears of bad blood test results (i.e. high levels of glycated hemoglobin), disease associated complications, having to switch to insulin treatment or that their condition would require hospital admittance (e.g. "I am afraid of bad test results."). This fear was most widely shared, indicated by 22.6% of patients. Crosstabulation showed this fear was associated with patients' age and diabetes type. Older patients (Fisher's exact test = 0.005) tended to be more afraid of a worsening condition (29.7% in the over 65 age group and 23.5% in the 34–65 age group) than younger patients (8.5% in the under 35 age group). Regarding diabetes type, patients with type II diabetes are more afraid of a worsening condition (29.1%)

than their type I counterparts (9.8%). These results may be because diabetes patients are aware that complications tend to appear more in older patients. At the same time, patients with type II diabetes know that a worsening condition means increased insulin resistance and maybe switching to insulin as opposed to oral antidiabetics. They do not wish for such a negative progression, so they may fear news in this regard.

Regarding interactions with the medical staff, 7% of patients fear these will not be pleasant. This involved patients' fears regarding inappropriate attitude of the medical staff (i.e. lack of patience, superficial interactions, lack of availability to provide explanations regarding the disease or the treatment prescribed, or medical staff scolding patients for bad test results). One patient indicated "I fear my doctor, how she interacts with me, she is not kind". A relationship was identified with patients' characteristics including age (Fisher's exact test = 0), income (Fisher's exact test = 0.013), and diabetes type (Fisher's exact test = 0.022). This fear was mostly present in younger patients (22% in the under 35 age group vs. 4.1% and 4% in the 36–64 and over 65 years old groups, respectively) and higher income patients (66.7% in patients earning over 1.000 euro, vs. 18.5% in those earning less). Patients with type II diabetes were less impacted by such interactions (4.5%) than patients with type I diabetes (11.6%). An explanation could be that these patients need more care and attention from their physician, as well as requiring more information about their disease. At the same time, they are more likely to experience anxieties and more difficulties in managing the disease. Therefore, they may fear unpleasant interactions as these will not help put them at ease. In the case of higher income patients, it is possible that because they have the means to access high quality services in their daily lives, they have lower tolerance for unpleasant interactions.

The lack of resources was feared by 8.5% of patients. This related to patients' fears they would not be able to access the treatment they needed (e.g. free prescriptions, blood tests, hospital admittance, quality consults etc.) because of lack of funding. Analyses showed a relationship with patients' age (Fisher's exact test = 0.009) in that older patients tended to be less fearful of this (2% in patients over 65 years old vs. 11.2% in those aged 35–64 and 11.9% in those under 35 years old). This may be because older people lived in times of scarcity so they may be more used to the lack of resources than their younger counterparts. A relationship was also present with regards to the length of the disease (Fisher's exact test = 0.006). Patients diagnosed less than 1 year before the study (12.5%) and those diagnosed between 6 and 15 years previously (13.9%) tended to fear lack of resources more than the other categories. One motive could be that patients diagnosed recently are still adjusting to the disease, so they need stability. At the same time, those diagnosed between 6 and 15 years could be seeing their disease progressing, so resources are important to them. We estimate patients having the disease for over 15 years have become accustomed to the system and its shortcomings and have developed coping mechanisms.

3.3 Patients Suggestions for Improving Ambulatory Care Services

After sharing their fears, patients were asked to indicate one aspect of ambulatory medical services they would change if they had the power to. Responses were investigated via text analyses and grouped in the following six categories: reducing waiting times, improved access to resources, more attention given to patients, improved service accessibility, more medical staff and more education. The latter was only indicated by 2.7% of patients so no further analyses were performed on this category.

Reducing waiting times before seeing the physician was suggested by 13% of respondents. No relationships were identified between this category and patient demographics suggesting this is an improvement wished by patients irrespective of their profile.

Improved access to resources was suggested by 11.3% of patients and involved better access to tests and treatments including (glycemic profiles, insulin pumps, specific medicines and tests for monitoring glucose levels). A relationship was identified between this category and patients' income levels (Fisher's exact test = 0.04), and complications associated with the disease (Fisher's exact test = 0.046). It was interesting to note that patients with higher income levels wished better access to treatments in a much higher proportion that those with lower income levels (66.7% in the over 1.000-euro group and only 23.3%). This may be because they are aware of their contributions to the medical system, so they expect access to medical resources in return. Patients with diabetes complications proposed a change in this regard in higher proportion (13.5%) than those not experiencing complications (6.4%). This was to be expected since patients with complications will require more medical attention in the form of tests, medication etc.

More attention given to patients referred to their desire to feel seen and heard during a consultation, to be allowed enough time to ask their questions and discuss their fears. Patients indicated "I would like more time to discuss with my doctor". This group category was indicated by 9.4% of patients and a relationship was identified with patients' age (Fisher's exact test = 0), and diabetes type (Fisher's exact test = 0). Younger patients (27.1% of the under 35 group vs. 11.6% for those over 35 years old), and patients with type I diabetes (21.4% vs. 3.2% for patients with type II diabetes) suggested this change the most. This is consistent with demographic data for patients indicating unpleasant interactions as their fear when accessing ambulatory medical care.

Improved service accessibility was indicated by 12.6% of patients and referred to better location, better waiting areas and more flexible scheduling. Patients suggested "better waiting conditions for receiving my recipe", "bigger clinic space", "adequate waiting areas" or "more flexible scheduling". A relationship was identified with the duration of the relationship with the current physician (Fisher's exact test = 0.0037), as well disease self-management (Fisher's exact test = 0.004). Patients with relationship durations of less than 5 years tended to suggest this improvement more (31.3%) than the rest (14.3%). An explanation could be that patients having a longer relationship with their physician have gotten used to the surroundings and the protocol and no longer pay much attention to shortcomings in this regard. Patients declaring themselves as somewhat careful (17%) or not careful (17,6%) in the management of their disease asked for better treatment facilities. This may be because they are in a worse condition when accessing medical care than patients attentive to disease management (5,4%).

More personnel referred to patients' desire for a higher number of medical staff available in the ambulatory care. Patients' answers suggested they wished for more staff in the hope of reduced waiting times and more time available for individual consults. The improvement was suggested by 8.5% of respondents and a relationship was identified with the length of the disease (Fisher's exact test = 0.02). Patients diagnosed up to five years prior to the study tended to suggest this improvement more (41.3%) than those diagnosed more than 6 years before the study (13%). Probably, patients having the disease for over five years have adjusted their expectations to the possibilities of the medical system and tend to ask less for things their experience showed were difficult to obtain.

4 Concusions

Patients' worst fear regarding their ambulatory care appointment was they would discover their condition had worsened. This was followed by increased waiting times, the lack of resources and unpleasant interactions with the medical staff. A relationship was identified between these fears and patients demographic data including age, diabetes type, income, and duration of the disease.

Patients' suggestions covered both systemic and unit-level improvements. Suggestions were consistent with their fears, indicating that respondents focused on solving acute issues they were facing, rather than focusing on ideal services. Patients suggested reduced waiting times, better access to resources, improved medical interactions and clinic facilities, as well as more personnel. It worths noting that two out of five categories of improvements refer to the medical staff which suggests patients do not feel comfortable with these interactions and associate the medical personnel, or lack thereof, with some of the shortcomings they encounter when accessing ambulatory medical care.

While systemic changes could be more difficult to implement, patients also indicated improvements that could be conducted at unit level. Such interventions could include more exact patient scheduling for their appointments (e.g. scheduling by time interval, not only by day) or workshops for developing communication skills among medical staff, improved waiting areas in the sense of accessibility etc. Further investigation into these categories is necessary to detail exact interventions, but we believe this list of improvement categories may represent a starting point for improving ambulatory care for diabetes patients.

Acknowledgement. This research was co-financed through the European Social Fund through Sectoral Operational Programme Human Resources Development 2007–2013, project number POSDRU159/1.5/S/142115 "Performance and Excellence in Doctoral and Postdoctoral Research in Romanian economics science domain".

Conflict of Interest. The authors declare they have no conflict of interest.

Informed Consent. All participants in the study were provided with an informed consent document detailing the purpose, method and duration of the study. Participation in the research was entirely voluntary.

References

1. World Health Organisation (WHO). https://www.who.int/health-topics/diabetes#tab=tab_1. Accessed 5 Sept 2020
2. International Diabetes Federation (IDF). https://www.idf.org/aboutdiabetes/what-is-diabetes/facts-figures.html. Accessed 5 Sept 2020
3. Boudreaux, E.D., O'Hea, E.L.: Patient satisfaction in the emergency department: a review of the literature and implications for practice. J. Emerg. Med. **26**, 13–26 (2003)
4. Taylor, C., Benger, J.R.: Patient satisfaction in emergency medicine. Emerg. Med. J. **21**, 528–532 (2004)
5. Jenkinson, C., Coulter, A., Bruster, S., Richards, N., Chandola, T.: Patients' experiences and satisfaction with health care: results of a questionnaire study of specific aspects of care. Qual. Satisfaction Health Care **11**, 335–339 (2002)
6. Saatchi, E., Tahmiscioglu, G., Bozdemir, N., Akpinar, E.: The well-being and treatment satisfaction of diabetic patients in primary care. Health Qual. Life Outcomes **8**, 67–75 (2010)
7. Redekop, W.K., Koomanschap, M.A., Stolk, R.P., Rutten, G.E.H.M., Wolffenbuttel, B.H.R., NIessen, L.W.: Health-related quality of life and treatment satisfaction in Dutch patients with type 2 diabetes. Diabetes Care **25**, 458–463 (2002)
8. Diț, L., Băban, A., Dumitrașcu, D.L.: Impactul severității bolii asupra satisfacției cu calitatea îngrijirii medicale la pacienții cu diabet de tip 2. Clujul Med. **85**, 573–577 (2012)
9. Field, A.: Discovering Statistics using IBM SPSS Statistics. Sage, Thousand Oaks (2018)

Are Electrical and Hybrid Vehicles Safe for Human Health?

Madalina-Alexandra Cotiu(✉), Anca Constantinescu-Dobra, and Claudia Steluta Martis

Faculty of Electrical Engineering, Technical University of Cluj-Napoca, Cluj-Napoca, Romania
madalina.cotiu@enm.utcluj.ro

Abstract. The past few decades have acknowledged a revival of producers and consumers' interest in electric vehicles. The aim of the paper is to analyze if batteries used in EV or HEV have a negative effect on human health, especially for the driver and passengers. Articles published in the last 20 years were considered.

The studies were chosen from three databases using specific keywords. A number of 33 articles and book chapters were assessed, but only 13 were taking into account in this study because the focus was on the healthcare for human beings and not on the designing, optimizing or testing EVs.

The problems related to the interference of electromagnetic field was the most investigated issue regarding EVs, hybrid and human health. The articles in the sample state that the highest mean of electromagnetic field was reached in moving condition vs. standing condition also, the higher the speed of the car is, the higher the fields are. The results demonstrate that EMF emissions can be substantial, especially at high-power transfer levels and misaligned conditions. There are many studies that support the safety of the EVs that can be compared with conventional cars. Although, few researchers request the need to re-draw fine limits of the existing guidelines in the fields, by including metrics in certain and specific conditions.

Keywords: Electric vehicles · Human health · Electromagnetic field

1 Introduction

1.1 EV Market Evidence

In the present context of increased interest directed to stopping climate change and reducing greenhouse gas emissions, electric vehicles (EVs) are becoming important triggers for boosting sustainable development trends. All major producers across the world have included electric vehicles in their model range. Furthermore, several new producers are entering the electric vehicles (EV) manufacturing industry drawn by constant estimates of positive market development for such vehicles. At present, transportation accounts for 26% of all primary energy consumption and automobiles are particularly important as they are generally dominating street traffic in most countries. Car sales are projected to register strong growth rates in the future and the forecast of EVs production states an increase with 37 million units per year in ten years

S. Vlad and N. M. Roman (Eds.): MEDITECH 2020, IFMBE Proceedings 88, pp. 424–431, 2022.
https://doi.org/10.1007/978-3-030-93564-1_47

[1]. At the same time, as indicated by the World Health Organization, exhaust emissions directly affect the humans' health, many inhabitants all over the world suffering from different respiratory diseases determined by low air quality. On the same line, there is an increasing concern of researchers along with the potential clients in providing or choosing safe EVs especially from the point of view of the interference of the magnetic field of the battery or the motors.

Electric vehicles present several benefits such as cutting-edge technology and connectivity, high energy efficiency, emissions reduction, small size, noise reduction, lower running cost than their conventional counterparts, as well as full city access. They are, at the same time, also confronted with several limitations such as limited driving range, charging infrastructure, cost issues and potential impact on the power system. Sales are expected to continue to grow, with electric vehicles reaching 21 million in 2030 and EVs expected to account for 70% of total EVs that same year [1]. At present, the penetration of EVs is still low, but rising with a sustained pace. According to [1], lately, on the global electric market there are approximatively 1 billion customers. The biggest demand was registered in China and US. In the year 2017 the hybrid and EV's market share surpassed 1% of global car sales. Two main factors were identified as driving EV market expansion: policy regulation aimed at curbing gas emissions and customer demand as potential consumers become more and more aware of EV benefits

The aim of the paper is to analyze if batteries used in EV or HEV have a negative effect on human health, especially for the driver and passengers.

2 Electric Vehicles and Human Health. Evidence from Literature Review

There are a lot of scenarios concerning the connection between the usage of the EV or hybrid cars and human health, especially because of the pipe gas reduction in urban/crowded area. Also, there are a lot of initiatives related to analyzing the carbon footprint of a conventional car in comparison to an EV. The generated electromagnetic fields (EMF) of EVs represent one concern of potential users.

Generally, the researches pointed out that the drivers and the passengers in a car are safe on long term exposure relatively to the level of EMFs inside a traditional or electric vehicles.

On one hand, the powertrains and some component of the electric systems, inverters or batteries are the main components inside of cars that can generate EMFs, or by the power cables from the electric system, inverters, EVs' batteries and cordless communication systems [3].

On the other hand, some studies reveal that EMFs, particular static magnetic fields (SMFs) can have some negative influences on the human health, by affecting the following: neurological coordination of eye-hand [4], remembrance and focus for short time, the optical function, and the responsiveness time. Several researches [3] have not found conclusive findings in their effort of investigating the correlation between EMF exposure on long periods of times and some diseases, as cancer or neurodegenerative disorders.

Therefore, in order to provide a safer environment for the passengers and drives, the International Commission on Non-Ionizing Radiation Protection (ICNIRP) has elaborated a guide containing protection principles against worming-up effects to EMF, exposed organism tissues or to nervous transmission [5].

2.1 Methodology

The authors assessed studies from three databases: Science Direct, Springerlink and IEEExplore, by using the following keywords:

- Human health,
- Electric vehicle,
- Hybrid vehicle,
- Electromagnetic field.

The authors defined as another constraint the publishing period of the research to be between years 2000 and 2020, because of the evolution of the constructive models of EVs.

A number of 33 articles and book chapters were assessed, but only 12 were taken into account in this syntheses study, because the focus was on the health care for humans and not on the designing, optimizing or testing EVs. We have to mention also that we eliminated the studies with the exclusive goal of investigation the human health under the influences of some equipment that exist in an EV, that is not a definitory for EV (the electric motors for air conditioning or heating for example). Moreover, if the study referred to a comparison between different types of cars (EV, hybrid, diesel, gasoline or gas fueled cars) in connection with human health, or a specific segment of patients (e.g. persons with cardiac diseases), it was included in the sample. The studies dealing with the health and less pollution because of the EV were also eliminated from the sample.

2.2 Findings

The objectives of those studies along with the methods used and study results are displayed in Table 1.

Table 1. The summary of the main objective, methods and findings of the articles in the sample.

Author	Objective	Sample/Method	Findings/Observations
1. Hareuveny R et al. [2]	Investigate electromagnetic fields (EMF) from diesel, gasoline and hybrid cars; Analysis over all four seats; Analysis over different types of driving	10 cars: three diesel, four gasoline, and three hybrid cars	Hybrid cars had the highest mean of MF levels; EMF higher for moving conditions compared to standing; EMF higher at 80 km/h compared to 40 km/h EMF in conventional car - stronger at front seats, EMF in hybrids - stronger at back seats

(*continued*)

Table 1. (*continued*)

Author	Objective	Sample/Method	Findings/Observations
2. Hsu et al. [7]	EFM in EV at different level of a human body	7 different electric cars, one hydrogen-powered car, two gasoline-fueled cars and one diesel-fueled car	Exposure at less than 2 percent of the non-ionizing radiation limit at head-height; The highest EMF- close to the ground of the EVs, near to the cells; The emissions grow up quickly during power-on mode in comparison with power off
3. Tell et al. [8]	EV and broadband magnetic fields	6 gasoline-powered vehicles and 8 were electric vehicles of various types	All fields measured in vehicles in the sample were not achieve the recommended/imposed limits by the International Commission on Non-Ionizing Radiation Protection (ICNIRP) and the Institute of Electrical and Electronics Engineers (IEEE)
4. Lennerz et al. [9]	EV, patients with pacemakers or defibrillators	108 persons, (in cars Nissan Leaf, BMW i3, VW eUp, Tesla S)	No modification in equipment function or programming was registered; The EMF identified – over the charging cable length, when is used energy at higher intensity; The EMF emission in the passengers' compartment was reduced
5. Yang L et al. [10]	Extremely low frequency of EMF in EVs, long-term (one year) surveillance; Investigations -at the front and back row seats in the cabin during acceleration and constant-speed driving modes	3EV	Extremely low frequency of EMF are approximately constant on long-term driving or regular maintenance; The necessity of periodically control extremely low frequency in EVs, especially after important crashes

(*continued*)

Table 1. (*continued*)

Author	Objective	Sample/Method	Findings/Observations
6. He et al. [11]	Static electric fields and human neuropsychological reactions and brain activity; Driving reaction time test	17 subjects	The static electric fields have no significant impact of the investigated subject from driving performance point of view and the cognitive capabilities; Was found a connection between beta sub-band of the EEGs and the reaction time of the investigated subjects
7. Erdem et al. [6]	Analyzing the wireless electric vehicle charging systems, pointing out the hurdles and safety of intermediate frequency EMF emissions for wireless electric vehicle charging applications	2 study cases, different wireless power transfer investigations in different energy levels	EMF emissions are even higher especially at high-power transfer levels and systems/ conditions are improperly aligned; -should be reduced below the limits identified in the ICNIRP 2010 guidelines
8. Lin J et al. [12]	Analyzing the impact of EMF on children in EVs	10 EV, different driving sessions	Considerably lower limit in than the boundary of ICNRP; "Although small children may be exposed to higher EMF strength, were much lower than that of adults due to their particular physical dimensions"
9. Xiaolin et al. [13]	Investigation of EMF impact in an EV, verified by wireless power transmission	A simulation model using an EV systems for cordless power transmission and a manequin with three levels of tissues	"The results of simulation show that the maximum electromagnetic radiation dose SAR under the magnetic resonance working state is 2.1693 * 10–13 W/Kg, which is lower than the safety limit of ICNIRP|"
10. Cheng et al. [14]	Influence on human body of a complex automotive electromagnetic environment	Car simulation model, through electromagnetic characteristics of antenna model	Automotive electromagnetic environment is almost safe to passengers
11. Sapunaru et al. [15]	Evaluating the generated level of radiated emissions in different operating modes	real radiated emissions tests in a semi-anechoic chamber on an EV	The differences between different operating modes, different location of the charging and different types of charging cables
12. Moreno-Torres et al. [3]	Assessing the EMF in the interior of EVs	EV, na	With appropriate design guidelines, it might be possible to make electric vehicles safe from the electromagnetic radiation point of view

The interference of EMF was the most investigated issue regarding EVs, hybrid cars and human health. The articles state that the highest mean of EMF was reached in moving condition vs. standing condition [2, 9]. Also, the higher the speed of the car is, the strong the fields are [2, 15]. In addition, EMF in hybrid cars is higher than in the conventional ones [3, 12, 14].

Referring to the specific place in a car, in conventional ones, EMF tend to be most powerful in front seats, while for hybrid cars, the persons from rear seats (especially in the right side) are most likely to be affected in a higher extent (with a percentage between 16–69%, depending on different state of driving) [2]. The highest indicator for EMF was detected along the charging cable, while the EV is charged with high currents (116.5 μT) [12]. However, all the measured indicators respect the international standards for producers.

Furthermore, in EVs, the highest indicator for EMF was registered at feet level (with 20% higher than chest or head level) inside of an EV's cabin [3]. Tell et al. [8] claimed that the magnetic fields reached the same level in gasoline fueled car like in EVs.

Linn et al. [12] examined small children exposure in an EV, giving the fact that their heads are near to the ground and, consequently to the battery of the EVs. The conclusions suggested children are exposed to a strong EMF, but the health impact is lower for them in comparison to an adult, because of their specific corporal size.

Yang et al. [10] demonstrate that for long periods of time (one year), the extremely low frequency of EMF does not modify, if the regular maintenance of the EV is properly done. Moreover, it is important the design process of the car in order to avoid components misalignment. At the same time, He et al. [11] investigated static magnetic field and find that drivability and the ability of brain to process the information are not significant influenced by EVs. The researchers still found a connection between the highest beta sub-band and slowness in response capability of the driver.

For drivers or passengers with pacemaker or defibrillator, EVs are completely safe, because physiological function of the body or the parameters of the attached dispositive were not affected or modified during the experiments.

In their research, Erdem et al. [6] consider ICNRIPF 2010 guide "conservative" because it does not integrate fully specific situations investigated in correlation with EMF power, especially for cordless EV charging systems. The results demonstrate that EMF emissions can be substantial, especially at high-power transfer levels and misaligned conditions. Closing the debate, Moreno-Torres et al. [3] sustained an EV can be safe from EMF perspective if the producers carefully monitor the design phase of an EV.

3 Conclusion

The present article does not intend to offer a final verdict in how safe are EVs and hybrid cars for the human body. The most assessed studies emphasize on meeting the existing standard for human health. The concern of EMF emissions is present in society, hence, this dimension is investigated in a great extent by the researchers.

The results support there is no reason for the driver and passengers to worry about electric cars by exposing them to high risks, even when the period of time spent in the cabin is long. There are many voices that support the safety of the EVs that can be compared with conventional cars. The warning side of the investigated article is related equally with the battery charging process of EVs that can be sometimes risky for human tissues because of the powerful emissions of EMF. Although, few researchers request the need to re-draw fine limits of the existing guidelines in the fields, by including metrics in certain and specific conditions.

Subsequently, the risks implied for driving an EV can be increased due to some drivers, such as:

- The short distance between equipment (e.g. traction drive) and passenger
- The longer time of exposure
- The aggressive style of driving
- The wireless speed charging

This issue remains however nuanced, even though the most evidences strengthen the less invasive impacts for those people who use an EV.

Conflict of Interest. The authors declare they have no conflict of interest.

References

1. IEA: Global EV Outlook 2019. https://www.iea.org/reports/global-ev-outlook-2019. Accessed 21 Sept 2020
2. Hareuveny, R., et al.: Characterization of extremely low frequency magnetic fields from diesel, gasoline and hybrid cars under controlled conditions. Int. J. Environ. Res. Public Health **12**, 1651–1666 (2015)
3. Moreno-Torres, C.P., Velez, P., Lafoz, M., Arribas, J.R.: Passenger exposure to magnetic fields due to the batteries of an electric vehicle. IEEE Trans. Veh. Technol. **65**, 4564–4571 (2016)
4. De Vocht, F., Stevens, T., Glover, P., Sunderland, A., Gowland, P., Kromhout, H.: Cognitive effects of head-movements in stray fields generated by a 7 Tesla whole-body MRI magnet. Bioelectromagn. **28**, 247–255 (2007)
5. International commission on non-ionizing radiation protection. Guidelines for limiting exposure to time-varying electric, magnetic, and electromagnetic fields (up to 300 GHz). Health Phys. **74**, 494–522 (2010)
6. Erdem, A., Mostak, M., Omer, C.O., Jason, P., Veda, G., Gui-Jia, S.: Review of safety and exposure limits of EMF in wireless electric vehicle charging applications. In: IEEE Transportation Electrification Conference&Expo, 23–26 June 2020 (2020). https://doi.org/10.1109/ITEC48692.2020.9161597
7. Hsu, J.: Magnetic fields in electric cars won't kill you. IEEE Spectrum (2014)
8. Tell, R.A., Sias, G., Smith, J., Sahl, J., Kavet, R.: ELF magnetic fields in electric and gasoline-powered vehicles. Bioelectromagn. **34**(2), 156–161 (2013). https://doi.org/10.1002/bem.21730
9. Lennerz, C., Horlbeck, L., Weigand, S.: Patients with pacemakers or defibrillators do not need to worry about e-Cars: an observational study. Technol. Health Care **28**(1), 1–12 (2019). https://doi.org/10.3233/THC-191891

10. Yang, L., et al.: Long-Term monitoring of extremely low frequency magnetic fields in electric vehicles. Int. J. Environ. Res. Public Health **16**, 3765 (2019). https://doi.org/10.3390/ijerph16193765
11. He, Y., Sun, W., Leung, P., Chow, Y.: Effect of static magnetic field of electric vehicles on driving performance and on neuro-psychological cognitive functions. Int. J. Environ. Res. Public Health **16**(18), 3382 (2019). https://doi.org/10.3390/ijerph16183382
12. Lin, J., Lu, M., Wu, T., Yang, L., Wu, T.N.: Evaluating extremely low frequency magnetic fields in the rear seats of the electric vehicles. Radiat. Prot. Dosim. **182**, 190–199 (2018)
13. Xiaolin, L., HaoKun, C., Yu-ting, S., Feixiang, G.Y.: Sun Electromagnetic Radiation Safety Verification of Electric Vehicle's Wireless Power Transmission based on Magnetic Resonance. In: Asia Energy and Electrical Engineering Symposium (2020). https://doi.org/10.1109/AEEES48850.2020.9121495
14. Cheng, Q., Zhong-Min, D.: Engineering, personal safety in electromagnetic environment of electric vehicle. Int. J. Sig. Process. Image Process. Pattern Recogn. **9**(6), 103–114 (2016). https://doi.org/10.14257/IJSIP.2016.9.6.09
15. Săpunaru, A., Ionescu, V., Popescu, M., Popescu, C.: Study of radiated emissions produced by an electric vehicle in different operating modes, environmental science. In: EV International Conference (2019). https://doi.org/10.1109/EV.2019.8893142

Using an Unsupervised Neural Network to Detect and Categorize Offensive Language in Social Media

Emil Stefan Chifu(✉), Viorica Rozina Chifu, and Ana-Maria Costea

Department of Computer Science, Technical University of Cluj-Napoca, Bariţiu 28, 400027 Cluj-Napoca, Romania
Emil.Chifu@cs.utcluj.ro

Abstract. The offensive language in present social media could harm the mental health of the minors. The paper aims to identify the offensive content in the posts published on the Twitter social network. More concrete, we categorize the text content of the tweets according to the characteristics they meet: offensive vs. non-offensive tweets, non-targeted tweets vs. tweets targeted on someone, and, more specifically, tweets targeted on an individual, on a group of people, or else targeted towards another category (i.e. towards an organization, a situation, an event, or an issue). This multi-level hierarchical categorization behaves like a top-down decision tree process that classifies the tweets against a tree like ontological taxonomy. This is an offensiveness taxonomy, which defines the above mentioned offensive tweet categories. We use an unsupervised neural network for this hierarchical categorization, as applied on the OLID (Offensive Language Identification Dataset) data set. The OLID dataset consists of tweets, and it was offered as benchmark at Task 6 of the SemEval 2019 competition, a task which actually inspired our paper.

Keywords: Categoriztion of offensive content · Text categorization · Unsupervised neural network · Offensiveness ontology

1 Introduction and Related Work

Although social media has become one of the most important ways to communicate, posts are often accompanied by abusive language that can have detrimental effects on users, and especially on the mental health of the minors. This phenomenon is a result of the ease with which anyone can now join a social media community. Fake accounts often manipulate the users.

The identification and classification of offensive content are usually modeled as a supervised classification problem [9], where systems are trained on posts annotated according to the presence of abusive form or offensive content. As opposed, our approach is unsupervised, and consequently it needs no annotated tweets. Specifically, unsupervised approaches do not need a corpus of tweets where each tweet is tagged with its proper category of offensive content. This is the main advantage over the supervised machine learning based methods.

S. Vlad and N. M. Roman (Eds.): MEDITECH 2020, IFMBE Proceedings 88, pp. 432–441, 2022.
https://doi.org/10.1007/978-3-030-93564-1_48

A good survey of the research on detection and categorization of offensive content from text is [9]. Among many other features, the authors of this survey mention the relatedness of this research field with the sentiment analysis. From this point of view, the approach in our current paper is similar with our previous work on the field of sentiment analysis [8].

The supervised approaches use traditional machine learning or deep learning. The traditional learning systems (the most used being SVM, Random Forests, Logistic Regression) work well when representing the text with n-gram vectors (bag-of-words, TF-IDF). Deep learning (CNN, LSTM) tends to use the text as represented by sequence order, with word embedding vectors. These are confirmed for instance in [5], where the authors describe their system as participant in the HASOC 2019 competition (Hate Speech and Offensive Content Identification in Indo-European Languages) [6]. The task in this competition is similar with Task 6 of the SemEval 2019 competition in the framework of the International Workshop on Semantic Evaluation 2019 [1].

Our paper aims to identify and categorize the offensive content in the Twitter social environment, being inspired from the above mentioned Task 6 of the SemEval 2019 competition. For this purpose, we use an unsupervised machine learning method; more specifically, the approach will use an unsupervised neural network as an offensive language classifier.

The data set provided for Task 6 of SemEval-2019 – i.e. the Offensive Language Identification Dataset (OLID) – consists of posts (tweets) extracted from the Twitter social network, where users have the freedom to express themselves freely. At a first glance, one can easily notice those offensive tweets, some directed at other people, organizations, and political groups, others accompanied by URLs and followed by words considered as vulgar. When classifying such tweets, we use the bag of words to represent the text of a tweet.

2 The Neural Network Based Classifier

We started from one of the tasks of the SemEval 2019 competition [1], and we wished to identify and categorize the offensive content in the posts on Twitter. In most cases, it can be difficult to recognize, among long, veiled posts, the intention of the author, which can later prove to be the initiator of an attack (on a person, a group, an organization etc.). Offensive content can occur in a variety of forms, from insults often considered harmless, to serious threats against clearly identified people, as well as manipulation, instigation of public opinion through vulgar language or hate speech, meant to encourage violence.

In our approach, the categorization of the offensive content in texts is ensured by an unsupervised neural network. The neural model used – Enrich-GHSOM [4] – is an extension of the Growing Hierarchical Self-organizing Maps (GHSOM) [3] (see Fig. 1), originating from the Kohonen Self-organizing Maps [2]. This model is applied on the OLID dataset (Offensive Language Identification Dataset) [10], which is actually a set of tweets. The tweets in the OLID dataset are annotated by following a three-level hierarchical annotation schema[1]: besides saying the content of a tweet is offensive or not (level 1), in case it is offensive the annotation also considers the type of offensive content (level 2), and furthermore the target of the insult (level 3). This hierarchical categorization is best represented as a taxonomic ontology, illustrated in Fig. 2.

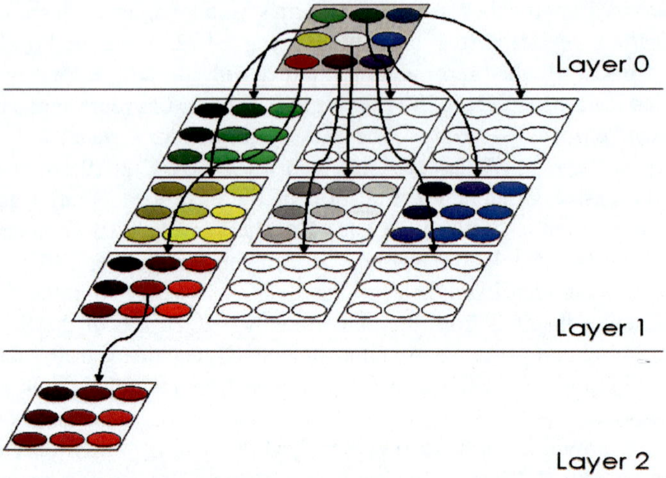

Fig. 1. The growing hierarchical self-organizing maps.

Task 6 of the SemEval 2019 competition – which inspired this paper – used the hierarchically annotated OLID data set consisting of tweets. In Task 6, the main idea was to extract and categorize tweets that use offensive language. This categorization should be hierarchical, i.e. it should comply with the hierarchical annotation of the OLID tweets, according to the taxonomy in Fig. 2. Actually, Task 6 of SemEval 2019 is divided into three sub-tasks, corresponding to the three depth levels of the taxonomic tree in Fig. 2:

- Sub-task A (level 1). *Offensive language identification*: NOT (not offensive) and OFF (offensive).
- Sub-task B (level 2). *Automatic categorization of the offensive content*: TIN (targeted insult) and UNT (untargeted).
- Sub-task C (level 3). *Identifying the target of the offensive language*: IND (when the target is an individual), GRP (the target is a group of people considered as a unity), and OTH (other, when the target is in neither of the first two categories, i.e. not IND, nor GRP).

[1] https://scholar.harvard.edu/malmasi/olid.

In our approach, this three-stage hierarchical classification is done in a single phase. The hierarchical classification of the tweets behaves like a top-down decision tree classifier against a predefined tree, where the predefined tree is set to be just the tree like ontology described above and illustrated in Fig. 2.

The embedding vectors of the tweets seen as short text documents propagate downwards starting from the root of the ontological taxonomy of tweet categories (illustrated in Fig. 2). This process behaves like a top-down decision tree classifier, being based on the vector similarity between the tweets that get classified on one hand and the concepts (i.e. the categories, the nodes) of the taxonomy on the other hand. Each tweet arrives finally at a node that defines (as discussed above) a category of tweets. The conclusion for such a classification of a tweet is that the tweet would belong to that category of tweets which is defined by the node. One notice is needed here: in any decision tree classifier the leaf nodes are the final result classes found by the classifier. We are interested in our approach as well to only classify the tweets against the leaf nodes of the offensiveness ontology in Fig. 2 (the nodes in blue color in Fig. 2). In other words, we combine the three sequential subtasks (A, B, and C) proposed in SemEval 2019 Task 6 into a single classification phase, as we only care about the final – the most fine-grained – category of offensive tweets. As such, we only measure the classification of the tweets in the leaves of the offensiveness ontology (Fig. 2). The leaves are the following: NOT, UNT, IND, GRP, and OTH.

We represent any tweet as a "classic" embedding vector, more exactly as a bag of words. The features of this vector are the numbers of occurrences of different words in the tweet seen as a short text document. In the experiments described in Sect. 3, the embedding vectors of the tweets have 4625 features each, since the vocabulary of the corpus of tweets used in the experiments consists of 4625 distinct word forms.

Equally like the tweets have an embedding vector, the nodes of the ontology (i.e. the ontology concepts) as well need a representation as a numeric feature vector. The process of top-down classification – as ensured by the Enrich-GHSOM neural network [4] –, and more exactly the decision to follow one of the children of the current ontological concept node depends on the Euclidean similarity between the embedding vector of the classified tweet on one hand and the vectors of the children nodes on the other hand. The most similar child node is the one chosen. As such, we have to define the feature vector of each ontological node. By looking at the ontology in Fig. 2, we only see some abstract concepts (categories), not anchored to any (more or less offensive) piece of text. Trying to establish this missing anchor though, we consolidated the offensiveness ontology, by attaching some additional leaf children to every leaf concept in Fig. 2. What used to be leaf nodes in Fig. 2 (the color blue nodes in Fig. 2) now become parents for the newly inserted leaves. The result is the ontology illustrated in Fig. 3.

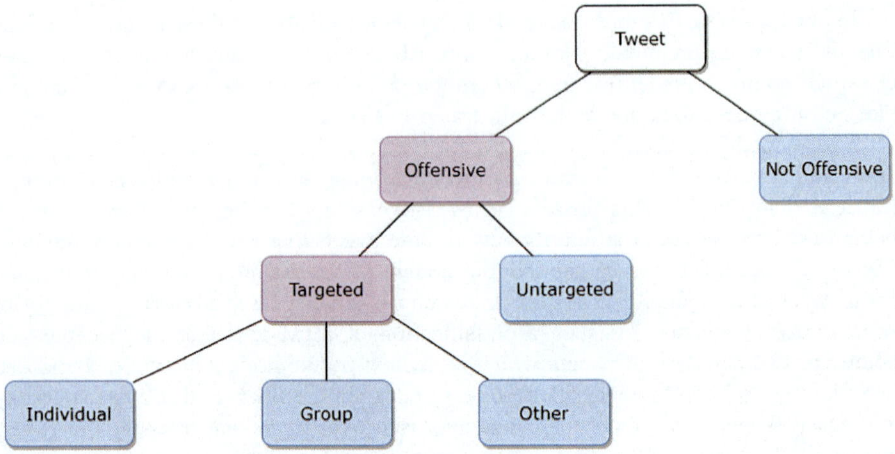

Fig. 2. The offensiveness taxonomic ontology.

Actually, we enriched the offensiveness ontology (Fig. 2) by choosing and considering some tweets which are representative for the five leaf categories in Fig. 2, and then simply defining these tweets as new leaf nodes inserted as children under what used to be leaf categories in Fig. 2. These representative tweets are able to provide an embedding vector powerful enough to represent the semantics of a leaf category. The feature vector of a leaf category is then computed as the *centroid* (average vector) or *category*[2] (unit-normalized sum vector) of its subtree (actually the *centroid* or *category* of all of its newly inserted tweet-leaf children, i.e. *centroid* or *category* of the embedding vectors of the tweets representative for the leaf category). As a consequence, we expect these new, representative, leaf nodes will have the power to suitably attract the correct, semantically similar tweets – tweets that are similar with each other and, more importantly, similar with the feature vector of one of the leaf categories in Fig. 2 i.e. just the leaf category into which we expect these tweets will get classified. We expect this to happen during the classification ensured by the Enrich-GHSOM neural network. For each of these newly inserted leaves of the taxonomy (each of which originating from a tweet), a genuine, true tweet – i.e. a real world tweet, as present in the OLID data set – should be used as representative.

The vector for the nonterminal category nodes of the ontology (i.e. the nodes colored mauve in Fig. 2) does not need to be anchored to any representative tweets (as newly inserted leaf children) from the corpus, as was the case for the leaf categories. Rather we consider for any intermediate category a vector built as a *centroid* or *category* of the vectors of all the nodes in the subtree of the category.

[2] We highlight always in italics the *category* vector as the unit normalized sum of vectors, to make a difference to the category of tweets, the latter being rather a node (concept) in our offensiveness ontology (in Fig. 2).

```
is_a(OFF, tweet)
is_a(NOT, tweet)
is_a(TIN, OFF)
is_a(UNT, OFF)
is_a(IND, TIN)
is_a(GRP, TIN)
is_a(OTH, TIN)
is_a(76669tweet#28, NOT)
is_a(45855tweet#484, NOT)
is_a(35940tweet#490, NOT)
is_a(36737tweet#506, NOT)
is_a(30899tweet#48, UNT)
is_a(80947tweet#70, UNT)
is_a(42196tweet#156, UNT)
is_a(10684tweet#186, UNT)
is_a(83681tweet#5, IND)
is_a(46444tweet#38, IND)
is_a(73516tweet#87, IND)
is_a(33394tweet#491, IND)
is_a(58995tweet#29, GRP)
is_a(40386tweet#72, GRP)
is_a(24040tweet#85, GRP)
is_a(59691tweet#495, GRP)
is_a(27158tweet#108, OTH)
is_a(15998tweet#59, OTH)
is_a(10918tweet#228, OTH)
```

Fig. 3. The offensiveness ontology (in Fig. 2) enriched with representative tweets as children for the leaf categories in Fig. 2.

3 Experimental Results

In our experiments, the data set on which the offensive content categorization is made is represented, as already mentioned, by OLID. The OLID dataset contains a number of 14,100 annotated tweets, divided into a training partition of 13,240 tweets and a test partition of 860 tweets. In the experimental evaluation, we applied our neural network based classifier on the 860 tweets of the test partition.

3.1 The Gold Standard

We exploited the hierarchical annotation of the OLID corpus, by considering the gold labels present in the OLID dataset files for the three hierarchical levels (layers) 1, 2, and 3 (corresponding to the three SemEval 2019 Task 6 subtasks A, B, and C respectively). The gold labels mapped to the 860 tweets in the OLID test set constitute for us a "gold standard" reference for evaluating the accuracy of the offensive content classification in our experimental setting. Actually we only take into account the five gold labels representing the five leaves in the offensiveness ontology in Fig. 2: NOT, UNT, IND, GRP, and OTH, since we are only interested in classifying the tweets against the leaf categories of the offensiveness ontology in Fig. 2, as already mentioned in Sect. 2. The distribution of these five leaf gold labels on the 860 tweets of the OLID test set is the following: 620 tweets are labeled NOT, 27 tweets are labeled UNT, 100 are labeled IND, 78 tweets are GRP, and 35 are OTH.

A fragment of the gold standard mapping is illustrated below, followed by the full text content of a tweet in each of the five leaf category labels. The tweets given subsequently in full text are marked with a '*' in this illustrative gold standard fragment:

```
tweet : leaf gold label
15923tweet#0 : OTH
27014tweet#1 : NOT *
30530tweet#2 : NOT
13876tweet#3 : NOT
46139tweet#64 : GRP *
34263tweet#15 : IND *
96874tweet#6 : NOT
27158tweet#108 : OTH *
78910tweet#8 : NOT
46363tweet#9 : NOT
68123tweet#10 : NOT
22452tweet#11 : NOT
15565tweet#12 : NOT
64376tweet#13 : NOT
42196tweet#156 : UNT *
```

27014tweet#1:

#ConstitutionDay is revered by Conservatives, hated by Progressives/Socialist/Democrats that want to change it.

46139tweet#64:

#Conservatism101 It's not about our disagreements with #Conservatives. Its that Conservatives can't debate honestly, and they have no integrity. Whatever gets them thru today, is all that matters to them. They're fundamentally dishonest people. URL

34263tweet#15:

#StopKavanaugh he is liar like the rest of the #GOP URL

27158tweet#108:

Amazon will ship live Christmas trees to your door — but will they stay and put on the damn lights? URL

42196tweet#156:

"#TheArchers I want to be brilliant at Karate so I can kick the shit out of everyone"""""

3.2 Evaluation Measures

The *accuracy* of a system that assigns offensive content categories to tweets quantifies the percentage of tweets correctly labeled by the system, out of the total number of tweets given as input to the system (the system being the offensiveness classifier):

$$accuracy = \frac{|correctly\_labeled\_tweets|}{|given\_tweets|} \quad (1)$$

For the experimental evaluation, we adopt the accuracy measure as described and defined above (Formula 1). We also used the *learning accuracy* metric [7], which is a semantic based accuracy measure, and, as such, suitable to our semantic oriented approach. "Near misses" in the classification are strictly counted as zero by the standard *accuracy* (according to formula 1), and leniently as non-zero by the *learning accuracy*. A chance is given by the *learning accuracy* metric, by judging that an intermediate category node chosen by the classifier is not that wrong when it is a generalization category (ancestor in the ontology) of the correct leaf category node. For instance, it is not so wrong to classify a tweet as "Targeted" (TIN - nonterminal category), as far as it is expected (according to our gold standard) to be "Individual" (IND, targeted towards an individual - leaf category). Consequently, in the experimental results illustrated in what follows, the *learning accuracy* will always have higher values than the standard *accuracy*.

3.3 Evaluation Results

The experimental evaluation consists in comparing the offensive content categorization as ensured by our classifier with the gold standard mapping described in Sect. 3.1. Actually, we measure the *accuracy* and *learning accuracy* of labeling the tweets with the correct, gold standard, leaf nodes. These results are illustrated in Table 1.

In the first column of Table 1 (*Experimental Setting*), we have the following abbreviations: (*i*) *F* means *flat*, i.e. the embedding vector of any tweet has as features flat counts (flat frequencies) of occurrences of different words in the tweet; (*ii*) *T* means *TFIDF*, i.e. the "term (word) frequency times inverse document frequency" weighting scheme has been applied on the flat count embedding vectors of the tweets; (*iii*) *C* and *K* mean *centroid* and *category* respectively, i.e. the two different ways of defining the vector associated to a nonterminal node of the offensiveness taxonomy (Fig. 2), as already mentioned in Sect. 2: *centroid* of the vectors of the leaves in the subtree of the node, and *category* of the vectors of the leaves in the subtree; (*iv*) *N* means *normalized*, i.e. the embedding vectors of the tweets are normalized to unit norm (as unit length vectors).

From Table 1, we can notice a tendency for the quality of offensive content categorization to become better when turning from *centroid* towards *category* as a vector representation for the nonterminal category nodes of the taxonomy in Fig. 2. Moreover, the best results (in boldface in Table 1) are achieved indeed with *category*. Keeping flat count embedding vectors for the tweets (instead of choosing the TF-IDF weighting scheme as applied on the flat counts) tends to lead to improvements in the two accuracy measures illustrated in Table 1. The best results achieved correspond to flat counts, together with *category*, as already mentioned (the *Experimental Setting* "FK" in Table 1, with the values of the two accuracy metrics in boldface).

Table 1. Accuracy of the offensive content categorization on tweets.

Experimental setting	Learning accuracy (%)	Accuracy (%)
TC	44.1	1.05
TK	43.34	0.35
TNC	43.31	0.12
TNK	43.52	0.47
FC	57.27	28.95
FK	**70.07**	**55.58**
FNC	43.31	0.12
FNK	42.28	20.93

Many of the strict *Accuracy* scores in Table 1 are very poor, and seem to be in disagreement with the pretty good *Learning Accuracy* scores achieved in the same *Experimental Setting* (i.e. in the same row in Table 1). The explanation is that there are very many "near misses" in the offensiveness category labels predicted by our classifier. For instance, the offensive content of a tweet is predicted as TIN instead of being correctly tagged (according to the gold standard) as IND. This is a "near miss" since the concept node TIN (targeted insult) is very close semantically to the node IND (insult targeted towards an individual). The two nodes are only one edge away to each other in the offensiveness ontology in Fig. 2; more exactly, category node IND is child (i.e. subconcept) of category node TIN.

4 Conclusions and Further Work

The paper presented a method for categorizing the offensive content in posts on Twitter that relies on an unsupervised neural network. For an unsupervised classifier, we achieved good accuracy results, as the classifier does not require annotated training data to be built for the given semantic domain, and as such it easily portable to other domains and even other languages. For instance, it can be used in order to detect and classify fake news in the medical domain, including the pandemics.

As further work, to improve the accuracy of labeling the offensive language, we will start from pre-trained word embeddings in order to build the vector for a tweet as the average vector of the pre-trained embedding vectors of the words that compose the tweet. We expect an improvement of the classification accuracy due to the much less sparse vectors built based on pre-trained word embeddings.

Acknowledgments. This work was supported in part by the Department of Computer Science, Technical University of Cluj-Napoca, Romania.

Conflicts of Interest. There is no conflict of interests between the authors.

References

1. Zampieri, M., Malmasi, S., Nakov, P., Rosenthal, S., Farra, N., Kumar, R.: SemEval-2019 Task 6: identifying and categorizing offensive language in social media (OffensEval). In: Proceedings of the 13th International Workshop on Semantic Evaluation, pp. 75–86 (2019)
2. Kohonen, T.: Self-organizing Maps. Springer Series in Information Sciences, 3rd edn., vol. 30. Springer, Berlin (2001). https://doi.org/10.1007/978-3-642-56927-2
3. Dittenbach, M., Merkl, D., Rauber, A.: Organizing and exploring high-dimensional data with the growing hierarchical self-organizing map. In: Wang, L. et al. (eds.) 1st International Conference on Fuzzy Systems and Knowledge Discovery, vol. 2, pp. 626–630 (2002)
4. Chifu, E.Ş., Leţia, I.A.: Self-organizing maps in web mining and semantic web. In: Matsopoulos, G.K. (ed.) Self-Organizing Maps, INTECH, pp. 357–380 (2010)
5. Bashar, M.A., Nayak, R.: QutNocturnalHASOC'19: CNN for hate speech and offensive content identication in Hindi language. In: Proceedings of the 11th annual meeting of the Forum for Information Retrieval Evaluation, pp. 237–245 (2019)
6. Mandl, T., et al.: Overview of the HASOC track at FIRE 2019: hate speech and offensive content identification in Indo-European languages. In: Proceedings of the 11th Forum for Information Retrieval Evaluation, pp. 14–17 (2019)
7. Grobelnik, M., et al.: Task description for PASCAL challenge. Evaluating ontology learning and population from text (2006)
8. Chifu, E.S., Letia, T.S., Chifu, V.R.: Unsupervised aspect level sentiment analysis using ant clustering and self-organizing maps. In: Proceedings of the 8th International Conference on Speech Technology and Human-Computer Dialogue (SpeD 2015), pp. 143–151, Bucharest, Romania (2015)
9. Schmidt, A., Wiegand, M.: A survey on hate speech detection using natural language processing. In: Proceedings of the Fifth International Workshop on Natural Language Processing for Social Media, pp. 1–10 (2017)
10. Zampieri, M., Malmasi, S., Nakov, P., Rosenthal, S., Farra, N., Kumar, R.: Predicting the type and target of offensive posts in social media. In: Proceedings of the Conference of the North American Chapter of the Association for Computational Linguistics: Hiuman Language Technologies, pp. 1415–1420 (2019)

Author Index

S. Vlad and N. M. Roman (Eds.): MEDITECH 2020, IFMBE Proceedings 88, pp. 443–444, 2022.
https://doi.org/10.1007/978-3-030-93564-1